ANTISENSE STRATEGIES

ANNALS OF THE NEW YORK ACADEMY OF SCIENCES
Volume 660

ANTISENSE STRATEGIES

Edited by Renato Baserga and David T. Denhardt

The New York Academy of Sciences
New York, New York
1992

⊗ The paper used in this publication meets the minimum requirements of American National Standard for Information Sciences–Permanence of Paper for Printed Library Materials, ANSI Z39.48-1984.

Cover: The figure was adapted from an illustration kindly provided by Synthecell Corp., Baltimore, Maryland.

Library of Congress Cataloging-in-Publication Data

Antisense strategies/edited by Renato Baserga and David T. Denhardt.
p. cm.—(Annals of the New York Academy of Sciences, ISSN 007-8923; v. 660)
Includes bibliographical references and index.
ISBN 0-89766-747-6 (cloth: alk. paper).—ISBN 0-89766-748-4 (pbk.: alk. paper)
1. Antisense nucleic acids—Congresses. I. Baserga, Renato. II. Denhardt, David T. III. Series.
[DNLM: 1. Antisense Elements (Genetics)—congresses. W1 AN626YL v. 660]
Q11.N5 vol. 660
[QP623.5.A58]
500 s—dc20
[574.87′322]
DNLM/DLC
for Library of Congress 92-49544
CIP

SP
Printed in the United States of America
ISBN 0-89766-747-6 (cloth)
ISBN 0-89766-748-4 (paper)
ISSN 0077-8923

ANNALS OF THE NEW YORK ACADEMY OF SCIENCES

Volume 660
October 28, 1992

ANTISENSE STRATEGIES[a]

Editors and Conference Organizers
RENATO BASERGA and DAVID T. DENHARDT

CONTENTS

Introduction. *By* RENATO BASERGA 1

Synthesis and Anti-HIV Activity of Oligoribonucleotides and Their Phosphorothioate Analogs. *By* SUDHIR AGRAWAL, J. Y. TANG, D. SUN, P. S. SARIN, and P. C. ZAMECNIK 2

Antisense Strategies to Characterize the Role of Genes and Oncogenes Involved in Myeloid Differentiation. *By* S. FERRARI, R. MANFREDINI, A. GRANDE, and U. TORELLI 11

Control of Gene Expression by Triple Helix-Forming Oligonucleotides. The Antigene Strategy. *By* CLAUDE HELENE, NGUYEN T. THUONG, and ANNICK HAREL-BELLAN 27

Specific Inhibition of Oncogene Expression *in Vitro* and *in Vivo* by Antisense Oligonucleotides. *By* LEN NECKERS, ANGELO ROSOLEN, BRIGID FAHMY, and LUKE WHITESELL 37

Variables Affecting Antisense RNA Inhibition of Gene Expression. *By* CAROL A. SCHERCZINGER, ADRIENNE A. YATES, and DAVID A. KNECHT.. 45

Modulation of c-*myc* Transcription by Triple Helix Formation. *By* EDITH H. POSTEL 57

Inhibition of Cell Cycle Progression by Antisense Oligodeoxynucleotides. *By* RENATO BASERGA, KRYSZTOF REISS, HANSJUERG ALDER, ZBIGNIEW PIETRZKOWSKI, and EWA SURMACZ 64

Mechanism of Action of Antisense RNA. Sometime Inhibition of Transcription, Processing, Transport, or Translation. *By* DAVID T. DENHARDT 70

Antisense RNA and p53 Regulation in Induced Murine Cell Differentiation. *By* SAADI KHOCHBIN, MARIE-PAULE BROCARD, DIDIER GRUNWALD, and JEAN-JACQUES LAWRENCE 77

Antisense Promoter Mapping. Inhibitory Methods of Transcriptional Analysis. *By* JEFFREY T. HOLT 88

Status of Ribozyme and Antisense-Based Developmental Approaches for Anti-HIV-1 Therapy. *By* JOHN A. ZAIA, SASWATI CHATTERJEE, K. K. WONG, DAVID ELKINS, NERIDA R. TAYLOR, and JOHN J. ROSSI 95

[a]This volume is the result of a conference entitled Antisense Strategies which was sponsored by the New York Academy of Sciences and held on January 12–15, 1992 in Philadelphia, Pennsylvania.

Epstein-Barr Viral Latency and Cell Immortalization as Targets for Antisense Oligomers. *By* JOSEPH S. PAGANO, GUSTAVO JIMENEZ, NANCY S. SUNG, NANCY RAAB-TRAUB, and JUNG-CHUNG LIN 107

Protooncogenes in the Regulation of Normal Hematopoiesis. An Antisense Approach. *By* BRUNO CALABRETTA, THOMAS SKORSKI, CEZARY SZCZYLIK, DANIELE CARACCIOLO, MAURO VALTIERI, and DONATELLA VENTURELLI 117

Reversal of Transformed Phenotypes by Antisense *fos*. *By* MATTHEWS O. BRADLEY, SUJATA MANAM, ANDREW R. KRAYNAK, WARREN W. NICHOLS, and BRIAN J. LEDWITH 124

Transferrinfection: A Highly Efficient Way to Express Gene Constructs in Eukaryotic Cells. *By* KURT ZATLOUKAL, ERNST WAGNER, MATT COTTEN, STEPHEN PHILLIPS, CHRISTIAN PLANK, PETER STEINLEIN, DAVID T. CURIEL, and MAX L. BIRNSTIEL 136

Antisense and Differentiation. *By* C. Thomas Caskey 154

Nonionic Oligonucleotide Analogs (Matagen™) as Anticodic Agents in Duplex and Triplex Formation. *By* P. O. P. TS'O, L. AURELIAN, E. CHANG, and P. S. MILLER 159

Therapeutic Applications of Antisense DNA in the Treatment of Human Leukemia. *By* ALAN M. GEWIRTZ 178

Myelin-Deficient Mutant Mice. An *in Vivo* Model for Inhibition of Gene Expression by Natural Antisense RNA. *By* JEAN-MARIE MATTHIEU, MIRJANA TOSIC, and ARTHUR ROACH 188

Antisense Oligodeoxynucleotides as Probes of T-Lymphocyte Gene Function. *By* DOUGLAS R. GREEN, HEGUANG ZHENG, and YUFANG SHI 193

Interference of Gene Expression in *Drosophila* by Antisense Ribosomal Protein Genes Integrated into the Germ Line. *By* MARCELO JACOBS-LORENA 204

Cell Cycle Effects of Microinjected Antisense Oligodeoxynucleotides to $p34^{cdc2}$ Kinase. *By* W. EDWARD MERCER, STEPHEN J. ULLRICH, MICHELLE T. SHIELDS, DAVID LIN, and HANSJUERG ALDER 209

Antisense Approaches to the Function of Glial Cell Proteins. *By* LINDA J. VAN ELDIK, STEVEN W. BARGER, and MICHAEL J. WELSH 219

Use of Antisense Oligomers to Study the Role of c-*jun* in G_1 Progression. *By* KENNETH J. SOPRANO, STEPHEN C. COSENZA, GLADYS YUMET, and DIANNE ROBERT SOPRANO 231

Modulation of Double-Stranded RNAs *in Vivo* by RNA Duplex Unwindase. *By* KAZUKO NISHIKURA 240

Antisense RNA. History and Perspective. *By* SIDNEY PESTKA 251

Poster Papers

Cleavage of Inhibin α Subunit mRNA by Engineered Ribozyme. *By* L. GE and A. W. SEITZ 263

Catalytic RNA Molecules and Their Cleavage of Viral mRNA. *By* SIEW PENG HO and DUSTIN H. O. BRITTON 265

Antisense RNA-Mediated Inhibition of the Replication of the Human Immunodeficiency Virus Type 1. Discussion of Critical Parameters. *By* GEORG SCZAKIEL, MICHAEL PAWLITA, KAROLA RITTNER, and MATTHIAS HOMANN 268

Inhibition of HIV-1 Replication by Novel Multitarget Ribozymes. *By* CHANG-JIE CHEN, AKHIL C. BANERJEA, KARL HAGLUND, GEORGE G. HARMISON, and MANFRED SCHUBERT 271

Gene Inhibition of HIV-1 Replication. A Comparative and Mechanistic Study. *By* WILLIAM JAMES, PAUL CRISELL, and ANDREW RHODES 274

Design of RNAs That Inhibit the Activated c-Ha-*ras* Gene in Mammalian Cells. *By* MAKOTO KOIZUMI and EIKO OHTSUKA 276

Internally Deleted Antigenomic Hepatitis Delta RNA Self-Cleaves Even in 18 M Formamide. *By* Y. PRASAD, M. G. BELINSKY, J. B. SMITH, and G. DINTER-GOTTLIEB 277

Inhibition of Processing of the Primary Transcript of the Gene Encoding Tissue Inhibitor of Metalloproteinases (TIMP) by Antisense TIMP RNA in Mouse 3T3 Cells. *By* BO FENG and DAVID T. DENHARDT 280

Antisense RNA to the Putative Tumor Suppressor Gene "Deleted in Colorectal Cancer" Transforms Fibroblasts. *By* KENNETH G. LAWLOR, NITIN T. TELANG, MICHAEL P. OSBORNE, ROEL Q. SCHAAPVELD, KATHLEEN R. CHO, BERT VOGELSTEIN, and RAMASWAMY NARAYANAN ... 283

Inhibition of the Transformed Phenotype of Carcinoma Cells That Contain Human Papillomavirus. *By* E. J. SHILLITOE and C. STEELE 286

Use of an Antisense RNA Expression System to Explore Functions of PKC Isoenzymes in T-Cell Activation. *By* J. FREIRE-MOAR, A. TSUTSUMI, and P. BARJA 288

A Novel Method to Modulate Desensitization and Truncation of LH Receptors Using Antisense Oligodeoxynucleotides. *By* ANTHONY P. WEST and BRIAN A. COOKE 291

Stable Transfection of Human Malignant Melanoma Cells with Basic Fibroblast Growth Factor Antisense cDNA. *By* ULLRICH GRAEVEN, ULRICH RODECK, ROBERTO WEINMANN, and MEENHARD HERLYN 293

Early Cell Growth Stimulation Is Inhibited by Casein Kinase II Antisense Oligodeoxynucleotides. *By* WALTER PYERIN, RAINER PEPPERKOK, WILHELM ANSORGE, and PETER LORENZ 295

Behavioral Change after Local Administration of Antisense Sequence for Progesterone Receptor mRNA in Female Rat Hypothalamus. *By* SONOKO OGAWA, URSULA E. OLAZÁBAL, and DONALD W. PFAFF 298

Liposomal Delivery of Antisense Oligodeoxynucleotides. Application to the Inhibition of the Multidrug Resistance in Cancer Cells. *By* ALAIN R. THIERRY and ANATOLY DRITSCHILO 300

Design, Synthesis, and Assay of Tetrapeptide-Acridine Mimics of Ribonuclease. *By* CHING-HSUAN TUNG and STANLEY STEIN 303

Chemical Modifications to Improve Uptake and Bioavailability of Antisense Oligonucleotides. *By* M. MANOHARAN, L. K. JOHNSON, D. P. C. MCGEE, C. J. GUINOSSO, K. RAMASAMY, R. H. SPRINGER, C. F. BENNETT, D. J. ECKER, T. VICKERS, L. COWSERT, and P. D. COOK 306

Separation of Synthetic Phosphorothioate Oligonucleotides from Phosphodiester-Defect Species by Strong-Anion Exchange HPLC. *By* B. JOHN BERGOT and GERALD ZON 310

Mechanistic Aspects of the Interaction of Polyanionic Oligodeoxynucleotides with HL60 Cells. *By* C. A. STEIN, J. TONKINSON, J. GERVASONI, L. M. ZHANG, S. KRISHNA, R. TAUB, S. FIELDS, T. DELOHERY, and S. ROTENBERG 313

Cell Physiology of Antisense TGF-β Oligomers in Hepatoma Cells. *By* MITSUHIRO INAGAKI, KAZUMI TOGAWA, BRIAN I. CARR, KRISHNA GHOSH, and JACK S. COHEN 315

Capped Oligodeoxynucleotide Phosphorothioates. Pharmacokinetics and Stability in Mice. *By* J. TEMSAMANI, J.-Y. TANG, and SUDHIR AGRAWAL 318

Ion Exchange HPLC Analysis of Oligoribonucleotides and Chimeric Oligoribo-oligodeoxyribonucleotides. *By* VALERI METELEV, KONRAD MISIURA, and SUDHIR AGRAWAL 321

Enantio- and *Meso-*DNA. *By* KOICHI SHUDO 324

Intracellular Oligo Targeting. A Theoretical Dissertation. *By* T. T. CHUANG 326

Cellular Uptake, Distribution, and Degradation of Phosphorothioate Oligonucleotides in *Euplotes crassus*. *By* L. L. THOMAS and C. M. PRICE 328

Sequence-Specific Activity of Antisense Oligonucleotides Conjugated to Poly(L-Lysine) Carriers. *By* GENEVIÈVE DEGOLS, JEAN-PAUL LEONETTI, and BERNARD LEBLEU 331

Inhibition of the Friend Retrovirus by Antisense Oligonucleotides. Indirect Evidence for the Necessity of Forced Cellular Uptake. *By* C. ROPERT, M. LAVIGNON, J. L. IMBACH, P. COUVREUR, and C. MALVY 334

Structure-Function Relation of the Hepatitis Delta Virus Genomic Ribozyme. *By* MARTA BLUMENFELD, GILBERT THILL, FRANCK LESCURE, and MARC VASSEUR 336

Antisense Oligonucleotides as Antiviral Agents. *By* JUDY COULSON and ALAN D. B. MALCOLM 339

Conference Summary. *By* DAVID T. DENHARDT 342

Index of Contributors 347

Subject Index 349

Financial assistance was received from:

Supporters

- GLAXO RESEARCH LABORATORIES
- JEFFERSON MEDICAL COLLEGE/THOMAS JEFFERSON UNIVERSITY

Contributors

- ABBOTT LABORATORIES
- AMERICAN CYANAMID COMPANY
- BIO 101 INC.
- BURROUGHS WELLCOME CO.
- HOFFMANN-LA ROCHE INC.
- LILLY RESEARCH LABORATORIES
- MERCK SHARP & DOHME RESEARCH LABORATORIES
- NATIONAL FOUNDATION FOR CANCER RESEARCH
- WYETH-AYERST RESEARCH

Introduction

RENATO BASERGA

Department of Microbiology
Thomas Jefferson University
Philadelphia, Pennsylvania 19107–5587

The purpose of this conference was to bring together scientists from disparate fields to discuss the various modalities of the antisense strategy. There is no question that the antisense strategy already occupies a prominent place of interest among both basic scientists and clinicians. Oligodeoxynucleotides complementary to RNAs have already shown their usefulness as a biological tool and their potential as a chemotherapeutic agent. The large number of laboratories that have used the antisense strategy to dissect delicate molecular problems and the many industrial companies that are interested in its development affirm that the antisense strategy is here to stay, at least for some time to come.

Therefore, this conference was organized not to extol its merits but, on the contrary, to highlight its problems. Anyone who has used antisense oligodeoxynucleotides in the laboratory knows that the results are not always consistent. Antisense oligos work well with certain RNA targets and poorly, or not at all, with others. Worse, one antisense sequence is effective in inhibiting the expression of a targeted gene product, and a sequence upstream or downstream from the first one is totally ineffective. In short, we do not know why certain antisense oligos are effective and others are not. This conference gives us some clues.

Even when antisense is effective, the doubt persists that it may not be specific for the targeted RNA. In selecting an antisense, a computer is always used to rule out the possibility that the same sequence is present in another RNA. But because the number of known RNA (and hnRNA) sequences is so small (probably 2% of the total), it is quite possible that the same sequence may be present in both the targeted and the unknown RNA. How can we convince ourselves that the effect of antisense is specific?

And, finally, how can we make the antisense strategy more effective? Modified oligodeoxynucleotides, ribozyme, formation of triple-helix DNA in the promoter region of a gene, use of vectors to direct the oligos to specific cells, these and other devices are continuously proposed to improve effectiveness and potential chemotherapeutic activity.

The participants were selected with these goals in mind. I can say that these goals have been achieved; the papers presented have gone a long way in clarifying some of these issues. The New York Academy of Sciences and the sponsors can be proud of the results of this conference.

Synthesis and Anti-HIV Activity of Oligoribonucleotides and Their Phosphorothioate Analogs

SUDHIR AGRAWAL,[a] J. Y. TANG,[a,d] D. SUN,[b]
P. S. SARIN,[c] AND P. C. ZAMECNIK[a]

[a] *Worcester Foundation for Experimental Biology*
222 Maple Avenue
Shrewsbury, Massachusetts 01545

[b] *Laboratory of Tumor Cell Biology*
National Cancer Institute
Bethesda, Maryland

[c] *Department of Biochemistry*
George Washington University Medical Center
Washington, DC

In previous reports, we demonstrated that replication of human immunodeficiency virus (HIV) could be inhibited by unmodified phosphodiester oligodeoxynucleotides complementary to HIV RNA.[1,2] However, a relatively short half-life of unmodified oligodeoxynucleotides in serum and in cells, due to the presence of nucleases, limits their potential usefulness *in vivo*. To overcome this limitation, we and others have studied internucleotide phosphate backbone modified oligonucleotides, such as methyl phosphonates,[3,4] phosphoramidates,[5] and phosphorothioates.[5–9] Several oligonucleotide conjugates such as oligonucleotide-cholesteryl conjugates[10] and oligonucleotide-polylysine conjugates[11] have also been studied. These analogs of oligodeoxyribonucleotides are resistant to nucleases and have generally increased hydrophobicity, which provides longer survival time *in vivo* and may increase cell membrane permeability. Oligodeoxynucleotide phosphorothioates are effective in inhibiting HIV replication in tissue culture (IC_{50}) at a concentration of approximately 1×10^{-7} M.[9] The anti-HIV activity of oligodeoxyribonucleotide phosphorothioate is found to be "relatively sequence independent" against freshly infected cells, but bear in mind that it is difficult for a computer to define a completely random 20-mer sequence against HIV and one that does not activate RNase H. There is, however, striking sequence-specific activity against HIV chronically infected cells.[6,7] The cholesteryl conjugates of oligodeoxyribonucleotide phosphorothioates were found to be more effective than their unconjugated counterparts, but cholesterol conjugation increases nonsequence-specific activity as well.[10] The inhibition of HIV replication by an oligodeoxyribonucleotide phosphorothioate is length dependent, a 25 mer being more active than 15–20 mers.[12] An increase in length to 30 mer or 35 mer had little or no enhanced effect on anti-HIV activity over that of the 25 mer.

Our continued interest in developing antisense oligodeoxynucleotides and their various analogs as anti-HIV agents has stimulated us to investigate oligoribonucleotides and their phosphorothioate analogs as anti-HIV agents.

[d] Present Address: Hybridon, Inc., One Innovation Drive, Worcester, MA 01605.

SYNTHESIS OF OLIGORIBONUCLEOTIDES AND THEIR PHOSPHOROTHIOATE ANALOG

Oligoribonucleotides were synthesized using the H-phosphonate approach, employing *t*-butyldimethylsilyl for 2′-hydroxyl protection and dimethoxytrityl for 5′-hydroxyl protection.[13,14] Syntheses were carried out on a large scale (5–10 μmol) employing the coupling cycle shown in TABLE 1, using an 8750 machine (Millipore, Bedford, MA). The advantage of an optimized coupling cycle is the consumption of only fivefold excess of nucleoside H-phosphonate for each coupling, with an average coupling efficiency of 95–97%. After assembly of the required sequence, the CPG-bound oligoribonucleoside H-phosphonate was oxidized to the corresponding oligoribonucleotide by treatment with iodine or to the corresponding phosphorothioate analog by treatment with sulfur. The CPG-bound oligoribonucleotide was then deprotected using the same condition as reported earlier.[13,14] The oligoribonucleotide and its phosphorothioate analog were purified by ion exchange chromatography (DEAE; DE-52, Whatman), using triethylammonium bicarbonate buffer (0.1–1 M gradient) followed by dialysis. Purified fractions of oligoribonucleotides and their phosphorothioate analogs were kinased at the 5′-end by ($\gamma^{32}P$) ATP and analyzed by gel electrophoresis using 20% polyacrylamide containing 7 M urea (FIG 1).

TABLE 1. Coupling Cycles for 5–10 μmol Scale Synthesis[a]

Step	Solvents/Reagents	Reaction Time (s)	Volume Consumed (ml)	Notes
1. Detritylation	3% dichloroacetic acid in dichloroethane	150	8	Flow 40 sec; wait 10 s; repeat 3 times
2. Washing	Acetonitrile	90	3–4	
3. Washing	Acetonitrile-pyridine (1:1, v/v)	56	2	Flow 5 s; wait 3 s; repeat 6 times
4. Coupling	Monomer (10 mM solution) and adamantane carbonyl chloride (100 mM solution) in acetonitrile-pyridine (1:1, V/V)	66	5–8[a]	Flow 0.3 s each of monomer and activator alternating 20 times; wait 10 s; repeat 2 times
5. Washing	Acetonitrile	90	4	Flow 10 s; wait 5 s; repeat 5 times

[a]When synthesis was carried out on a 5-μmol scale, the flow was about 5 ml during the coupling step. Thus, the monomer consumption was in fivefold excess.

Oligoribonucleotides 1 and 3 (TABLE 2) were digested with snake venom phosphodiesterase (SVPD) and analyzed for base composition on reversed phase HPLC. Results were in agreement with the theoretical base composition value (data not shown). Oligoribonucleotide phosphorothioate 2 and 4 (TABLE 2) were also digested using the same condition. The rate of digestion was determined by measurement of hyperchromicity at 260 nm. Phosphorothioate oligoribonucleotides were found to be digested with SVPD, with a relative $t_{1/2}$ about 10^2–10^3 times longer than that of the phosphodiester oligoribonucleotides.

THERMAL STABILITY OF THE DUPLEXES

The thermal stability of 20-mer duplexes containing RNA-RNA, RNA-DNA, RNA-RNA (phosphorothioate), and RNA-DNA (phosphorothioate) was evaluated by determination of ultraviolet-absorbance-temperature profiles. The normalized curves are shown in FIGURE 2.

All four duplexes showed cooperative melting, and the change in hyperchromicity was about 20% in the case of RNA hybridizing to RNA or DNA and about 17% in the case of the same RNA hybridizing to the phosphorothioate analog of RNA or DNA. The highest Tm value (44°C) was observed for the RNA-RNA duplex followed by RNA-DNA (41°C), RNA-RNA (phosphorothioate) (37°C), and RNA-DNA (phosphorothioate) (33°C). The Tm of duplexes involving RNA-RNA or RNA-DNA, to some extent, also depend on the base composition of RNA and DNA sequences. Phosphorothioate analogs of RNA and DNA generally had lower Tms than did their phosphodiester counterparts when hybridized to complementary normal RNA sequences.

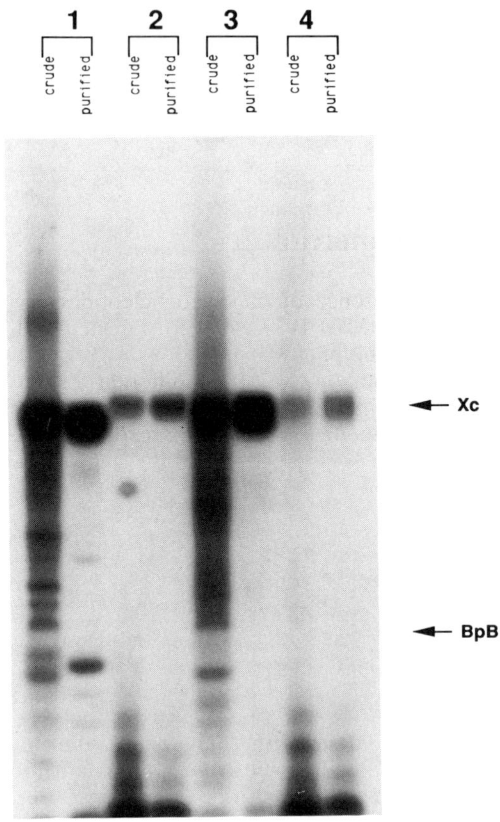

FIGURE 1. Autoradiogram of 20% polyacrylamide gel (7 m urea) derived from ^{32}P-5′-phosphorylated crude and purified, 20-mer oligoribonucleotide and its phosphorothioate analog. *Lane 1,* oligomer 1; *lane 2,* oligomer 2; *lane 3,* oligomer 3; and *lane 4,* oligomer 4.

TABLE 2. Oligoribonucleotide Assembled

No.	Sequence (5′ → 3″)	Phosphate/ Backbone Modification	Scale of Synthesis μmol	Average Coupling Yield (%)	Isolated Yield (%)
1	ACA CCC AAU UCU GAA AAU GG	Phosphodiester	8.0	95.5	34
2	ACA CCC AAU UCU GAA AAU GG	Phosphorothioate	5.9	97.4	32
3	CCA UUU UCA GAA UUG GGU GU	Phosphodiester	7.8	95.0	31
4	CCA UUU UCA GAA UUG GGU GU	Phosphorothioate	5.8	96.7	32
5	CGA UAC GAU ACG AUA CGA UU	Phosphodiester	6.0	97.1	30
6	CGA UAC GAU ACG AUA CGA UU	Phosphorothioate	5.9	96.9	31

ASSAY FOR HIV-1 INHIBITION

Inhibition of HIV-1 replication in the presence of oligoribonucleotides was studied, as described earlier. In short, H-9 or MOLT-3 cells were infected with HIV-1 (HTLV IIIb and HTLV IIIc) with a multiplicity of infections of 0.5 to 1. Oligoribonucleotides were added only once to the tissue culture at different concentrations either at 0 hours (simultaneously with the virus) or at 24 or 48 hours after the addition of virus to cells. With the 24- and 48-hour additions of virus before the

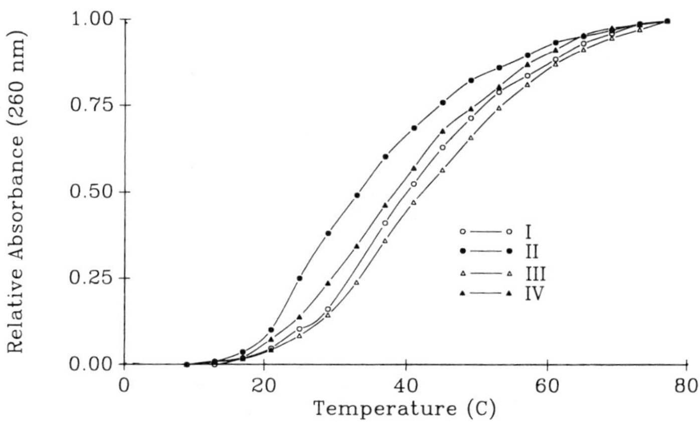

FIGURE 2. Melting curves of the duplexes of oligoribonucleotide, oligomer 3 (TABLE 2), with (I) oligoribonucleotide 1, (II) oligodeoxynucleotide, ACA CCC AAT TCT GAA AAT GG, (III) phosphorothioate oligoribonucleotide, oligomer 2 (TABLE 2), and (IV) phosphorothioate oligodeoxyribonucleotide ACA CCC AAT TCT GAA AAT GG. All duplexes were formed in a 1:1 mixture in phosphate buffer (100 mM NaCl, pH 7.4). The samples were heated to 70°C and then cooled to 10°C at a rate of 2°C per minute. Melting curves were measured as hyperchromicity at 260 nm from 10°C to 80°C (1°C increase per minute).

addition of the oligoribonucleotide, excess virion particles were washed off. The cells were incubated in culture medium containing RPMI 1640, 10% *v/v* fetal bovine serum, 2 μM glutamine, and 250 μg of gentamicin per milliliter at 37°C. After 96 hours, the cells and supernatant were examined for the level of HIV-1 expression by measuring syncytia, the viral antigens p17 and p24, reverse transcriptase, and also cell viability. When oligoribonucleotides were added at 24 or 48 hours after viral infection, the experiment was continued for 96 hours thereafter before terminating.

RESULTS

Anti-HIV Activity of Oligoribonucleotides

Three oligoribonucleotides containing unmodified phosphodiester internucleotide linkages were screened against HIV-1. The oligomers were 20 mers complementary to the splice acceptor site of HIV-1 tat gene (5358), oligomer 1 (TABLE 2), the sense sequence of the splice acceptor site of HIV-1 tat gene (5358), oligomer 3 (TABLE 2), and a mismatched control oligomer 5 (TABLE 2). All three oligomers showed similar dose-dependent anti-HIV activity when added to tissue cultures simultaneously with the virus. When added to tissue cultures at 24 or 48 hours postdosing, they failed to show any anti-HIV activity, even at higher concentrations (FIG. 3). Similar results were obtained with oligodeoxynucleotides containing unmodified phosphodiester internucleotide linkages.[7]

Anti-HIV Activity of Oligoribonucleotide Phosphorothioates

The oligoribonucleotide phosphorothioates, which were screened for their anti-HIV activity, were also 20 mers in length. The oligomer sequences were complementary to the splice acceptor site of HIV-1 tat (5358), oligomer 2 (TABLE 2), or the sense sequence of splice acceptor site of HIV-1 tat (5358), oligomer 4 (TABLE 2), and a mismatched control, oligomer 6 (TABLE 2). When added simultanously with the virus, all three oligoribonucleotide phosphorothioates showed similar anti-HIV activity. Inhibition of HIV-1 replication was dose dependent. When oligoribonucleotide phosphorothioates were added at 24 hours postinfection, only the complementary oligoribonucleotide phosphorothioate showed anti-HIV activity. The inhibitory dose of complementary oligoribonucleotide phosphorothioate was higher than that observed on simultaneous addition. The complementary oligoribonucleotide phosphorothioate was also effective in inhibiting HIV-1 replication when added 48 hours after infection.

Comparative Anti-HIV Activity of Phosphorothioate Analogs of Oligodeoxyribonucleotide and Oligoribonucleotide

The phosphorothioate analogs of oligodeoxyribonucleotide and oligoribonucleotide, 20 mers, complementary to the splice site of HIV-1 tat (5358), were screened for their comparative anti-HIV activity in tissue culture under similar conditions. The oligomers were added to 24-hour infected cells, and the level of HIV-1 expression was monitored after 96 hours by measurement of syncytia formation and expression of viral antigens p17 and p24. All three parameters showed a similar level of HIV-1 expression. Both oligomers showed dose-dependent inhibition. The oligodeoxyribonucleotide phosphorothioate was more effective than the oligoribonucleotide phosphorothioate in inhibiting HIV replication (FIG. 5).

The IC_{50} of antisense oligodeoxyribonucleotide phosphorothioate in the previously infected cells was about 1×10^{-6} M compared to the IC_{50} of antisense oligoribonucleotide phosphorothioate of 5×10^{-6} M.

DISCUSSION

A study has been made of the inhibitory efficacy of antisense oligoribonucleotides and their phosphorothioate analogs against HIV in tissue culture. The investigation was carried out for two main reasons: (1) the synthons used for assembling oligoribonucleotides are less expensive than the synthons for oligodeoxyribonucleotide synthesis; and (2) the duplex formed between antisense oligoribonucleotides and RNA is more stable than the duplex between antisense oligodeoxyribonucleotide and RNA.

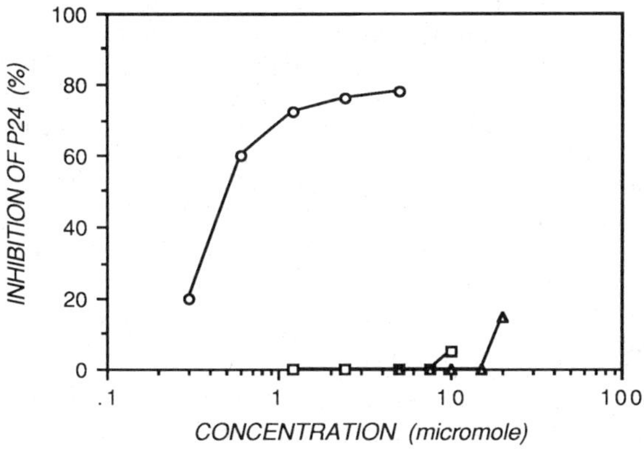

FIGURE 3. Oligoribonucleotide 1, complementary to the tat splice acceptor site of HIV, inhibited replication of HIV when added to the cell culture simultaneously with the virus (○). The inhibition level of HIV-1 expression monitored by syncytia formation and p17 and p24 expression showed similar results. This figure shows the p24 expression level as assayed by immunofluorescence. When the same or an even higher concentration of oligoribonucleotide was added to cells that were infected for 24 (□) or 48 hours (△), no inhibition was observed.

Oligoribonucleotides containing normal phosphodiester internucleotide linkages were synthesized using H-phosphonate chemistry. With an optimized coupling cycle on an automated DNA synthesizer, synthesis was carried out on a 5–10 μmol scale. The average coupling efficiency, based on dimethoxytrityl measurement, was 96–97%. The duplex stability study showed that a duplex of an oligoribonucleotide and its complementary oligoribonucleotide is more stable than a duplex of oligodeoxyribonucleotide and a complementary oligoribonucleotide (FIG. 2), as measured by the melting temperature. Three oligoribonucleotides were tested for their anti-HIV activity. One oligomer was antisense to the tat splice acceptor site of HIV-1, the second oligomer was the sense strand of the tat splice acceptor site of HIV-1, and the third oligomer was mismatched. All three sequences showed similar anti-HIV activity when added to uninfected cells at the same time as HIV-1. However, when added to postinfected cells (24 or 48 hours), only the antisense oligoribonucleotide showed marginal anti-HIV activity, but at a higher concentration. Similar results were observed with oligodeoxyribonucleotides containing a phosphodiester backbone.[7] The inability of these oligomers to inhibit HIV in already infected cells is most probably due to limited half-life and lack of *in vivo* persistence.

We have also studied phosphorothioate analogs of oligoribonucleotides. These analogs were synthesized using H-phosphonate chemistry followed by oxidation of the CPG-bound oligoribonucleoside H-phosphonate with a mixture of elemental sulfur, carbon disulfide, pyridine, and triethylamine. The Tms of the oligoribonucleotide phosphorothioates were lower than those of their oligoribonucleotide phosphodiester counterparts. However, the Tm was higher than that of the oligodeoxyribonucleotide phosphorothioate. Oligoribonucleotide phosphorothioates were more resistant to exonucleases than were oligoribonucleotide phosphodiesters.

Three oligoribonucleotide phosphorothioates were studied for their anti-HIV activity in tissue culture. The oligomer sequences were the same, as mentioned, except for containing phosphorothioate internucleotide linkages. All three oligoribonucleotide phosphorothioates showed a similar level of anti-HIV activity when added to tissue cultures simultaneously with the virus. Inhibition of HIV replication was found to be dose dependent. When added to postinfected cells (24 or 48 hours after infection), only complementary oligoribonucleotide phosphorothioates showed anti-HIV activity (FIG. 4). The other two oligoribonucleotide phosphorothioates failed to show inhibition of HIV-1 replication. The effective concentration of the antisense oligoribonucleotide phosphorothioate was higher as compared to zero hour infection. In comparison, oligoribonucleotide phosphorothioate was more effective than its phosphodiester counterpart.

We also studied the comparative anti-HIV activity of oligodeoxyribonucleotide phosphorothioate and oligoribonucleotide phosphorothioate. Anti-HIV activity was compared when the oligomers were added to cells that were already infected for 24 hours. The dose required to inhibit HIV replication (IC_{50}) by oligoribonucleotide

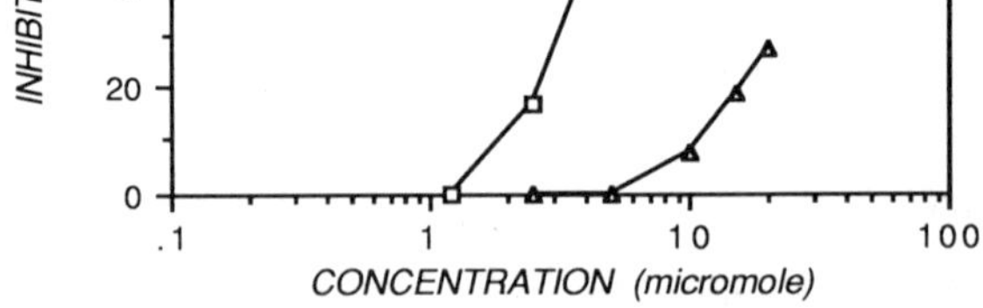

FIGURE 4. Oligoribonucleotide phosphorothioate 2, complementary to tat splice acceptor site of HIV, inhibited HIV replication. When added to cell cultures simultaneously with the virus (○) or even to cells that were infected for 24 (□) or 48 hours (△), inhibition was dose dependent. The mismatched phosphorothioate analog of oligoribonucleotide 6 and sense oligoribonucleotide 4 showed no inhibitory effect when added to cells already infected.

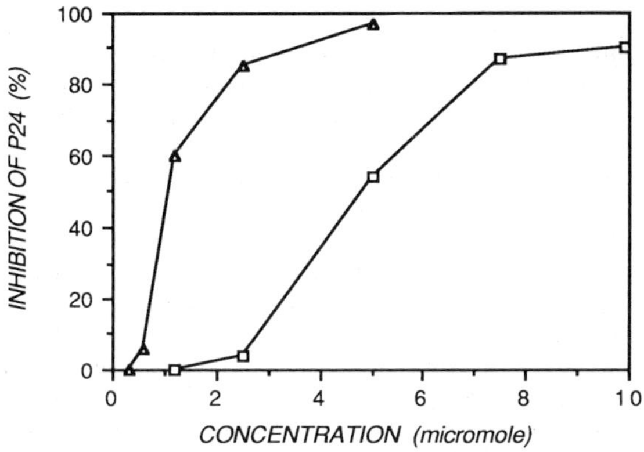

FIGURE 5. Comparative inhibition of HIV-1 expression by an antisense oligoribonucleotide phosphorothioate and an antisense oligoribonucleotide phosphorothioate. Both sequences were complementary to the tat splice acceptor site of HIV-1. The oligonucleotides were added to cells that were already infected for 24 hours. Both oligonucleotides showed a dose-dependent effect. Oligodeoxyribonucleotide phosphorothioate was about five times more active than the corresponding oligoribonucleotide phosphorothioate.

phosphorothioate was approximately five times greater than that required by oligodeoxyribonucleotide phosphorothioate (FIG. 5). A possible explanation for the superiority of antisense oligodeoxyribonucleotide phosphorothioate over its ribo-counterparts in HIV inhibitory potency is the activity of RNase H in degrading the mixed hybrid in the translation step of protein synthesis. This enzyme is inactive against double-stranded RNAs, and therefore there is no synergistic effect of inhibition by hybridization arrest plus RNase H cleavage of the target RNA strand. We and others have observed similar lesser inhibitory activity of oligonucleotide analogs that do not activate RNase H such as methylphosphonate,[3] phosphoramidate,[5] and 2′-OMe oligoribonucleotide phosphorothioates.[18]

It may be, however, that the RNA series of either antisense or sense, nuclease-resistant oligomers may have advantages in test systems other than HIV. It is therefore worth evaluating complementary oligoribonucleotides in other biological systems in which genetic information readout is being studied. This point of view is particularly appropriate in efforts to induce inhibition at the replication eye or transcription eye stage, at which both antisense and sense strands of the eye would appear to be appropriate targets for blocking movement of the eye in the 5′ → 3′ sense direction.

CONCLUSION

The goal of this research is principally to develop new compounds for the treatment of HIV. This approach to chemotherapy stems from earlier work which showed that replication and expression of *Rous sarcoma* virus could be inhibited in tissue culture by hybridization of an oligonucleotide complementary to a segment of the RSV genome.[15,16]

Initially, we synthesized and tested unmodified oligonucleotides, complementary to a conserved segment of the viral genome, which showed promise as inhibitors of

HIV replication and expression in tissue culture.[1,2] Degradation of the oligomers by intracellular enzymes, however, made it desirable to make nuclease-resistant modifications.[7–9] The IC_{50} was enhanced from the 10^{-5} to the 10^{-7}–10^{-8} molar range in this way. We recently explored the toxicity and pharmacokinetics of the oligodeoxynucleotide phosphorothioates[17] with the aim in particular of developing an antisense oligonucleotide-based chemotherapeutic approach for AIDS.

REFERENCES

1. ZAMECNIK, P. C., J. GOODCHILD, Y. TAGUCHI, & P. S. SARIN. 1986. Proc. Natl. Acad. Sci. USA **83:** 4143.
2. GOODCHILD, J. G., S. AGRAWAL, M. P. CIVEIRA, P. S. SARIN, D. SUN & P. C. ZAMECNIK. 1988. Proc. Natl. Acad. Sci. USA **85:** 5507.
3. SARIN, P. S., S. AGRAWAL, M. P. CIVEIRA, J. GOODCHILD, T. IKEUCHI & P. C. ZAMECNIK. 1988. Proc. Natl. Acad. Sci. USA **85:** 7448.
4. ZAIA, J. A., J. J. ROSSI, G. J. MURADAWA, P. A. SPALLONE, D. A. STEPHENS, B. E. KAPLAN, R. ERITZA, R. B. WALLACE & E. M. CANTIN. 1988. J. Virol. **62:** 3914.
5. AGRAWAL, S., J. GOODCHILD, M. P. CIVEIRA, A. H. THORNTON, P. S. SARIN & P. C. ZAMECNIK. 1988. Proc. Natl. Acad. Sci. USA **85:** 7079.
6. MATSUKURA, M., G. ZON, K. SCHINOZUKA, M. ROBERT-GUROFF, T. SHIMADA, C. A. STEIN, H. MITSUYA, F. WONG-STAAL, J. S. COHEN & S. BRODER. 1989. Proc. Natl. Acad. Sci. USA **86:** 4244.
7. AGRAWAL, S., T. IKEUCHI, D. SUN, P. S. SARIN, A. KONOPKA, J. MAIZEL & P. C. ZAMECNIK. 1989. Proc. Natl. Acad. Sci. USA **86:** 7790.
8. AGRAWAL, S. & P. S. SARIN. 1991. *In* Advanced Drug Delivery Reviews. R. Juliano, ed. **6**(3):251–270. Elsevier. Amsterdam.
9. AGRAWAL, S. 1991. *In* Prospects for Antisense Nucleic Acid Therapy of Cancer and AIDS, E. Wickstrom, ed. 143–158. Wiley-Liss. NY.
10. LETSINGER, R. L., G. ZHANG, D. SUN, T. IKEUCHI & P. S. SARIN. 1989. Proc. Natl. Acad. Sci. USA **86:** 6553.
11. STEVENSON, M. & P. IVERSON. 1989. J. Gen. Virol. **70:** 2673.
12. AGRAWAL, S., M. ZAMECNIK, P. S. SARIN & P. C. ZAMECNIK. 1991. *In* Gene Regulation by Antisense RNA and DNA. R. P. Erickson & J. Izant, ed.: 273–283. Raven Press. NY.
13. OGILVIE, K. K., N. USMAN, K. NICHOGHOSIAN & R. J. CEDERGREN. 1988. Proc. Natl. Acad. Sci. USA **85:** 5764–5768.
14. AGRAWAL, S. & J.-Y. TANG. 1990. Tet. Lett. **31:** 7541–7544.
15. ZAMECNIK, P. C. & M. L. STEPHENSON. 1978. Proc. Natl. Acad. Sci. USA **75:** 280–284.
16. STEPHESON, M. L. & P. C. ZAMECNIK. 1978. Proc. Natl. Acad. Sci. USA **75:** 285–288.
17. AGRAWAL, S., J. T. TEMSAMANI & J.-Y. TANG. 1991. Proc. Natl. Acad. Sci. USA **88:** 7595–7599.
18. SHIBAHARA, S., S. MUKAI, H. MORISAWA, H. NAKASHIMA, S. KOBAYASHI & N. YAMAMOTO. 1989. Nucleic Acids Res. **17:** 239.

Antisense Strategies to Characterize the Role of Genes and Oncogenes Involved in Myeloid Differentiation[a]

S. FERRARI,[b] R. MANFREDINI, A. GRANDE, AND U. TORELLI

Experimental Hematology Center
II Medical Clinic
University of Modena
Via del Pozzo 71, Policlinico
41100 Modena, Italy

Several genes that are associated with cell proliferation[1,2] and differentiation[3] have been isolated. The definition of "cell cycle-related or differentiation-related genes" is simply based on the concept that mRNA and/or protein abundance of these genes increases in these two functional states of the cell.[4] Hundreds of genes have been characterized using very different biological systems and technical strategies such as differential screening of cDNA libraries[5] or, more recently, using subtraction libraries[6] starting from polyadenylated mRNA molecule populations extracted from cells in different functional states such as proliferating, quiescent, or differentiated cells. One approach is to use as a biological model temperature-sensitive mutant cell lines that can be arrested, depending on permissive or nonpermissive temperature, in different phases of the cell cycle progression.[7] With this biological system, genes involved in the cell cycle progression have been isolated.[8] Remarkable progress in the study of the molecular mechanisms underlying cell proliferation and differentiation has been made with the molecular cloning and characterization of cytokines and their receptors[9] as well as lineage-specific transcriptional factors[10] such as the GATA 1 gene which plays an important role in erythroid differentiation[11] or other binding proteins for myeloid differentiation.[12] It is clearly very important to distinguish between gene products that are simply associated with a functional state from gene products that restrict and perhaps regulate that function, in other words, genes whose expression is essential for the proliferative or differentiative genetic programs. In fact, these genes can play a regulatory role because of their interference with the several signal transduction pathways underlying both proliferation and differentiation.[13] That a gene product is restricted to a cell function has been demonstrated with very different technical approaches based on different methodologies such as microinjection of antibodies or synthetic antisense RNA,[14] treatment with antisense oligomers,[15,16] transfection of constructs expressing high levels of antisense RNA under the control of strong and often inducible promoters,[17] transgenic mice,[18] and more recently homologous recombination.[19] One opportunity offered by these methods is to obtain specific and possibly complete inhibition of a gene product at the level of either the genetic locus[20] or its transcribed product (precursor mRNA,

[a]This work was supported by a grant from A.I.R.C. (Associazione Italiana per la Ricerca sul Cancro).

[b]Address for reprint requests: Dr. Sergio Ferrari, Experimental Hematology Center, Second Medical Clinic, University of Modena, Via del Pozzo 71-Policlinico, 41100 Modena, Italy.

mature mRNA, and protein) and to look at the biological consequences of the specific gene inactivation.[21] As pointed out before, these experimental approaches allow us to discriminate gene products that are simply associated with a specific functional state from gene products that limit and perhaps regulate a cell function, such as proliferation, differentiation, or development. The method used in our laboratory is the antisense oligodeoxynucleotide approach which is based on the concept of using precursor or mature mRNA as the primary target. The hybrid duplex formed between RNA and the complementary oligomer will lead basically to translational arrest, but also, according to the experimental design, to a block of the splicing of precursor RNA molecules or to inhibition of gene expression because of a triple DNA helix formation.[20,21] The continuous evolution of this method has led to technical improvement, because it is possible to synthesize modified oligodeoxynucleotides, which are more stable, and conjugate these oligomers with different chemical compounds that can improve uptake in the cells of these molecules.[22] With this method it has been reported that oncogenes such as c-fos, c-myc, c-myb, c-abl, and bcl2 and normal genes such as CDC2, PCNA, DNA polymerase alfa, bFGF, CSF1, and IL6 play an essential role in cell proliferation, because their inactivation inhibits the proliferative capacity of several different cell lines and of normal hematopoietic cells (TABLE 1). It is currently very difficult to establish a hierarchy of these genes in the molecular mechanism underlying the cell cycle progression. These results offer new insights into the molecular mechanisms underlying and controlling the prolifer-

TABLE 1. Genes and Oncogenes Whose Effect on Cell Proliferation Is Inhibited by Antisense Sequences

Target Gene	Cells	Biological Effects	References
C-FOS	3T3 fibroblasts	Inhibition of cell proliferation, inhibition of transformation or reversion	80,81
C-MYC	T lymphocytes HL-60 ST-486 (B.L.C.L.)	Inhibition of cell proliferation of either T lymphocytes or HL-60 myeloid cells which are induced to spontaneous differentiation; growth inhibition of a tumor cell line derived from a Burkitt's lymphoma	52,53,82,83
N-MYC	CHP-100 (N.E.C.L.) LAN-5	Reduction of proliferative activity in human neuroepithelioma and neuroblastoma cell lines	84,85
C-MYB	T lymphocytes HL-60 ML-3 KG-1 Bone marrow cells	Inhibition of proliferative activity of myeloid cells in culture, normal myeloid hemopoietic precursors, and mitogen-stimulated T lymphocytes	16,33,34,86

TABLE 1. *Continued*

Target Gene	Cells	Biological Effects	References
PCNA (cyclin)	BALB 3T3	Inhibition of proliferative activity	87
C-ABL	K-562 HL-60 Bone marrow cells	Partial inhibition of cell proliferation capacity and of myeloid colony formation	88
CDC-2	T lymphocytes	Inhibition of cell proliferation of mitogen-stimulated T lymphocytes	82
Butyril-cholin-esterase	Murine bone marrow cells	Selective inhibition of megacariocytopoiesis	89
Pro-thymosin alpha	Human myeloma cells	Inhibition of human myelomatous cells in culture	90
S100 beta	C6 cells	Inhibition of proliferation of rat glioma C6 cell line; morphological changes and cytoskeletal alterations are also observed	91
BCL-2	Leukemic 697 cells	Inhibition of this lymphomatous cell line with massive cell death	92
B-FGF	Human melanoma cells	Inhibition of cell proliferation and decrease in the clonogenicity capacity of these cells in semisolid media	93
DNA polymerase beta	3T3 cells	Alteration of the growth behavior of this fibroblastic murine cell line	94
CSF-1	Murine monocytes	Inhibition of proliferation	95
IL-1 alpha	Human endothelial cells	Treatment with IL-1 alpha antisense prevents these cells from becoming senescent, thus increasing their proliferative capacity	96
IL-6	U-266	Treatment with IL-6 antisense inhibits proliferation of this myelomatous cell line	97
H-RAS	T24 bladder carcinoma cells	Inhibition of T24 cell proliferation	98
BCR-ABL	CML (chronic myeloid leukemia cells)	Suppression of chronic myelogenous leukemia blast cell colony formation	99

ative pathways. Knowledge of the molecular mechanisms underlying and controlling the cell differentiation pathways is, on the contrary, poor, but we believe that the study of the function of oncogene products will provide new insights into the genetic control of cell differentiation[23] and programmed cell death (apoptosis) which seems to be the physiological mechanism of death of a terminally differentiated cell.[24–27] During differentiation, the cell cycle is characterized by a progressive increase in the length of its G1 and S phases, and terminally differentiated cells are arrested mainly in the G1 phase of the cycle. We have studied two particular aspects of human myeloid cell proliferation and differentiation. The first experimental approach was to establish which correlations exist between the proliferative and differentiative pathways in the HL-60 myeloid cell lines. The second experimental approach was to distinguish between genes that are simply associated and genes or oncogenes that control myeloid differentiation induced *in vitro* in HL-60 cells. In fact, the two oncogenes we studied, c-myb and c-fes, were found by Northern blot analysis to be clearly expressed in human normal myeloid cells and in several leukemic myeloid cell lines several years ago.[28,29] Furthermore, the c-myb protooncogene is expressed in the late G1 phase of the cell cycle of PHA-stimulated T lymphocytes as described.[30] Changes in c-myb and c-fes protooncogene expression that occur during *in vitro* differentiation of myeloid cells are remarkably different, as already described by both Northern blot analysis[28,29] and *in situ* hybridization.[31] The c-myb oncogene mRNA abundance drops to undetectable levels after a few hours of treatment of HL-60 cells with several differentiation inducers. This behavior, which is common to several other cell-cycle-related genes, may be associated with the inhibition of cell proliferation, which is the consequence of the biological effect of all differentiation inducers. A completely different pattern of expression was shown by the c-fes protooncogene. In fact, the induction to granulocytic differentiation of HL-60 cells obtained with retinoic acid (RA) or dimethylsulfoxide (DMSO) causes a clear increase in the level of expression of this oncogene, whose mRNA is therefore abundant in the granulocytic[28] but not in the monocytic or macrophagic[32] HL-60 cells. The role played by c-myb in myeloid differentiation is still uncertain. It has been shown that an antisense c-myb oligomer inhibits proliferation of myeloid cell lines.[33] However, c-myb involvement in myeloid differentiation is also suggested by two observations: (1) c-myb function seems to be preferred for the proliferation of intermediate hematopoietic progenitors, whereas it seems less critical for the proliferation of early progenitors[34]; (2) c-myb transfection suppresses macrophage and erythroid differentiation.[35,36] The role of c-myb oncogene will be discussed in detail by Calabretta in another paper in this volume.

CORRELATION BETWEEN PROLIFERATIVE CAPACITY AND DIFFERENTIATION POTENTIAL IN MYELOID CELLS *IN VITRO*

The observations made in our laboratory that acute leukemic blast cells are arrested mainly in the G1 phase of the cell cycle and that they lose the capability to progress through the G1-S transition have led us to postulate a "pathology of the cell cycle"[37,38] probably determined by several posttranscriptional alterations in gene expression occurring in blast cell populations.[39–42] Furthermore, blast cells are incapable of carrying out the genetic program leading to terminal differentiation, so that arrest occurs at different stages.[43,44] To explore the molecular mechanisms underlying a particular biological behavior of leukemic cells, particularly the possible relation between the proliferative and differentiative pathways, we studied the ability of G1-arrested HL-60 myeloid cells to differentiate *in vitro* using different inducer agents. It is important to specify that the HL-60 cells were obtained several years ago at the seventh passage directly from the Tumor Cell Biology Laboratory (NCI, Bethesda, MD)[45] and they were not a variant HL-60 clone.[46] In fact, this myeloid, M2 type,[47] cell line has become the archetype model for studying the regulation of hematopoietic growth and differentiation even though they are leukemic cells with several chromosomal abnormalities. These cells are capable of granulocytic, macrophagic, and monocytic differentiation depending on the differentiation inducer used.[48–51] We want to emphasize that several HL-60 variant cells are described to show that growth arrest and expression of the differentiated phenotype can be uncoupled, indicating separate regulatory pathways. These cells allow *in vitro* study of the control of cellular proliferation and differentiation along the myelomonocytic pathways at the cellular, molecular, and genomic levels. It has been shown that synthetic oligodeoxynucleotides complementary to c-myc and c-myb protooncogenes can inhibit the proliferative capacity of these cells. These specific AS aODN drastically reduce the growth fraction of these cells which in untreated cultures is about 45% with a doubling time of 24 hours. To inhibit the proliferative capacity of these cells we used an anti-c-myb specific oligomer instead of an antisense c-myc, as the c-myc inactivation of cell proliferation also causes spontaneous differentiation of these cells.[52,53] After 5 days of treatment of HL-60 cells with an antisense c-myb oligomer, more than 90% of the cells were arrested in the G1 phase of the cell cycle as evaluated by flow cytometry. At variance from the cell treated with c-myc antisense oligomer, these cells did not show spontaneous differentiation. These G1-arrested HL-60 cells were treated with different inducer agents such as DMSO and RA to trigger granulocytic differentiation, TPA to induce mono/macrophagic differentiation, and vitamin D3 to induce monocytic differentiation. In all experiments the S or AS c-myb oligomer was present in cultures at a concentration of about 14 μM. Unexpectedly, when we treated the G1-arrested HL-60 cells with DMSO or RA, we were unable to induce terminal granulocytic differentiation, but the cells, even after 8 days of treatment, had the monocytic morphology and prevalent expression of monocytic specific markers, particularly butyrate esterase, accompanied by a marked decrease in the expression of chloroacetate esterase, specific for granulocytes (FIG. 1). These results strongly suggest that in complete inhibition of cell proliferation, myeloid HL-60 cells lose the capability to differentiate along the granulocytic differentiation pathway, raising several questions about the mechanisms of action of the specific granulocytic differentiation inducers. It is possible that down-modulation of the expression of retinoic acid receptors induced by the G1 arrest of cell proliferation is responsible for the lack of activity of this inducer. Furthermore, our experiments suggest that the proliferative and differentiative genetic programs must be carried out simultaneously to push the cell in the granulocytic terminal differentiation pathway. The treatment of normal bone marrow cells with an AS c-myb oligomer allows the formation of monocytic/macrophagic colonies in methyl cellulose or agar dishes but not the formation of granulocytic colonies (personal communication with Bruno Calabretta). The spontaneous differentiation observed in HL-60 cells after inhibition of cell proliferation with a c-myc AS oligomer is controversial. A report by Wickstrom *et al.*[52] indicates that the differentiated cells are monocytes, whereas Holt *et al.*[53] indicate that the spontaneously differentiated cells are granulocytes. These discrepancies can be due to the different HL-60 cells used for these experiments. On the contrary, G1-arrested HL-60 cells can differentiate, as expected, to mono/macrophagic cells when treated with TPA and to a pure monocytic population when treated with vitamin D3. Our results show that G1-arrested cells can differentiate only along the monocytic and macrophagic pathways, and not along the granulocytic pathway.[54,55] To attain granulocytic differentiation apparently requires the simultaneous activation of both the proliferative and the differentiative metabolic pathways. Several questions arise from our observations: (1) What is the role of the protooncogene c-myb in terminal differentiation? We are studying now whether simple proliferation arrest is responsible for the different differentiation potential of HL-60 cells or whether the genetic pathway activated by the product of this protooncogene is involved specifically in granulocytic differentiation. (2) Is a down-modulation of RA receptor in G1-arrested HL-60 cells responsible for the failure of granulocytic differentiation in these cells? We are now tackling the question by comparing the number of RA receptors and vitamin D3 receptors in these G1-arrested cells. (3) Are our results a particular biological behavior of the leukemic HL-60 cell line? In other normal biological models, such as normal human bone marrow cells, the antisense c-myb oligomer

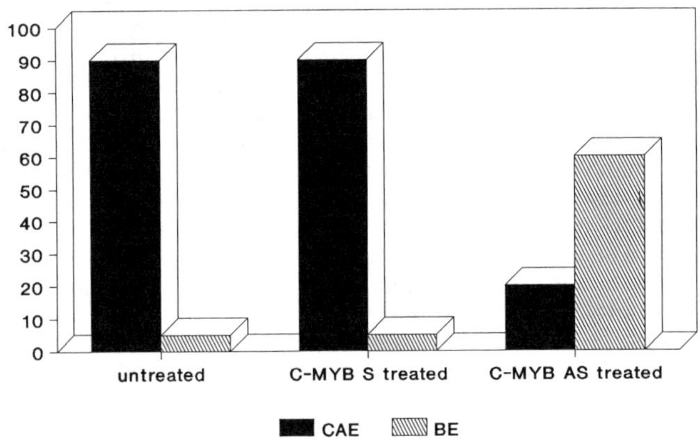

FIGURE 1. Effect of the c-myb antisense (AS) oligodeoxynucleotide on the expression of naphthol AS-D chloroacetate esterase (CAE), specifically detectable in granulocytic cells, and alfa-naphthylbutyrate esterase (BE), specifically detectable in monocytic cells, after complete inhibition of HL-60 cell proliferation and treatment of G1 arrested cells with RA-DMSO for 5 days. The increase of the BE enzyme activity in the c-myb AS treated cells is evident. These data confirm the monocytic differentiation of these c-myb AS treated cells. In the untreated and c-myb sense (S) treated HL-60 cells the CAE enzymatic activity after 5 days of RA-DMSO treatment is prevalent confirming the granulocytic differentiation of these cells. The percentage of positive cells was evaluated by counting at least 200 cells. Purified peripheral blood granulocytes or monocytes were used as positive controls for cytochemical assay.

treatment allows the formation of monocytic/macrophagic colonies only. In conclusion, we can speculate from the data obtained that during myeloid differentiation, different genetic programs are activated by the different inducer agents and the possibility of preferentially activating one of these differentiation programs depends on the functional state of the precommitted myeloid cell.

ROLE OF C-FES PROTOONCOGENE IN MYELOID DIFFERENTIATION

Several oncogenes that encode proteins with tyrosine-kinase activity are involved in hematopoietic development, particularly in the proliferation and differentiation of hematopoietic early progenitors. As mentioned in the introduction, several protooncogene products regulate the proliferative pathway also in hematopoietic cells (TABLE 1). It is our belief that the study of the function of oncogenes will provide critical insights into the molecular mechanisms underlying cell differentiation. Mammalian cells contain a gene equivalent to the avian c-fps which was originally identified as the viral oncogene of the Snyder-Theilen feline sarcoma virus (ST-FeSV) and hence is known as c-fes; despite the difference in nomenclature, c-fps and c-fes correspond to a single, common genetic locus[56,57] mapped on chromosome 15q25-q26.[58] This protooncogene is of particular interest mainly for two reasons: (1) The avian c-fps gene is involved in hematopoiesis, as the transcription and synthesis of the $p98^{c\text{-}fps}$ protein is largely restricted to hematopoietic tissue such as bone marrow.[59] Furthermore, the v-fps oncogene encodes for the transforming Fujinami avian sarcoma virus (FSV) protein ($p140^{gag\text{-}fps}$) which is characterized by an intrinsic tyrosine-specific protein kinase activity capable of both autophosphorylation and phosphorylating exogenous protein substrates.[60] (2) A survey of protooncogene expression showed that both c-fes mRNA and its encoded protein $p92^{c\text{-}fes}$ are restricted to myeloid hematopoietic cells.[28,61–63] We, in fact, first showed by Northern blot analysis and by an *in situ* hybridization technique performed on normal bone marrow specimens that c-fes mRNA is myeloid specific, and with these techniques it is not detectable in lymphoid normal and leukemic cells.[28,31] (3) we showed also that c-fes mRNA expression is remarkably increased in granulocytic terminal differentiated HL-60 cells, so that we can define c-fes as a granulocytic differentiation associated gene. (4) Together with c-abl, c-fes is an oncogenic tyrosine kinase that is not exclusively cell membrane associated, but can be detected in the cytoplasm of the cell. We further know that the carboxyl-terminus of c-fes protein is unrelated to that of the src family.[64] (5) The 5′ cDNA region of the c-fes oncogene is different from the 5′ regions of the other tyrosine kinases encoded by other oncogenes.[65] We therefore selected this specific c-fes region to synthetize an 18-mer antisense oligomer to inactivate c-fes mRNA in HL-60 cells. To obtain a better understanding of the role of the c-fes oncogene in myeloid differentiation, we studied the ability of HL-60 myeloid cells to differentiate under the effects of several inducers after practically complete inhibition of c-fes function using an antisense c-fes oligomer complementary to the 5′ region of the c-fes mRNA. The method using antisense oligodeoxynucleotides to selectively inactivate gene products is well described. Firstly, we evaluated the growth kinetics of HL-60 cells after several days of treatment with the c-fes antisense oligomer at a concentration in the culture medium of 10 μM. Control experiments were carried out without oligomer treatment and with sense c-fes oligomer treatment. Nonmodified oligomers were added at 24-hour intervals based on the stability of the oligomer in the culture medium, and we estimate that the oligomer concentration was at least 10 μM in all cultures. No inhibition of HL-60 cell proliferation was observed after 5 days of antisense c-fes oligomer treatment, because the fraction of the cell population in each phase of the cell cycle, as evaluated by flow cytometric analysis, remained unmodified after this treatment. The lack of relevance of the c-fes mRNA and protein to cell proliferation was confirmed by the disappearance of any detectable amount of the specific mRNA by the RT-PCR technique after 70–80 hours of treatment. In all experiments performed using different oligomer preparations and HL-60 cell cultures, the growth fraction of HL-60 cells remained unmodified by the c-fes antisense oligomer when compared with that of untreated and sense oligomer-treated cultures. Furthermore, in all experiments the cell viability after 120 hours of treatment with c-fes antisense and sense oligomers was higher than 90–95%. Therefore, after this period of oligomer treatment, the culture medium was discarded and fresh medium containing the

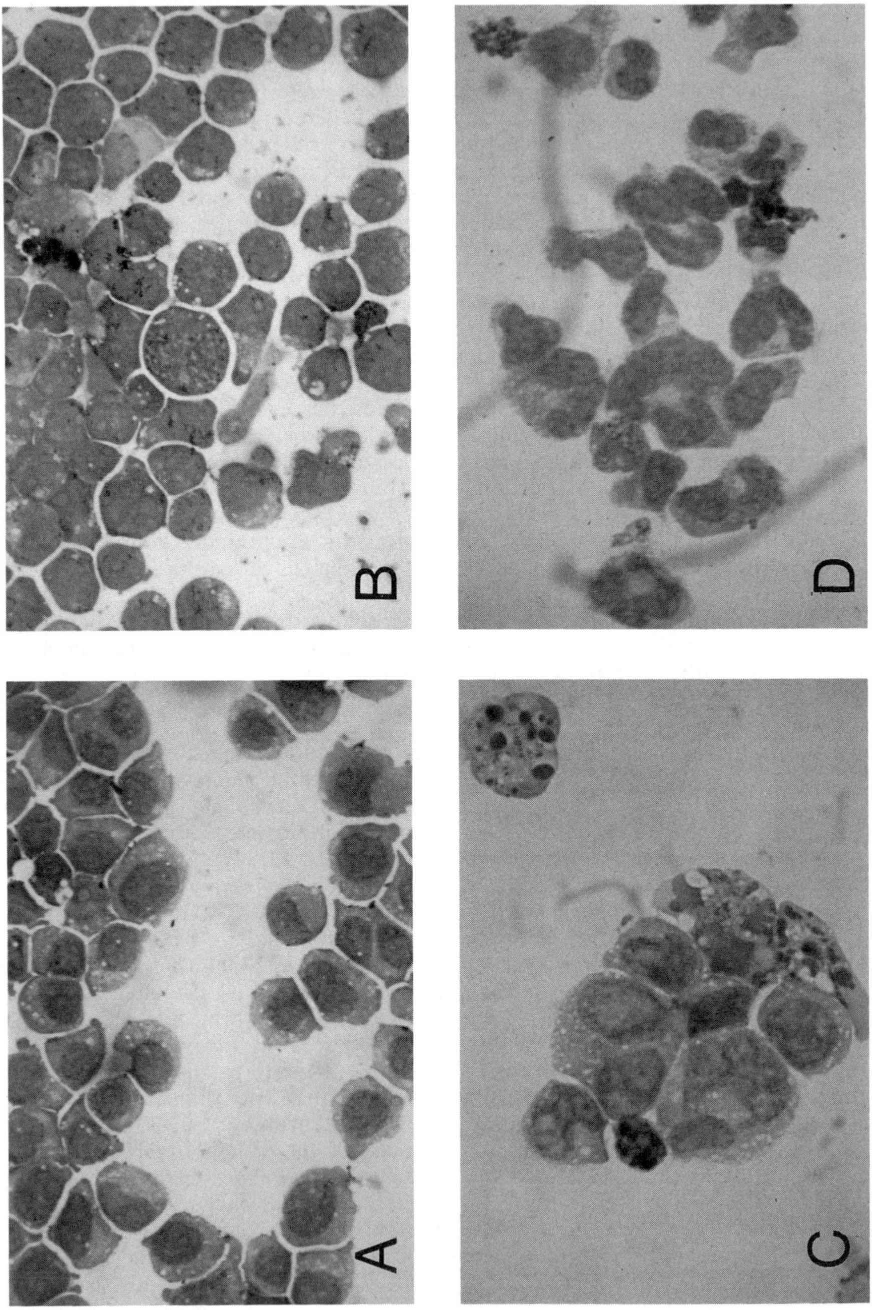

FIGURE 2.

different inducer agents was added to the cultures. The cells were then harvested after 24 hours of treatment with TPA, four days of treatment with vitamin D3, and at 24-hour intervals for 5 days after treatment with RA-DMSO. The morphology of the cells in the various experiments performed was evaluated on cytospin preparations stained with May-Grunwald-Giemsa. Cytochemistry studies were performed using standard conditions. A small aliquot of the cells was used for RNA extraction using standard techniques. The morphology of the cells after incubation with TPA and vitamin D3 indicates that these cells were triggered in the terminal differentiation monocytic pathway. Furthermore, in several experiments performed with the c-fes antisense oligomer in TPA-treated cells, we never observed adherence to the plastic flask wall, which is a well-known characteristic of cells terminally differentiated to macrophages. This result has allowed us to postulate an interference of the c-fes protooncogene with the expression of the different adhesion molecules.[66,67] Different behavior was observed when the HL-60 cells were induced to differentiate with RA-DMSO after almost complete inhibition of the c-fes mRNA. The morphology of the HL-60 cells treated as just described was studied at 24-hour intervals after the addition of the inducers. The morphology with respect to proliferating HL-60 cells was almost unmodified, as these cells maintained the characteristics of immaturity but the cytochemical character of monocytic differentiation. However, the most surprising finding was that a proportion of cells increasing from about 20% at 24 hours to more than 90% after 80–90 hours died, showing pyknoses and karyorrhexis. The small proportion of surviving cells had a monocytoid morphology. Cell death was clearly due to the inability of the HL-60 c-fes antisense oligomer-treated cells to activate the genetic program underlying the granulocytic differentiation pathway, because the same antisense oligomer had no toxic effect in HL-60 cells induced to differentiate with TPA or vitamin D3. Control cultures carried out without oligomers or with c-fes sense oligomer differentiated normally to granulocytes when treated with RA-DMSO, and cell viability after 5 days of treatment remained higher than 95% (FIG. 2). Overall, our experiments lead to the conclusion that granulocytic differentiation, besides c-myb conditioned proliferation, requires the activity of $p90^{c\text{-}fes}$ protein kinase encoded by the c-fes protooncogene, an activity that cannot be substituted by other kinases, as the specific inactivation of this oncogene causes a complete block of granulocytic differentiation determined by premature cell death.[54,68]

POSSIBLE INTERFERENCE OF C-FES PROTOONCOGENE WITH THE MECHANISMS OF PROGRAMMED CELL DEATH (APOPTOSIS)

Our investigation showed that HL-60 cells induced to differentiate by granulocytic inducers after complete inhibition of the c-fes protooncogene die. The morphol-

FIGURE 2. (**A**) Morphology of untreated proliferating HL-60 cells. (**B**) Morphology of c-fes AS oligomer treated HL-60 cells. The antisense oligomer treatment was performed for 5 days at a concentration of 10 μM. The viability, as evaluated by the tripan bleu exclusion test, was higher than 95%. After this period of oligomer treatment the c-fes mRNA was not detectable with the RT-PCR reaction. The distribution of the cells in the different phases of the cell cycle, as evaluated by flow cytometric analysis, was the same of that of the untreated cells. (**C**) morphology of the cells described in picture B after 36 hours of treatment with RA-DMSO in the presence of the c-fes AS oligomer. It is evident that several cells are dead with the characteristic morphology described in the text. (**D**) morphology of the cells described in picture B after 5 days of treatment with RA-DMSO in the presence of c-fes S oligomer. The cells are capable, as expected, to reach the end point of granulocytic differentiation.

ogy of these dead cells is more compatible with the characteristics of programmed cell death (apoptosis) than with the characteristics of necrotic cell death (resulting from injury, infection, or other external stress). In fact, the majority of cells treated with c-fes antisense oligomer together with RA-DMSO showed the classic features of apoptosis: pyknotic chromatin, nuclear fragmentation, vacuolated cytoplasm, plasma membrane ruffling, and apoptotic bodies.[69] In our laboratory preliminary results obtained with Southern blot analysis performed on DNA extracted from these cells showed a characteristic "DNA ladder" pattern resulting from the cleavage of nuclear DNA in internucleosomal regions. These preliminary but indicative results have led us to perform more detailed investigations on these dead cells using the electron microscope. Furthermore, these preliminary results confirm that inhibition of granulocytic cell differentiation in HL-60 cells treated with a c-fes antisense oligomer is not a toxic effect. Therefore, whereas most oncogenes probably contribute to neoplasia by promoting cell proliferation, cancer may also be induced through the activation of genes whose products allow the survival of a cell that should otherwise die.[25] So far, we have no evidence that c-fes is an anti-apoptotic gene in normal cells, but we have strong evidence of the basic role of this protooncogene in the genetic program underlying the granulocytic differentiation pathway.

OTHER GENE AND PROTOONCOGENE PRODUCTS INVOLVED IN MYELOID DIFFERENTIATION

Other interesting results come from the work of Wu and collaborators[70] who demonstrated that c-fms protooncogene has an important role in HL-60 cells during the mono/macrophagic differentiation induced by phorbol-12-myristate-13-acetate (PMA). They found that the differentiation of HL-60 cells into macrophages was accompanied by transcription of the c-fms protooncogene. This oncogene was identified as the transforming gene of the McDonough strain of feline sarcoma virus, which also transforms fibroblasts *in vitro.* The cellular homolog, called c-fms, is the receptor for CSF1, which is a factor involved in the differentiation and perhaps proliferation of monocytes and macrophages. After treatment with a c-fms antisense oligomer, PMA- and hrGM-CSF–induced macrophage differentiation was significantly inhibited, whereas neither vitamin D3-induced monocytic nor DMSO-induced granulocytic differentiation was inhibited. These investigators concluded that HL-60 cells differentiate into monocytes and macrophages along two different pathways; one is involved in the action of the c-fms oncogene and the other is not. Several other gene products are associated, that is, their mRNA and protein are induced in myeloid differentiation. For example, transcription of c-jun is increased several times when proliferating HL-60 cells are induced to mono/macrophagic differentiation by TPA or monocytic differentiation by vitamin D3 (1,25-dihidroxy vitamin D3).[71] Another gene involved in myeloid differentiation is protein kinase C, which is induced during treatment of HL-60 cells with TPA but not after treatment of these cells with RA.[72] Other genes and oncogenes are involved in myeloid differentiation. Particularly interesting are the genes belonging to the "src family." All these genes share high homology, and some of them, such as hck, lck, blk, c-fes, ltk, c-fgr, c-src, and protein kinase C, are strictly associated in their expression with myeloid differentiation (TABLE 2).[74–80] We do not have any evidence that these genes or oncogenes play a regulatory role in the myeloid differentiation pathway, but their tyrosine kinase activity can be specifically involved, as we demonstrated for the c-fes protooncogene, in one of the different metabolic pathways underlying granulocytic, monocytic, or macrophagic differentiation.

CONCLUSIONS

The results of our experiments support the concept that differentiation, like proliferation, may be genetically dissected, and this observation reveals different orders of function of involved genes. As a matter of fact, we have to distinguish several different conditions: (1) Some genes conditioning the cell cycling, such as c-fos, c-myc, c-myb, and Jun D, must be shut off to permit the occurrence of the entire differentiation program. (2) Other genes have to be activated to allow a single specific differentiation program; this is the case of transcriptional factor such as GATA 1 and others. (3) A specific differentiation pathway may require the increased abundance of already transcribed genes; the inactivation of some of them inhibits in different ways the genetic program of a specific pathway (for instance, inhibition of granulocytic differentiation by c-fes and inhibition of macrophage differentiation by c-fms). (4) The inhibition of some other genes whose expression increases during differentiation does not interfere with the specific differentiation pathway, but may deprive the differentiated cell of one of its specific functions. This is the case, for instance, of MPO or lactoferrin genes in granulocytic cells. One conclusion we can draw from these observations is that different differentiation end points, such as that

TABLE 2. Genes and Oncogenes Encoding for Protein Tyrosine Kinases That Are Differentiation Related

Gene	Differentiation Pathway	Cell Lines	References
HCK	Myeloid	HCK expression increases during differentiation of HL-60 cells induced by DMSO or TPA and differentiation of ML-1 cells induced by TPA	73,74
LCK	Lymphoid	LCK is highly and specifically expressed in normal cells of lymphoid lineage and is also expressed in lymphoid neoplastic cell lines and nonlymphoid neoplastic cell lines especially of colon tumors	75
BLK	Lymphoid	Normal and neoplastic B lymphoid cells	76
C-FES	Myeloid	C-FES expression increases many times in granulocytic differentiation of HL-60 cells induced by RA and in normal bone marrow granulopoiesis	28,31
LTK	Leukemic cells	LTK mRNA is expressed with high specificity by leukemic cells but not by other neoplastic cells or normal hematopoietic cells	77
C-FGR	Myeloid	This gene is expressed in normal and leukemic monocytes and in B-cell malignancies, especially those that are EBV positive	79
C-SRC	Myeloid	Induction of monocytic and granulocytic differentiation of HL-60 cells by TPA and DMSO, respectively, is associated with activation of $pp60^{src}$ tyrosine kinase activity	78
PKC	Myeloid	PKC expression is increased in HL-60 cells treated with RA	72

of myeloid cells, are reached through specific and separately regulated genetic programs. It seems very likely that different cytokines play an important role in modulating the different pathways.

REFERENCES

1. DENHARDT, D. T., D. R. EDWARDS & C. L. J. PARFETT. 1986. Gene expression during the mammalian cell cycle. Bichim. Biophys. Acta **865:** 83–125.
2. FERRARI, S. & R. BASERGA. 1987. Oncogenes and cell cycle genes. BioEssays **7:** 9–13.
3. SHIOSAKA, T. & Y. TANAKA. 1989. Expression of selected genes and oncogenes in differentiated HL-60 cells and primary cells from human leukemias. Anticancer Res. **9:** 1249–1264.
4. BASERGA, R. 1985. The Biology of Cell Reproduction. Harvard University Press. Cambridge, Massachusetts/London England.
5. HIRSCHHORN, R. R., P. ALLER, Z. YOUAN, C. W. GIBSON & R. BASERGA. 1984. Cell-cycle-specific cDNAs from mammalian cells temperature sensitive for growth. Proc. Natl. Acad. Sci. USA **81:** 6004–6008.
6. SCHNEIDER, C., R. M. KING & L. PHILIPSON. 1988. Genes specifically expressed at growth arrest of mammalian cells. Cell **54:** 787–793.
7. MEISS, H. K. & C. BASILICO. 1972. Temperature sensitive mutants of BHK 21 cells. Nature New Biol. **239:** 241–244.
8. GRECO, A., M. ITTMAN & C. BASILICO. 1987. Molecular cloning of a gene that is necessary for G1 progression in mammalian cells. Proc. Natl. Acad. Sci. USA **84:** 1565–1569.
9. METCALF, D. 1989. The molecular control of cell division, differentiation commitment and maturation in haemopoietic cells. Nature **339:** 27–30.
10. DYNAN, W. S. & R. TJIAN. 1985. Control of eukaryotic messenger RNA synthesis by sequence-specific DNA-binding proteins. Nature **316:** 774–778.
11. PEVNY, L., M. C. SIMON, E. ROBERTSON, W. H. KLEIN, S. F. TSAI, V. D'AGATI, S. H. ORKIN & F. COSTANTINI. 1991. Erythroid differentiation in chimeric mice blocked by a targeted mutation in the gene for transcription factor GATA-1. Nature **349:** 257–260.
12. STEIN, G., J. LIAN, J. STEIN, R. BRIGGS, V. SHALHOUB, K. WRIGHT, U. PAULI & A. VAN WIJNEN. 1989. Altered binding of human histone gene transcription factors during the shutdown of proliferation and onset of differentiation in HL-60 cells. Proc. Natl. Acad. Sci. USA **86:** 1865–1869.
13. SHARP, P. A. 1989. Gene regulation and oncogenes. Cancer Res. **49:** 2188–2194.
14. HYLAND, J. K., C. M. ROGERS, E. M. SCOLNIK, R. B. STEIN, R. ELLIS & R. BASERGA. 1985. Microinjected ras family oncogenes stimulate DNA synthesis in quiescent mammalian cells. Virology **141:** 333–336.
15. STEIN, C. A. & J. S. COHEN. 1988. Oligodeoxynucleotides as inhibitors of gene expression: A review. Cancer Res. **48:** 2659–2668.
16. CALABRETTA, B. 1991. Inhibition of protooncogene expression by antisense oligodeoxynucleotides: Biological and therapeutic implications. Cancer Res. **51:** 4505–4510.
17. COLMAN, A. 1990. Antisense strategies in cell and developmental biology. J. Cell Sci. **97:** 399–409.
18. KATSUKI, M., M. SATO, M. KIMURA, M. YOKOYAMA, K. KOBAYASHI & T. NOMURA. 1988. Conversion of normal behavior to shiverer by myelin basic protein antisense cDNA in transgenic mice. Science **241:** 593–595.
19. LOZANNE, A. D. & J. SPUDICH. 1987. Disruption of the Dictyostelium myosin heavy chain gene by homologous recombination. Science **236:** 1088–1091.
20. MAHER, L. J., III, P. B. DERVAN & B. J. WOLD. 1990. Kinetic analysis of oligodeoxynucleotide-directed triple-helix formation on DNA. Biochemistry **29:** 8820–8826.
21. DER KROL, A. R., J. N. M. MOL & A. R. STUITJE. 1988. Modulation of eukariotic gene expression by complementary RNA or DNA sequences. BioTechniques **6:** 958–976.
22. ZON, G. 1988. Oligonucleotide analogues as potential chemotherapeutic agents. Pharmaceutical Res. **5:** 539–549.
23. BOTTINGER, D. 1989. Interaction of oncogenes with differentiation programs. Curr. Topics Microbiol. Immunol. **147:** 31–77.

24. MARTIN, S. J., J. G. BRADLEY & T. G. COTTER. 1990. Hl-60 cells induced to differentiate towards neuthophils subsequently die via apoptosis. Clin. Exp. Immunol. **79:** 448–453.
25. YONISH-ROUACH, E., D. RESNITZKY, J. LOTEM, L. SACHS, A. KIMKI & M. OREN. 1991. Wild-type p53 induces apoptosis of myeloid leukemic cells that is inhibited by interleukin-6. Nature **352:** 345–347.
26. HOCKENBERY, D., G. NUNEZ, C. MILLIMAN, R. D. SCREIBER & S. KORSMEYER. 1990. Bcl-2 is an inner mitochondrial membrane protein that blocks programmed cell death. Nature **348:** 334–336.
27. WILLIAMS, G. T., C. A. SMITH, E. SPOONCER, T. M. DEXTER & D. R. TAYLOR. 1990. Haemopoietic colony stimulating factors promote cell survival by suppressing apoptosis. Nature **343:** 76–79.
28. FERRARI, S., U. TORELLI, L. SELLERI, A. DONELLI, D. VENTURELLI, L. MORETTI & G. TORELLI. 1985. Expression of c-fes oncogene occurs at detectable levels in myeloid but not lymphoid cell populations. Br. J. Haematol. **59:** 21–25.
29. FERRARI, S., U. TORELLI, L. SELLERI, A. DONELLI, D. VENTURELLI, L. MORETTI & G. TORELLI. 1985. Study of the levels of expression of two oncogenes c-myc and c-myb in acute and chronic leukemias of both lymphoid and myeloid lineage. Leukemia Res. **9:** 833–842.
30. TORELLI, G., L. SELLERI, A. DONELLI, S. FERRARI, G. EMILIA, D. VENTURELLI, L. MORETTI & U. TORELLI. 1985. Activation of c-myb expression by phytohaemagglutinin stimulation in human T lymphocytes. Mol. Cell. Biol. **5:** 2874–2877.
31. EMILIA, G., A. DONELLI, S. FERRARI, U. TORELLI, L. SELLERI, P. ZUCCHINI, L. MORETTI, D. VENTURELLI, G. CECCHERELLI & G. TORELLI. 1986. Cellular levels of mRNA for c-myc, c-myb, and c-fes onc-genes in normal myeloid and erythroid precursor of human bone marrow: An *in situ* hybridization study. Br. J. Haematol. **62:** 287–292.
32. LANFRANCONE, L., P. MANNONI, M. J. PEBUSQUE, A. CARE, C. PESCHLE, F. GRIGNANI & P. G. PELICCI. 1989. Expression pattern of c-fes oncogene mRNA in human myeloid cells. Int. J. Cancer Sup. **4:** 35–38.
33. ANFOSSI, G., A. M. GEWIRTZ & B. CALABRETTA. 1989. An oligomer complementary to c-myb-encoded mRNA inhibits proliferation of human myeloid leukemia cell lines. Proc. Natl. Acad. Sci. USA **86:** 3379–3383.
34. CARACCIOLO, D., D. VENTURELLI, M. VALTIERI, C. PESCHLE, A. M. GEWIRTZ & B. CALABRETTA. 1990. Stage-related proliferative activity determined c-myb functional requirements during normal human hematopoiesis. J. Clin. Invest. **85:** 55–61.
35. DURBAN, E. M. & D. BOETTIGER. 1981. Differential effects of transforming avian RNA tumor viruses on avian macrophages. Proc. Natl. Acad. Sci. USA **78:** 3600–3604.
36. CLARKE, M. F., J. F. KUKOWSKA-LATALLO, E. WESTIN, M. SMITH & E. PROCOWNICK. 1988. Constitutive expression of a c-myb cDNA blocks Friend murine erythroleukemia cell differentiation. Mol. Cell. Biol. **8:** 884–892.
37. TORELLI, U. & S. FERRARI. 1987. Oncogenes and cell cycle genes and proliferation arrest in acute myeloid leukemia. *In* Genotypic, Phenotypic and Functional Aspects of Haematopoiesis. F. Grignani, F. M. Martelli, D. Y. Mason, eds. Vol. **41:** 123–131. Serono Symposia Publications. Raven Press. New York, NY.
38. FERRARI, S., E. TAGLIAFICO, G. CECCHERELLI, L. SELLERI, B. CALABRETTA, A. DONELLI, P. TEMPERANI, M. SARTI, S. SACCHI, G. EMILIA, G. TORELLI & U. TORELLI. 1989. Myeloperoxidase gene expression in acute and chronic myeloid leukemias: Relationship to the expression of cell cycle related genes. Leukemia **3:** 423–430.
39. FERRARI, S., R. MANFREDINI, E. TAGLIAFICO, E. ROSSI, A. DONELLI, G. TORELLI & U. TORELLI. 1990. Noncoordinate expression of S6, S11 and S14 ribosomal protein genes in leukemic blast cells. Cancer Res. **50:** 5825–5828.
40. FERRARI, S., E. TAGLIAFICO, R. MANFREDINI, A. GRANDE, E. ROSSI, P. ZUCCHINI, G. TORELLI & U. TORELLI. 1992. Abundance of the primary transcript and its processed product of growth-related genes in normal and leukemic cells during proliferation and differentiation. Cancer Res. **52:** 11–16.
41. FERRARI, S., E. TAGLIAFICO, M. D'INCA, G. CECCHERELLI, R. MANFREDINI, L. SELLERI, A. DONELLI, S. SACCHI, G. TORELLI & U. TORELLI. 1990. Ratios between the abundance of messenger RNA and the corresponding protein of two growth-related genes, c-myc and vimentin, in leukemic blast cells. Cancer Res. **50:** 1988–1991.
42. FERRARI, S., M. T. MARIANO, E. TAGLIAFICO, M. SARTI, G. CECCHERELLI, L. SELLERI, F.

MERLI, F. NARNI, A. DONELLI, G. TORELLI & U. TORELLI. 1988. Myeloperoxidase gene expression in blast cells with lymphoid phenotype in cases of acute lymphoblastic leukemia. Blood **72:** 873–877.

43. FEARON, E. R., P. J. BURKE, B. A. ZEHNBAUER, B. VOGELSTEIN & C. A. SCHIFFER. 1986. Differentiation of blast cells in acute nonlymphocytic leukemia. N. Engl. J. Med. **315:** 1488.
44. RIFKIND, R. A. 1986. Acute leukemia and cell differentiation. N. Engl. J. Med. **315:** 56–57.
45. COLLINS, S. J., R. C. GALLO & R. E. GALLAGHER. 1977. Continuous growth and differentiation of human myeloid leukemic cells in suspension culture. Nature **270:** 347–349.
46. YEN, A. 1990. HL-60 cells as a model of growth control and differentiation: The significance of variant cells. Hematol. Rev. **4:** 5–46.
47. DALTON, W. T., M. J. AHEARN, K. B. MCCREDIE, E. J. FREIREICH, S. A. STASS & J. M. TRUJILLO. 1988. HL-60 cell line was derived from a patient with FAB-M2 and not FAB-M3. Blood **71:** 242–247.
48. COLLINS, S. J., F. W. RUSCETTI, R. E. GALLAGHER & R. C. GALLO. 1978. Terminal differentiation of human promyelocytic leukemia cells induced by dimethylsulfoxide and other polar solvents. Proc. Natl. Acad. Sci. USA **75:** 2458–2462.
49. COLLINS, S. J., K. A. ROBERTSON & L. MUELLER. 1990. Retinoic acid-induced granulocytic differentiation of HL-60 myeloid leukemia cells is mediated directly through the retinoic acid receptor (RAR-a). Mol. Cell. Biol. **10:** 2154–2163.
50. ROVERA, G., D. SANTOLI & C. DAMSKY. 1979. Human promyelocytic leukemia in culture differentiate into macrophage-like cells when treated with forbol esters. Proc. Natl. Acad. Sci. USA **76:** 2779–2783.
51. KASUKABE, T., Y. HONMA, M. HOZUMI, T. SUDA & Y. NISHII. 1987. Control of proliferating potential of myeloid leukemia cells during long-term treatment with vitamin D3 analogues and other differentiation inducers in combination with antileukemic drugs: *In vitro* and *in vivo* studies. Cancer Res. **47:** 567–572.
52. WICKSTROM, E. L., T. A. BACON, A. GONZALES, D. L. FREEMAN, G. H. LYMAN & E. WICKSTROM. 1988. Human promyelocytic leukemia HL-60 cell proliferation and c-myc protein expression are inhibited by an antisense pentadecadeoxynucleotide targeted against c-myc mRNA. Proc. Natl. Acad. Sci. USA **85:** 1028–1032.
53. HOLT, J. T., R. L. REDNER & A. W. NIENHUIS. 1988. An oligomer complementary to c-myc mRNA inhibits proliferation of HL-60 promyelocytic cells and induces differentiation. Mol. Cell. Biol. **8:** 963–973.
54. FERRARI, S., A. DONELLI, R. MANFREDINI, M. SARTI, R. RONCAGLIA, E. TAGLIAFICO, E. ROSSI, G. TORELLI & U. TORELLI. 1990. Differential effects of c-myb and c-fes antisense oligodeoxynucleotides on granulocytic differentiation of human myeloid leukemia HL60 cells. Cell Growth Diff. **1:** 543–548.
55. ROVERA, G., N. OLASHAW & P. MEO. 1980. Terminal differentiation in human promyelocytic leukaemic cells in the absence of DNA synthesis. Nature **284:** 69–70.
56. HUNTER, T. 1989. Oncogene products in the cytoplasm: The protein kinases. *In* Oncogenes and the Molecular Origins of Cancer. Robert A. Weinberg, ed.: 147–173. Cold Spring Harbor Laboratory Press. Cold Spring Harbor, NY.
57. ROEBROEK, A. J. M., J. A. SCHALKEN, S. VERBEEK, A. M. W. VAN DEN OUWELAND, C. ONNEKINK, H. P. J. BLOEMERS & W. J. M. VAN DE VEN. 1985. The structure of the human c-fes/fps proto-oncogene. EMBO J. **4:** 2897–2903.
58. DALLA-FAVERA, R., G. FRANCHINI, S. MARTINOTTI, F. WONG-STAHL, R. C. GALLO & C. M. CROCE. 1982. Chromosomal assignment of the human homologues of feline sarcoma virus and avian myeloblastosis oncogenes. Proc. Natl. Acad. Sci. USA **79:** 4714–4717.
59. SAMARUT, J., B. MATHEY-PREVOT & H. HANAFUSA. 1985. Preferential expression of the c-fps protein in chicken macrophages and granulocytic cells. Mol. Cell. Biol. **5:** 1067–1072.
60. PAWSON, T., J. GUYDEN, T. H. KUNG, K. RADKE, T. GILMORE & G. S. MARTIN. 1980. A strain of Fujinami sarcoma virus which is temperature-sensitive in protein phosphorylation and cellular transformation. Cell **22:** 767–775.
61. FELDMAN, R. A., J. L. GABRILOVE, J. P. TAM, M. A. S. MOORE & H. HANAFUSA. 1985.

Specific expression of the human cellular fps/fes encoded protein NCP92 in normal and leukemic myeloid cells. Proc. Natl. Acad. Sci. USA **82:** 2379–2383.
62. Smithgall, T. E., G. Yu & R. I. Glazer. 1988. Identification of the differentiation associated p93 tyrosine protein kinase of HL-60 leukemia cells as the product of the human c-fes locus and its expression in myelomonocytic cells. J. Biol. Chem. **263:** 15050–15055.
63. Greer, P., V. Maltby, J. Rossant, A. Bernstein & T. Pawson. 1990. Myeloid expression of the human c-fps/fes proto-oncogene in transgenic mice. Mol. Cell. Biol. **10:** 2521–2527.
64. Parsons, J. T. & M. J. Weber. 1989. Genetics of src: Structure and functional organization of a protein tyrosine kinase. Curr. Top. Microbiol. Immunol. **147:** 79–127.
65. Alcalay, M., F. Antolini, W. Van de Ven, L. Lanfrancone, F. Grignani & P. G. Pelicci. 1990. Characterization of human and mouse c-fes cDNA clones and identification of the 5′ end of the gene. Oncogene **5:** 267–275.
66. Arnaut, M. A. 1990. Structure and function of the leukocyte adhesion molecules CD11/CD18. Blood **75:** 1037–1050.
67. Chiang, M. Y., H. Chan, M. A. Zounes, S. M. Freier, W. F. Lima & C. F. Bennett. 1991. Antisense oligonucleotides inhibit intercellular adhesion molecule 1 expression by two distinct mechanisms. J. Biol. Chem. **266:** 1862–1871.
68. Ferrari, S., R. Manfredini, A. Grande, G. Torelli & U. Torelli. 1992. Proliferation and differentiation arrest and survival in leukemic blast cells. *In* Aging and Cellular Defense Mechanisms. C. Franceschi, G. Crepaldi, V. J. Cristofalo, L. Masotti, J. Vijg, eds. Ann. N.Y. Acad. Sci. In press.
69. Wyllie, A. H., R. G. Morris, A. L. Smith & D. Dunlop. 1984. Chromatin cleavage in apoptosis: Association with condensed chromatin morphology and dependence on macromolecular synthesis. J. Pathol. **142:** 67–77.
70. Wu, J., J. Q. Zhu, K. K. Han & D. X. Zhu. 1990. The role of *c*-fms oncogene in the regulation of HL-60 cell differentiation. Oncogene **5:** 873–877.
71. Sherman, M. L., R. M. Stone, R. Datta, S. H. Bernstein & D. W. Kufe. 1990. Transcriptional and post-transcriptional regulation of c-jun expression during monocytic differentiation of human myeloid leukemic cells. J. Biol. Chem. **265:** 3320–3323.
72. Hashimoto, K., A. Kishimoto, H. Aihara, I. Yasuda, K. Mikawa & Y. Nishizuka. 1990. Protein kinase C during differentiation of human promyelocytic leukemia cell line, HL-60. FEBS Letts. **263:** 31–34.
73. Ziegler, S. F., J. D. Marth, D. B. Lewis & R. M. Perlmutter. 1987. Novel protein-tyrosine kinase (hck) preferentially expressed in cells of hematopoietic origin. Mol. Cell. Biol. **7:** 2276–2285.
74. Quintrell, N., R. Lebo, H. Varmus, J. M. Bishop, M. J. Pettenati, M. M. Le Beau, M. O. Diaz & J. D. Rowley. 1987. Identification of a human gene (HCK) that encodes a protein-tyrosine kinase and is expressed in hemopoietic cells. Mol. Cell. Biol. **7:** 2267–2275.
75. Sartor, O., F. S. Gregory, N. S. Templeton, S. Pawar, R. M. Perlmutter & N. Rosen. 1989. Selective expression of alternative lck mRNA in human malignant cell lines. Mol. Cell. Biol. **9:** 2983–2988.
76. Dymecki, S. M., J. E. Niederhuber & S. V. Desiderio. 1990. Specific expression of a tyrosine kinase gene, blk, in B lymphoid cells. Science **247:** 332–336.
77. Maru, Y., H. Hirai & F. Takaku. 1990. Human ltk: Gene structure and preferential expression in human leukemic cells. Oncogene Res. **5:** 199–204.
78. Barnekow, A. & M. Gessler. 1986. Activation of $pp^{60c\text{-}src}$ kinase during differentiation of myelomonocytic cells *in vitro.* EMBO J. **5:** 701–705.
79. Ley, T. J., N. L. Connolly, S. Katamine, M. S. C. Cheah, R. M. Senior & K. C. Robbins. 1989. Tissue specific expression and development regulation of the human fgr proto-oncogene. Mol. Cell. Biol. **9:** 92–99.
80. Nishikura, K. & J. M. Murray. 1987. Antisense RNA of protooncogene c-fos blocks renewed growth of quiescent 3T3 cells. Mol. Cell. Biol. **7:** 639–649.
81. Holt, J. T., T. V. Gopal, A. D. Moulton & A. W. Nienhuis. 1986. Inducible production of c-fos antisense RNA inhibits 3T3 cell proliferation. Proc. Natl. Acad. Sci. USA **83:** 4794–4798.
82. Furukawa, Y., H. Piwnica-Worms, T. J. Ernst, Y. Kanakura & J. D. Griffin. 1990.

CDC2 gene expression at the G1 to S transition in human T lymphocytes. Science **250:** 805–808.
83. McManaway, M. E., L. M. Neckers, S. L. Loke, A. A. Al-Nasser, R. T. Redner, B. T. Shiramizu, W. L. Goldsschmidts, B. E. Huber, K. Bhatia & I. T. Magrath. 1990. Tumor-specific inhibition of lymphoma growth by an antisense oligodeoxynucleotide. Lancet **335:** 808–811.
84. Rosolen, A., L. Whitesell, N. Ikegaki, R. H. Kennett & L. M. Neckers. 1990. Antisense inhibition of single copy M-myc expression results in decreased cell growth without reduction of c-myc protein in neuroepithelioma cell line. Cancer Res. **50:** 6316–6322.
85. Negroni, A., S. Scarpa, A. Romeo, S. Ferrari, A. Modesti & G. Raschella. 1991. Decrease of proliferation rate and induction of differentiation by a MYCN antisense DNA oligomer in a human neuroblastoma cell line. Cell Growth Diff. **2:** 511–518.
86. Venturelli, D., S. Travali & B. Calabretta. 1990. Inhibition of T-cell proliferation by a MYB antisense oligomer is accompanied by selective down-regulation of DNA polymerase alfa expression. Proc. Natl. Acad. Sci. USA **87:** 5963–5967.
87. Jaskuski, D., J. K. de Riel, W. E. Mercer, B. Calabretta & R. Baserga. 1988. Inhibition of cell proliferation by antisense oligodeoxynucleotides to PCNA cyclin. Science **240:** 1544–1546.
88. Caracciolo, D., M. Valtieri, D. Venturelli, C. Peschle, A. M. Gewirtz & B. Calabretta. 1989. Lineage-specific requirement of c-abl function in normal hematopoiesis. Science **245:** 1107–1110.
89. Patinkin, D., S. Seidman, F. Eckstein, F. Benseler, H. Zakut & H. Soreq. 1990. Manipulations of cholinesterase gene expression modulate murine megakaryocytopoiesis *in vitro.* Mol. Cell. Biol. **10:** 6046–6050.
90. Sburlati, A., R. E. Manrow & S. L. Berger. 1991. Prothymosin alfa antisense oligomers inhibit myeloma cell division. Proc. Natl. Acad. Sci. USA **88:** 253–257.
91. Selinfreund, R. H., S. W. Barger, M. J. Welsh & L. J. Van Eldik. 1990. Antisense inhibition of glial S100 beta production results in alterations in cell morphology, cytoskeletal organization, and cell proliferation. J. Cell. Biol. **111:** 2021–2028.
92. Reed, J. C., C. Stein, C. Subasinghe, S. Haldar, C. M. Croce, S. Yum & J. Cohen. 1990. Antisense-mediated inhibition of BCL2 protooncogene expression and leukemic cell growth and survival: Comparisons of phosphodiester and phosphorothioate oligodeoxynucleotides. Cancer Res. **50:** 6565–6570.
93. Becker, D., C. B. Meier & M. Herlyn. 1989. Proliferation of human malignant melanomas is inhibited by antisense oligodeoxynucleotides targeted against basic fibroblast growth factor. EMBO J. **8:** 3685–3691.
94. Zmudzka, B. Z. & S. H. Wilson. 1990. Deregulation of DNA polymerase-beta by sense and antisense RNA expression in mouse 3T3 cells alters cell growth. Somatic Cell Mol. Genet. **16:** 311–320.
95. Birchenall-Roberts, M. C., C. Ferrer, D. Ferris, L. A. Falk, J. Kasper, G. White & F. W. Ruscetti. 1990. Inhibition of murine monocyte proliferation by a colony stimulating factor-1 antisense oligodeoxynucleotide. J. Immunol. **145:** 3290–3296.
96. Maier, J. A. M., P. Voulalas, D. Roeder & T. Maciag. 1990. Extension of the life-span of human endothelial cells by interleukin-1 alfa antisense oligomer. Science **249:** 1570–1574.
97. Schwab, G., C. B. Siegall, L. A. Aarden, L. M. Neckers & R. P. Nordan. 1991. Characterization of an interleukin-6-mediated autocrine growth loop in the human multiple myeloma cell line, U266. Blood **77:** 587–593.
98. Saison-Behomoaras, T., B. Tocque, I. Rey, M. Chassignol, N. T. Thuong & C. Helene. 1991. Short modified antisense oligonucleotides directed against Ha-ras point mutation induce selective cleavage of the mRNA and inhibit T24 cells proliferation. EMBO J. **10:** 1111–1118.
99. Szczylik, C., T. Skorski, N. C. Nicolaides, L. Manzella, L. Malaguarnera, D. Venturelli, A. M. Gewirtz & B. Calabretta. 1991. Selective inhibition of leukemia cell proliferation by BCR-ABL antisense oligodeoxynucleotides. Science **253:** 562–564.

Control of Gene Expression by Triple Helix-Forming Oligonucleotides

The Antigene Strategy[a]

CLAUDE HELENE

Laboratoire de Biophysique
INSERM U. 201 - UA. 481
Muséum National d'Histoire Naturelle
43, rue Cuvier
75231 Paris Cedex 05, France

NGUYEN T. THUONG

Centre de Biophysique Moléculaire
CNRS
1A, avenue de la Recherche Scientifique
45071 Orléans Cedex 02, France

ANNICK HAREL-BELLAN

Laboratoire d'Immunologie
URA - CNRS 1156
Institut Gustave Roussy
94805 Villejuif, France

Oligonucleotides can be used to control gene expression according to several strategies. In the *antisense* approach the oligonucleotide is targeted to a messenger RNA, thereby inhibiting translation, or to a viral RNA (or single-stranded DNA) where it can control RNA or DNA replication or reverse transcription. An oligodeoxynucleotide is able to induce cleavage of the target RNA via activation of an endogenous RNase H. An oligoribonucleotide may also induce catalytic cleavage of RNA in the absence of any protein (*ribozyme*). Antisense RNAs and ribozymes can be synthesized within cells via genetic constructs. In the *sense* approach the oligonucleotide is used to bind a protein: a single-stranded oligoribonucleotide can be used to trap viral proteins, such as tat and rev from HIV; a double-stranded oligonucleotide can act as a trap for a transactivating factor involved in the control of transcription. In the *antigene* strategy the oligonucleotide binds to a specific sequence of DNA, thereby forming a triple helix that may inhibit transcription of the target gene. This review presents the latest developments of this new strategy to control gene expression. Reviews of the other strategies are presented elsewhere.[1,2]

[a] This work was supported in part by Rhône-Poulenc-Rorer and by the Agence Nationale de Recherche contre le Sida (ANRS).

TRIPLE HELIX FORMATION BY OLIGONUCLEOTIDES BOUND TO THE MAJOR GROOVE OF DNA

Base Triplets

A homopurine·homopyrimidine sequence of duplex DNA can be recognized by an oligonucleotide containing the usual bases (FIG. 1). Recognition involves the formation of two hydrogen bonds with the purines of Watson-Crick base pairs within the major groove of the double helix. Thymine, cytosine, and guanine can adopt two different orientations; they will be called "Hoogsteen" and "reverse Hoogsteen" by analogy with the hydrogen-bonding scheme discovered by Hoogsteen in co-crystals of A and T derivatives (FIG. 1). In contrast, adenine and inosine can form two hydrogen bonds with an A.T base pair in a single orientation. It should be noted that in order to form two hydrogen bonds with G, cytosine must be protonated. Therefore, triplets involving $C^+ \times G.C$ are more stable at acidic pH. Methylation at C-5 of cytosine shifts the pH range where triplexes can form to higher values, probably as a result of

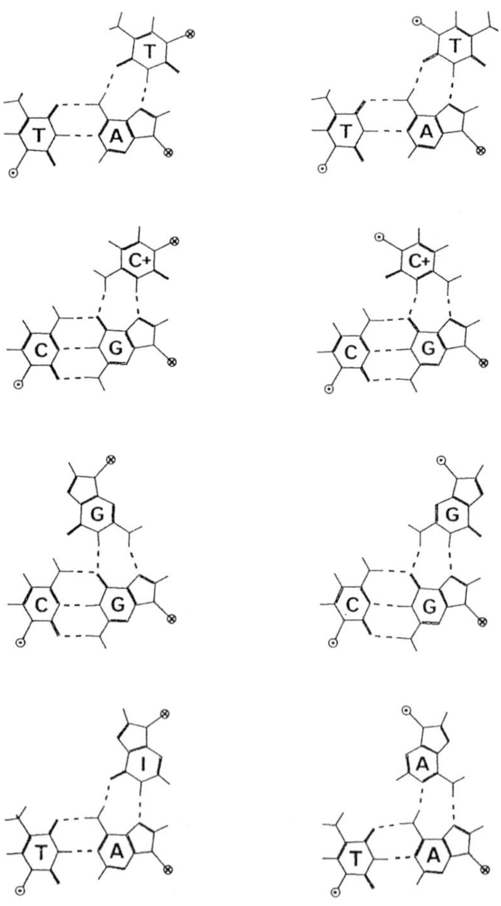

FIGURE 1. Base triplets formed by Watson-Crick A.T and G.C. base pairs with T, protonated C (C^+), G, A, and I. If bases in the third strand are in the anti conformation with respect to the sugar ring, the triplets in the *left column* correspond to a parallel orientation of the third strand with respect to the homopurine sequence of DNA target; those in the *right column* correspond to an antiparallel orientation. If bases in the third strand were in the syn conformation, the orientation would be reversed. For the T × A.T triplets the left-hand one corresponds to the so-called Hoogsteen hydrogen-bonding pattern; the right-hand one to reverse Hoogsteen.

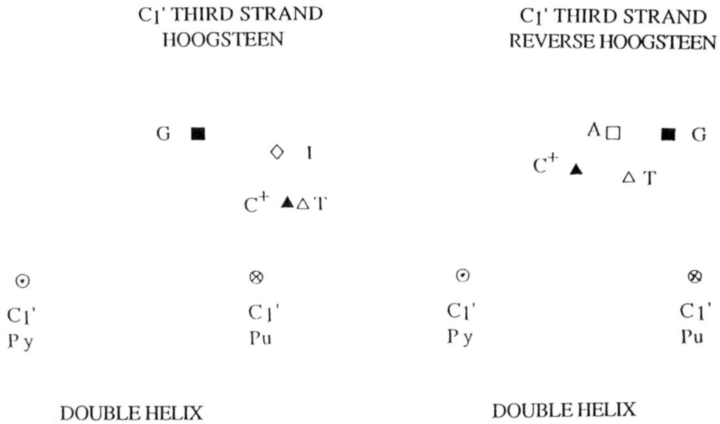

FIGURE 2. Position of C 1′ atoms in the plane of base triplets. The C 1′ atoms of the Watson-Crick strands are fixed in order to show the variations in position of the C 1′ atom of the third strand of the triple helix. Hoogsteen and reverse Hoogsteen refer to the type of hydrogen bonds formed by the third base with the Watson-Crick base pair. They correspond to the left and right columns of FIGURE 1, respectively.

hydrophobic interactions between the methyl substituents of 5-methylcytosine and thymine which form a helical spine of hydrophobic groups.[3,4]

An important parameter governing the stability of triple helices (besides the intrinsic energy of triplet formation) is the energy penalty associated with the distortion of the phosphodiester backbone when one moves from one triplet to the next one. This is illustrated in FIGURE 2 where the position of the C1′ atom of the nucleotide in the third strand is given with respect to the C1′ atoms of the two nucleotides involved in Watson-Crick hydrogen bonding. Only two triplets are isomorphous: T × A.T and C^+ × G.C in the Hoogsteen configuration. Therefore, only homopyrimidine oligonucleotides can form a third strand without backbone distortion. All other triplets (including T × A.T and C^+ × G.C in the reverse Hoogsteen configuration) are not isomorphous. Any combination of these triplets will be associated with an energy penalty due to backbone distortion.

Third-Strand Orientation

The orientation of the third strand will be parallel to the homopurine target strand if the base triplets are in the Hoogsteen configuration and if nucleotides in the third strand have the anti conformation. Third-strand orientation is antiparallel to the homopurine strand when reverse Hoogsteen hydrogen bonding is involved with anti conformation. Changing the conformation of the nucleosides from anti to syn will change the orientation of the third strand.

Energy calculations suggest that Hoogsteen triplets T × A.T, C^+ × G.C, and G × G.C are more stable than are those involving reverse Hoogsteen hydrogen bonding.[5] For oligonucleotides containing only T and C, the parallel orientation (with respect to the homopurine sequence) will always be favored. However, the distortion of third strand backbone is less in the reverse Hoogsteen configuration for T × A.T and G × G.C triplets (FIG. 2). Consequently the orientation of oligonucleotides containing G and T is expected to depend on the sequence[5]: a high number of GpT or TpG steps will be detrimental to parallel binding (with respect to the homopurine sequence) in the Hoogsteen configuration as a result of the higher energy penalty associated with backbone distortion. Therefore, oligo (G,T) with a large number of GpT and TpG steps will prefer an antiparallel orientation even though reverse Hoogsteen hydrogen bonding is not the lowest energy triplet configuration for both T × A.T and G × G.C triplets.

In contrast, an oligo (G,T) with few GpT and TpG steps will bind more strongly in a parallel orientation. We have exploited this observation to design oligonucleotides that will bind to two consecutive homopurine·homopyrimidine sequences where the two homopurine sequences belong to different strands of the double helix.[5] The oligo (G,T) switches from one strand to the other without any interruption of base triplet formation. This allows extension of the number of DNA sequences that can be recognized by triple helix-forming oligonucleotides. Other possibilities for switching from one strand of the double helix to the other exist, but they involve a linker between two oligonucleotide sequences and leave unrecognized one or several base pairs between the two target sequences.[6,7]

Specificity of Triple Helix Formation

Mismatches in short oligonucleotide duplexes destabilize the hybrids. This destabilization has been exploited in the case of antisense oligonucleotides to discriminate between two messenger RNAs that differ by a single point mutation, as in ras oncogene activation.[8]

When a homopurine sequence in DNA is interrupted by a pyrimidine, destabilization of triplex structures is observed. The stability of triple helices has been estimated from the extent of cleavage induced by oligonucleotide-EDTA-Fe conjugates[9] and determined from Tm measurements.[10] Measurements of the superhelical twist required to induce intramolecular triplex formation (H-DNA) have also been used to determine the energy penalty associated with mismatches in triplexes.[11] The change in temperature of half dissociation (Tm) of the third strand from its duplex target depends on the nature and location of the mismatch.[10] The most destabilizing triplet in our experiments was C × T.A. A T.A base pair in the center of Pu.Py base pairs can be recognized by G in the third strand if both neighbors of T are adenines on each side,[9] but a guanine on either (or both) side(s) of T does not allow recognition of T.A by G.[10] As expected, a terminal mismatch is less detrimental to triplex stability than is a central mismatch.

Kinetics and Thermodynamics of Triplex Formation

The association of an oligonucleotide with duplex DNA to form a triplex is much slower than is duplex formation.[12,13] Measurements of activation energies have led to the conclusion that the rate-limiting step (nucleation) involves the formation of 3 to 5 base triplets before zippering occurs. A high negative activation enthalpy is associated with the nucleation process.[13] Dissociation of a third strand from a duplex is also a slow process as compared to duplex dissociation. When formed, a triplex can be rather long-lived.

These kinetic phenomena have to be taken into account when competition experiments are carried out between a triple helix-forming oligonucleotide and another DNA ligand, such as a protein.

The stability of triplexes depends on base sequence and on pH if cytosines are involved in the third strand (as a result of the requirement for cytosine protonation). Thermodynamic studies are presently underway in several laboratories to determine thermodynamic parameters associated with each base triplet in its appropriate environment.

Cations stabilize triple helices as expected because three negatively charged backbones are involved in triplex formation. Polycations such as Mg^{2+} or $Spermine^{4+}$ are more efficient than are monocations such as Na^+. In the presence of Mg^{2+} or $Spermine^{4+}$, Na^+ ions destabilize triple helical structures. It should be noted that formation of G × G.C reverse Hoogsteen base triplets requires the presence of Mg^{2+}; that of A × A.T base triplets appears to be favored by Zn^{2+}.

α-*Oligonucleotides*

Oligonucleotides synthesized with the α-anomers of nucleotide units instead of natural β-anomers are assumed to adopt the opposite orientation. (In the α-anomeric configuration the base is on the same side as the 3′-OH with respect to the deoxyribose, whereas in the natural β-anomers the base is on the same side as the 5′-OH). An oligo-α-T_n was shown to bind in a parallel orientation with respect to the homopurine sequence, that is, the same orientation as the corresponding oligo-β-T_n.[14] If in both cases thymidines have an anti conformation, this result implies that the T × A.T triplets adopt the reverse Hoogsteen configuration for the α-anomers and the Hoogsteen configuration for the β-anomers. Oligo-α-deoxynucleotides containing both C and T bind in an antiparallel orientation, while the corresponding oligo-β-deoxynucleotides bind in a parallel orientation with respect to the homopurine sequence.[15] Therefore, in both cases the Hoogsteen configuration is preferred for both triplets T × A.T and C^+ × G.C, as expected because these Hoogsteen triplets are isomorphous. Combination of bases other than C and T in the third strand have not yet been investigated for α-oligonucleotides.

Stabilization of Triple Helical Structures

As mentioned, substitution of 5-methylcytosine to cytosine increases the stability of triplex structures obtained with oligonucleotides containing C and T. This holds true for α-oligonucleotides as well as natural β-oligomers. The pK of cytosine is only slightly shifted (0.2 pH units) by methylation at the C-5 position. Therefore, it appears more likely that additional stabilization is provided by hydrophobic interactions between the methyl groups of 5-methylcytosine and thymine. When the third strand wraps around the double helix within the major groove, a helical "spine" of methyl groups is formed.

Strong stabilization of triplex structures was obtained when the third-strand oligonucleotide was substituted by an intercalating agent.[4] The effect was greater when substitution occurred at the 5′ end. The substituent (an acridine derivative) was shown to be intercalated at the triplex-duplex junction. Intercalation provides additional binding energy that stabilizes the triple helical complex. Similar stabilization was observed when a homopyrimidine oligo-α-deoxynucleotide was substituted at its 3′ end.[15] As the α-oligomer was bound in a reverse orientation as compared to the corresponding β-oligomer, intercalation took place at the same triplex-duplex junction as with the 5′ substituted β-oligomer. A 16-bp homopurine·homopyrimidine sequence repeated twice in the HIV proviral genome was chosen as a target for these oligonucleotides.[15]

When a psoralen derivative was attached to the 5′ end of an oligopyrimidine, cross-linking of the two strands of the duplex was observed at the triplex-duplex junction when the complex was UV irradiated.[16] This cross-linking is much more efficient when a 5′TpA3′ step is present on the homopurine-containing strand.

CONTROL OF TRANSCRIPTION BY TRIPLEX-FORMING OLIGONUCLEOTIDES

Published experiments aimed at inhibiting gene transcription by triplex-forming oligonucleotides are still rather limited (in contrast to the numerous examples of antisense inhibition of mRNA translation).

Several mechanisms exist by which triple helix formation can alter gene transcription:

1. Triple helix formation within the promoter region can change DNA conformation and therefore alter the rate and efficiency of RNA polymerase initiation. This can lead to either activation or inhibition of transcription.

2. Oligonucleotide binding to a DNA sequence overlapping a transcription factor binding site may inhibit its transactivating capacity.

3. Triplex formation within or adjacent to the region where RNA polymerase binds may inhibit transcription initiation even if RNA polymerase and transcription factors are still bound to the promoter.

4. Oligonucleotide binding downstream of the RNA polymerase recognition site might inhibit progression of the transcription machinery along the DNA and therefore block RNA elongation.

Inhibition of Transcription by Inhibition of Transcription Factor Binding

Cooney *et al.*[17] first described that a 28nt-long oligonucleotide bound to a purine-rich sequence (but not homopurine) in the c-myc gene promoter inhibited transcription *in vitro.* The same group later reported that the same oligonucleotide could inhibit c-myc gene transcription in cells in culture. (As a matter of fact, the oligonucleotide was bound to an overlapping sequence in the reverse orientation as compared to the original proposal).[18] They also provided evidence for triple helix formation within cells by demonstrating that a DNase hypersensitive site located within the target sequence was protected in cells incubated with the oligonucleotide. It was proposed that the 28nt-long oligonucleotide inhibited c-myc mRNA synthesis via competition with a transcription factor recognizing part or all of the triplex target.[18]

Subunit α of the interleukin-2 receptor (IL2-Rα) was also chosen as a target for triplex-forming oligonucleotides. A 30% inhibition of IL2-Rα expression was observed when cells were incubated in the presence of a 34nt-long purine-rich oligonucleotide expected to bind in an antiparallel orientation with respect to a purine-rich sequence within the promoter region.[19] Grigoriev *et al.*[20] used a 15-mer homopurine oligonucleotide that binds to a 15-bp homopurine·homopyrimidine sequence in the promoter region of the IL2-Rα gene (FIG. 3). This target sequence overlaps by 3 bp the recognition sequence of the NF-KB transcription factor. Triplex formation was shown to inhibit transcription factor binding *in vitro.* The stability of the triplex was strongly increased by substituting 5-methylcytosine to cytosine and by covalent attachment of an intercalating agent at the 5′ end of the oligonucleotide. Oligonucleotide-intercalator conjugates exhibited much higher affinity due to additional binding energy provided by intercalation at the duplex-triplex junction.[4]

The IL2-Rα gene promoter was fused to the bacterial reporter chloramphenicol acetyltransferase (CAT) gene. To demonstrate that the effects of triplex-forming oligonucleotides on CAT gene expression were due to triple helix formation, a mutant of the target sequence was constructed by inverting three base pairs within the homopurine · homopyrimidine sequence (FIG. 3). Footprinting studies showed

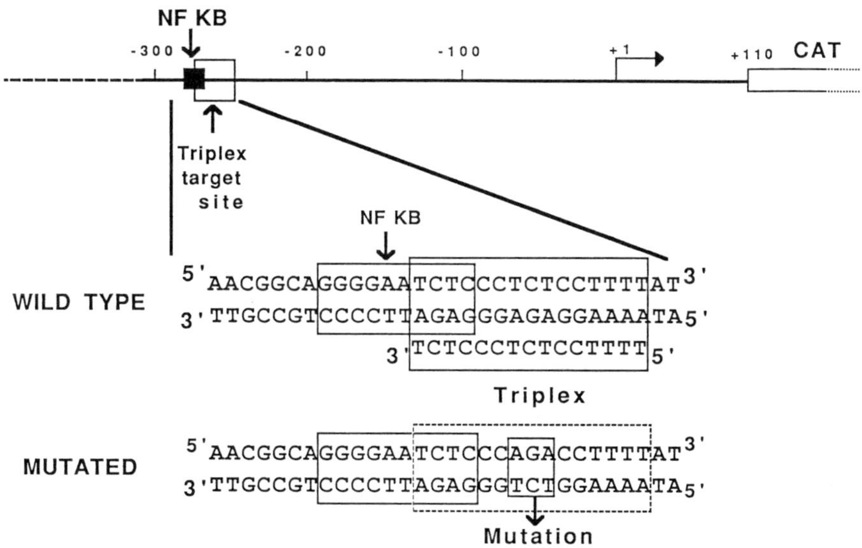

FIGURE 3. The promoter region of the gene coding for subunit **α** of the interleukin-2 receptor (IL2-R**α**) has been cloned upstream of the chloramphenicol acetyltransferase (CAT) gene. The binding site for the transcription factor NF-KB and the site for triple helix formation are boxed. The mutant promoter used as a control does not allow for triple helix formation. The triplex-forming 15-mer oligonucleotide containing thymine and 5-methylcytosine (shown below the boxed sequence of the wild-type promoter) was substituted at its 5′-end by either an acridine derivative to stabilize the triple helix (refs. 4 and 15) or a psoralen derivative to photo-induce a cross-link between the two strands of the duplex at the 5′TpA3′ step on the 3′-side of the homopyrimidine target sequence (ref. 16). See references 20 and 21 for experimental results.

that the triplex-forming 15 mer did not bind to the mutant sequence. This mutation, however, did not affect NF-KB-dependent transcription of the IL2-Rα-CAT construct.

When the IL2-Rα-CAT construct was transfected into human lymphocytes and exposed to the 15 mer containing T and ^{m}C, no inhibition was observed on either the wild-type (WT) or the mutated promoter. However, when the 15 mer was 5′-substituted by acridine, a strong reduction in CAT gene expression was seen with the WT sequence but not with the mutant. An acridine-substituted 11 mer had no effect on either construct, demonstrating that the effect of Acr-15-mer was not due to degradation products such as acridine itself.

When the 15 mer was 5′-substituted with psoralen, we could easily show *in vitro* that the two strands of the plasmid carrying the WT promoter were cross-linked under UV irradiation, whereas no cross-linking was detected with the mutant. We first irradiated the plasmids *in vitro* and then transfected the cross-linked plasmids into recipient cells; CAT gene expression was strongly reduced with the WT construct, whereas no inhibition was seen with the mutant. In another set of experiments the plasmids were first transfected into cells in the presence of pso-15-mer, and then the transfected cells were irradiated. The plasmids were extracted and cut with restriction enzymes that generate a 2,400-bp fragment expected to contain the cross-link. After denaturation of the fragments those that are cross-linked migrate as dimers, because the two strands remain covalently attached to each other. The results clearly showed that the WT construct was cross-linked at the expected position, whereas the mutant was not. When CAT expression was measured in irradiated cells, the correlation between cross-linking and inhibition of CAT gene expression was very good. The experiments with the mutant clearly showed that the observed effects were due to triple helix formation at the expected target site.

Control of Transcription by Repressor Oligonucleotides

A 13-mer oligonucleotide was bound to a homopurine · homopyrine sequence adjacent to or slightly overlapping the RNA polymerase binding site on the *Escherichia coli* β-lactamase gene promoter.[22] Both RNA polymerase and the oligonucleotide could bind to the same DNA fragment, as shown by footprinting experiments. Triplex formation inhibited transcription initiation in a concentration- and temperature-dependent manner. However, this inhibition was only transiently observed. Dissociation of the oligonucleotide from its target sequence allowed fast initiation of transcription in the presence of the four NTPs. As reported above, dissociation and association reactions are slow processes for intermolecular triple helix formation from a duplex and a single-stranded oligomer. Transcription from the β-lactamase promoter is faster (when RNA polymerase is bound to its promoter before adding the four NTPs) than association of the oligonucleotide at the concentrations used in these studies. Therefore, only a temperature-dependent delay was seen in the synthesis of the full transcripts.

On the contrary, a time-independent inhibition was obtained when the oligonucleotide was covalently linked to a psoralen derivative and the complex irradiated before the start of transcription. Covalent linkage of the oligonucleotide-psoralen conjugate occurred on the nontranscribed strand. (No cross-link between the two DNA strands was made due to the absence of a *5′*TpA*3′* sequence at the triplex-duplex junction.) Nevertheless, transcription initiation was irreversibly arrested.

The 13-mer triple-helix–forming oligonucleotide targeted to the β-lactamase gene behaves as a repressor of transcription. An analogy can be drawn with the *lac* repressor which binds to a region overlapping the RNA polymerase binding site but does not prevent polymerase binding to the promoter sequence.[23] When the repressor is tightly bound, transcription initiation does not occur. In the presence of an inducer the dissociation rate of the repressor is increased and transcription begins.

CONCLUSION

Formation of a triple helix resulting from oligonucleotide binding to the DNA double helix offers new possibilities to control gene expression at the transcriptional level. Two modes of action have been documented until now: (1) the oligonucleotide can behave as a negative regulatory factor by inhibiting the binding of a transcription-activating protein to its recognition sequence; (2) the oligonucleotide can act as a repressor when it binds to a site adjacent to or overlapping the binding site of RNA polymerase in the downstream region of the polymerase binding site. Additional modes of action can be expected, such as (1) alteration of DNA conformation with subsequent effects on transcription kinetics and efficiency even when the oligonucleotide binding site does not overlap any binding site of a component of the transcription machinery; and (2) transcription arrest when the oligonucleotide is bound within the transcribed region but not adjacent to the polymerase bound to its promoter. Oligonucleotide-intercalator conjugates might prove to be strong stoppers of RNA chain elongation, especially when the site of intercalation is located upstream of the triple helix and encountered first by the RNA polymerase moving along its template.

It should be kept in mind that triplex-forming oligonucleotides might also interfere with DNA replication.[24] When the oligonucleotide carries a reactive group, DNA alterations that are induced site-specifically are expected to be recognized by repair enzymes. Experiments are presently underway to analyze the mechanisms and biological consequences of DNA modifications induced on DNA within living cells by triple-helix–forming oligonucleotides.

Triple helices can be formed on single-stranded nucleic acids. A homopurine sequence can be recognized by two oligonucleotides, one forming Watson-Crick base pairs, the other one forming Hoogsteen hydrogen bonds with the Watson-Crick double helix. These two oligonucleotides can be linked together at either one[25] or both ends.[26] The stability of the triple helix formed by such molecules is strongly increased as compared to the oligonucleotide forming only Watson-Crick base pairs. Nearly all hydrogen-bonding possibilities of the purines in the target sequences are used for recognition.[25] Therefore, triple helices might also find applications in the *antisense* field by allowing strong complexes to be formed on selected homopurine sequences of a messenger RNA.

ACKNOWLEDGMENTS

We wish to thank all colleagues who have contributed to the work described in this review. Their names and contributions can be found in the reference list.

REFERENCES

1. HÉLÈNE, C. & J. J. TOULMÉ 1990. Biochim. Biophys. Acta **1049:** 99–125.
2. HÉLÈNE, C. 1991. Eur. J. Cancer **27:** 1466–1471.
3. THOMAS, J. P. & P. DERVAN. 1989. J. Am. Chem. Soc. **111:** 3059–3061.
4. SUN, J. S., J. C. FRANÇOIS, T. MONTENAY-GARESTIER, T. SAISON-BEHMOARAS, V. ROIG, N. T. THUONG & C. HÉLÈNE. 1989. Proc. Natl. Acad. Sci. USA **86:** 9198–9202.
5. SUN, J. S., T. DE BIZEMONT, G. DUVAL-VALENTIN, T. MONTENAY-GARESTIER & C. HÉLÈNE. 1991. C. R Acad. Sci. Paris, t. 313, Série III: 585–590.

6. Horne, D. A. & P. Dervan. 1990. J. Am. Chem. Soc. **112:** 2435–2437.
7. Ono, A., C. N. Chen & L. S. Kan. 1991. Biochemistry **30:** 9914–9921.
8. Saison-Behmoaras, T., B. Tocqué, I. Rey, M. Chassignol, N. T. Thuong & C. Hélène. 1991. EMBO J. **10:** 1111–1118.
9. Griffin, L. C. & P. Dervan. 1989. Science **254:** 967–971.
10. Mergny, J. L., J. S. Sun, M. Rougée, T. Montenay-Garestier, F. Barcelo, J. Chomilier & C. Hélène. 1991. Biochemistry **30:** 9791–9798.
11. Belotserkowskii, B. P., A. G. Veselkov, S. A. Filippov, V. N. Dobrynin, S. M. Mirkin & M. D. Frank-Kamenetskii. 1990. Nucleic Acids Res. **18:** 6621–6624.
12. Maher, L. J., P. B. Dervan & B. J. Wold. 1990. Biochemistry **29:** 8820–8826.
13. Rougée, M., B. Faucon, J. L. Mergny, F. Barcelo, C. Giovannangeli, T. Montenay-Garestier & C. Hélène. 1992. Biochemistry. Submitted.
14. Le Doan, T., L. Perrouault, D. Praseuth, N. Habhoub, J. L. Decout, N. T. Thuong, J. Lhomme & C. Hélène. 1987. Nucleic Acids Res. **15:** 7749–7760.
15. Sun, J. S., C. Giovannangeli, J. C. François, R. Kurfurst, T. Montenay-Garestier, T. Saison-Behmoaras, N. T. Thuong & C. Hélène. 1991. Proc. Natl. Acad. Sci. USA **88:** 6023–6027.
16. Takasugi, M., A. Guendouz, M. Chassignol, J. L. Decout, J. Lhomme, N. T. Thuong & C. Hélène. 1991. Proc. Natl. Acad. Sci. USA **88:** 5602–5606.
17. Cooney, M., G. Czernuszewicz, E. H. Postel, S. J. Flint & M. E. Hogan. 1988. Science **241:** 456–459.
18. Postel, E. H., S. J. Flint, D. J. Kessler & M. E. Hogan. 1991. Proc. Natl. Acad. Sci. USA 1991. **88:** 8227–8231.
19. Orson, F. M., D. W. Thomas, W. M. McShan, D. J. Kessler & M. E. Hogan. 1991. Nucleic Acids Res. **19:** 3435–3441.
20. Grigoriev, M., D. Praseuth, P. Robin, A. Hemar, T. Saison-Behmoaras, A. Dautry-Varsat, N. T. Thuong, C. Hélène & A. Harel-Bellan. 1992. J. Biol. Chem. **267:** 3389–3395.
21. Grigoriev, M., D. Praseuth, P. Robin, A. L. Guyesse, N. T. Thuong, C. Hélène & A. Harel-Bellan. 1992. Submitted.
22. Duval-Valentin, G., N. T. Thuong & C. Hélène. 1992. Proc. Natl. Acad. Sci. USA **89:** 504–508.
23. Hélène, C. & G. Lancelot. 1982. Prog. Biophys. Molec. Biol. **39:** 1–68.
24. Birg, F., D. Praseuth, D. Zérial, N. T. Thuong, U. Asseline, T. Le Doan & C. Hélène. 1990. Nucleic Acids Res. **18:** 2901–2908.
25. Giovannangeli, C., T. Montenay-Garestier, M. Rougée, M. Chassignol, N. T. Thuong & C. Hélène. 1991. J. Am. Chem. Soc. **113:** 7775–7777.
26. Kool, E. T. 1991. J. Am. Chem. Soc. **113:** 6265–6266.

Specific Inhibition of Oncogene Expression *in Vitro* and *in Vivo* by Antisense Oligonucleotides

LEN NECKERS, ANGELO ROSOLEN, BRIGID FAHMY, AND LUKE WHITESELL

Clinical Pharmacology Branch
National Institutes of Health
National Cancer Institute
Bethesda, Maryland 20892

To understand the role of individual genes in regulating biological processes, one must be able to interfere specifically with either their expression or their function. Although monoclonal antibodies have proven very useful in studying cell surface proteins, the specific inhibition of intracellular proteins in viable cells is a much more difficult problem. The goal of antisense technology is to develop small oligonucleotides, plasmids, or retroviral vectors that can be introduced easily into viable cells to specifically inhibit gene products. This report describes our use of antisense DNA and RNA to study the role of the *NMYC* protooncogene in neuroectodermal cell growth and differentiation and to begin to assess the potential of *in vivo* antisense oligonucleotide administration as gene-targeted therapy.

NMYC is a DNA-binding phosphoprotein of unknown function. Unlike its homolog, *MYC*, *NMYC* expression is temporally regulated and is limited to immature T and B lymphocytes, fetal tissues, and various tumors of neuroendocrine origin, such as neuroblastoma, neuroepithelioma, retinoblastoma, small cell lung carcinoma, and Wilms' tumor.[1–3] In neuroblastoma, *NMYC* copy number is consistent in tumor biopsies obtained from individual patients over time, whereas *NMYC* expression may vary. Thus, *NMYC* gene amplification is an intrinsic property of a subset of these tumors and does not occur with progressive disease if it is not present at diagnosis. However, the development of progressive disease is likely to occur in patients with amplified or overexpressed *NMYC*.[4] The transient and localized expression of *NMYC* during the early stages of organogenesis, together with the presence of conserved motifs characteristic of transcription factors, suggests that *NMYC* may regulate gene expression in the developing organism.[5–8]

CONSEQUENCES OF *NMYC* SUPPRESSION *IN VITRO*

Sixteen to 20 hours after the addition of 100 μM 15 mer complementary to the translation start site of *NMYC* mRNA, *NMYC* protein is significantly reduced.[9] During this same time, DNA synthesis, as measured by labeled BrdU incorporation, is also reduced. The daily addition for 5 days of 100 μM antisense, but not sense, oligomer results in a reduction in the number of viable cells on day 6. Yet the level of *MYC* protein, highly expressed in these cells, does not change throughout the course of the experiment. To explain how the growth of these cells could be affected without altering *MYC* expression, we need to backtrack a bit and look more closely at how such neuroectodermal tumors grow *in vitro*.

Many neuroectoderm-derived tumors, such as neuroblastoma, display two distinct morphologies *in vitro*[10,11] (FIG. 1). These two morphologies persist for many passages, over a period of years, suggesting that the presence of both supplies a growth advantage to the culture. It is not clear if these two cell types can interconvert, or transdifferentiate, between each other or if both morphologic types arise from a low abundance, regenerating stem cell present in the cultures. What is

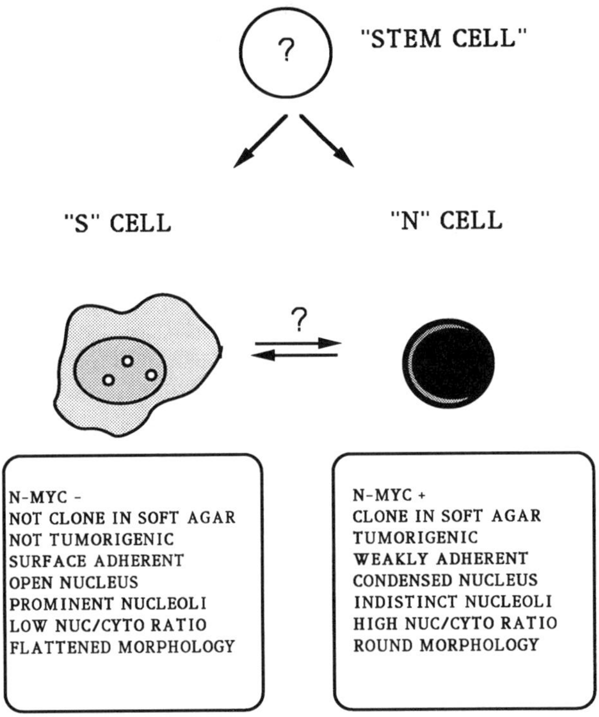

FIGURE 1. Many neuroectoderm-derived cell lines consist of two morphologic types *in vitro*. The "S" cells are epithelial in nature, whereas the "N" cells are neuroblastic in nature. The two cell types may interconvert between themselves or arise individually from a low-abundance precursor cell also present in the culture.

intriguing about the stable coexistence of two morphologies *in vitro* is that one morphologic type, termed "S" cells because they are substrate adherent, consists of flat, epithelial-like cells that are contact inhibited and neither clone in soft agar nor are tumorigenic in nude mice. The other morphologic type, termed "N" cells, consists of round, neuritic-like cells that are not contact inhibited, grow in soft agar, and form tumors in nude mice.

We documented that a single copy *NMYC* neuroectoderm-derived cell line, the neuroepithelioma CHP100, also stably expresses two morphologies in culture.[9,12]

These two cell types can be separated from each other by limiting dilution and maintained over the short term as monomorphic clones. As with other neuroectoderm-derived cell lines, the S clones do not form colonies in soft agar, whereas the N clones do.[12] Also characteristic of these subclones are distinct expression patterns of neuroendocrine antigens. Thus, although all morphologic variants express S-100, only the neuroblastic clones demonstrate reactivity for secretogranin I and neuron-specific enolase (NSE). In addition, on examining *NMYC* protein expression, we observed that the neuroblastic cells express *NMYC,* whereas the epithelial cells do not.[12] Although the factors that regulate such morphologic variation are not known, it is of interest that even in an *NMYC*-amplified cell line, S cells express little or no *NMYC,* while N cells are highly *NMYC* positive.[13]

We wondered if differential *NMYC* expression is in any way involved in the maintenance of the two morphologies *in vitro* and if morphologic heterogeneity was advantageous to growth. When we looked at the morphology of antisense treated cells, we observed a subtle, but reproducible change: following the addition of *NMYC* antisense, fewer small N cells were present. If elimination of one morphology from the culture reduced the growth rate of the remaining cells, we might have an explanation for how *NMYC* antisense was affecting growth without altering *MYC* level. In fact, careful analysis of the growth rates of pure S or N cultures, isolated by repeated single cell cloning of wild-type cultures, revealed that both pure cultures grow more slowly than do mixed wild-type cultures.[9,12]

To characterize the effects of *NMYC* on the maintenance of morphologic heterogeneity, hinted at by the oligo experiments, required more sustained inhibition of *NMYC* in many more cells than was achievable with exogenously added oligos. Stable transfection of cultured cells with an antisense vector allows for continuous generation of antisense RNA. However, antisense RNA must be generated in large excess to be effective. Even with the strongest eukaryotic promoters, conventional integrating vectors are rarely able to suppress an amplified gene. Episomal replicons have been developed as a means of amplifying vector copy number.[14] We used the episomally replicating vector pREP3 (obtained from Mark Tyckocinski at Case Western Reserve University), which relies on the presence of the Epstein-Barr virus origin of replication and nuclear antigen 1 (EBNA I) to confer episomal replication capability. As an added advantage, upon removal of selective antibiotic pressure (hygromycin), episomal replicons are readily lost from the culture by dilution, and their potential antisense effects can be reversed.

Episomally derived *NMYC* antisense inhibits *NMYC* protein expression in CHP100 cells and confirms the results obtained with exogenously added antisense oligomers. In antisense transfected cultures in the presence of hygromycin, only large *NMYC*-negative cells are present. If hygromycin is removed from the culture, both N-type cells and *NMYC* reactivity reappear within 14 days.[12] This cannot be due to clonal selection, because removal of hygromycin from the culture results in concomitant reappearance of both N-like cells and *NMYC* reactivity. Likewise, it cannot be due to the presence of hygromycin, because sense-transfected cells in the presence of hygromycin display S and N morphologies as well as *NMYC* reactivity. These results were confirmed in the *NMYC*-amplified neuroblastoma line IMR32. Western blot analysis of sense and antisense transfected cells reveals that the *NMYC* episomal vector in the presence of hygromycin is able to markedly diminish *NMYC* reactivity and that this effect is reversible upon removal of hygromycin. In addition, *NMYC* suppression results in a loss from culture of N-type cells with concomitant accumulation of S-like cells. Removal of hygromycin leads to a rapid return of populations of mixed morphology. Again, *MYC* expression is not affected by *NMYC* antisense RNA

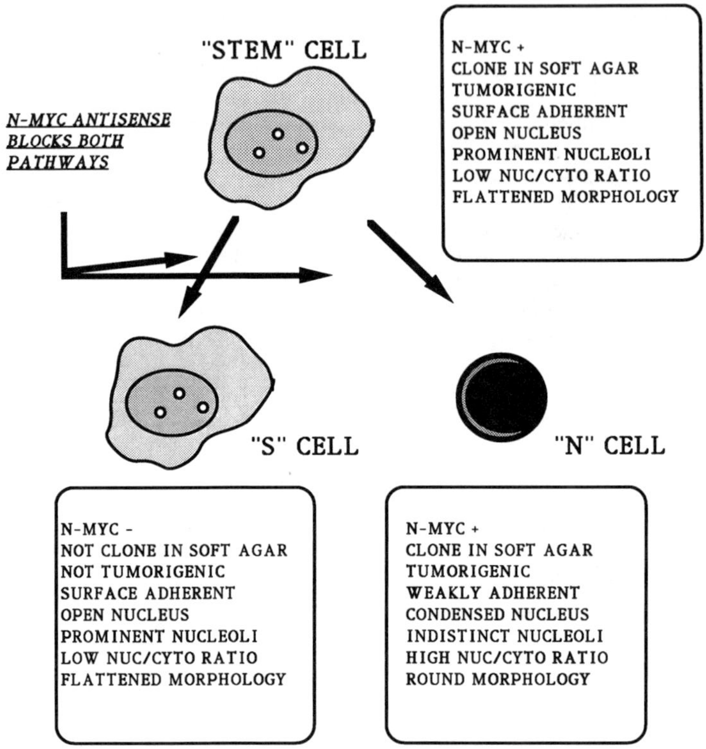

FIGURE 2. *In vitro NMYC* suppression appears to result in suppression of differentiation of an *NMYC*-positive, low-abundance precursor cell expressing characteristics of both "N" and "S" cells.

generation, and yet antisense transfected cultures grow more slowly than do wild-type cultures. Once hygromycin is removed, the growth rate returns to normal.

Thus, both the oligo and the vector experiments suggest that *NMYC* suppression alters the morphologic heterogeneity of neuroectodermal cultures, resulting in loss of *NMYC*-expressing N-type cells and a concomitant reduction in growth rate of the overall culture. We have reason to believe, however, that *NMYC* antisense suppressed cells, although morphologically similar, are not equivalent to S cells. The two groups of cells express several divergent characteristics: (1) Although most monotypic N cultures clone in soft agar to the same degree as do wild-type cultures, S cultures are poorly clonogenic. On the other hand, *NMYC*-suppressed cultures clone readily in soft agar. (2) *NMYC*-suppressed cultures still form tumors in nude mice, although at a slower rate than do wild-type cells. (3) *NMYC*-suppressed cultures are phenotypically distinct from either N or S monotypic cultures.

These data lead us to hypothesize that *NMYC* suppression results in the accumulation of a monomorphic stem cell population that, although morphologically resembling S cells, is of neither S nor N phenotype (FIG. 2). As release from *NMYC* suppression leads to rapid reappearance of both S and N cells in culture, we hypothesize that low level *NMYC* expression occurs in a stem cell population, and that this expression is required to generate both S and N cells from the stem cells. Why *NMYC* expression is turned off in S cells as opposed to N cells is a question of great interest and may lead to an understanding of the role of *NMYC* in the progression of neuroendocrine tumors.

CONSEQUENCES OF *NMYC* SUPPRESSION *IN VIVO*

To extend these *in vitro* observations to the modulation of protooncogene expression *in vivo* and to explore the use of antisense oligos as gene-targeted pharmacologic agents, we have developed an athymic murine tumor xenograft model in which oligo solutions are continuously perfused via subcutaneously implanted microosmotic pumps[15] (FIG. 3). Such administration prevents degradation of bulk oligo while allowing controlled, continuous release of relatively small quantities of drug into surrounding tissue.

Oligo-treated tumors were grown and harvested in athymic mice. Two-week constant subcutaneous perfusion of unmodified phosphodiester oligos at the rate of 2.5 nmol per hour resulted in no obvious whole animal toxicity (i.e., altered water or food intake, weight loss, or lethargy). *NMYC* gene expression in tumor xenografts from these oligo-perfused animals was assessed by immunoblotting. A marked reduction in *NMYC*-specific signal was observed in lysates of antisense-perfused tumors compared to lysates of sense-perfused tumors. The decrease in *NMYC* expression following antisense perfusion is specific for the targeted gene, because, as seen previously *in vitro,*[9,12] expression of the closely related *MYC* gene remains unchanged in samples of both sense- and antisense-perfused tumors.[15] Because the *MYC* antibody used for this blot does not cross-react with murine *MYC,*[16] equivalent amounts of human tissue must have been lysed in sense- and antisense-perfused tumors. Thus, the decrease in *NMYC* signal signifies decreased gene expression and not simply cell loss or dilution by murine stromal components. *In vivo* antisense inhibition of *NMYC* expression in tumor xenografts also results in loss of expression of secretogranin I, as was evident following *in vitro NMYC* suppression.[12]

Although a decline in *NMYC* expression has been associated with retinoic acid-stimulated differentiation in neuroblastoma,[17] the multiple activities of this drug prevent a simple cause-and-effect interpretation of such data. We previously showed that *in vitro* antisense inhibition of CHP100 *NMYC* expression decreases growth rate without reducing *MYC* expression.[9] This decrease in growth rate may be due to inhibition of the normal transdifferentiation process rather than to a direct antiproliferative effect.[12] Thus, *NMYC*-suppressed cells (containing an *NMYC* antisense expression vector) continue to form tumors in nude mice with good efficiency, but these tumors grow more slowly than do those formed by nonsuppressed cells. Similarly, continuous perfusion of CHP100 cells *in vivo* with antisense oligo also does

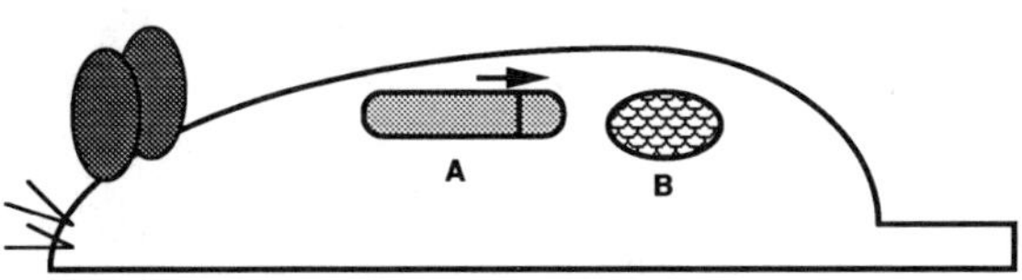

FIGURE 3. Microosmotic pumps (**A**) are implanted subcutaneously into athymic mice followed by subcutaneous injection of tumor cells (**B**). Oligo diffuses out of the pump at a constant rate in the direction of the *arrow*. Tumor grows subcutaneously as a discrete mass (**B**).

not abrogate tumor formation, but instead leads to significantly smaller tumors.[15] In several experiments in which tumors were not excised for 1 week following pump expiration, no significant difference was observed between tumor mass in sense- and antisense-treated animals, demonstrating that the effect on tumor growth was reversible. At the same time, reappearance of *NMYC* protein was observed in tumor tissue taken from these animals. Thus, although *NMYC* amplification has been associated with poor prognosis and tumor progression in human neuroblastoma, and although *NMYC* down-regulation is associated with retinoic acid-induced cellular differentiation, *NMYC* alone may not be an appropriate focus for further development of gene-targeted antitumor therapy, because its specific inhibition *in vivo* does not lead to loss of tumorigenicity or cell death.

In concordance with the multihit hypothesis of oncogenesis, it is possible that the interplay of several positive and negative oncogenes is required to maintain the tumorigenicity of CHP100. Simultaneous, or even sequential, inhibition of such multiple genes could lead to loss of the tumorigenic phenotype. As our understanding of the precise role played by various growth-associated genes in the malignant phenotype develops, more of these targets are likely to become apparent. Nonetheless, the results presented here demonstrate that *in vivo* administration of an unmodified oligo can specifically modulate protooncogene expression with the same sequelae that are seen *in vitro.* Thus, not only does expression of the target gene decline (without decline in expression of a closely related gene), but also the phenotype and biologic behavior of the perfused tumors is altered as well.

USE OF ANTISENSE OLIGONUCLEOTIDES AS ANTISENSE INHIBITORS OF GENE EXPRESSION

We have sought to overcome some of the obstacles associated with *in vivo* administration of oligos by avoiding the need for their systemic application. Successful systemic administration of oligos has been hindered by several technical difficulties, including oligonucleotide instability in biological fluids and the rapid clearance of systemically administered oligonucleotides.[18–21] Add to these considerations the observation that oligo uptake is a property of most, if not all, cells,[22–23] and it is apparent that systemic delivery of a nontargeted antisense reagent in sufficient concentration to its intended site is currently a very difficult task. Given the relatively low potency of unmodified antisense oligos, certain antisense-based therapeutic strategies requiring systemic administration may not prove effective, even if stability and bulk production problems can be overcome. Furthermore, systemic administration of DNA, modified to enhance stability or cellular uptake, must be viewed with particular caution, because eventual degradation of these oligos could lead to ultimate incorporation of modified nucleotides into cellular DNA, with a significant potential for mutagenicity. This led us to test the *in vivo* efficacy of unmodified phosphodiester oligos in the model presented here.

Our data suggest that the current generation of oligos can be developed as clinically efficacious therapeutics in well-defined clinical settings where systemic administration is not required. Localized continuous perfusion of antisense oligos does effectively modulate gene expression in an animal. The model we have described can be useful for *in vivo* efficacy screening of both new target sequences and novel oligo modifications as well as for determining *in vivo* toxicities of antisense reagents. In addition, this model can be used to test the role of target genes in *in vivo* tumor growth. For example, our current data clearly demonstrate that sole inhibition of tumor cell *NMYC in vivo* is not sufficient to abrogate tumorigenesis. At the same time, other clinically relevant models suggest themselves. Studies are underway to explore the use of intrathecal perfusion of oligos for the treatment of leptomeningeal tumors in the athymic rat and topical administration to hyperproliferative epidermis in a murine model of psoriasis.

REFERENCES

1. ALT, F., R. DEPINHO, K. ZIMMERMAN, E. LEGOUY, K. HATTON, P. FERRIER, A. TESFAYE, G. YANCOPOULOS & P. NISEN. 1986. The human myc gene family. Cold Spring Harbor Symp. Quant. Biol. **51:** 931–941.
2. DILDROP, R., K. ZIMMERMAN, R. DEPINHO, G. YANCOPOULOS, A. TESFAYE & F. ALT. 1988. Differential expression of myc family genes during development: Normal and deregulated *NMYC* expression in transgenic mice. Curr. Top. Microbiol. Immunol. **141:** 100–109.
3. DILDROP, R., A. MA, K. ZIMMERMAN, E. HSU, A. TESFAYE, R. DEPINHO & F. ALT. 1989. IgH enhancer-mediated deregulation of *NMYC* gene expression in transgenic mice: Generation of lymphoid neoplasias that lack *MYC* expression. EMBO J. **8:** 1121–1128.
4. BRODEUR, G. M., C. FONG, M. MORITA, R. GRIFFITH, F. A. HAYES & R. A. SEEGER. 1989. Molecular analysis and clinical significance of *NMYC* amplification and chromosome 1p monosomy in human neuroblastomas. *In* Advances in Neuroblastoma Research. A. E. Evans, G. J. D'Angio, A. G. Knudson & R. C. Seeger, eds.: 3–15. Alan R. Liss Publishers. New York.
5. ZIMMERMAN, K. A., G. D. YANCOPOULOS, R. G. COLLUM, R. K. SMITH, N. E. KOHL, K. A. DENIS, M. M. NAU, O. N. WITTE, D. TORAN-ALLERAND, C. E. GEE, J. D. MINNA & F. ALT. 1986. Differential expression of myc family genes during murine development. Nature **319:** 780–783.
6. JACOBOVITZ, A., M. SCHWAB, J. BISHOP & G. MARTIN. 1985. Expression of NMYC in teratocarcinoma stem cells and mouse embryos. Nature **318:** 188–191.
7. COLLUM, R. & F. ALT. 1990. Are myc proteins transcription factors? Cancer Cells **2:** 69–75.
8. NAKAJIAMA, H., M. IKEDA, M. TSUCHIDA, S. NISHIMURA & Y. TAYA. 1989. Inactivation of the *NMYC* gene product by single amino acid substitution of leucine residues located in the leucine-zipper region. Oncogene **4:** 999–1002.
9. ROSOLEN, A., L. WHITESELL, N. IKEGAKI, R. H. KENNETT & L. M. NECKERS. 1990. Antisense inhibition of single copy NMYC expression results in decreased cell growth without reduction of *MYC* protein in a neuroepithelioma cell line. Cancer Res. **50:** 6316–6322.
10. ROSS, R. A., B. A. SPENGLER & J. L. BIEDLER. 1983. Coordinate morphological and biochemical interconversion of human neuroblastoma cells. J. Natl. Cancer Inst. **71:** 741–747.
11. CICCARONE, V. C., B. A. SPENGLER, M. B. MEYERS, J. L. BIEDLER & R. A. ROSS. 1989. Phenotypic diversification in human neuroblastoma cells: Expression of distinct neural crest lineages. Cancer Res. **49:** 219–225.
12. WHITESELL, L., A. ROSOLEN & L. M. NECKERS. 1991. Episome-generated NMYC antisense RNA restricts the differentiation potential of primitive neuroectodermal cell lines. Mol. Cell. Biol. **11:** 1360–1371.
13. MICHITSCH, R. W., J. L. BIEDLER & P. W. MELERA. 1988. Modulation of NMYC expression, but not tumorigenicity, accompanies phenotypic conversion of neuroblastoma cells in prolonged culture. *In* Advances in Neuroblastoma Research. A. E. Evans, G. J. D'Angio, A. G. Knudson & R. C. Seeger, eds.:103–120. Alan R. Liss Publishers, New York.
14. HAMBOR, J. E., C. A. HAUER, H. SHU, R. K. GROGER, D. R. KAPLAN & M. L. TYKOCINSKI. 1988. Use of an Epstein-Barr virus episomal replicon for antisense RNA-mediated gene inhibition in a human cytotoxic T-cell clone. Proc. Natl. Acad. Sci USA **85:** 4010–4014.
15. WHITESELL, L., A. ROSOLEN & L. M. NECKERS. 1991. *In vivo* modulation of *NMYC*

expression by continuous perfusion with an antisense oligonucleotide. Antisense Res. Dev. **1:** 343–350.

16. EVAN, G., G. LEWIS, G. RAMSEY & J. BISHOP. 1985. Isolation of mononuclear antibodies specific for human *MYC* proto-oncogene product. Mol. Cell. Biol. **5:** 3610–3616.
17. THIELE, C. J., C. P. REYNOLDS & M. A. ISRAEL. 1985. Decreased expression of NMYC precedes retinoic acid-induced morphological differentiation of human neuroblastoma. Nature **313:** 404–406.
18. CHEM, T.-L., P. S. MILLER, P. O. P. T'SO & O. M. COLVIN. 1990. Disposition and metabolism of oligodeoxynucleoside methylphosphonate following a single i.v. injection in mice. Drug Metab. Dispos. **18:** 815–818.
19. AGRAWAL, S., J. TEMSAMANI & J. Y. TANG. 1991. Pharmacokinetics, biodistribution, and stability of oligodeoxynucleotide phosphorothioates in mice. Proc. Natl. Acad. Sci. USA **88:** 7595–7599.
20. GOODCHILD, J., B. KIM & P. C. ZAMECNIK. 1991. The clearance and degradation of oligodeoxynucleotides following intravenous injection into rabbits. Antisense Res. Dev. **1:** 153–160.
21. ZON, G. 1989. Pharmaceutical considerations. *In* Oligodeoxynucleotides: Antisense Inhibitors of Gene Expression. J. S. Cohen, ed.: 233–248. Macmillan Press. London.
22. LOKE, S. L., C. A. STEIN, X. H. ZHANG, K. MORI, M. NAKANISHI, C. SUBASINGHE, J. S. COHEN & L. M. NECKERS. 1989. Proc. Natl. Acad. Sci. USA **86:** 3474–3478.
23. YAKUBOV, L. A., E. A. DEEVA, V. F. ZARYTOVA, E. M. IVANOVA, A. S. RYTE, V. L. YURCHENKO & V. V. VLASSOV. 1989. Mechanism of oligonucleotide uptake by cells: Involvement of specific receptors? Proc. Natl. Acad. Sci. USA **86:** 6454–6458.

Variables Affecting Antisense RNA Inhibition of Gene Expression[a]

CAROL A. SCHERCZINGER, ADRIENNE A. YATES, AND DAVID A. KNECHT[b]

Department of Molecular and Cell Biology
University of Connecticut
Storrs, Connecticut 06269

Antisense inhibition of gene expression has become a widely used method for specifically interfering with gene function. Inhibition of particular target genes has been achieved using injected RNAs, expressed RNAs, and oligodeoxynucleotides.[1,2] In the eukaryotic ameba *Dictyostelium discoideum,* transformation with a portion of the myosin heavy chain II (MHCII) gene in antisense orientation resulted in a stable cell line severely inhibited for MHCII gene expression.[3] Transformed cells were shown to accumulate the vector-derived antisense RNA transcript in excess, with a concomitant 200-fold reduction in the amount of MHCII protein present. In addition, MHCII mRNA was undetectable by RNA blot analysis. The inhibited cells were impaired in cytokinesis, generating giant multinucleated progeny. Additionally, they were unable to complete the normal developmental sequence initiated upon starvation; they could not complete morphogenesis, although they expressed development-specific markers.[4]

Despite the wide application of antisense RNA technology, there is little actual information on the mechanism of inhibition or on critical parameters for experimental success. Numerous undocumented failures are not suitable for publication, making the rate of success difficult to determine. To generate some guidelines for future experiments, we have systematically studied some of the variables that might affect the successful inhibition of expression from a nuclear transcribed antisense gene. In particular, the size of the antisense fragment, its location in the RNA transcript, the portion of the gene from which it is derived, and the type of vector used to express the antisense gene were analyzed.

MATERIALS AND METHODS

Dictyostelium axenic strain AX4 was transformed by the $CaPO_4$-DNA coprecipitation method. All transfected strains were maintained in axenic broth medium containing 15 μg/ml G418. Methods for Western blot analysis, Northern blot analysis, fixation, and staining were all performed according to standard methods described previously.[3,5] The antibody for Western blot analysis is My11, a mouse monoclonal antibody that reacts with MHCII near the COOH-terminus of the protein. The probes for Northern blots were single-stranded RNA riboprobes made using the Gemini transcription system (Promega Biotech Inc.).

[a]This work was supported by grants from the NIH (GM40599), the Muscular Dystrophy Association, and the American Cancer Society (ACSIN152E).

[b]To whom correspondence should be addressed.

RESULTS

Vector Construction

A series of transformation vectors was constructed to assess the relative contribution of several parameters to successful inhibition of gene expression by antisense RNA. In each case, a fragment of the myosin heavy chain II gene (MHCII) was cloned into an expression vector in both the sense (+) and antisense (−) orientations relative to the promoter driving RNA transcription.

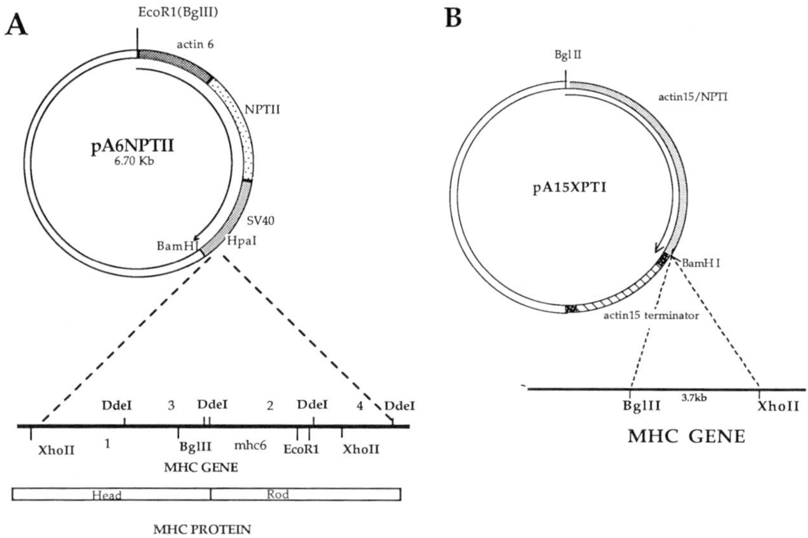

FIGURE 1. Transformation vectors. All fragments were cloned in both the sense (+) and antisense (−) orientations. **(A)** Fragments of the *Dictyostelium discoideum* myosin heavy chain II gene cloned into transformation vector pA6NPTII. The promoter (actin 6), selectable marker gene (NPTII), and terminator (SV40) are noted. *Arrow* indicates direction of transcription. mhc8+/9− = entire 8.0-kb genomic fragment cloned into the *Bam*HI site; 2+/−; 3+/−; 4+/− = *Dde*I fragments cloned into the *Hpa*I site; sizes are 2.0, 1.6, and 1.5 kb, respectively; 1+/− = 1.6 kb 5′ *Xho*II/*Bgl*II fragment cloned into the *Bam*HI site; mhc6+/− = 3.7 kb *Bgl*II/*Xho*II fragment cloned into the *Bam*HI site; A6B+/− = 3.7 kb *Bgl*II/*Xho*II fragment cloned into the origin of the vector. The cloning site was change from an *Eco*RI to a *Bgl*II site by the addition of linkers. **(B)** mhc1+/mhc2− = 3.7-kb *Bgl*II/*Xho*II fragment cloned into the *Bam*HI site of transformation vector pA15XPTI. The promoter/selectable marker gene (actin 15/NPTI) and terminator are noted. *Arrow* indicates the direction of transcription.

Four approximately equal sized fragments spanning the MHCII gene were subcloned into the transformation vector pA6NPTII (1+/−;2+/−;3+/−;4+/−) to determine if different portions of the gene vary in their ability to inhibit gene expression (FIG. 1A). The vector pA6NPTII consists of the *Dictyostelium* actin 6 promoter driving transcription of a bacterial neomycin phosphotransferase selectable marker gene, followed by SV40 termination sequences. As the SV40 terminator is not completely efficient in *Dictyostelium* cells, transcription initiation at the actin 6 promoter generates two transcripts: a 2.0-kb drug resistance transcript that terminates within the SV40 sequences, and a plasmid-sized transcript that terminates in the actin 6 promoter.[6] Thus, myosin fragments cloned downstream of the terminator (*Bam*HI site) will be included in the readthrough transcript, whereas sequences cloned upstream (*Hpa*I site) will be present on both transcripts.

To test whether the position of the antisense sequence on the plasmid-derived transcript is a contributing factor for inhibition, the 3.7-kb *Bgl*II/*Xho*II fragment used in the original antisense experiments with this gene was cloned into the origin of pA6NPTII (A6B+/−). This construct places the myosin sequences at the end of the long readthrough transcript, including both NPT and pBR322 sequences. The same 3.7-kb *Bgl*II/*Xho*II fragment was also cloned into the *Bam*HI site of pA6NPTII(mhc6+/−) and pA15XPTI (mhc1+/2−) (FIG. 1B). This vector contains the *Dictyostelium* actin 15 promoter, the neomycin selectable marker gene, and the *Dictyostelium* actin 15 termination sequences. As this homologous terminator is very efficient, the only RNA derived from this vector is the NPT-antisense myosin fusion transcript.[7]

The size of the antisense RNA sequence homologous to the endogenous mRNA might also affect the extent of inhibition of gene expression. The 3.7-kb *Bgl*II/*Xho*II fragment gave complete inhibition of expression of MHCII mRNA.[3] A 396-bp *Eco*RI fragment contained within the 3.7-kb fragment was tested as well as the entire 8.0-kb genomic fragment (mhc8+; mhc9−). In addition, the *Dde*I fragment 2 described above (FIG. 1) is a 2-kb subfragment of the 3.7-kb *Bgl*II/*Xho*II.

Portion of the Gene Used to Produce Antisense RNA

All plasmids were individually introduced into axenic *Dictyostelium discoideum* cells by calcium phosphate coprecipitation, and stable transformants were selected for resistance to G418. Vector DNA without MHCII sequences was used as a control for nonspecific effects of transformation and/or plasmid integration. Both pA6NPTII and pA15XPTI were previously shown to integrate into the genome in a tandem array of high copy number.[6,7] Transformed cells began to show the giant multinucleated morphologic characteristic of antisense myosin mutants after 1–2 weeks of selection. This phenotype results from a defect in cytokinesis in cells lacking MHCII protein. Not all cell lines accumulated giant multinucleated cells to the same extent, suggesting differences in inhibition of MHCII gene expression between them.

To directly determine the extent of MHCII inhibition, protein extracts were prepared from transformed cell populations and examined by Western blot analysis using an anti-MHCII monoclonal antibody (FIG. 2). Accumulation of MHCII protein varied widely in the cell lines analyzed. Transformants containing the 3.7-kb fragment (mhc6−, A6B−, mhc2−) showed extreme reduction, whereas the subclone transformants accumulated MHCII protein to different levels, ranging from undetectable (2−) to virtually wild type (3−). Cells transformed with the fragments derived from the 3′ end of the gene (2− and 4−) had considerably less MHCII protein than did those derived from the 5′ end of the gene (1− and 3−). MHCII protein accumulation was also reduced in cells transformed with the entire genomic fragment (mhc9−), although not to the degree seen in the 3′ end subclones.

Clonal cell lines were derived from the transformed populations by limiting dilution. Clonal derivatives were extremely uniform with respect to MHCII protein levels; all clones accumulated the protein to the same extent as did the populations (data not shown). Therefore, the differences in MHCII protein accumulation between the transformed cell lines cannot be due to mixed populations of inhibited and noninhibited cells. We conclude that although the choice of transformation vector (mhc6−; mhc2−) and the relative position of the antisense sequence (A6B−) do not affect inhibition of MHCII gene expression, the efficiency of antisense RNA inhibition can be affected by the portion of the gene used to generate the antisense sequence. Fragments representing the 3′ end of the gene (2−;4−) are much more effective than the 5′ fragments (1−; 3−).

RNA isolated from the clonal cell lines was analyzed for the presence of MHC mRNA and antisense RNA transcripts. Each produced vector-derived antisense RNA, and the pattern of MHCII mRNA accumulation reflected that of the protein (data not shown). Therefore, antisense RNA inhibition in this system is at the level of mRNA accumulation. It is not known if this decrease in steady-state mRNA levels is the result of a decrease in transcription rate or mRNA stability.

Size of the Antisense Sequence

The size of the antisense sequence also had an effect on the extent of inhibition. Both the 3.7-kb fragment and the 2-kb Dde fragment derived from it (2 in FIG. 1) give

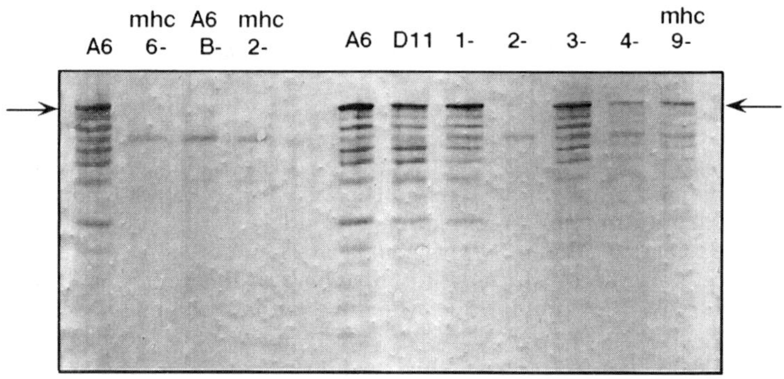

FIGURE 2. Western analysis of MHCII protein in antisense-transformed cell populations. Protein extracts (10 μg) were stained with an anti-MHCII monoclonal antibody. Cell line designations are the same as in FIGURE 1. A6 is a control transformation with pA6NPTII. *Arrow* indicates the position of the 243-kD MHCII protein.

complete inhibition of expression. When a 396-bp *Eco*RI fragment contained within the *Dde*I fragment was tested, no inhibition was observed (FIG. 3, lane 3). In addition, the vector containing the entire 8-kb MHCII sequence gave only partial inhibition (FIG. 2, mhc9). Therefore, the size of the antisense sequence, as well as the portion of the gene used is important, but the sequence context may also affect the result. The 3.7-kb fragment is effective when associated with pBR322 or NPT sequences, but not when in the context of the rest of the myosin gene.

Transformed cells were selected using a concentration of 10 μg G418/ml. When the amount of drug was increased, cells expressing more of the NPT protein and mRNA were selected. Selection for increased amounts of the antisense RNA also occurred, because the two transcripts are linked. Induction occurs as a result of increased transcription; no increase in copy number of vector DNA was seen (data not shown). When cells expressing the antisense 396-bp *Eco*RI fragment were

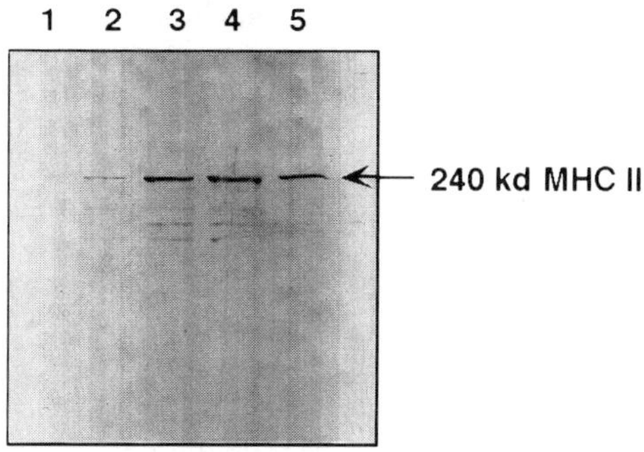

FIGURE 3. Increased inhibition by drug induction. Cells containing the 396-bp *Eco*RI antisense fragment and cells containing vector alone were selected for growth in 50 and 250 μg G418/ml. Protein extracts were prepared and MHCII protein immunostained on a Western blot. *Lane 1,* antisense 250 μg/ml; *lane 2,* antisense 50 μg/ml; *lane 3,* antisense 10 μg/ml; *lane 4,* control 250 μg/ml; *lane 5,* control 50 μg/ml.

exposed to increasing amounts of G418, MHCII expression was gradually reduced until apparently complete inhibition was achieved at 250 μg G418/ml (FIG. 3, lanes 1 and 2). This result indicates that the extent of inhibition with antisense RNA might be due to a combination of the amount of antisense RNA, its sequence context, and the particular portion of the coding sequence transcribed.

Reversion of Antisense Inhibition

Some of the cell lines used in these experiments have been maintained in culture for extended periods of time. Periodically, cultures of cell lines that had previously undetectable levels of MHCII began to reexpress the protein. A variety of cell lines was produced by limited dilution cloning. Some cells still did not accumulate measurable MHCII protein, others expressed nearly normal levels, and a third class had low but detectable amounts of protein (FIG. 4). Northern blots of RNA extracted from these cells were probed to identify the cause of this reversion. The renewed accumulation of MHCII protein was correlated with the reappearance of MHCII

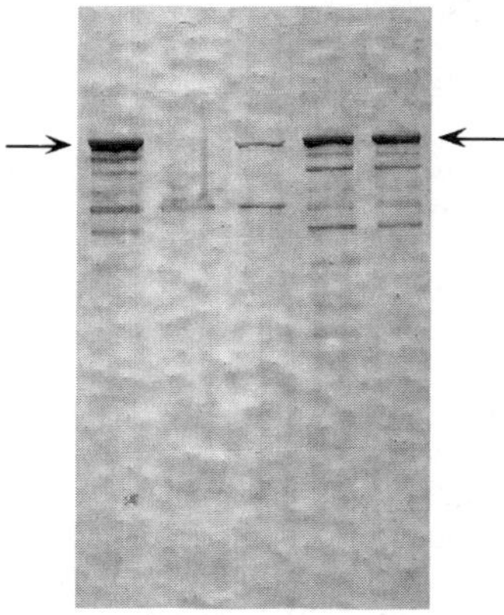

FIGURE 4. Reversion of antisense-inhibited cells. Clonal derivatives from a population of cells transformed with pA6MHC6- which showed reexpression of MHCII. Protein extracts from each clone were Western blotted and immunostained with anti-MHCII monoclonal antibody. *Arrow* indicates the position of the MHCII protein.

mRNA and decreased accumulation of antisense RNA (FIG. 5). The decrease in the amount of the 10.4-kb antisense transcript was correlated with an increase in the amount of the 2.0-kb NPT transcript terminating in the SV40 sequence. The most likely explanation for this phenomenon is that the cells have increased the efficiency of the SV40 termination site so that the readthrough transcription of the antisense RNA is reduced.

Sense Orientation Transformants

All transformations were performed with sense and antisense orientation clones in order to investigate the orientation dependence of inhibition. Surprisingly, cells transformed with sense orientation clones also began to accumulate giant cells. When these populations were stained with the DNA binding dye Hoechst and examined by fluorescence microscopy, they were found to contain a mixture of multinucleated and normal cells (FIG. 6A). Cell lines containing each of the sense orientation clones had giant multinucleated cells, but those containing the 3.7-kb *Bgl*II/*Xho*II fragment were the most dramatic. This phenotype is usually associated with a lack of MHCII protein as seen in the antisense mutants as well as in cells in which the MHCII gene was deleted by homologous recombination.[8]

MHCII protein accumulation was examined in clonal derivatives of 3.7-kb sense orientation transformants (FIG. 6B). These cells accumulated 50–75% of wild-type MHCII protein levels, as seen by comparison with the titration of control extract. Inhibition of MHCII gene expression was more noticeable at the mRNA level in the sense transformants. Northern blot analysis revealed substantial reduction in the steady-state levels of MHCII mRNA with a concomitant accumulation of vector-derived sense transcripts (data not shown).

In one clone (pA15mhc+1) no MHCII protein was detected with the monoclonal antibody. When this extract was stained with polyclonal antiserum, a 180-kd band was detected. The size of this band is consistent with the production of a truncated peptide generated by homologous insertion of the transformation vector into the

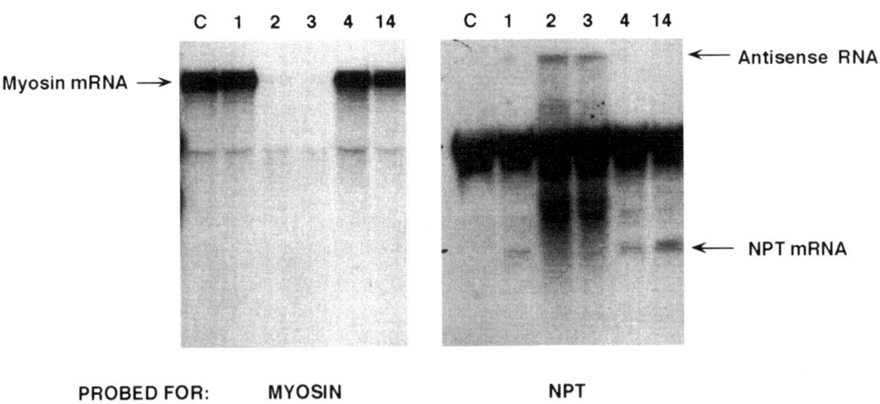

FIGURE 5. Northern analysis of revertants. Total RNA was analyzed for the presence of MHCII mRNA and antisense RNA. Duplicate filters were probed with single-stranded RNA probes to detect the two transcripts. Note that the clones that show myosin mRNA correspond to those showing reduced antisense RNA and increased MHCII protein on the Western blot in FIGURE 4.

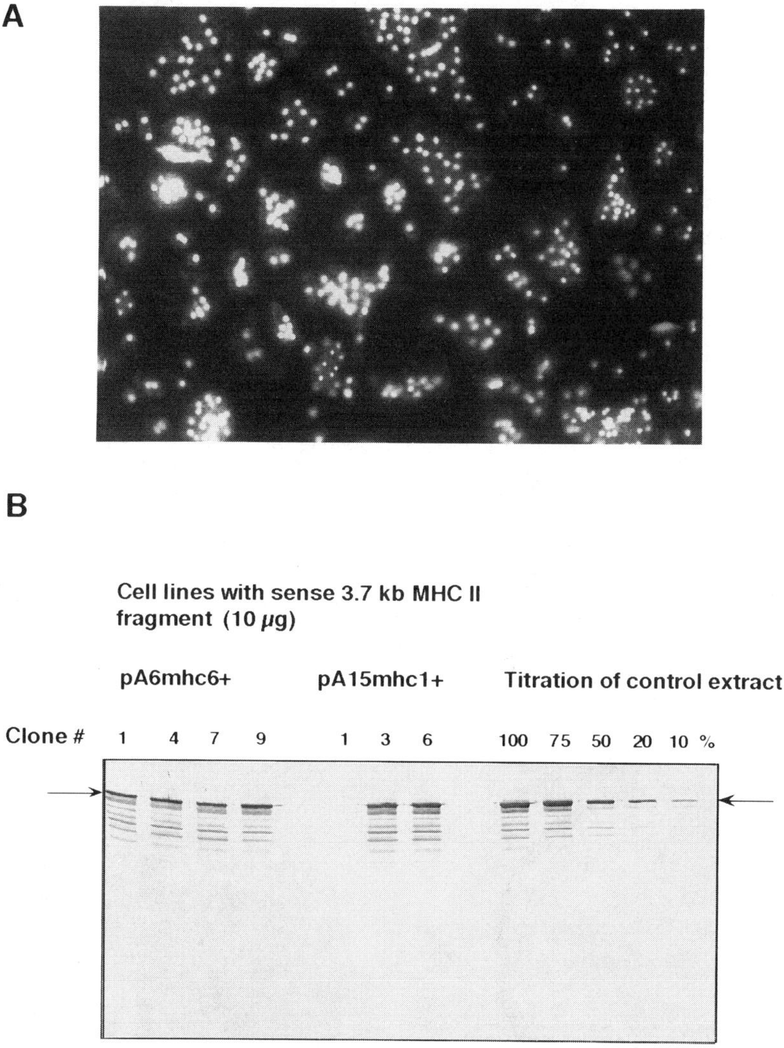

FIGURE 6. Sense orientation transformants. **(A)** Hoechst 33258 fluorescence micrograph showing giant multinucleated morphology of mhc1+ transformed cells. **(B)** Western analysis of MHCII protein in clonal derivatives of sense-transformed populations. 10 μg of extract was immunostained with an anti-MHCII monoclonal antibody. Cell line designation and the individual clone number are indicated above the lane. Control extract is from pA6NPTII transformed cells and titration begins at 10 μg. *Arrows* indicate the position of MHCII protein.

chromosomal copy of the MHCII gene. None of the other vectors used was ever found to give homologous recombination into the genome.

Considerable variation was found in the amount of MHCII protein among clonal derivatives of sense transformants, unlike the antisense clones, which were quite uniform. Some clones had a reduced level of MHCII protein and accumulated giant multinucleated cells, whereas others appeared to be completely wild type. No sense clone was found to give the virtually complete inhibition noted in the most effective antisense cell lines. However, the 50% reduction seen in some of the clones (e.g., mhc6+4) was as good or better than the less effective antisense clones (1−; 3−; mhc9−).

To determine if expression of the vector-derived MHCII sequences is necessary for sense inhibition, the 3.7-kb *Bgl*II/*Xho*II fragment was cloned into an unexpressed region of the transformation vector pA15TX. This vector is similar to pA15XPTI, except the *Bam*HI cloning site is located downstream of the actin 15 terminator (FIG. 7A). The transcript produced upon initiation at the actin 15 promoter should therefore terminate before reaching the myosin sequences. The 3.7-kb fragment was cloned in both orientations with respect to the actin 15 promoter (TXS+; TXA−), although orientation should be irrelevant as the myosin sequences should not be expressed. Cells transformed with both TXS+ and TXA− became giant and multinucleated. Analysis of MHCII protein levels in the transformed populations showed a reduction of approximately 50% for both orientations (FIG. 7B). There are several possible explanations for this result. The simplest is that expression of the MHCII sequences on the transformation vector is not necessary for inhibition of gene expression. Alternatively, there could be a small amount of readthrough transcription of the actin 15 terminator. We are currently analyzing RNA from clonal derivatives of these transformants to determine if the cells are accumulating vector-derived transcripts containing the MHCII sequences.

Preliminary evidence from other experiments in our laboratory suggested a third possibility: the A15 and A6 promoters might be capable of initiating transcription in a bidirectional manner. To explore this possibility, RNA was isolated from populations of cells transformed with pA6NPTII, pA15XPTI, or pA15TX. This RNA was

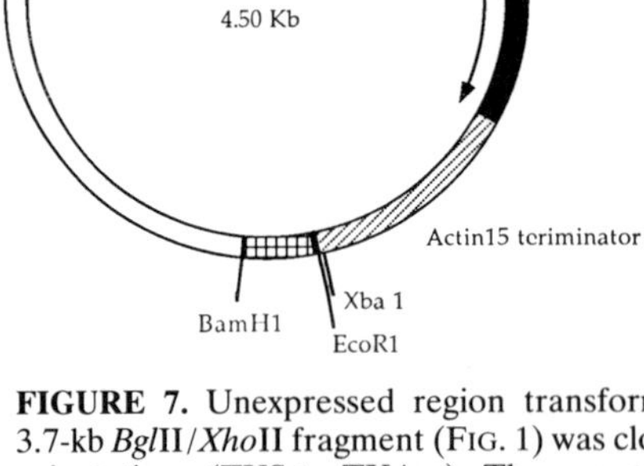

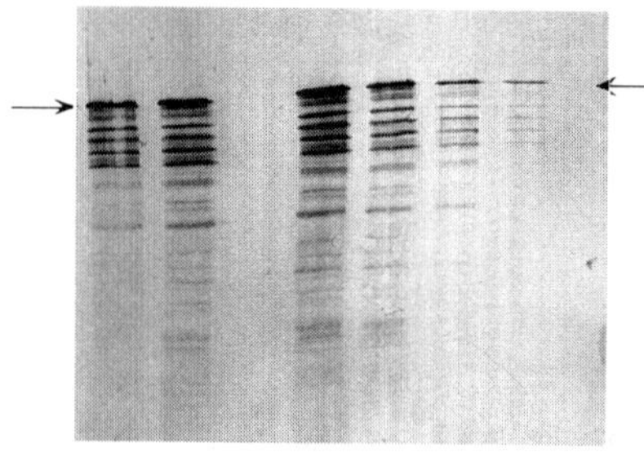

FIGURE 7. Unexpressed region transformants. (**A**) Transformation vector pA15TX. The 3.7-kb *Bgl*II/*Xho*II fragment (FIG. 1) was cloned into the *Bam*HI site in the sense and antisense orientations (TXS+; TXA−). The promoter, selectable marker gene, and terminator are noted. As the *Bam*HI site is located downstream of the terminator, the myosin sequences should not be included on the transcript initiated at the actin 15 promoter (*arrow*). (**B**) Western analysis of MHCII protein in unexpressed region transformants. 10 μg of protein was immunostained with anti-MHCII monoclonal antibody. Cell line designations are as in **A.** Control extract is pA6NPTII, and titration begins at 10 μg. *Arrow* indicates MHCII protein.

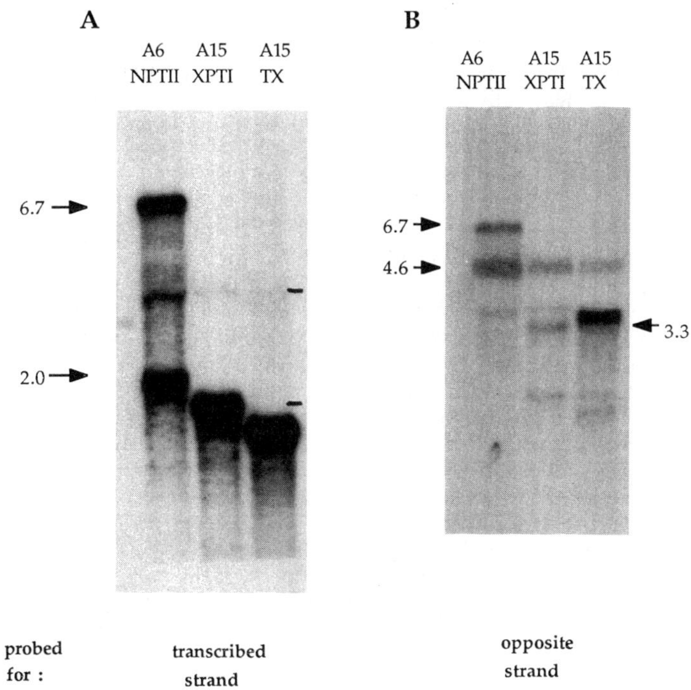

FIGURE 8. Northern analysis of transcripts in vector-transformed cell lines. 10 μg of total cell RNA was analyzed using strand-specific RNA probes. Vector is indicated above the lane. Panel A was probed for sense NPT sequences. Panel B was probed for pBR322 sequences transcribed in the opposite orientation.

probed for transcripts in the predicted transcription orientation as well as for transcripts initiated in the other direction using strand-specific RNA probes (FIG. 8). The strongly hybridizing bands in panel A correspond to the predicted drug resistance transcripts for each vector, and the larger band in the pA6NPTII lane is the readthrough transcript. A number of bands were detected when the same RNA samples were probed for the opposite strand of the plasmid. The largest band in the pA6NPTII lane is the same size as the readthrough transcript in the sense orientation (~6.7 kb), suggesting that this transcript initiates on the opposite strand and terminates in the actin 6 promoter. The smaller band (~4.6 kb) is in the size range predicted for initiation on the opposite strand and termination within the SV40 terminator. The largest band in both the pA15XPTI and pA15TX lanes is of approximate vector size (~4.6 kb) and again suggests transcription initiation on the opposite strand and termination within the promoter. The more strongly hybridizing smaller band (~3.3 kb) in the pA15TX sample is of a size consistent with termination within the actin 15 terminator. Preliminary data would therefore suggest that the *Dictyostelium* actin 6 and actin 15 promoters in these transformation vectors are capable of promoting transcription bidirectionally.

This finding suggests that inhibition of MHCII gene expression by sense orientation clones might, in fact, be caused by transcription of the other strand of the transformation vector, resulting in the production of antisense MHCII sequences.

We are continuing to analyze the RNAs accumulated by the sense transformants to determine the nature of inhibition by sense sequences. Although it is possible that the other strand of the transformation vector is being transcribed to produce an antisense RNA in these cells, we believe that sense orientation inhibition might be the cumulative result of this and other factors to be discussed.

DISCUSSION

Antisense RNA has become a widely used tool for analysis of gene function and holds great promise for therapeutic use in the future. Despite a large body of literature in which antisense RNA has been used, still little is known about the mechanism of inhibition, primarily because of the difficulty in studying the process *in vivo.* It is generally assumed that the mechanism of inhibition by nuclear-derived antisense RNA entails sequence-specific hybridization of antisense RNA transcript to the target mRNA in the nucleus. The point at which this event happens relative to transcription termination, assembly of a ribonuclear protein particle (RNP), processing of the 3′ end, polyadenylation, splicing, or nuclear export is unknown. It is possible that there is a particular step at which mRNA is most accessible to form a duplex. It seems probable that hybridization would occur prior to RNP assembly, because RNA accessibility within the RNP is likely to be limited.

With this idea in mind, we set out to determine the sequence specificity and context that give the most efficient antisense inhibition of gene expression in the case of MHCII mRNA in *Dictyostelium.* It is clear from our results that many factors affect the extent of inhibition achieved. An interplay of the amount of antisense and the length of the sequence used are important factors, as would be expected if antisense inhibition is viewed as a concentration-driven hybridization reaction. However, other factors are clearly involved. The fact that the 3.7-kb *Bgl*II/*Xho*II fragment derived from the tail region of the myosin gene and a subclone of this region were the most effective fragments was not predictable. This region of the gene gave complete inhibition no matter what transformation vector was used or the position of the antisense sequence within the RNA transcript. The only situation in which it was not completely effective was when it was in the context of the entire antisense MHCII gene.

There are several reasons why this fragment might be the most effective. First, the folding of this RNA might make it particularly accessible for hybridization to its mRNA complement. However, this fragment functioned when at the end of a 5.7-kb fusion with NPT I sequence, in the middle of a 10-kb fusion with NPT II and pBR322 sequence, and at the end of a fusion with NPT II and pBR322. This result implies that the 3.7-kb domain had to fold into the appropriate conformation with little influence from the rest of the RNA molecule. Until more is known about how RNA and RNP molecules fold *in vivo* and the accessibility of RNA sequences in RNP molecules, it will be difficult to assess the validity of this proposal.

Another possibility that we have considered is that the effectiveness is determined by how much cross-hybridization occurs with other sequences in the cell. It is presumed that to obtain effective inhibition, there must be an excess of antisense RNA over mRNA in the nucleus. If a particular antisense sequence is titrated out by other mRNAs, there might be less inhibition of the target mRNA. The extent of cross-reaction with other RNAs is impossible to determine without some knowledge of all of the homologous genes in the cell. As it is unclear what level of specificity is tolerated in the formation of sense-antisense hybrids, this problem is difficult to address. However, for clinical application of this technology, it is important to know that no other genes are being affected. We are addressing this question currently by examining the inhibition of other myosin-related genes (myosin Is) in cells expressing different antisense fragments of the MHCII gene. As these myosin I genes vary in their homology to MHCII, we may see cross-inhibition with some fragments.

The inhibition of MHCII expression by sense constructs was unexpected and may again demonstrate how little we know about the mechanism of antisense inhibition. Since we first observed this phenomenon, several other groups have published similar observations.[9–12] Now termed "sense co-suppression," numerous explanations have been put forward to explain these unusual findings. Our first concern was that the sense transcript was producing a small amount of peptide, and that this peptide was interfering with myosin assembly or function. Although theoretically possible, we have been unable to detect any vector-derived peptides in these cells. Inhibition of accumulation of MHCII mRNA also argues that the problem occurs pretranslationally. Cameron and Jennings[12] have argued that perhaps the sense RNA and the mRNA can form a hybrid. Although RNA modeling computer programs can generate a duplex between RNAs, we doubt that this is any more likely with endogenous mRNA than any other mRNA in the cell. Cells presumably have mechanisms to prevent RNA-RNA duplex formation unless the driving force is great enough. However, their data suggesting a requirement for transcription of the sense RNA in order to observe co-suppression are very compelling. It is also possible that the presence of the extra gene copies in the nucleus is sufficient to give co-suppression through some interaction of the genes or titration of nuclear factors. In plants, the phenomenon is highly variable from plant to plant and even in different cells or tissues of the same plant. Thus, it is hard to investigate this phenomenon except on a cell-by-cell basis. We have added to the puzzle by showing that the promoters used to drive antisense RNA can also initiate transcription in the opposite direction. Therefore, unintended antisense transcripts could be produced in the nucleus. It is also possible that the termination sequences are not 100% efficient or that transcription proceeds beyond this region before processing adds polyA at an upstream site. A transient sense or antisense RNA would then be produced in the nucleus. This unintended antisense transcript might inhibit accumulation of its complementary mRNA. Alternatively, it is possible that the sense (or antisense) sequences might exert their effect at the DNA level, interfering with transcription by hybridizing to the endogenous gene. Some combination of these ideas is likely to account for co-suppression.

SUMMARY

We have sought to determine what variables affect the extent of inhibition of gene expression by stable nuclear-derived antisense RNAs. Myosin heavy chain gene II (MHCII) expression in *Dictyostelium* was used as a model system, because previous results have shown that nearly complete inhibition of expression can be achieved under appropriate conditions. Various fragments of the myosin gene were inserted into several transformation vectors in both sense and antisense orientation, and the effects on expression of protein and RNA from the endogenous MHC II gene were assayed. The results indicate that the critical factor was the particular fragment of the gene used to produce the antisense RNA. Some fragments produced complete inhibition of expression, whereas others gave only slight inhibition. The fragments that produced the greatest inhibition were from the tail region of the gene. In addition, cells were capable of overcoming the inhibition while still expressing the antisense RNA. We hypothesize that the three-dimensional topology of the antisense and sense RNAs determines their accessibility for interstrand hybridization.

Transformation with sense fragments of the myosin gene also caused inhibition of the endogenous myosin gene. The inhibition seen with sense expression is not as dramatic as that for antisense, but it had the same phenotypic consequences for the cells. This phenomenon, which has now been documented in other systems, may be mechanistically similar to inhibition by antisense RNA in our system.

REFERENCES

1. Takayama, K. M. & M. Inouye. 1990. Antisense RNA. Crit. Rev. Biochem. Mol. Biol. **25:** 155–184.
2. Helene, C. & J. J. Toulme. 1990. Specific regulation of gene expression by antisense, sense and antigene nucleic acids. Biochim. Biophys. Acta **1049:** 99–125.
3. Knecht, D. & W. Loomis. 1987. Antisense RNA inactivation of myosin heavy chain gene expression in *Dictyostelium discoideum.* Science **236:** 1081–1086.
4. Knecht, D. & W. Loomis. 1988. Developmental consequences of the lack of myosin heavy chain in *Dictyostelium discoideum.* Dev. Biol. **128:** 178–184.
5. Sambrook, J., E. F. Fritsch & T. Maniatis. 1989. Molecular Cloning: A Laboratory Manual. Cold Spring Harbor Laboratory Press. Cold Spring Harbor, NY.
6. Knecht, D., S. Cohen, W. Loomis & H. Lodish. 1986. Developmental regulation of *Dictyostelium discoideum* actin gene fusions carried on low-copy and high-copy transformation vectors. Mol. Cell Biol. **6:** 3973–3983.
7. Cohen, S. M., D. Knecht, H. F. Lodish & W. F. Loomis. 1986. DNA sequences required for expression of a *Dictyostelium* actin gene. EMBO J. **5:** 3361–3366.
8. Manstein, D. J., M. A. Titus, L. A. De & J. A. Spudich. 1989. Gene replacement in *Dictyostelium:* Generation of myosin null mutants. EMBO J. **8:** 923–932.
9. Smith, C. J., C. F. Watson, C. R. Bird, J. Ray, W. Schuch & D. Grierson. 1990. Expression of a truncated tomato polygalacturonase gene inhibits expression of the endogenous gene in transgenic plants. Mol. Gen. Genet. **224:** 477–481.
10. Napoli, C., C. Lemieux & R. Jorgensen. 1990. Introduction of a chimeric chalcone synthase gene into petunia results in reversible co-suppression of homologous genes *in trans.* Plant Cell. **2:** 279–289.
11. van der Krol, A. R., L. A. Mur, M. Beld, J. N. M. Mol & A. R. Suite. 1990. Flavonoid genes in petunia: Addition of a limited number of gene copies may lead to suppression of gene expression. Plant Cell. **2:** 291–299.
12. Cameron, F. H. & P. A. Jennings. 1991. Inhibition of gene expression by a short sense fragment. Nucleic Acids Res. **19:** 469–475.

Modulation of c-*myc* Transcription by Triple Helix Formation[a]

EDITH H. POSTEL

Department of Molecular Biology
Princeton University
Princeton, New Jersey 08546–1014

Transcription of eukaryotic genes is known to be mediated by *trans*-acting protein factors that bind to various *cis*-acting DNA control elements, thereby serving as regulatory signals for transcriptional activation of the gene. One known control element of the human c-*myc* oncogene, termed a nuclease-hypersensitive element (NHE),[1] lies approximately −142 to −115 base pairs from the P1 promoter and plays an important role in c-*myc* transcription *in vivo* and *in vitro*.[1–4] This NHE regulatory region is the site of interaction of a number of likely regulatory proteins,[1,2,5,6] of which at least one, PuF, is a positively acting transcription initiation factor.[1]

Several properties of the NHE c-*myc* DNA segment suggest the presence of an atypical secondary structure that may be important in the transcriptional regulation of the c-*myc* gene.[1,4,5,7,8] The unusually high homopurine-homopyrimidine strand asymmetry[6] of this DNA segment may cause structural perturbations that can lead to hypersensitivity to the endonucleases DNaseI[3] and Sl[7] and to numerous single-strand specific chemical reagents.[11] Additionally, the observation that the −115 NHE region can assume a DNA secondary structure that is in equilibrium between at least two alternative helix conformations[7] has been the basis of its selection as a site capable of forming a stable triple-stranded complex.[4,9,10]

Thus, based on the prediction that the −115 NHE region might be a high affinity site for triplex formation and the fact that the NHE is a regulatory region for ligand binding, a 27-base long oligonucleotide (PU1) was synthesized to assess both its capacity to bind to the −115 target sequence and the possible biological significance of such site-selective oligonucleotide binding.[4] That work showed by DNaseI footprinting methods that the PU1 oligomer bound tightly to its duplex target sequence *in vitro* and that upon addition to cell free HeLa nuclear extracts, PU1 was capable of blocking c-*myc* transcription initiation in a dose-dependent manner and with apparent high specificity.

The first evidence that triplex formation between the PU1 oligonucleotide and the NHE element can also occur *in vivo* was demonstrated by DNaseI mapping[9] of nuclei isolated from HeLa cells following cellular uptake of the PU1 oligonucleotide.[9] Such site-selective triplex formation at the NHE sequence was presumed to have caused the significant reduction of c-*myc* mRNA synthesis seen upon PU1 oligonucleotide treatment, under circumstances in which unrelated oligonucleotides did not elicit the same effect.

Recent evidence, moreover, suggests that triplex formation *in vitro* between a small DNA fragment as the duplex target site and PU1 and other triplex-forming oligonucleotide homologs inhibits binding of the PuF factor[1] to the duplex target, as

[a] This work was funded by grants from The New Jersey Cancer Commission and the National Institutes of Health.

assessed by electrophoresis mobility-shift assays (E. Postel, R. Durland, and M. Hogan, unpublished results).

This paper examines with greater resolution the inhibitory effect of triplex-forming oligonucleotides on transcription from the dual c-*myc* promoters *in vitro,* and in particular, on PuF-initiated transcription in a reconstituted system that consists of partially purified RNA polymerase II and the PuF factor.[1] It is shown here that PuF factor-mediated transcription from both the P1 and P2 c-*myc* promoters is inhibited, in a dose-dependent manner, by oligonucleotides capable of triplex formation with the NHE target sequence, whereas oligonucleotides that do not form triplexes with this target site do not elicit the same effect. Together with previous observations,[1,4,9,10] these new data confirm the hypothesis put forth earlier[1,4,9] that oligonucleotide-directed triple-helix formation at the human c-*myc* NHE regulatory element is inhibiting sequence-specific DNA-protein interactions necessary for proper transcriptional activation of the c-*myc* gene.

RESULTS AND DISCUSSION

*Inhibition of c-*myc *Transcription* in Vitro *by Triplex Formation*

PU1 is a 27-base long purine-rich oligonucleotide (FIG. 1) that has been shown to bind in a stable fashion and with very high selectivity to a c-*myc* target within the NHE control region.[4] The work suggested that the gross secondary structure of the complex might resemble a colinear triplex in which the oligonucleotide bound parallel relative to the more purine-rich strand of the duplex target stabilized by TAT, AAT, and GGC triplets.[4] Subsequent work has shown, however, that PU1 binds in an antiparallel orientation, giving rise to an AGC and a GAT triplet mismatch.[10]

This same study[4] showed that upon the addition of PU1 to HeLa cell nuclear extracts, c-*myc* transcription initiation was blocked in a dose-dependent manner and with apparent specificity, as judged by the lack of a measurable effect with the complementary oligonucleotide (PY1) that does not bind to the NHE region. Transcription in this early work was assessed by runoff analysis, employing linearized plasmid templates where *in vitro* synthesized RNA was glyoxylated and then sized on agarose gels. In those nuclear extracts the majority of RNA synthesized was from the P2 initiation site, the main c-*myc* transcription start site *in vitro.*[1] Subsequently, a quantitative RNase protection assay was developed, allowing assessment of the relative strengths of the dual P1 and P2 promoters, which are separated by 161 base pairs and hence give rise to two different-sized transcripts.[1,9] It is now possible to prepare nuclear extracts and chromatographic fractions thereof that are capable of initiating transcription from the P1 start as well, albeit at a lower rate than either P2 RNA synthesis *in vitro* or P1 RNA relative to P2 RNA *in vivo.* (Compare for example, FIG. 3, lanes 19 and 20.) We do not yet understand the reasons why the P2 promoter is used more efficiently *in vitro,* but it may be related to loss, during extraction, of additional factors needed for P1 specific transcription.

The transcription factor PuF, described previously,[1] is essential for accurate and efficient *in vitro* initiation of c-*myc* transcription. PuF has been partially purified by chromatography of HeLa cell nuclear extracts through Affigel-heparin-agarose (H.2 fraction) and DEAE-sepharose columns (D.4 fraction).[1] DNA mobility shift and methylation interference assays located the PuF factor binding domain at −142 to −115 base pairs, with respect to the P1 promoter. These findings suggested that the PuF binding domain and the triplex-forming sequences determined earlier[4] might involve the same control element and that PuF and the oligonucleotide PU1 might compete for the same target sequence when present together in transcription assays.

To determine directly whether the PU1 oligonucleotide can inhibit PuF-mediated transcription, the oligonucleotide was mixed with supercoiled template DNA, and after incubation to allow for triplex formation, transcription was initiated by the addition of the appropriate chromatographic fractions and standard transcription components.[1,4] (See also FIG. 2 legend.) FIGURE 2 and other data not shown

A

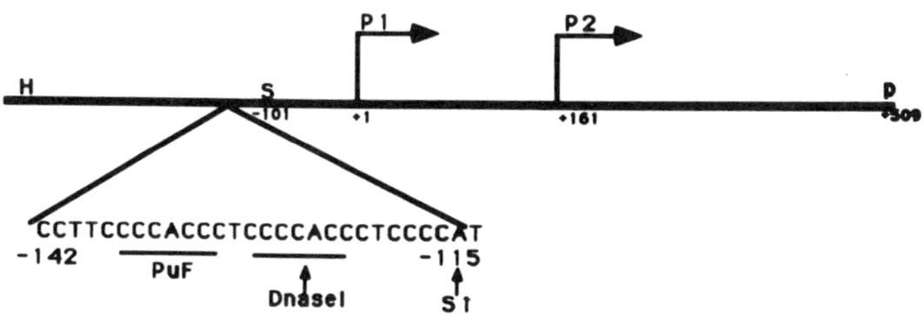

B

HUMAN C-MYC P1 PROMOTER TARGET SITE:

```
            -142                        -115
5'-TCTCCTCCCCACCTTCCCCACCCTCCCCACCCTCCCCATAAGCGCCCCTGAT-3'
3'-AGAGGAGGGGTGGAAGGGGTGGGAGGGGTGGGAGGGGTATTCGCGGGGACTA-5'
```

OLIGONUCLEOTIDES USED FOR TRIPLEX FORMATION:

```
5'-GTGGTGGGGTGGTTGGGGTGGGTGGGGTGGGTGGGG-3'  GT36
    5'-TGGGGAGGGTGGGGAGGGTGGGGAAGG-3'        PU1
    5'-CCTTCCCCACCCTCCCCACCCTCCCCA-3'        PY1
   5'-GTTTTTGGGTGTTGTGGGTGTGTGTGGTT-3'      HIV2C
```

FIGURE 1. **(A)** Partial map of the human c-*myc* gene showing the two major c-*myc* promoters P1 and P2, the DNaseI-hypersensitive site at around −125,[3] a major S1 cleavage site,[7] and the sequence of the −115 to −142 NHE region with respect to the P1 promoter. The PuF transcription factor binding sequence is underlined. **(B)** DNA sequence of the −115 to −142 duplex target site also including some flanking sequences. Sequences of the triplex-forming PU1 and GT36 oligonucleotides, with respect to their binding orientation to the duplex target, and control PY1 and HIV oligonucleotides are indicated last.

here indicate that the triplex-forming PU1 oligonucleotide completely inhibited transcription at concentrations around 400 nM PU1, whereas the complementary, nonbinding PY1 oligomer did not achieve a comparable effect, even at a 10-fold greater concentration.

Whereas FIGURE 2A shows the effect of the two oligonucleotides on P2 RNA synthesis, FIGURE 2B, a longer exposure of the same autoradiogram, shows a roughly similar inhibitory effect of PU1 on P1 RNA synthesis, although in other experiments (e.g., FIG. 3) P1 RNA appears to be somewhat more sensitive to inhibition.

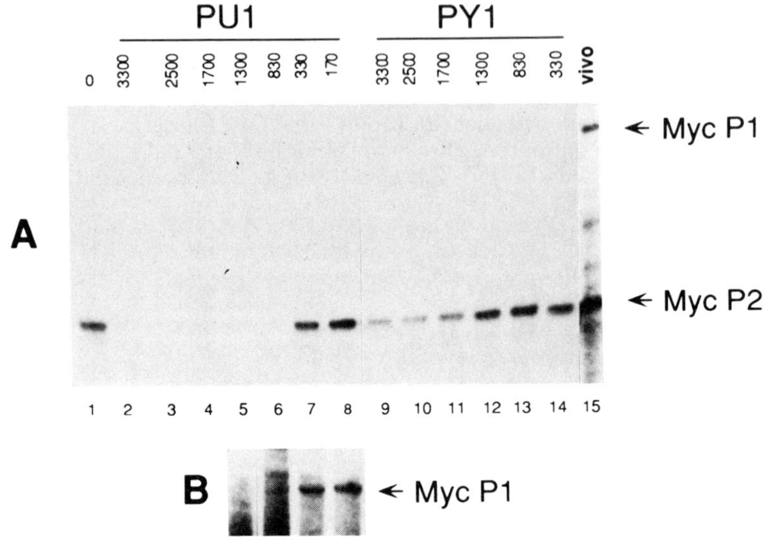

FIGURE 2. Inhibition of PuF-mediated c-*myc* transcription *in vitro* by the triplex-forming PU1 oligonucleotide. 250 ng of the 9.6-kb supercoiled plasmid template RV101[1] was incubated at 22°C for 10 minutes with oligonucleotides at concentrations (nM) indicated above each lane. A reaction mixture consisting of the PuF-rich D.4 fraction (40 μg/ml), the RNA polymerase II-rich C1.25 fraction (100 μg/ml), in the presence of 13.4 mM HEPES, 67 mM KCl, 5 mM $MgCl_2$, 14% glycerol, 2% polyvinyl-alcohol, 0.1 mM EDTA, 0.1 mM EGTA, 0.5 mM DTT, 1 μg/ml each of the protease inhibitors pepstatin A, antipain, and leupeptin, 0.2 mM PMSF, and 250 μM each of unlabeled ATP, CTP, GTP, and UTP, was added in a total volume of 30 μl. After incubation for 60 minutes at 30°C, RNA was purified and analyzed in an RNAse protection assay as described previously.[1] **(A)** PU1 and PY1 titrations showing P2 RNA made *in vitro* (*lanes 1–14*) and *in vivo* synthesized (10 μg) total cytoplasmic RNA purified from HeLa cells[9] (*lane 15*). *Lane 1* is a control with no oligonucleotide added. **(B)** Effect of PU1 on c-*myc* P1 RNA synthesis. Longer exposure of top portion of the autoradiogram shown in **A,** *lanes 5–8.*

Titration of the PU1 oligomer presented in FIGURE 2 shows that approximately 10–20 times more oligonucleotide is needed to completely inhibit PuF-mediated RNA synthesis in this reconstituted system than was necessary to achieve the same degree of inhibition in unfractionated crude nuclear extracts.[4] This result is consistent with the observed 20–50-fold enrichment of the PuF factor in the D.4 fraction, as assessed by DNA-binding mobility shift assays,[1] and is further support for the argument that triplex formation specifically competes with PuF binding.

The effect of additional oligonucleotides that do (GT36) and do not (HIV) form triplexes[10] with the same target sequences was also assessed. FIGURE 3 shows again that PU1 and also its longer homolog, GT36, inhibited PuF-mediated c-myc RNA synthesis, whereas control oligomers that do not form triplexes (PUY and HIV) did not inhibit transcription under the conditions of this assay. GT36, a longer version of PU1, has a greater capacity *in vitro*[10] for site-selective triplex formation than does PU1, which is in variance with their roughly similar transcriptional effects observed here. The origin of this discrepancy is unknown, but it may result from the fact that oligonucleotide binding affinity is measured under a different set of conditions[10] than is transcriptional activity.[1] A similar reduction in the binding constant differential between PU1 and GT36 was also observed in the PuF/triplex binding competitions, in which both oligomers appeared to be equally effective inhibitors of PuF binding in the PuF assay system (E. Postel, R. Durland, and M. Hogan, unpublished results).

As stated, the ratio of P1/P2 RNA synthesized *in vitro* is approximately 1:10, whereas this ratio is closer to 1:1 *in vivo.* (See FIG. 3, lanes 1, 2, 19, and 20.) As found *in vivo*[9] and discussed below, synthesis of P1 RNA appeared to be more sensitive to oligonucleotide treatment, consistent with genetic evidence[2] of the relative importance of this region to c-*myc* promoter utilization.

Thus, it is clear from these data that oligonucleotides that can bind to the NHE target sequence to form triple helices, competitively inhibit PuF from binding within the same target, blocking transcriptional activation of the gene.

*Modulation of Steady-State c-***myc** *mRNA Levels in Cultured HeLa Cells by an Oligonucleotide That Forms a Triplex* **in Vitro**

In view of the functional importance of the −142 to −115 NHE sequence of the c-*myc* gene and the fact that the PU1 oligonucleotide can bind to it *in vitro* to form a triplex and can thus inhibit transcription of the gene, it is clearly relevant to determine if PU1 can also modulate transcription of c-*myc in vivo.* Unmodified oligonucleotides, including PU1, can readily enter HeLa cells in culture, whereupon they penetrate the nuclei and remain stable for several hours.[9] We have used resistance to DNaseI cleavage as a method to map binding of PU1 in the nucleus, as the hypersensitive site within the NHE[4] would become insensitive to digestion if it were engaged in a triplex structure.[4,9] Nuclei extracted from HeLa cells that had been treated with PU1 displayed inhibition of DNaseI cleavage at the NHE target sequence, whereas untreated or PY oligomer-treated nuclei remained sensitive to the enzyme.[9] This result provided the first indication of triplex formation *in vivo.*[9]

To determine if PU1 binding to its target site in the nucleus caused transcription to falter, we also measured steady-state mRNA synthesis in these oligonucleotide-

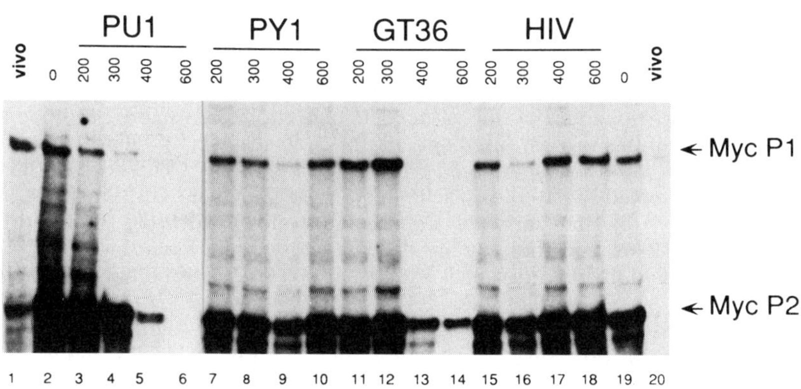

FIGURE 3. Inhibition of PuF-mediated c-*myc* transcription *in vitro* by triplex-forming oligonucleotides PU1 and GT36. Oligonucleotide titrations as in FIGURE 2, with oligonucleotide concentrations (nM) in the transcription assay shown above each lane. *Lanes 1 and 20, in vivo* RNA; *lane 2–6,* PU1; *lanes 7–10,* PY1; *lanes 11–14,* GT36; *lanes 15–19,* HIV. Reaction conditions were as described in FIGURE 2.

treated cells. As can clearly be seen in FIGURE 4, PU1, but not the control oligonucleotides PY1 and HIV, caused the reduction of c-*myc* steady-state mRNA in these treated cells. This effect was more pronounced on the synthesis of P1 RNA (FIG. 4), an observation that is consistent with previous deletional analysis showing that the region encompassing the NHE sequences is more important to P1 than to P2 transcription.[2]

SUMMARY AND CONCLUSIONS

The human c-*myc* oncogene promoter was used as a model with which to study the mechanism of action of oligodeoxyribonucleotides targeted to a gene regulatory region. The nuclease-hypersensitive element, NHE, lying −115 bp from the P1 promoter of the human c-*myc* gene, is known to be required in *cis* for transcription of the gene from both P1 and P2 promoters (FIG. 1). Inhibition of c-*myc* transcription

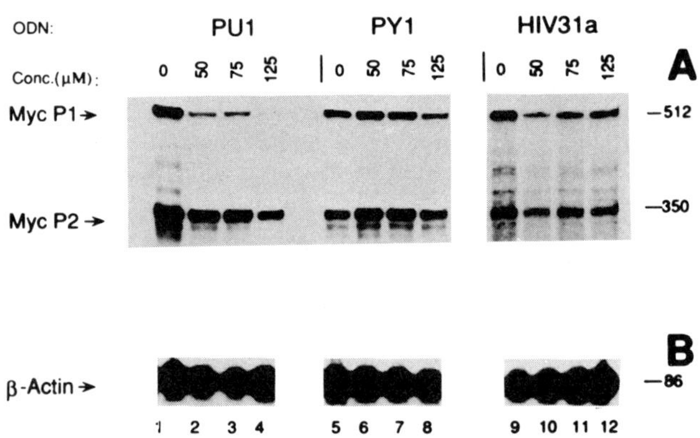

FIGURE 4. *In vivo* repression of c-*myc* mRNA synthesis by PU1 (*lanes 1–4*). Controls PY1 (*lanes 5–8*) and HIV31A (*lanes 9–12*) oligonucleotides show no measurable effect. Partially reproduced from Postel *et al.*[9]

by an oligonucleotide designed to bind to NHE by triplex formation has been observed in a cell-free transcription assay.[4] Using a reconstituted transcription system with the semipurified PuF transcription factor whose site of interaction resides within the NHE, it is shown here that the oligonucleotide inhibits PuF-mediated transcription.

These findings, together with data presented elsewhere showing that: (1) PU1 binds to cloned DNA fragments to form a colinear triplex[4]; (2) PU1 inhibits transcription in nuclear extracts[4]; (3) triple helix formation inhibits the binding of PuF to its target NHE element in an *in vitro* binding competition assay (E. Postel, R. Durland, and M. Hogan, submitted); (4) triplex formation at the NHE target site can occur in living HeLa cells treated with the triplex-forming PU1 oligomer,[9] and (5) c-*myc* mRNA synthesis in these treated cells is repressed,[9] clearly support the proposed model[1,4,9] in which the oligonucleotide targeted against the c-*myc* NHE promoter region binds to form a triplex, thereby blocking access to the regulatory protein PuF. This results in promoter-sensitive repression of transcriptional activation of the c-*myc* gene. The potential for manipulation of gene expression by oligonucleotides targeted to a DNA sequence of the c-*myc* oncogene promoter and other gene promoters is clear.

ACKNOWLEDGMENTS

I wish to thank my collaborators, M. Hogan and J. Flint, for continued support.

REFERENCES

1. Postel, E. H., S. Mango & S. J. Flint. 1989. Mol. Cell. Biol. **9:** 5123–5133.
2. Hay, N. J., M. Bishop & D. Levens. 1987. Genes Dev. **1:** 659–671.
3. Siebenlist, U., L. Henninghousen, J. Battey & P. Leder. 1984. Cell **37:** 381–391.
4. Cooney, M., G. Czernuszewicz, E. H. Postel, S. J. Flint & M. E. Hogan. 1988. Science **241:** 456–459.
5. Davis, T. L., A. B. Firulli & A. J. Kinniburgh. 1989. Proc. Natl. Acad. Sci. USA **88:** 9682–9686.
6. Wells, R. D., D. A. Collier, J. C. Hanvey, M. Shimizu & F. Wohlrab. 1988. FASEB J. **2:** 2939–2949.
7. Boles, C. & M. Hogan. 1987. Biochemistry **26:** 367–376.
8. Mirkin, S. M., V. I. Lyamichev & K. N. Drushlyak. 1987. Nature **330:** 495–497.
9. Postel, E. H., S. J. Flint, D. J. Kessler & M. E. Hogan. 1991. Proc. Natl. Acad. Sci. USA **88:** 8227–8231.
10. Durland, R. H., D. J. Kessler, S. Gunnell, M. D. Duvic, B. M. Pettitt & M. E. Hogan. 1991. Biochemistry **30:** 9246–9255.
11. Postel, E.H., J. Flint, R. Cox & S. Mirkin. 1992. J. Cell. Biochem. Suppl **16E:** 134.

Inhibition of Cell Cycle Progression by Antisense Oligodeoxynucleotides[a]

RENATO BASERGA, KRYSZTOF REISS, HANSJUERG ALDER, ZBIGNIEW PIETRZKOWSKI, AND EWA SURMACZ

Jefferson Cancer Institute
Jefferson Medical College
Philadelphia, Pennsylvania 19107-5541

Our laboratory has been interested for many years in the use of targeted inhibitors for studying the control of cell proliferation in animal cells. Our first attempt[1] demonstrated that microinjection of an antibody to the p53 protein inhibited serum-stimulated DNA synthesis in 3T3 cells. Such a result seems difficult to explain today in light of the recent findings that the p53 gene product may actually act as an antioncogene, but the experiment was recently repeated with the same cells, with the same antibody, and with the same results.[2] Regardless of the explanation, the experiment was correct.

The advantage of antisense oligodeoxynucleotides (AOD) is that they do not have to be microinjected, and we used them for the first time to show that proliferating cell nuclear antigen (PCNA) is required for cell cycle progression in 3T3 cells.[3] The growth inhibitory effect of AOD to PCNA RNA was confirmed with another cell type in a different laboratory.[4]

Insulin-like growth factor I (IGF-1) is an important growth factor that is necessary for the proliferation of fibroblasts and fibroblast-like cells.[5–7] Certain hemopoietic cell lines are also capable of growing in IGF-1 only or in insulin at supraphysiological concentrations, in which conditions insulin activates the IGF-1 receptor.[8] IGF-1 is also important in human development[9] and in the maturation of *Xenopus* oocytes.[10] Recently, Pietrzkowski *et al.*[11] found that constitutively overexpressed cDNAs for IGF-1 and its receptor completely abrogated any requirement for exogenously added growth factors. In 3T3 cells constitutively expressing only the IGF-1 receptor cDNA, the addition of IGF-1 is sufficient to support growth. Under these conditions, the receptors for PDGF and EGF, the other two growth factors that play a role in the growth of 3T3 cells, are not activated. Previously, Travali *et al.*[12] had shown that the protooncogene c-myb induced the expression of IGF-1 mRNA as well as the IGF-1 receptor mRNA.[13]

Here, we summarize our studies with AOD to the IGF-1 receptor RNA, aimed at two goals: (1) to confirm the crucial role of the IGF-1 receptor in the control of cell proliferation; and (2) to generate a tentative sequence of events in gene expression during the cell cycle.

[a]This work was supported by grants GM33694, CA53484, and AG00378 from the National Institutes of Health, Bethesda, Maryland.

METHODS AND MATERIALS

These are described in previous papers from our laboratory and include the cell lines used (BALB/c3T3 and p6 cells, HL-60 cells), RNA extraction and blotting, probes used, and so forth.[11–14] p6 cells are BALB/c3T3 cells constitutively expressing the human IGF-1 receptor cDNA.[12] Peripheral blood mononuclear cells (PBMCs) were prepared and stimulated as previously described.[15] Serum-free medium is DMEM containing ferrous sulfate (1.0 μM) in lieu of transferrin. As an attachment factor, we used either poly-D-lysine or serum proteins.

Oligodeoxynucleotides were synthesized as previously described. However, in more recent experiments we used phosphorothioate oligodeoxynucleotides.[16]

Phosphorothioate oligonucleotides were synthesized on the Applied Biosystems DNA synthesizer model 394 using tetraethylthiuran disulfide (TETD). TETD in acetonitrile converts internucleotide cyanoethyl phophite to the phosphorothioate triester within 15 minutes at room temperature. Phosphorothioate oligonucleotides were cleaved and deprotected at the completion of synthesis with concentrated ammonia and finally precipitated with ethanol.

TABLE 1. Effect of Antisense Oligodeoxynucleotides on the Growth of HL-60 Cells[a]

Treatment	Percentage of Labeled Cells
None	75
Control oligo (80 μg + 40 μg)	60
Antisense oligo to IGF-1 receptor RNA (80 μg + 40 μg)	20

[a]Exponentially growing HL-60 cells were treated as indicated. Twenty-four hours after treatment, the cells were incubated with (^{3}H)-thymidine for 24 hours. Autoradiographs were prepared by standard procedures.

RESULTS

Inhibition of DNA Synthesis and Cell Growth by Antisense Oligodeoxynucleotides to IGF-1 Receptor RNA

The addition of AOD to IGF-1 receptor RNA to exponentially proliferating HL-60 cells inhibits their entry into S phase and cell division (TABLE 1). At the concentrations used (80 μg/ml), a sense oligo had essentially no effect. A similar result was obtained with p6 cells, the 3T3 derivatives that constitutively overexpress the IGF-1 receptor RNA and protein[12] and grow in serum-free medium with the sole addition of IGF-1. The addition of AOD to IGF-1 receptor RNA to p6 cells growing in serum-free medium plus IGF-1 (100 ng/ml) resulted in 50% inhibition of growth. For instance, in one experiment in which 4×10^4 cells/cm^2 were plated in every dish, the number of cells after 5 days was 22×10^4 in controls, 21×10^4 in cultures treated with sense oligos, and 12×10^4 in cultures treated with AOD (always per cm^2).

These two experiments conclusively demonstrate that expression of IGF-1 receptors is necessary for the proliferation of HL-60 and 3T3 cells. The fact that p6 cells grow in serum-free medium with the sole addition of IGF-1 suggests that the IGF-1 receptor is not only necessary but also sufficient.

Role of the IGF-1 Receptor in the Response of Peripheral Blood Mononuclear Cells to Phytohemagglutinin

We summarize here our studies on human PBMC stimulated with phytohemagglutinin (PHA). A typical experiment is shown in FIGURE 1, which is a densitometry reading of a reverse transcriptase-polymerase chain reaction (RT-PCR) of RNAs from PHA-stimulated PBMC. Amplification products were displayed on a Southern blot, which was hybridized to the appropriate probes. PBMC were stimulated with PHA for 48 hours. In panel a, cells were treated with AOD to c-myb, and the extracted RNA was amplified with the appropriate amplimers for IGF-1 receptor RNA. The first (flat) peak is from cells treated with AOD, and the second and third peaks from cells treated with a mismatched oligo, or untreated, respectively. Panel b

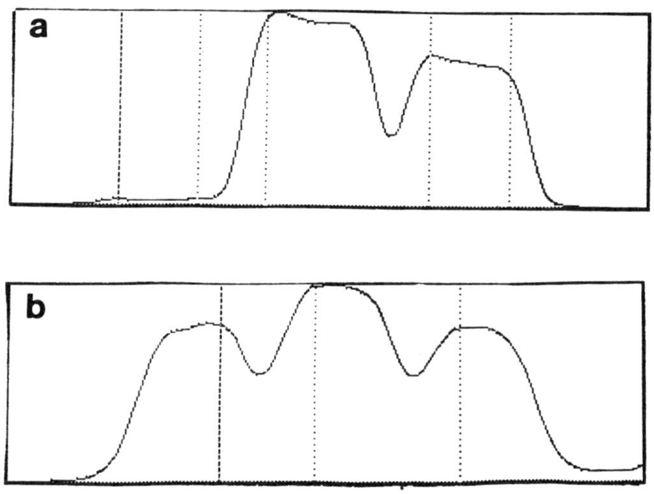

FIGURE 1. Reciprocal effects of antisense oligodeoxynucleotides to c-myb and IGF-1 receptor RNAs. Peripheral blood mononuclear cells (PBMCs) were mitogen stimulated for 48 hours. Densitometer reading of RT-PCR amplification products hybridized to the appropriate probes.[19] (**a**) In the presence of antisense oligos to c-myb (flat peak), sense and no oligos (last two peaks, respectively); RT-PCR[19] for IGF-1 receptor RNA. (**b**) In the presence of antisense oligos to IGF-1 receptor RNA, sense and untreated, respectively; RT-PCR for c-myb RNA.

shows a similar experiment, but in reverse: the AOD were against IGF-1 R RNA and the amplification products are from c-myb mRNA. AOD to c-myb inhibit the expression of IGF-1 receptor RNA, but AOD to IGF-1 receptor RNA have no effect on the expression of c-myb.

AOD to IGF-1 receptor RNA inhibit DNA synthesis, and cell proliferation in HL-60 cells (see above) and also in PBMC. We previously reported that the protooncogene c-myb induces transcription of the IGF-1 receptor gene.[13] Using the antisense strategy mentioned here, we attempted to establish a temporal order in the expression of cell cycle genes as well as the dependency from each other. The rationale is as follows: if AOD to mRNA from gene A inhibits the expression of gene B, whereas AOD to RNA from gene B does not inhibit the expression of gene A, it can reasonably be assumed that gene A is upstream of gene B and that gene B

TABLE 2. Effect of Antisense Oligodeoxynucleotides on the Expression of Cell Cycle Genes[a]

Antisense Used	Gene Product Tested	Effect
c-myc	c-myb	Decrease
c-myb	IGF-1 R	Decrease
IGF-1 R	c-myb	No change
IGF-1 R	PCNA	Decrease
IGF-1 R	DNA synthesis	Decrease
IGF-1 R	cdc2	Decrease

[a] All antisense studies used as controls either sense oligos or oligos with a randomly generated sequence having the same nucleotide composition as the antisense oligo. Most of these results are unpublished data (Reiss *et al.*, in preparation) on HL-60 cells or peripheral blood mononuclear cells stimulated with PHA and IL-2.

expression is somehow regulated by the product of gene A. We studied this problem in HL-60 cells and in PHA-stimulated PBMC, using AOD to either c-myb or IGF-1 receptor RNAs and RT-PCR to identify the RNAs. The results of our experiments are summarized in TABLE 2. From our data and a previous report by Gewirtz and Calabretta,[17] we outlined a temporal sequence of cell cycle events in which the expression of each gene is dependent on the expression of a previous gene or genes. This sequence is depicted in FIGURE 2.

DISCUSSION

The use of antisense oligodeoxynucleotides has given us a valuable tool in elucidating the mechanism(s) of cell cycle progression. AOD, like antibodies, can be targeted to specific gene products, but, whereas antibodies have to be microinjected to be effective (except for antibodies to cell surface proteins), AOD can simply be added to the culture medium, an obvious advantage in terms of both speed and requisite skills.

Two important questions have to be considered in experiments with AOD: (1) How valid are the results we obtain with them? and (2) What kind of information do they provide?

The validity of results obtained with AOD is by no means accepted universally by the scientific community. We are not referring here to the well-known lack of reliability of AOD; sometimes they work, and sometimes they don't. The reasons for

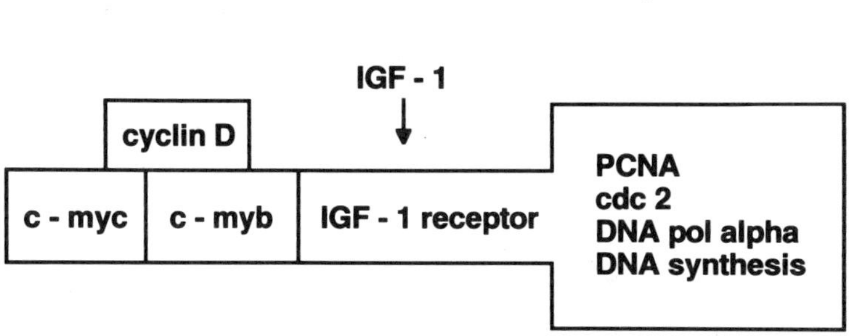

FIGURE 2. A tentative sequence of cell cycle gene expression in mitogen-stimulated lymphocytes. See text for explanation.

this lack of reliability are discussed elsewhere in this book. But let us assume that a given AOD works, that is, that it consistently inhibits cell proliferation. Are we sure that its effect on cell proliferation is due to inhibition of the targeted gene product? We introduce into cells a sequence antisense to a given RNA; but just as antibodies can cross-react, a sequence can interact with more than one RNA. The use of sense oligodeoxynucleotides is a necessary, but insufficient control. The abrogation of function (in our case, the growth response to IGF-1) is, of course, comforting, but it still is possible that growth arrest may be due to inhibition of another RNA of similar partial sequence but different from the IGF-1 receptor RNA. A decrease in the number of IGF-1 binding sites would add more credibility to our interpretation, but in our opinion the most convincing evidence would be the demonstration that by increasing stepwise the amount of IGF-1 receptor RNA, the inhibitory effect of the AOD would correspondingly decrease.

Assuming that the inhibitory effect of an AOD is actually specific, what information can it give us? I have to reiterate here what I stated on a previous occasion[18] that an inhibitory effect of an AOD or an antibody simply indicates that the targeted gene product is *required* for cell proliferation. It does not say whether or not that gene product is *growth regulatory* in the sense of determining and controlling the extent of cell proliferation. To make a rough but effective example, amino acids are required for cell proliferation, but they really are not growth regulatory. Increasing the amount of amino acids in the medium will not induce a G_0 cell to enter the cell cycle. In other words, to be growth regulatory, something should be *both necessary* and *sufficient.* At the present moment, we do not know any cellular gene that is both necessary and sufficient for cell proliferation.

Even with these qualifications, AOD can be usefully employed in determining the sequence of events in mitogen-stimulated cells, as we have done in this report (FIG. 2). For instance, if an AOD to the IGF-1 receptor RNA inhibits the expression of PCNA but not the expression of c-myb, it can reasonably be concluded that PCNA mRNA levels are dependent on a functional IGF-1 receptor, but c-myb mRNA levels are not. If, in addition, AOD to c-myb inhibit the expression of IGF-1 receptor RNA, it can reasonably be concluded that the c-myb gene product is temporally upstream of the IGF-1 receptor, whereas the PCNA gene product is temporally downstream. This approach should turn out to be very useful in elucidating the temporal sequence of events that occur when quiescent cells are stimulated to proliferate.

SUMMARY

We have used the antisense strategy to study the role of certain genes in cell cycle progression. In particular, we used antisense oligodeoxynucleotides to study: (1) the role of the IGF-1 receptor in the control of cell proliferation; and (2) the sequence of gene expression during the cell cycle. Our results can be summarized as follows: (1) the activation of the IGF-1 receptor by its ligand, IGF-1, is an obligatory step in the proliferation of fibroblasts and hemopoietic cells; and (2) the expression of DNA synthesis genes, such as PCNA, DNA polymerase α, and cdc2, is dependent on the expression of previous genes. A tentative temporal order is: c-myc > c-myb > IGF-1 receptor > DNA synthesis genes.

REFERENCES

1. MERCER, W. E., D. NELSON, A. B. DELEO, L. J. OLD & R. BASERGA. 1982. Proc. Natl. Acad. Sci. USA **79:** 6309–6312.

2. DEPPERT, W., G. BUSCHHAUSEN-DENKER, T. PATSCHINSKY & K. STEINMEYER. 1990. Oncogene **5:** 1701–1706.
3. JASKULSKI, D., J. K. DERIEL, W. E. MERCER, B. CALABRETTA & R. BASERGA. 1988. Science **240:** 1544–1546.
4. LIU, Y. C., R. L. MARRACCINO, P. C. KENG, R. A. BAMBARA, E. M. LORD, W. C. CHOU & S. B. ZAIN. 1989. Biochemistry **28:** 2967–2974.
5. SCHER, C. D., R. C. SHEPHARD, H. N. ANTONIADES & C. D. STILES. 1979. Biochem. Biophys. Acta **560:** 217–241.
6. STILES, C. D., G. T. CAPONE, C. D. SCHER, N. H. ANTONIADES, J. J. VAN WYK & W. J. PLEDGER. 1979. Proc. Natl. Acad. Sci. USA **76:** 1279–1283.
7. PHILLIPS, P. D. & V. J. CRISTOFALO. 1988. Exp. Cell Res. **175:** 395–403.
8. FLIER, J. S., P. USHER & A. C. MOSES. 1986. Proc. Natl. Acad. Sci. USA **83:** 664–668.
9. VAN WYK, J. J., L. E. UNDERWOOD, A. J. DE ERCOLE, D. R. CLEMMONS, W. J. PLEDGER, W. R. WHARTON & E. B. LEOF. 1981. *In* The Biology of Normal Human Growth. :223–239. Raven Press. New York.
10. SADDLER, S. E. & J. L. MALLER. 1987. J. Biol. Chem. **262:** 10644–10650.
11. PIETRZKOWSKI, Z., M. YOSHINOUCHI, R. LAMMERS, M. LIMARDO, P. D. PHILLIPS, A. ULLRICH & R. BASERGA. 1992. Cell Growth & Diff. **3:** 199–205.
12. TRAVALI, S., K. REISS, A. FERBER, S. PETRALIA, W. E. MERCER, B. CALABRETTA & R. BASERGA. 1991. Mol. Cell. Biol. **11:** 731–736.
13. REISS, K., A. FERBER, S. TRAVALI, P. PORCU, P. D. PHILLIPS & R. BASERGA. 1991. Cancer Res. **51:** 5997–6000.
14. SURMACZ, E., P. NUGENT, Z. PIETRZKOWSKI & R. BASERGA. 1992. Exp. Cell Res. **199:** 275–278.
15. CALABRETTA, B., L. KACZMAREK, W. MARS, D. OCHOA, C. W. GIBSON, R. R. HIRSCHHORN & R. BASERGA. 1985. Proc. Natl. Acad. Sci. USA **82:** 4463–4467.
16. STEIN, C. A. & J. S. COHEN. 1988. Cancer Res. **48:** 2659–2668.
17. GEWIRTZ, A. & B. CALABRETTA. 1988. Science **242:** 1303–1306.
18. BASERGA, R. 1990. Cancer Res. **50:** 6769–6771.
19. RAPPOLEE, D. A., A. WANG, D. MARK & Z. WERB. 1989. J. Cell. Biochem. **29:** 1–11.
20. CHANG, C. D., L. OTTAVIO, S. TRAVALI, K. E. LIPSON & R. BASERGA. 1990. Mol. Cell. Biol. **7:** 3289–3296.

Mechanism of Action of Antisense RNA

Sometime Inhibition of Transcription, Processing, Transport, or Translation[a]

DAVID T. DENHARDT

Department of Biological Sciences
Rutgers University
Piscataway, New Jersey 08855

Research performed in this and other laboratories is reviewed here to identify unifying themes relevant to understanding how antisense RNA functions to reduce gene expression in eukaryotic cells. The review is restricted to experimental systems in which antisense (AS) RNA is generated by an artificial (mini)gene stably incorporated into the genome and capable of producing an AS transcript either constitutively or inducibly. Depending upon the system, inhibition is sometimes observed at the level of splicing of the primary transcript, transport of the mature mRNA into the cytoplasm, or translation of the mRNA. I am aware of only one report of transcription suppression by AS RNA, and that is the decreased transcription of the *c-myc* gene observed by Yokoyama and Imamoto[1] in HL60 cells expressing AS RNA complementary to the coding sequence of *c-myc;* these results are intriguing, because the *c-myc* promoter appeared to be inhibited by AS RNA targeted to sequences downstream of the promoter itself.

INHIBITION AT THE LEVEL OF SPLICING

Khokha *et al.*[2] described two mouse cell lines, LA1 and LA7, that expressed substantially different amounts of AS TIMP RNA (essentially the complete cDNA), yet were down-modulated for TIMP expression to similar extents, 30–60% depending upon the experiment. LA1 had approximately 100 copies of the AS construct incorporated into the genome, whereas LA7 had only 1. FIGURE 1 shows the structure of the AS construct and a Southern blot analysis of the genomic DNA; transcription was driven by the inducible (at least in some constructs) mouse metallothionein I promoter, and processing signals were provided at the 3′ end by a 3′ segment of the gene encoding human growth hormone. LA1 contained an amount of AS RNA in the cytoplasm comparable to a typical modestly abundant mRNA; LA7 did not contain detectable (by Northern blotting) cytoplasmic AS TIMP RNA.[2] The levels of cytoplasmic TIMP mRNA and secreted TIMP protein were each reduced 30–50% in both cell lines despite the great difference in concentration of cytoplasmic AS RNA. Because down-modulation of TIMP expression was comparable in LA1 and LA7, we tentatively conclude that inhibition of translation of TIMP mRNA by cytoplasmic AS TIMP RNA was not the major contributor to down-modulation.[3] There was also no inhibition of transcription of the gene.[4]

[a]Research described herein was supported by NIH CA50183 and the Bureau of Biological Research, Rutgers University.

To determine why TIMP mRNA abundance was reduced in the cytoplasm of these two AS TIMP RNA-containing cell lines, we investigated the amount of unprocessed and processed (sense) TIMP RNA in the nucleus. These studies revealed that although nuclear mRNA levels were diminished to about the same extent as were cytoplasmic mRNA levels, there was a striking increase in the amount of unprocessed primary transcript.[4] This was determined from the abundance of RNA from the first intron, as measured by a reverse transcriptase-coupled PCR reaction. The intron cannot be part of a duplex structure formed with the AS RNA, because the AS RNA is derived from the cDNA. These observations suggest that in these cells under these conditions AS TIMP RNA interferes with the processing of the primary transcript.

Kelly *et al.*[5] noted that in human placenta there is an excess of RNA complementary to (i.e., antisense) the mRNA encoding gonadotropin-releasing hormone (GnRH). They also noted that the ratio of precursor unprocessed GnRH RNA to processed mRNA was 10-fold larger than the 0.01–0.02 ratio observed for some other genes. This finding is consistent with the hypothesis that in these cells the naturally occurring AS GnRH RNA impairs processing of the primary GnRH transcript. Another example of a naturally occurring AS RNA was described by Munroe and Lazar[6] in the rat *erbA*α locus. The *erbA*α*1* transcript encodes the thyroid hormone receptor. The *erbA*α*2* transcript encodes a protein related to but different from the thyroid hormone receptor in its COOH-terminal region; it is longer than the *erbA*α*1* transcript and its 3′ end overlaps with an AS transcript (*Rev-erbA*α) that encodes a protein of unknown function. A fivefold excess of *Rev-erbA*α inhibits splicing of the *erbA*α*2* transcript, increasing the ratio of *erbA*α*1* to *erbA*α*2* transcripts

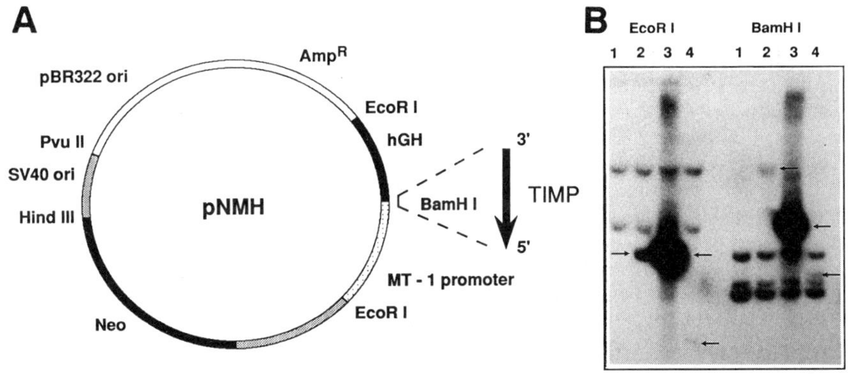

FIGURE 1. Analysis by Southern blotting of the DNA from MC2, 16S1, LA1, and LA7 cells.[2,3,19] **(A)** Structure of the construct used to make the transfected lines, which were initially selected for G418 resistance. **(B)** Southern blot of genomic DNA restricted with *Eco*RI (*left*) and *Bam*HI (*right*). *Lanes 1 to 4* are from MC2, 16S1, LA1, and LA7, respectively. MC2 is a control G418-resistant line revealing the restriction pattern of the endogenous TIMP gene. 16S1 is a cell line with the TIMP cDNA in a sense orientation. LA1 and LA7 are two lines with the TIMP cDNA in an AS orientation. The intensities of the autoradiographic signals suggest that there is one copy of the TIMP cDNA minigene in LA7 and many copies in LA1. *Arrows* indicate the positions of new bands in the 16S1, LA1, and LA7 lines.

produced in an *in vitro* splicing reaction. The investigators propose that inhibition of splicing *in vivo* by an AS RNA may play a physiologically important role in controlling the expression of the thyroid hormone receptor encoded by this gene.

INHIBITION AT THE LEVEL OF TRANSPORT

Kim and Wold[7] studied the inhibition of the expression of the *Herpes simplex* virus (HSV) thymidine kinase (TK) gene, which contains no introns, by AS TK RNA generated in stably transfected mouse L cells. Cell lines producing different amounts of AS RNA were constructed by attaching the TK gene (specifically 225 bases from the 3′ end in a reverse orientation) to the 3′ end of the dihydrofolate reductase gene (DHFR), so that when the DHFR gene was amplified by culturing the cells in the presence of increasing concentrations of methotrexate, the sequences encoding AS TK were also amplified. The AS TK RNA linked to the 3′ end of the DHFR transcript was able to reduce TK expression more or less in proportion to the extent of amplification of the DHFR gene. The TK mRNA was retained in the nucleus in a duplex structure, leading these investigators to conclude that the reduction in TK expression was the consequence of inhibition of the transport of mRNA out of the nucleus. The 1.1-kb TK mRNA did not appear to be degraded at an enhanced rate as a consequence of its forming a duplex structure with the 225-base AS RNA. Different clones were found to be down-regulated for TK expression to different extents (0.11–0.78 was the range) for unknown reasons, perhaps having to do with the location of the AS gene in the genome.

Prochownik *et al.*[8] made similar observations on the inhibition of *c-myc* expression by AS RNA in Friend murine erythroleukemia cells. They placed a full-length mouse *c-myc* cDNA in reverse orientation immediately upstream of a DHFR-encoding sequence and down-stream of an SV40 early promoter. Methotrexate-induced amplification of the DHFR gene generated AS myc RNA 10–25 times the level of the endogenous *c-myc* mRNA. Although total cellular *c-myc* mRNA levels were not significantly reduced, a higher proportion was retained in the nucleus.

A naturally occurring example of apparent AS RNA-mediated inhibition of export of a transcript from the nucleus is found in the myelin-deficient (mld) mouse. The *mld* mutation results in duplication and inversion of a major portion of the gene encoding myelin basic protein, so that an AS RNA transcript is generated from the inverted sequence immediately upstream of the normal gene, which for the most part is still transcribed largely from its own promoter. Okano *et al.*[9] reported that AS *mld* RNA was more abundant in the nucleus than in the cytoplasm of the cells from the *mld* mouse brain, and a portion of the AS RNA was sequestered in a nuclease-resistant duplex form. As reviewed by Matthieu *et al.*,[10] there is evidence that transcriptional readthrough and transcriptional interference may also contribute to the down-regulation of expression, but most of the effect is due to a posttranscriptional inhibition of processing and/or transport.

INHIBITION AT THE LEVEL OF TRANSLATION

A set of plasmid constructs was made by Chang and Stoltzfus[11] to ask if AS RNA directed at the Rous sarcoma virus *env* gene was more effective in inhibiting expression from transcripts requiring splicing than transcripts not requiring splicing. These experiments were performed by calcium phosphate-mediated cotransfection of pairs of plasmids encoding sense and AS transcripts into quail cells carrying an RSV genome lacking the *env* gene and thus dependent upon expression of the *env* gene from the transfected DNA for the production of focus-forming virus. Expression of *env* function from the genomic plasmid (encoding a transcript requiring splicing) and the cDNA plasmid (encoding a transcript not requiring splicing) was equally well inhibited by an AS *env* RNA. The investigators concluded that inhibition by AS RNA did not obligatorily act at the level of splicing, and they suggested that inhibition of translation was the most reasonable explanation, although a block to export from the nucleus or enhanced degradation of the mRNA could not be excluded.

Sullenger *et al.*[12] studied inhibition of Moloney murine leukemia virus replication by AS *gag* RNA or *pol* RNA produced as a tRNA chimera incorporated into the U3 element of the 3′ long terminal repeat (LTR) of a retroviral vector, a so-called "double-copy" vector. Short AS sequences were introduced into the 3′ end of a $tRNA^{met}$ gene in place of the sequence that normally mediates processing of the 3′ end of the tRNA. Transcription was driven at high levels from the tRNA promoter by RNA polymerase III, and high levels of cytoplasmic AS RNA were generated. Antisense *gag* RNA in the chimeric molecule substantially reduced the synthesis of viral GAG proteins while having no effect on the cytoplasmic level of *gag* mRNA. Hence, the investigators concluded that in this system AS *gag* RNA was acting to block translation of the message. Antisense *pol* RNA, although present at comparably high levels, had much less of an effect on virus replication and protein synthesis, perhaps, the authors suggest, because of secondary structure in the *pol* RNA.

INHIBITION WHEN ANTISENSE RNA IS APPARENTLY LESS ABUNDANT THAN THE TARGET RNA

A number of reports describe significant down-regulation of expression in circumstances in which the steady-state levels of mRNA substantially exceed the steady-state levels of (cytoplasmic) AS RNA. Thus, Nishikura and Murray[13] constructed an AS c-fos expression vector using the glucocorticoid-inducible mouse mammary tumor virus (MMTV) promoter to drive expression of the 5′ flanking region and first exon of c-fos. Cell lines containing one or more of these constructs made an AS RNA that was able in some cases to reduce the extent of c-fos expression (both mRNA and protein) even though the amount of detectable AS RNA was much below the amount of c-fos mRNA. Sheehy *et al.*[14] used AS RNA to the polygalacturonase mRNA to suppress synthesis of this enzyme in ripening tomatoes. Constitutive synthesis of the AS RNA under the control of the cauliflower mosaic virus 35S promoter caused a substantial reduction in mRNA and enzyme levels despite the greater abundance of mRNA than AS RNA.

These two reports and ours on TIMP described above are representative of other reports in the literature (e.g., Holt *et al.*[15] and Sun *et al.*[16]) of a lack of correlation between the amount of AS RNA apparently produced and its effect on the cell. There are several obvious possibilities to explain these discrepancies, mostly centering around what happens in the nucleus during the synthesis of the target mRNA. The actual amounts of AS RNA in the nucleus and cytoplasm are set by the various rate constants that determine the appearance and disappearance of AS RNA. The cartoon in FIGURE 2 illustrates this process; the equations in the legend reveal that the steady-state level of AS RNA in the cytoplasm can be more or less than the concentration in the nucleus depending upon the relative rates of transport of the AS RNA out of the nucleus and its stability in the nucleus and cytoplasm. In principle,

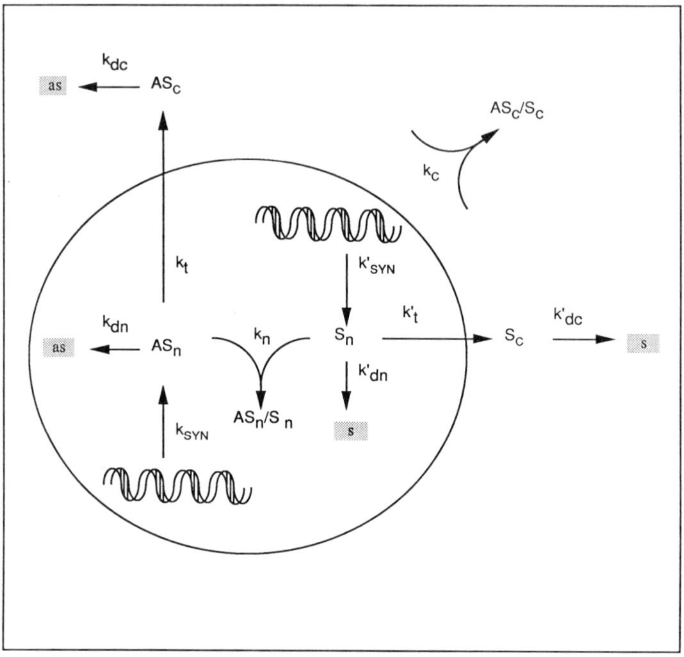

FIGURE 2. A simple model to indicate the dependence of the nuclear and cytoplasmic abundances of AS RNA on the rate of transcription (plus any processing) of the gene, transport of the RNA out of the nucleus, degradation of the RNA in both the nucleus and cytoplasm, and rate of hybridization. $[AS_n]$ and $[AS_c]$ are the concentrations of AS RNA in the nucleus and cytoplasm, respectively. [S] is likewise the concentration of the target sense transcript, and the comparable rate constants are primed. k_{SYN} is the (1st order) rate of synthesis of AS RNA; k_t is the rate constant for transport out of the nucleus; k_{dn} is the rate constant for degradation in the nucleus; k_{dc} is the rate constant for degradation in the cytoplasm; k_n and k_c are the (2nd order) rate constants for formation of duplex RNA in the nucleus and cytoplasm, respectively. *Shadowed, lower case terms* indicate degraded species.

$$\frac{d[AS_N]}{dt} = k_{SYN} - k_t[AS_N] - k_{dn}[AS_N] - k_n[AS_N][S_N]$$

$$= k_{SYN} - (k_t + k_{dn} + k_n[S_N])[AS_N]$$

$$\frac{d[AS_C]}{dt} = k_t[AS_N] - k_{dc}[AS_C] - k_c[AS_c][S_c]$$

Under steady-state conditions there is no change in the concentration of S and AS RNA with time. Hence:

$$k_{SYN} = (k_t + k_{dn} + k_n[S_N])[AS_N] \qquad [AS_N] = \frac{k_{SYN}}{k_t + k_{dn} + k_n[S_N]}$$

$$k_t[AS_N] = (k_{dc} + k_c[S_c])[AS_C] \qquad [AS_C] = \frac{k_t[AS_N]}{k_{dc} + k_c[S_c]}$$

$$= \frac{k_t\, k_{SYN}}{(k_{dc} + k_c[S_c])(k_t + k_{dn} + k_n[S_N])}$$

very effective antisense RNA-mediated down-modulation of expression of a target mRNA can be obtained even when the target mRNA is of high abundance and found in the cytoplasm at higher levels than the AS RNA.

Sequestration of portions of the transcript by various proteins probably influences the efficiency of AS RNA-mediated inhibition. Location of the AS gene in the nucleus relative to the target gene is likely to be important, as also would be the rate of transcription and the stability and "diffusibility" of the AS transcript. It is important for the AS transcript to make contact with its cognate mRNA before the mRNA is transported to the cytoplasm. Cell lines established from different transfection events obviously will have the AS gene in different locations in the chromosome. Given that the nucleus is structured, it is plausible to argue that the events that happen to place the AS gene near the target gene will yield lines that show the greatest amount of down-modulation. Because this is a kinetic phenomenon, there is no necessary relation between the absolute amount of AS RNA in the different constructs and its impact on expression of the target gene.

If the AS RNA cannot access the sense RNA in the nucleus, then it will have to act in the cytoplasm if it is going to have an effect. As the volume of cytoplasm is usually larger than the nucleus, it might be expected that down-modulation in expression as the result of events in the cytoplasm is of marginal significance. In any case the extent of down-regulation will depend on both the ratio of abundances of the complementary RNA species and the amount of secondary structure, or lack thereof, in the hybridizing regions. The possibility that association with proteins, such as ribosomes, could also interfere with antisense/sense annealing is yet another confounding factor.

NEGATIVE EXPERIMENTS IN WHICH ANTISENSE RNA HAD NO EFFECT

At least two publications address failure in specific instances of AS RNA to suppress expression of a targeted gene. In one case, a large excess of full-length AS RNA to a chloramphenicol acetyltransferase (CAT) "minigene" previously introduced into CV1 cells was unable to inhibit CAT expression at the mRNA or protein level.[17] In the second case, a 10- to 20-fold excess of AS H-*ras* RNA (the first exon) was unable to revert the transformed phenotype of H-*ras* transformed NIH 3T3 cells.[18] Although these failures are representative of many unpublished disappointments in this field, they should not deter future efforts to use this approach. It is now well established that many variables influence the effectiveness of a specific construct, and because we do not yet know what general principles govern regulation by AS RNA, it is best to try several alternative sequence elements simultaneously.

FIGURE 2 (continued) These equations reveal that the steady-state concentration of (available, single-stranded) AS RNA in the nucleus is determined by the ratio of the rate of synthesis to the sum of the rates of degradation, transport, and sequestration in a duplex form. If transport is inefficient and degradation minimal, then high concentrations of AS RNA in the nucleus will be found either free or in a duplex form. The concentration of (available, single-stranded) AS RNA in the cytoplasm is directly proportional to the rate of transport out of the nucleus and inversely proportional to the rate of degradation and hybridization with complementary mRNA; it can be less than or greater than the concentration of AS RNA in the nucleus. Note that the assumption that the concentration of these species is uniform throughout the nucleus and cytoplasm is only an approximation; the value of k_N will likely depend on the relative geographic locations of the AS minigene and the target gene.

There is no necessary correlation between the length of the sequence and its effectiveness.

SUMMARY

Anyone considering the use of AS RNA, generated endogenously, to inhibit gene expression should plan to generate several independent transfectants with nonoverlapping sequences; strategies that maximize both the transcription rate and the stability of the AS RNA are obviously desirable. Reasons why different results are obtained in different systems or with different constructs likely include the specific nucleotide sequence under investigation, the location of the AS gene in the nucleus relative to the endogenous gene, and the rate-limiting step in the expression of the target gene. Splicing may not be necessary, but an efficient polyadenylation signal likely is. Employment of a ribozyme-mediated strategy, discussed by various investigators in this volume, may be beneficial. There is no reason at present to conclude that any gene, however abundant its transcript might be, is inherently recalcitrant to AS-mediated down-regulation of expression.

ACKNOWLEDGMENT

I thank Dr. Bo Feng for her various contributions to this work.

REFERENCES

1. Yokoyama, K. & F. Imamoto. 1987. Proc. Natl. Acad. Sci. USA **84:** 7363–7367.
2. Khokha, R., P. Waterhouse, S. Yagel, P. K. Lala, C. M. Overall, G. Norton & D. T. Denhardt. 1989. Science **243:** 947–950.
3. Denhardt, D. T., R. Khokha, S. Yagel, C. M. Overall & R. S. Parhar. 1992. *In* Matrix Metalloproteinases and Inhibitors. H. Birkedal-Hansen, Z. Werb, H. Welgus & H. Van Wart, eds. Matrix, Suppl. **1:** 281–285.
4. Feng, B. & D. T. Denhardt. 1992. Ann. N.Y. Acad. Sci., this volume.
5. Kelly, A. C., A. Rodgers, K.-W. Dong, N. X. Barrezueta, M. Blum & J. L. Roberts. 1991. DNA Cell Biol. **10:** 411–421.
6. Munroe, S. H. & M. A. Lazar. 1991. J. Biol. Chem. **266:** 22083–22086.
7. Kim, S. K. & B. J. Wold. 1985. Cell **42:** 129–138.
8. Prochownik, E. V., J. Kukowska & C. Rodgers. 1988. Mol. Cell. Biol. **8:** 3683–3695.
9. Okano, H., J. Aruga, T. Nakagawa, C. Shiota & K. Mikoshiba. 1991. J. Neurochem. **56:** 560–567.
10. Matthieu, J.-M., M. Tosic & A. Roach. 1992. Ann. N.Y. Acad. Sci., this volume.
11. Chang, L.-J. & C. M. Stoltzfus. 1985. Mol. Cell. Biol. **5:** 2341–2348.
12. Sullenger, B. A., T. C. Lee, C. A. Smith, G. E. Ungers & E. Gilboa. 1990. Mol. Cell. Biol. **10:** 6512–6523.
13. Nishikura, K. & J. M. Murray. 1987. Mol. Cell. Biol. **7:** 639–649.
14. Sheehy, R. E., M. Kramer & W. R. Hiatt. 1988. Proc. Natl. Acad. Sci. USA **85:** 8805–8809.
15. Holt, J. T., T. V. Gopal, A. D. Moulton & A. W. Nienhuis. 1986. Proc. Natl. Acad. Sci. USA **83:** 4794–4798.
16. Sun, T. J., P. J. M. Van Haastert & P. N. Devreotes. 1990. J. Cell Biol. **110:** 1549–1554.
17. Kerr, S. M., G. R. Stark & I. M. Kerr. 1988. Eur. J. Biochem. **175:** 65–73.
18. Salmons, B., B. Groner, R. Friis, D. Muellener & R. Jaggi. 1986. Gene **45:** 215–220.
19. Khokha, R. & D. T. Denhardt. 1987. Anticancer Res. **7:** 653–660.

Antisense RNA and p53 Regulation in Induced Murine Cell Differentiation

SAADI KHOCHBIN, MARIE-PAULE BROCARD, DIDIER GRUNWALD, AND JEAN-JACQUES LAWRENCE[a]

Laboratoire de Biologie Moléculaire du Cycle Cellulaire
Unité INSERM 309
Département de Biologie Moléculaire et Structurale
Centre d'Etudes Nucléaires de Grenoble
38041 Grenoble Cedex, France

Cell differentiation is the center of important modifications of the expression state of the genome. A crucial step in this process is the regulation of nuclear proteins putatively involved in the control of gene expression, namely, those of presently unknown functions, such as certain nuclear oncoproteins, suppressor proteins, or even special histone subtypes.

The murine erythroleukemia (MEL) cell line is a virus-transformed erythroid cell precursor immobilized at an early stage in the erythroid differentiation process. The cells can resume a differentiation program under the action of a variety of chemical agents such as hexamethylene-bis-acetamide (HMBA).[1] Modulation of nuclear oncoproteins has been described during the early stage of induced differentiation,[2,3] namely, the critical period of precommitment in which important cell cycle modifications are observed, presumably in relation to the onset of a new gene expression program. Murine melanoma cells (line B16) are also able to undergo induced differentiation, but with different cell cycle parameter variations. We have been interested in examining the role of the p53 protein during the resumption of these two differentiation program upon the action of inducers, as evidence indicates that p53 is involved in the control of cell proliferation[4–6] as well as in the regulation of gene expression.

p53 expression is subjected to modulation during cell differentiation.[7–9] This regulation is controlled at a posttranscriptional level by processes that involve the induced synthesis of RNA, but not that of protein.[10] This paper summarizes our efforts to identify such RNAs and their possible role in cell differentiation.

MATERIALS AND METHODS

Cell Line and Culture. MEL cells and B16 melanoma cells were maintained in culture and induced to differentiate as described previously.[10,12]

RNA Preparation and Analysis. RNA was prepared and analyzed essentially as described previously.[10]

(^3H)-Uridine Pulse-Chase Experiments. 5×10^8 exponentially growing MEL cells were labeled in 50 ml of culture medium supplemented with 5.4 mCi (^{3}H)-uridine (43 Ci/mmol) for 2 hours. Cells were then rinsed with phosphate-buffered saline.

[a] Address for proofs and correspondence: Jean-Jacques Lawrence, DBMS/BMCC, INSERM U309, CEN-G, BP 85X, 38041 Grenoble Cedex, France.

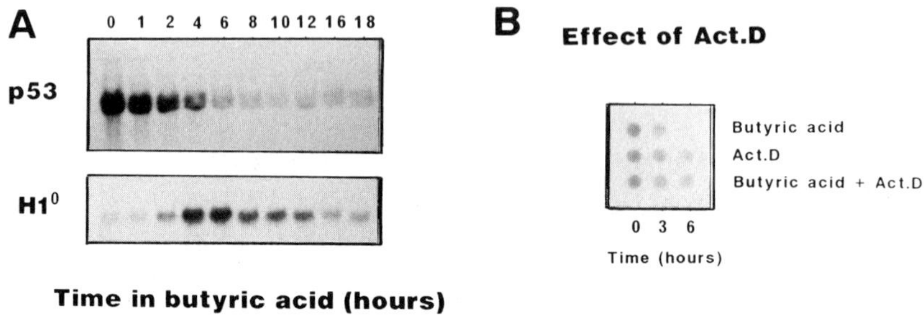

FIGURE 1. **(A)** Variation of p53 mRNA during induced differentiation of B16 cells. B16 cells were treated for varying periods with 5 mM butyric acid, and RNA was extracted from the cells at the indicated times. 50 μg of purified RNA from each sample was analyzed by Northern blot. The blots were probed with ^{32}P-labeled p53 and H1^0 cDNA, respectively. **(B)** Inducer-mediated down-regulation of p53 mRNA is blocked by actinomycin D. 10 μg of purified RNA obtained from cells after treatment for the indicated time with 5 mM butyric acid, 5 μg/ml of actinomycin D, or both were used to perform a dot blot analysis. The blots were then probed with ^{32}P-labeled p53 cDNA.

solution and transferred to normal (^{3}H)-uridine-free culture medium in the presence or absence of 4 mM HMBA. Cell samples were taken every 2 hours during the chase period, and nuclear RNA was prepared as described previously.[11] RNA was resuspended in standard hybridization medium, and hybridization with immobilized p53 cDNA on nitrocellulose filters was carried out for 48 hours. Filters were then washed and counted.

RESULTS

Down-Regulation of p53 mRNA during Induced Differentiation of B16 Cells

The decrease of p53 observed during the induction of differentiation of MEL cells is regulated at a posttranscriptional level and requires the induced synthesis of RNA.[11] We also found that down-regulation occurred in the nuclear compartment. To check for the general nature of this process, we investigated another system that is inducible for differentiation. As shown in FIGURE 1A, p53 mRNA decreased rapidly when B16 melanoma cells were treated with 5 mM butyric acid, and the decay followed a kinetics comparable to that observed during the HMBA-induced differentiation of MEL cells. During the same period, H1^0 mRNA, used as a differentiation marker,[12] increased rapidly.

In B16 cells, as in MEL cells, actinomycin D abolished the inducer-mediated decay of p53 mRNA (FIG. 1B), and cycloheximide treatment of the cells had no effect (not shown), indicating that a similar mechanism governs the down-regulation of p53 mRNA content in both cellular systems.

Visualization of p53 pre-mRNA Maturation Intermediates during Induced Differentiation of the B16 Melanoma Cell Line

During induced differentiation of MEL cells, we observed a simultaneous accumulation of high molecular weight p53 pre-mRNA maturation intermediates and that of a 1.3-kb RNA species that hybridized to a restricted part of the first intron of the p53 gene.[11]

To visualize these RNA species in B16 cells, Northern blots of RNAs extracted at various times of induction were probed with the *Pst*I-*Hind*III p53 first-intron fragment.[11] As in MEL cells, this probe recognized both the 1.3-kb RNA species and the high molecular weight p53 pre-mRNA intermediates (FIG. 2A). Contrary to what was observed in MEL cells, the 1.3-kb RNA did not accumulate during induction. (See below for further analysis of this RNA species.)

The accumulation of high molecular weight intermediates was still observed when an antisense probe, depleted of the B1 repetitive element, was used instead of

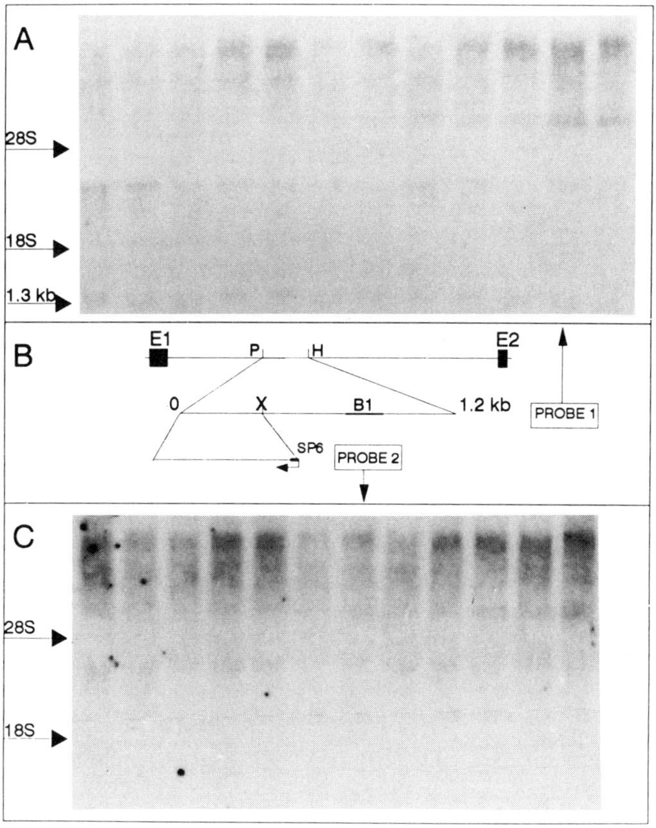

FIGURE 2. Use of p53 intronic probes to visualize high molecular weight RNA(s) and the 1.3-kb RNA species. **(A)** Northern blots of RNA purified from cells treated for the indicated times by 5 mM butyric acid were probed by ^{32}P-labeled intronic DNA fragment (probe 1), shown in **B. (B)** Probes of the first p53 intron used for the Northern blots displayed in **A** and in **C**; the B1 repetitive element is represented as a *black box; letters* represent the position of various restriction enzyme sites, as described previously (Bienz *et al.,* 1984). **(C)** The same Northern blot as in **A** was probed by a ^{32}P-labeled riboprobe transcribed from the indicated fragment (probe 2), using the SP6 promoter and an *in vitro* transcription system.

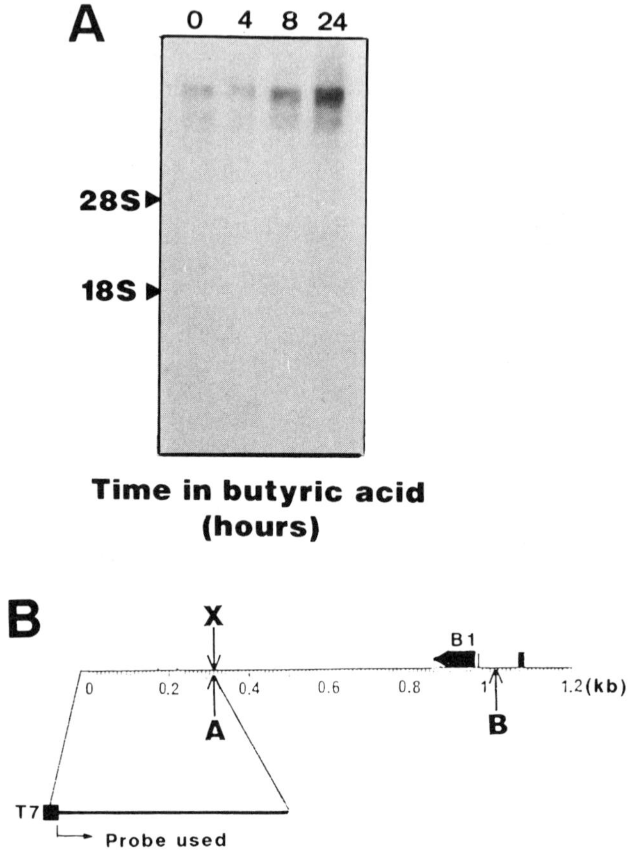

FIGURE 3. Accumulation of high molecular weight antisense RNA during induced differentiation. Northern blot of RNA extracted from cells treated for indicated times by 5 mM butyric acid (**A**) was probed by a ^{32}P-labeled riboprobe which is transcribed *in vitro* from the same fragment as in FIG. 2C but in the opposite orientation (**B**).

the full probe (FIG. 2B), ruling out a putative B1-mediated recognition of these high molecular weight RNA species.

High Molecular Weight Antisense p53 RNA Precursors Also Accumulate during Induced Differentiation

We then used a sense ^{32}P-labeled intronic probe depleted of the B1 repetitive element, and we found that an antisense RNA species accumulated in the high molecular weight fraction only (FIG. 3). This observation is consistent with the results obtained with induced MEL cells.[11]

Posttranscriptional Regulation of p53 mRNA Accumulation Is Controlled at the Level of pre-mRNA Maturation

The results presented indicate that a modification of p53 pre-mRNA maturation occurs during B16 and MEL cell-induced differentiation. To better understand the mechanism controlling the down-regulation of p53 mRNA, we performed actinomycin D chase experiments over a longer period than that usually used. A biphasic decrease of p53 mRNA was observed (FIG. 4B); in the first 2–3 hours of the chase, the transcript levels decreased slowly after which a sharp decay of the p53 mRNA was observed. The kinetics of the decay at this stage was very close to that associated with the early decrease of p53 mRNA upon treatment of the cells with the inducer. A delay of 2 hours occurred between the early inducer-mediated decrease of p53 mRNA and the later, rapidly decreasing phase of p53 mRNA upon actinomycin D treatment, although overall transcriptional activity ceased within the first 15 minutes of this treatment. This latency period is due to the significant amount of p53

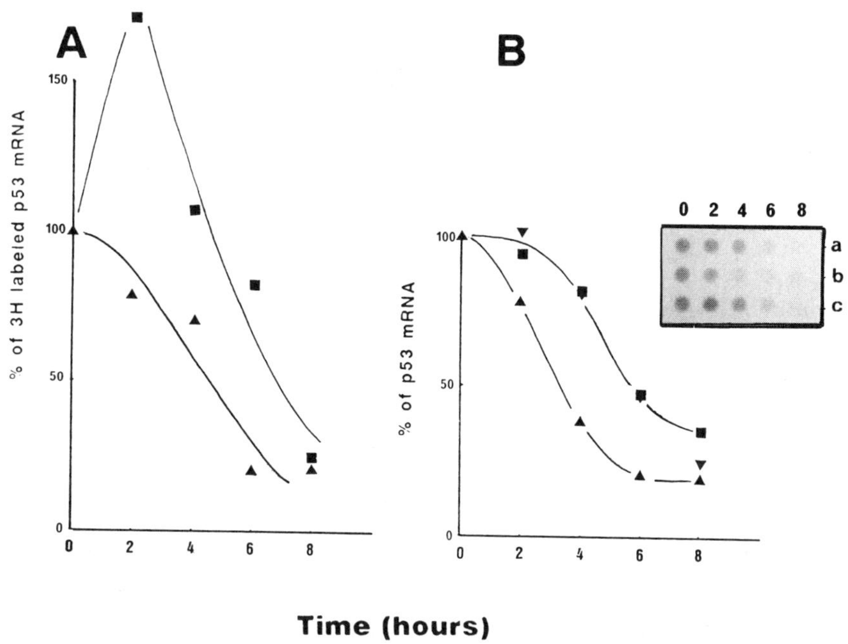

FIGURE 4. p53 mRNA down-regulation is controlled at the level of pre-mRNA maturation. **(A)** MEL cells were labeled with (^{3}H)-uridine for 2 hours, and the cells were then transferred to a (^{3}H)-uridine-free medium (time 0) in the presence (▲) or absence (■) of 4 mM HMBA. Nuclear RNA was prepared after the indicated times of chase and hybridized to a p53 cDNA-containing nitrocellulose filter. After hybridization, filters were washed and counted and normalized to time 0 (881 cpm). **(B)** MEL cells were treated by 4 mM HMBA (▲) or 5 μg/ml actinomycin D (■) or both (▼) for the indicated times. RNA was purified and subjected to dot blot analysis. The blot was probed with ^{32}P-labeled p53 cDNA; **(a)** actinomycin D alone, **(b)** HMBA alone, and **(c)** HMBA + actinomycin D.

pre-mRNA that has been committed to the maturation process, so that the delay observed represents the time necessary to exhaust the pre-mRNA splicing compartment and/or to inhibit splicing upon the action of actinomycin D. The inducer-mediated decrease of p53 mRNA is therefore due to rapid blockage of the maturation process of p53 pre-mRNA.

To exclude any artifact resulting from use of actinomycin D, we performed a (^{3}H)-uridine pulse-chase experiment. Uninduced cells were labeled with (^{3}H)-uridine for 2 hours, and the chase was carried out in the presence or absence of inducer. FIGURE 4A shows that in uninduced cells the amount of labeled nuclear p53 mRNA increases up to 2 hours of chase and then decreases with kinetics similar to that observed in the rapidly decreasing phase of actinomycin D treatment. In induced cells, the decrease of nuclear (^{3}H)-labeled p53 mRNA occurred almost immediately with a kinetics similar to that observed in uninduced cells. These results confirm that the maturation of p53 pre-mRNA is rapidly blocked when cells are treated by inducer.

Accumulation of Small RNA Species

We previously observed that during the induced differentiation of MEL cells, there is a simultaneous accumulation of high molecular weight p53 pre-mRNA maturation intermediates and a 1.3-kb RNA species that hybridized to a restricted part of the first intron of the p53 gene.[11] To visualize these RNA species in B16 cells, Northern blots of RNAs extracted at various times after induction were probed with the *Pst*I-*Hin*dIII p53 first-intron fragment.[11] As in MEL cells, this probe revealed both the 1.3-kb RNA species and the high molecular weight p53 pre-mRNA intermediates (FIG. 2A). However, contrary to its behavior in MEL cells, the 1.3-kb RNA species was not induced in B16 cells, but it decreased as the induction proceeded. We reported previously that this intronic probe contained a B1 repetitive element in an antisense orientation with respect to the p53 promoter. On the basis of the same cartographic procedure, we found that the homology between the intronic probe and the 1.3-kb RNA species is restricted to the B1 repetitive sequence (FIG. 5). The function of this RNA species is currently unknown, but it is clearly associated with the action of the inducer, and it behaves differently depending on the type of cell line being induced.

In the same way, we identified a B2 repetitive element present downstream of the 3′ region of the p53 gene (FIG. 6A).[13] We used a sense probe made from this element and found that it was able to reveal two sets of RNA species in the cells, one of high and the other of low molecular weight. Only an antisense probe could detect the lowest molecular weight species (FIG. 6B, 2) in the nucleus, suggesting that the antisense-oriented B2 containing RNAs cannot survive RNA processing and are no longer present in the low molecular weight RNAs. FIGURE 6C shows the variation of both species upon induction of differentiation. The cytoplasmic component decreased, whereas the nuclear component showed a biphasic variation, probably in relation to the variation in the cell cycle parameters observed during the same period.[13]

DISCUSSION

The amount of p53 decreases upon differentiation of normal or transformed cells.[7–9,14,15] In light of the recent findings on the properties of p53, particularly with

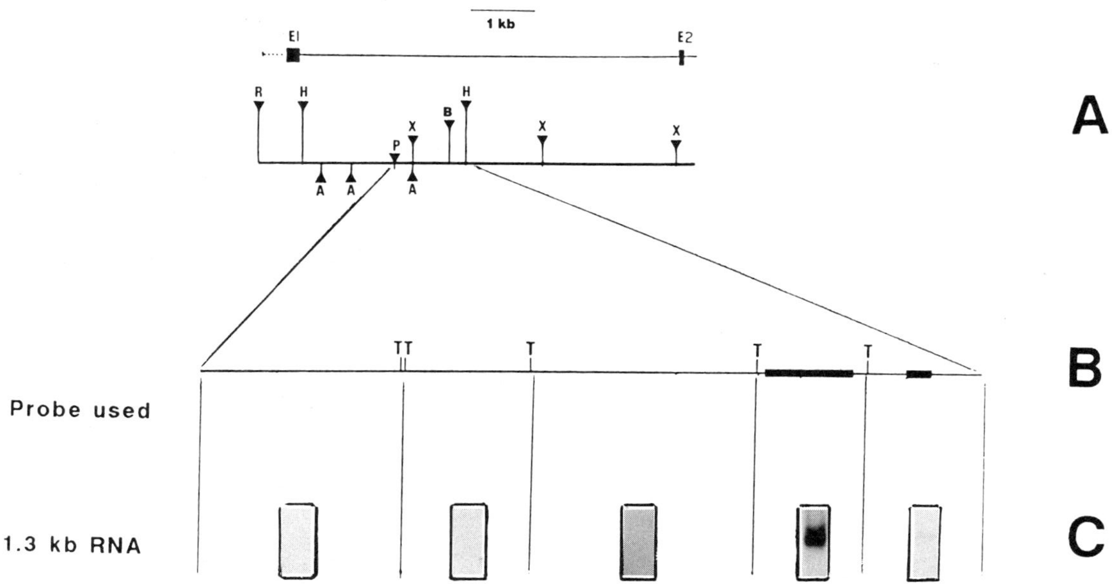

FIGURE 5. Homology of the *Pst*I-*Hind*III intronic fragment with the 1.3-kb RNA species. The *Pst*I-*Hind*III fragment of the first intron of p53 gene (**A**) was cut as indicated in **B** by *Taq*I, and fragments were ^{32}P-labeled and used to probe strips of Northern blots of RNAs extracted from MEL cells induced for 24 hours by 4 mM hexamethylene bisacetamide (HMBA) (**C**). *Black boxes* represent the identified repetitive elements, the *large box* for the B1 unit and the *small box* for the sequence homologous to the 3′ part of the 7SL coding gene.[11]

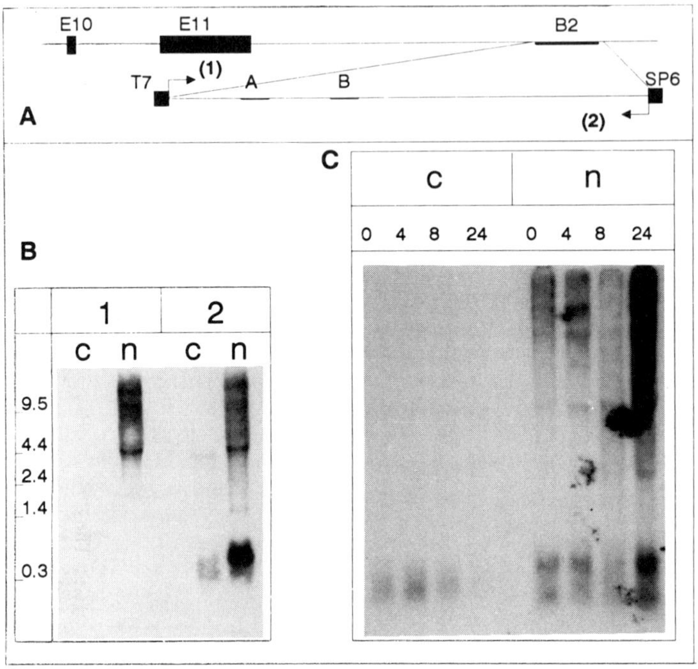

FIGURE 6. A B2 repetitive element present ≃2 kb down-stream of the coding region of the p53 gene (**A**) was used to probe total RNAs extracted from MEL cells. (**B**) Cytoplasmic (c) and nuclear (n) RNA was probed with sense (1) or antisense (2) probes. (**C**) Variation of B2 RNAs upon induction of differentiation of MEL cells; RNA was purified from cytoplasmic (c) or nuclear fractions (n) at the indicated times of induction (hours).

respect to its role in the control of cell proliferation, it is essential to know if the regulation of p53 during differentiation represents a general phenomenon and if the kind of regulation found in MEL cells can be extrapolated to other cellular systems. We studied the variation of p53 in B16 melanoma cells, another transformed cell line that can be induced to differentiate. We found that p53 mRNA decreases rapidly upon induction of differentiation. This decrease does not occur if the cells are induced in the presence of actinomycin D, which indicates that down-regulation requires transcriptional activity; cycloheximide experiments showed that it does not require protein synthesis. This regulation process in MEL cells therefore appears very similar to the process we observed in the B16 melanoma cell line. The posttranscriptional control of p53 mRNA production in cells inducible for differentiation occurs in several other systems as well. Although less documented, it seems to take place in an F9 teratocarcinoma cell line[9] as well as in chicken embryonic cells during development.[15] This suggests that such a mechanism might generally be employed for regulating the expression of p53 messenger RNA during differentiation, development, or both.

We previously identified the nucleus as the site of the regulation,[11] and we show in the present study that the down-regulation of p53 mRNA correlates with the accumulation of high molecular weight pre-mRNA maturation intermediates. This result suggests that in B16 cells as well as in MEL cells, a particular step in the processing of the p53 pre-mRNA is blocked, which in turn leads to the accumulation of mRNA precursors and to the disruption of the production of mature p53 mRNA. To confirm this hypothesis, we carried out actinomycin D chase experiments and (^{3}H)-uridine pulse-chase experiments. These experiments could readily be performed with MEL cells because of their high steady-state levels of p53 mRNA (3 to 4 times as much p53 mRNA in MEL cells as in B16 cells), as this increases considerably the ability to detect p53 RNA intermediates in a (^{3}H)-uridine pulse-chase experiment. We found that the decay of p53 mRNA upon actinomycin D treatment was a biphasic event; an apparent stability of p53 mRNA can be observed in up to 3–4 hours of actinomycin D treatment, after which p53 mRNA levels decrease with a half-life closely related to that observed when cells are treated by inducer. Thus, a latency period of 2 hours is required before p53 mRNA starts to decay. We postulate that this delay is due to the presence of a significant amount of p53 pre-mRNA committed into the splicing machinery. To investigate this notion and to confirm the data from the actinomycin D chase experiments, we performed (^{3}H)-uridine pulse-chase experiments. These data show that in uninduced cells (^{3}H)-labeled p53 mRNA production continued for at least 2 hours after removal of (^{3}H)-uridine. But when the chase was performed in the presence of inducer, a rapid decay of (^{3}H)-labeled p53 mRNA was observed. These data are in agreement with those obtained by quantitative dot blot hybridization, and they show that the inducer stopped the course of p53 mRNA production at a stage nearer to the mature mRNA than at the stage affected by actinomycin D., that is, p53 gene transcription. FIGURE 7 summarizes these data.

To look for RNA species capable of interfering with p53 pre-mRNA maturation, we used several genomic probes derived from the p53 gene, extending from the first exon up to the 3′ flanking region. This experimental approach allowed us to identify several RNA species that were modulated during the induction of differentiation.

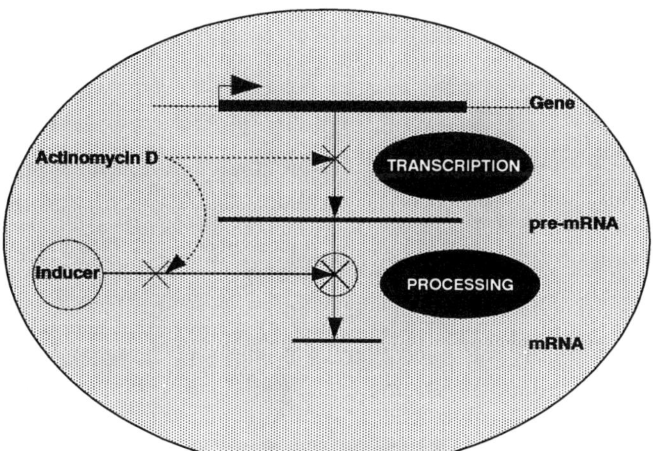

FIGURE 7. Schematic representation of the regulation steps of p53 mRNA expression during induction of differentiation of murine cells. Down-regulation of the p53 mRNA does not take place in the absence of transcriptional activity. Actinomycin D blocks both the transcriptional activity of the p53 gene and also the inducer-mediated transcription required for blockade of the maturation process of p53 pre-mRNA.

First, we visualized a high molecular weight RNA species, in an antisense orientation with respect to the p53 promoter and which accumulates as the induced differentiation proceeds. This finding may be related to the existence of a family of antisense RNAs in the bulk of hnRNA. Indeed, the presence of such antisense RNA was reported for c-myc[2,16–18] and c-myb.[19–21] These RNAs may have a rapid turnover as do the majority of hnRNAs, and their accumulation during differentiation may reflect a decrease in the turnover rate upon the action of the inducer. We cannot state at this time if this finding is insignificant or if it reflects a general regulation process of biological importance.

We also identified a 1.3-kb RNA species, which was recognized by a specific probe, a *Pst*I-*Hind*III fragment, originating from the first intron of the p53 gene. This RNA was in an antisense orientation with respect to the p53 gene and contained a B1 repetitive element.[11] We find that contrary to MEL cells, the 1.3-kb RNA does not accumulate during the induction of B16 cells, but rather it decreases. It is thus not likely to be involved in the maturation process of p53 pre-mRNA. This result led us to reassess the nature and the role of this RNA species in MEL cells. We found that the complementarity of this RNA with the intronic probe, which originally allowed its identification, was actually restricted to the B1-containing element of the probe. It is clear, however, that this particular B1-containing RNA species is subject to modulation during induced differentiation, in a manner different in MEL cells than in B16 cells, the biological significance of which is presently unknown. The same is true for the B2-containing RNAs that we identified by the use of a probe originating from the downstream region of the p53 gene which contains the B2 consensus sequence. Both high and low molecular weight "B2-RNAs" were subject to modulation during the induced differentiation of MEL cells. The low molecular B2-RNAs were in a sense orientation, whereas high molecular weight B2-RNAs existed in both orientations. This suggests an interesting model according to which only the sense B2 RNAs would survive the maturation process leading to the production of B2 small RNAs. The biological role of such small RNAs is presently unknown; an interesting hypothesis has been formulated by Clemens,[22] according to which B2-RNAs would play a role in regulating the concentration of certain RNAs in the cytoplasm, namely, a subset of those few that are involved in the control of cell proliferation. Our observation that B2-RNAs are modulated along with the induced differentiation of MEL cells and, more precisely, in an apparent relation with the modulation of the cell cycle parameters[13] reinforces this hypothesis.

Finally, we reassessed our earlier interpretation of the results on the RNase protection assay performed in a previous study,[11] in which protection of the PstI-*Hind*III intronic fragment was observed over at least 600 base pairs. Protection of this intronic p53 probe is likely to have been mediated by complementary antisense RNA present in the high molecular weight fraction. Our data has led us to speculate about the possible function of such high molecular weight antisense RNAs; by being able to hybridize to sense pre-mRNAs, they would be able to interfere with the regular maturation process and thus stop the production of mature mRNA. Whether or not this process is involved in p53 mRNA regulation during differentiation has yet to be demonstrated.

SUMMARY

p53 expression is strongly modulated during the process of induced differentiation, at the same time as both cell cycle and genetic expression become modulated, giving rise to a commitment to terminal differentiation. We took advantage of two murine cell lines inducible for differentiation, an erythroleukemia and a melanoma cell line, to outline common features of the regulation of p53 expression during the differentiation process.

We found that p53 mRNA decreased early after induced differentiation and that regulation was controlled at a posttranscriptional level. Our data showed that this regulation affects p53 pre-mRNA maturation. Because, in both systems used, actinomycin D treatment abolished the inducer-mediated decrease of p53 mRNA, we looked for induced RNAs potentially involved in this process. Using different parts of the p53 gene and flanking regions as probes, we identified three RNA species whose expression is modulated during induced differentiation. A first species is made of high molecular weight RNAs that accumulate in the nuclear compartment and seem to represent antisense transcripts of the p53 gene. A second species, 1.3-kb long, was found to accumulate in the nucleus of induced MEL cells and was homologous to a restricted part of the first intron of the p53 gene due to the presence of a B1 repetitive element in an antisense orientation with respect to the p53 pre-messenger RNA. Finally, a family of B2-containing RNAs was observed in both cytoplasmic and nuclear compartments. The variation in the amounts of sense and antisense RNAs, respectively, suggested an interesting speculative model for the maturation of B2-containing pre-messenger RNAs.

REFERENCES

1. Marks, P. A. & R. A. Rifkind. 1978. Ann. Rev. Biochem. **47:** 419–448.
2. Mechti, N., M. Piechaczyk, J. M. Blanchard, L. Marty, A. Bonnien, P. Jeanteur & B. Lebleu. 1986. Nucl. Acids Res. **14:** 9653–9666.
3. Lachman, H. M., G. Cheng & A. I. Skoultchi. 1986. Proc. Natl. Acad. Sci. USA **83:** 6480–6484.
4. Finlay, C. A., P. W. Hinds & A. J. Levine. 1989. Cell **57:** 1083–1093.
5. Eliyahu, D., A. Raz, P. Gruss, D. Givol & M. Oren. 1984. Nature **312:** 646–649.
6. Michalovitz, D., O. Halevy & M. Oren. 1990. Cell **62:** 671–680.
7. Shen, D. W., F. X. Real, A. B. Deleo, L J. Old, P. A. Marks & R. A. Rifkind. 1983. Proc. Natl. Acad. Sci. USA **80:** 5919–5922.
8. Dony, C., M. Kessel & P. Gruss. 1985. Nature **317:** 636–639.
9. Richon, V. M., R. G. Ramsay, R. A. Rifkind & P. A. Marks. 1989. Oncogene **4:** 165–173.
10. Khochbin, S., E. Principaud A. Chabanas & J. J. Lawrence. 1988. J. Mol. Biol. **200:** 55–64.
11. Khochbin, S. & J. J. Lawrence. 1989. EMBO J. **8:** 4107–4114.
12. Chabanas, A., E. Khoury, P. Goeltz, P. Foussard, R. Gjerset, B. Dod, H. Eisen & J. J. Lawrence. 1985. J. Mol. Biol. **183:** 141–151.
13. Khochbin, S., D. Grunwald, E. Bergeret & J. J. Lawrence. 1991. Biochem. Biophys. Res. Com. **175:** 91–97.
14. Rogel, A., M. P. Popliker, G. G. Webb & M. Oren. 1985. Mol. Cell Biol. **5:** 2851–2855.
15. Louis, J. M., V. W. Mc Farland, P. May & P. Mora. 1988. Biochim. Biophys. Acta **950:** 395–402.
16. Nepveu, A. & K. B. Marcu. 1986. EMBO J. **5:** 2859–2865.
17. Bentley, D. L. & M. Groudine. 1986. Nature **321:** 702–706.
18. Kindy, M. S., J. E. McCormack, A. J. Bucker, A. R. Levine & G. E. Sonenshein. 1987. Mol. Cell Biol. **7:** 2857–2862.
19. Bender, T. P., C. B. Thompson & W. M. Kuehl. 1987. Science **237:** 1473–1476.
20. Watson, R. J. 1988. Mol. Cell Biol. **8:** 3938–3942.
21. Todokoro, K., R. J. Watson, H. Higo, H. Amanuma, S. Kuramochi, H. Yanagisawa & Y. Ikawa. 1988. Proc. Natl. Acad. Sci. USA **85:** 8900–8904.
22. Clemens, M. J. 1987. Cell **49:** 157–158.

Antisense Promoter Mapping

Inhibitory Methods of Transcriptional Analysis[a]

JEFFREY T. HOLT

Departments of Cell Biology and Pathology
Vanderbilt University
Nashville, Tennessee 37232

Antisense methods have been employed to inhibit the function of a number of genes that were ultimately shown to be transcription factors including the protooncogenes c-fos,[1–6] c-myc,[7–15] and c-myb.[16–18] Inhibition of many of these genes results in growth inhibition, suggesting that expression of these genes is required for the growth of both normal and transformed cells (reviewed in ref. 1). The discovery that these genes represent transcription factors has allowed antisense studies to extend beyond growth inhibition, employing antisense methods as a tool to study the functional role of specific transcription factors during cellular proliferation. Overexpression studies are important in analysis of transcriptional control, but they have some serious potential drawbacks. An example is the phenomenon of squelching, in which sequestration of cofactors creates the illusion that transcriptional activators are repressors.[19] Analysis of both overexpression and antisense data should provide increased assurance that the transcriptional mechanism being studied is a physiologically relevant one.

Growth factors induce rapid but transient expression of a subset of genes termed "immediate early genes" which are induced in the absence of protein synthesis.[20–25] In fact, the protooncogene c-fos is superinduced by cycloheximide,[20,25] suggesting that a newly transcribed gene product may function as a repressor. The simplest model to explain these results proposes that growth factors induce c-fos transcription, resulting in translation of the c-fos protein (Fos) that subsequently represses c-fos transcription.[26] This model explains both the transient induction and the effects of cycloheximide on c-fos expression. Overexpression of Fos protein inhibits the c-fos promoter, providing support for this model.[27–38]

If Fos protein functions as a repressor, then antisense inhibition of c-fos expression should "superinduce" c-fos expression, resulting in both increased and prolonged transcription of the c-fos promoter. To test this hypothesis we performed nuclear run-on studies in stably transfected cell lines that express steroid-inducible anti-fos RNA[2] and produce a 90% decrease in levels of Fos protein as measured by immunofluorescence[39] and immunoprecipitation (J. T. Holt, manuscript submitted). The results indicated that induction of anti-fos RNA produced a 20-fold increase in c-fos transcription (FIG. 1). Steroid treatment of control transfectants had no effect on c-fos transcription, indicating that the observed alteration in transcription was due to the anti-fos RNA and not to the steroid.[39] Induction of anti-fos RNA resulted in both increased transcription and prolonged transcription of the c-fos gene. Strong transcription of the c-fos gene could still be observed 4 hours after serum stimulation following anti-fos RNA induction, whereas c-fos transcription is not defected after the first hour in normal and control cells.[39]

[a]This work was supported by National Institutes of Health grant R01CA49052. Dr. Holt is a Pew Scholar and is supported by the Pew Memorial Trust.

ANTISENSE PROMOTER MAPPING

Because the nuclear run-on studies demonstrated that inhibition of c-fos expression increases c-fos transcription, we analyzed the expression of a series of Fos/promoter CAT constructs by transient transfection into the inducible anti-fos clones. The results showed that induction of anti-fos RNA produced a 10-fold increase in Fos promoter-regulated CAT activity. This "antisense activation" was maximal at 1 μg of transfected promoter/CAT target DNA and decreased with increasing concentration of target DNA, as would be expected during titration of a repressor.[26] Nuclease protection assays demonstrated that the increased CAT activity resulted from an increase in correctly initiated CAT mRNA. Steroid treatment of a "sense" control clone did not alter CAT activity, and transfection of an antisense resistant c-fos expression vector reversed the antisense effect.[39] Once the method had been documented, it was employed to map the region of the c-fos promoter which is repressed by c-fos protein. The results demonstrated that the serum response element was the site of initial repression, whereas a larger upstream region was required for prolonged repression by Fos protein (FIG. 2).

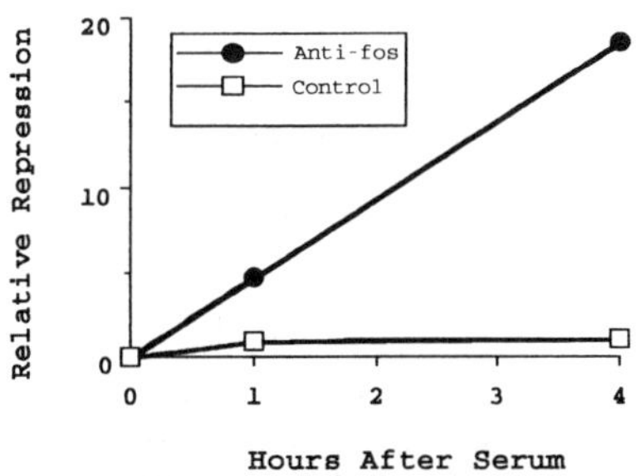

FIGURE 1. Standardized transcription of c-fos in antisense and control "sense" cells relative to that of the constitutively transcribed cyclophilin gene. Autoradiographs from nuclear run-on experiments were analyzed by densitometry to determine the relative level of c-fos transcription in the presence and absence of steroid (which induces antisense RNA). Modification of FIGURE 1 from Robinson-Benion *et al.*[39]

The previously described studies employed antisense methods to demonstrate a transcriptional effect of Fos protein: repression of the c-fos promoter. The success of these studies suggested that antisense methods might be employed to identify and clone transcriptional target genes. Overexpression of c-fos protein has been employed to identify target genes that are regulated by Fos,[40] but this approach has some serious problems. The most important concern is that Fos protein is unable to bind DNA unless it heterodimerizes with another basic zipper protein such as Jun or ATF (reviewed in refs. 41 and 42). Consequently overexpression of Fos will not induce target genes unless the appropriate partner protein is present. Antisense inhibition eliminates this problem, because it can theoretically identify genes that require coinduction of Fos and an unknown binding partner. Overexpression of Fos can also result in transformation, leading to the cloning of genes that result from transformation and not from the normal transcriptional effects of Fos protein. This should not be a problem with an antisense cloning method.

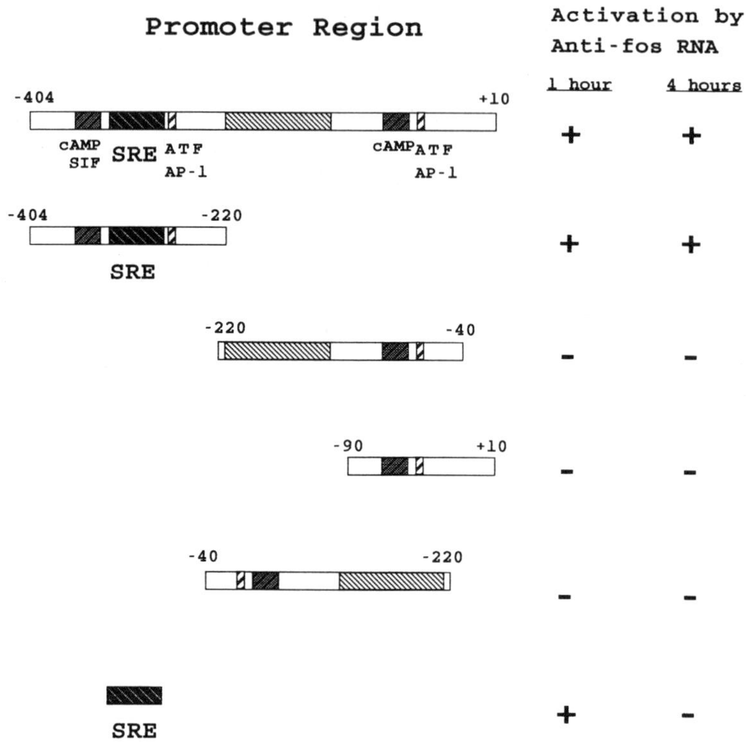

FIGURE 2. Map of promoter elements that are activated by anti-fos RNA (repressed by Fos protein). The left portion is a schematic drawing of the c-fos promoter region linked to CAT for each deletion plasmid. Note that the second, third, fifth, and last vectors (including SREcat) are fused to minimal tk promoters to provide a TATA element. The plus sign indicates that induction of anti-fos RNA in antisense cells resulted in increased fos/promoter CAT activity at the given time interval following quiescence. Activation was scored as positive if in multiple experiments a greater than fivefold induction occurred following the addition of dexamethasone to cells expressing inducible anti-fos RNA. Minus sign means the steroid treatment produced no significant ($<20\%$) change in CAT activity. Adapted from FIGURE 5 in Robinson-Benion *et al.*[39]

ANTISENSE CLONING STRATEGY

To identify physiologically relevant transcriptional targets of Fos protein, we have developed an antisense cloning method. This strategy employs our transfectants that express steroid-inducible anti-fos RNA as reagents to identify Fos target genes. A cDNA library was prepared from fibroblasts that were quiescent for 48 hours and then serum stimulated for 4 hours with 10% serum. Total RNA was prepared from these cells following lysis in guanidinium isothiocyanate and centrifugation on a cesium chloride step-gradient. Polyadenylated RNA (obtained by two rounds of oligo-dT selection) from these cells was appropriately linkered and cloned into lambda ZAP; this cDNA library contains 3×10^7 recombinants. The library was then differentially screened with two probes, one derived from 4-hour serum-stimulated fibroblasts, and a second derived from 4-hour serum-stimulated fibroblasts expressing anti-fos RNA. This antisense cloning strategy is outlined in FIGURE 3. We have already screened 200,000 recombinants, and four have been plaque purified through tertiary screening and subcloned by phagemid rescue. All four of these subclones lack significant homology to each other as judged by cross-hybridization studies. The cDNAs have been employed to probe Northern blots of serum-stimulated RNAs from anti-fos clones and control clones, and all are induced by serum unless anti-fos RNA is present. Cycloheximide blocks serum induction of all four cDNAs, suggesting that they are indirectly induced (require transcription of a new gene before they can be transcribed). This effect of steroids is not nonspecific, because steroid treatment of untransfected cells or control "sense" cells does not inhibit serum induction of these cDNAs. These clones are currently being further characterized to determine the molecular mechanisms through which Fos protein regulates their expression.

Although many studies describe the use of antisense methods to inhibit specific genes and disrupt cellular physiology, fewer studies have used antisense to dissect molecular mechanisms. Antisense methods of transcriptional analysis avoid many of the pitfalls inherent in overexpression studies and are particularly suited for analysis of factors whose dimerization partners are unknown. This review summarizes our previous studies employing antisense promoter mapping to identify the site where serum-induced endogenous Fos protein represses the Fos promoter[39] and describes a new antisense cloning method to identify transcriptional targets for Fos. After an investigator has created and characterized antisense reagents that demonstrate an important functional role for the gene product (e.g., the protein is required for cell growth or differentiation), it would be helpful if these same reagents could be

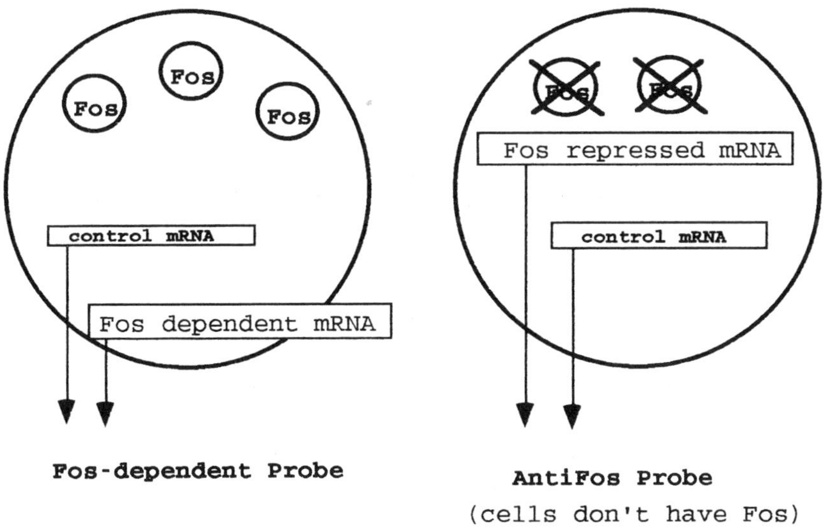

FIGURE 3. Illustration of antisense cloning method. RNA from the left cell (*circle*) was obtained 4 hours after serum restimulation of quiescent fibroblasts. RNA from the right cell was obtained 4 hours after serum restimulation of quiescent cells expressing inducible anti-fos RNA. Fos-dependent cDNAs are in the Fos-dependent probe but not the anti-fos probe. Library made from Fos-dependent mRNAs. Most cDNAs (control) are in both probes.

employed to analyze the molecular events that are governed by the protein. The judicious use of antisense promoter mapping and antisense cloning methods may augment the usefulness of antisense research by expanding these methods into studies of molecular mechanisms.

SUMMARY

We have employed antisense methods to study the transcriptional functions of c-fos protein (Fos). Clones expressing inducible anti-fos RNA have been employed to inhibit c-fos expression, resulting in activation of c-fos transcription by inhibiting its normal repressor function. The sites of negative regulation by Fos have been mapped using this antisense mapping method which demonstrates that the serum response element represents the major site of repression by endogenous c-fos protein. A similar strategy (antisense cloning) has been employed to clone four target genes that are Fos dependent. These cDNAs encode mRNAs that are rapidly induced by serum (although this induction is blocked by cycloheximide) but are blocked by induction of anti-fos RNA. These inhibitory methods of studying transcription factor function are extremely useful for transcription factors (like Fos) that require cooperation with other factors to modulate gene transcription.

REFERENCES

1. Holt, J. T. 1989. Growth inhibition by antisense nucleic acids. *In* Discoveries in Antisense Nucleic Acids. C. L. Brakel, ed. :81–94. Gulf Publishing. Houston, Texas.
2. Holt, J. T., T. V. Gopal, A. D. Moulton & A. W. Nienhuis. 1986. Inducible production of c-fos antisense RNA inhibits 3T3 cell proliferation. Proc. Natl. Acad. Sci. USA **83:** 4794–4798.
3. Nishikura, K. & J. M. Murray. 1987. Antisense RNA of proto-oncogene c-fos blocks renewed growth of quiescent 3T3 cells. Mol. Cell. Biol. **7:** 639–649.
4. Mercola, D., A. Rundell, J. Westwick & S. A. Edwards. 1987. Antisense RNA to the c-fos gene: Restoration of density-dependent growth arrest in a transformed cell line. Biochem. Biophys. Res. Comm. **147:** 288–296.
5. Levi, B. Z. & K. Ozato. 1988. Constitutive expression of c-fos antisense RNA blocks c-fos gene induction by interferon and by phorbol ester and reduces c-myc expression in F9 embryonal carcinoma cells. Genes Dev. **2:** 554–566.
6. Edwards, S. A., A. Y. Rundell & E. D. Adamson. 1988. Expression of c-fos antisense RNA inhibits the differentiation of F9 cells to parietal endoderm. Dev. Biol. **129:** 91–102.
7. Griep, A. E. & H. Westphal. 1988. Antisense myc sequences induce differentiation of F9 cells. Proc. Natl. Acad. Sci. USA **85:** 6806–6810.
8. Harel-Bellan, A., D. K. Ferris, M. Vinocur, J. T. Holt & W. L. Farrar. 1988. Specific inhibition of c-myc protein biosynthesis using an antisense synthetic deoxynucleotide in human T lymphocytes. J. Immunol. **140:** 2431–2435.
9. Heikkila, R., G. Schwab, E. Wickstrom, S. L. Loke, D. H. Pluznik, R. Watt & L. M. Neckers. 1987. A c-myc antisense oligonucleotide inhibits entry into S phase but not progress from G0 to G1. Nature **328:** 445–449.
10. Holt, J. T., R. L. Redner & A. W. Nienhuis. 1988. An oligomer complementary to c-myc mRNA inhibits proliferation of HL-60 promyelocytic cells and induces differentiation. Mol. Cell. Biol. **8:** 963–973.
11. Pietenpol, J. A., J. T. Holt, R. W. Stein & H. L. Moses. 1990. Transforming growth factor Beta-1 suppression of c-myc gene transcription: Role in inhibition of keratinocyte proliferation. Proc. Natl. Acad. Sci. USA **87:** 3758–3762.
12. Prochownik, E. V., J. Kukowska & C. Rodgers. 1988. C-myc antisense transcripts

accelerate differentiation and inhibit G1 progression in murine erythroleukemia cells. Mol. Cell. Biol. **8:** 3683–3695.

13. WICKSTROM, E. L., T. A. BACON, A. GONZALEZ, D. L. FREEMAN, G. H. LYMAN & E. WICKSTROM. 1988. Human promyelocytic leukemia HL-60 cell proliferation and c-myc expression are inhibited by an antisense pentadecadeoxynucleotide targeted against c-myc mRNA. Proc. Natl. Acad. Sci. USA **85:** 1028–1032.
14. WICKSTROM, E. L., T. A. BACON, A. GONZALEZ, G. H. LYMAN & E. WICKSTROM. 1989. Anti-c-myc DNA increases differentiation and decreases colony formation by HL-60 cells. In Vitro Cell Dev. Biol. **25:** 297–301.
15. YOKOYAMA, K. & F. IMAMOTO. 1987. Transcriptional control of the endogenous myc protooncogene by antisense RNA. Proc. Natl. Acad. Sci. USA **84:** 7363–7367.
16. VENTURELLI, D., S. TRAVALI & B. CALABRETTA. 1990. Inhibition of T-cell proliferation by a myb antisense oligomer is accompanied by selective downregulation of DNA polymerase alpha expression. Proc. Natl. Acad. Sci. USA **87:** 5963–5967.
17. GEWIRTZ, A. W., G. ANFOSSI, D. VENTURELLI, S. VALPREDA, R. SIMS & B. CALABRETTA. 1989. G1/S transition in normal human T-lymphocytes requires the nuclear protein encoded by c-myb. Science **245:** 180–183.
18. GEWIRTZ, A. W. & B. CALABRETTA. 1988. A c-myb antisense oligodeoxynucleotide inhibits normal human hematopoiesis *in vitro.* Science **242:** 1303–1306.
19. LEVINE, M. & J. L. MANLEY. 1989. Transcriptional repression of eukaryotic promoters. Cell **59:** 405–408.
20. LAU, L. F. & D. NATHANS. 1987. Expression of a set of growth-related immediate early genes in BALB/c 3T3 cells: Coordinate regulation with c-fos or c-myc. Proc. Natl. Acad. Sci. USA **84:** 1182–1186.
21. ALMENDRAL, J. M., D. SOMMER, H. MACDONALD-BRAVO, J. BURCKHARDT, J. PERERA & R. BRAVO. 1988. Complexity of the early genetic response to growth factors in mouse fibroblasts. Mol. Cell. Biol. **8:** 2140–2148.
22. CHAVRIER, P., U. JANSSEN-TIMMEN, M.-G. MATTEI, M. ZERIAL, R. BRAVO & P. CHARNAY. 1989. Structure, chromosome location, and expression of the mouse zinc-finger gene Krox-20: Multiple gene products and coregulation with the proto-oncogene c-fos. Mol. Cell. Biol. **9:** 787–797.
23. COCHRAN, B. H., A. C. REFFEL & C. B. STILES. 1983. Molecular cloning of gene sequences regulated by platelet-derived growth factor. Cell **33:** 939–947.
24. GREENBERG, M. E. & E. B. ZIFF. 1984. Stimulation of 3T3 cells induces transcription of the c-fos proto-oncogene. Nature **311:** 433–438.
25. GREENBERG, M. E., A. L. HERMANOWSKI & E. B. ZIFF. 1986. Effect of protein synthesis inhibitors on growth factor activation of c-fos, c-myc, and actin gene transcription. Mol. Cell. Biol. **6:** 1050–1057.
26. SASSONE-CORSI, P., J. C. SISSON & I. VERMA. 1988. Transcriptional autoregulation of the proto-oncogene fos. Nature **334:** 314–319.
27. GREENBERG, M. E., Z. SIEGFRIED & E. B. ZIFF. 1987. Mutation of the c-fos dyad symmetry element inhibits serum inducibility *in vivo* and nuclear regulatory factor binding *in vitro.* Mol. Cell. Biol. **7:** 1217–1225.
28. DESCHAMPS, J., F. MEIJLINK & I. M. VERMA. 1985. Identification of a transcriptional enhancer element upstream from the proto-oncogene fos. Science **230:** 1174–1177.
29. GIUS, D., X. CAO, F. J. RAUSCHER III, D. R. COHEN, T. CURRAN & V. P. SUKHATME. 1990. Transcriptional activation and repression by Fos are independent functions: The C terminus represses immediate early gene expression via CarG elements. Mol. Cell. Biol. **10:** 4243–4255.
30. KONIG, H., H. PONTA, U. RAHMSDORF, M. BUSCHER, A. SCHONTHAL, H. J. RAHMSDORF & P. HERRLICH. 1989. Autoregulation of fos: The dyad symmetry element as the major target of repression. EMBO J. **8:** 2559–2566.
31. LUCIBELLO, F. C., C. LOWAG, M. NEUBERG & R. MULLER. 1989. Transrepression of the mouse c-fos promoter: A novel mechanism of fos-mediated trans-regulation. Cell **49:** 741–752.
32. RIVERA, V. M., M. SHENG & M. E. GREENBERG. 1990. The inner core of the serum

response element mediates both the rapid induction and subsequent repression of c-fos transcription following serum stimulation. Genes. Dev. **4:** 255–268.
33. SHAW, P. E., S. FRISCH & A. NORDHEIM. 1989. Repression of c-fos transcription is mediated through p67SRF bound to the SRE. EMBO J. **8:** 2567–2574.
34. SHAW, P. E., H. SCHROTER & A. NORDHEIM. 1989. The ability of a ternary complex to form over the serum response element correlates with serum inducibility of the human c-fos promoter. Cell **56:** 563–572.
35. SUBRAMANIAM, M., L. J. SCHMIDT, C. E. CRUTCHFIELD III & M. J. GETZ. 1989. Negative regulation of serum-responsive elements. Nature **340:** 64–66.
36. TREISMAN, R. 1985. Transient accumulation of c-fos RNA following serum stimulation requires a conserved 5′ element and c-fos 3′ sequences. Cell **42:** 889–902.
37. WILSON, T. & R. TREISMAN. 1988. Fos C-terminal mutations block down-regulation of c-fos transcription following serum stimulation. EMBO J. **7:** 4193–4202.
38. KAMATA, N. & J. T. HOLT. 1992. Inhibitory effect of myristylation on transrepression by FBR (gag-fos) protein. Mol. Cell. Biol. **12:** 876–882.
39. ROBINSON-BENION, C., N. KAMATA & J. T. HOLT. 1991. Antisense mapping of the c-fos promoter: Role of the serum response element. Antisense Res. Dev. **1:** 21–34.
40. SUPERTI-FURGA, G., G. BERGERS, D. PICARD & M. BUSSLINGER. 1991. Hormone-dependent transcriptional regulation and cellular transformation by Fos-steroid receptor fusion proteins. Proc. Natl. Acad. Sci. USA **88:** 5114–5118.
41. ABATE, C. & T. CURRAN. 1990. Encounters with Fos and Jun on the road to AP-1. Sem. Cancer. Biol. **1:** 19–26.
42. HOLT, J. T. 1992. Fos and Jun: Transcriptional regulation of cell growth and tumorigenesis. *In* Oncogenes II. C. Benz & E. Liu, eds. Kluwer Academic Publishers, Norwell, MA. In press.

Status of Ribozyme and Antisense-Based Developmental Approaches for Anti-HIV-1 Therapy[a]

JOHN A. ZAIA,[b] SASWATI CHATTERJEE,[b] K. K. WONG,[c]
DAVID ELKINS,[d] NERIDA R. TAYLOR,[d]
AND JOHN J. ROSSI[d]

[b]*Division of Pediatrics and* [c]*Medicine*
City of Hope National Medical Center
Duarte, California 91010–0269

[d]*Division of Biology*
Bechman Research Institute of the City of Hope
Duarte, California 91010–0269

The observation that the complementary interaction of nucleotide sequences exists in nature for the regulation of cellular functions led to the concept of "antisense" as a biological entity.[1] Antisense is any DNA or RNA that inhibits a gene function by methods that require complementary base-pairing to the genetic target. In the last 15 years, it has been shown that the expression of antisense RNA as well as the exogenous introduction of antisense DNA can affect a variety of cell functions.[2–5]

The observation that RNA has autocatalytic function that is mediated in part by complementary base-pairing has led to a second generation of antisense which is called ribozyme. Ribozyme is an RNA molecule having catalytic enzyme activity that cleaves RNA in a sequence-dependent manner, thereby inactivating the RNA. There are several types of ribozymes. (For a review see ref. 6 and FIG. 1). The first of these, from the *Tetrahymena* group one intervening sequence, was shown by Zaug and Cech[7] to be essential for cleaving an intron from messenger RNA. A second ribozyme was noted by Altmann and co-workers to function as the RNA component of RNAseP in the maturation of tRNAs.[8] RNAseP consists of both RNA and protein moieties that function together to complete the 5′ cleavage of pre-tRNAs. In this situation, the RNA component can function independently, but the 14-kD protein moiety is thought to create a microenvironment that optimizes the catalytic reaction (FIG. 1). A third type of ribozyme, the hammerhead ribozyme, is the one that exists in certain viroids and virusoids as well as in *Neurospora.* These ribozymes require specific flanking sequences that target the ribozyme and bring together an active center, causing cleavage at that specific site. Synthetic ribozyme can be made synthetically, and the use of ribozyme for inhibition of HIV infection will be discussed in detail here.

The fourth type, the hairpin ribozyme, was discovered by Hampel and Tritz[11] in the negative strand of satellite RNA of tobacco ring spot virus. Here the minimal sequence required for catalytic activity consists of a 15-nucleotide RNA catalyst and a 14-nucleotide RNA target substrate.[12] As shown in FIGURE 1, the presence of four helices maintains a secondary configuration, and helices I and II are formed between the substrate and the ribozyme by virtue of antisense complementation. Helices III

[a]This work was supported by US Public Health Service grant AI25959–06 and AI29329.

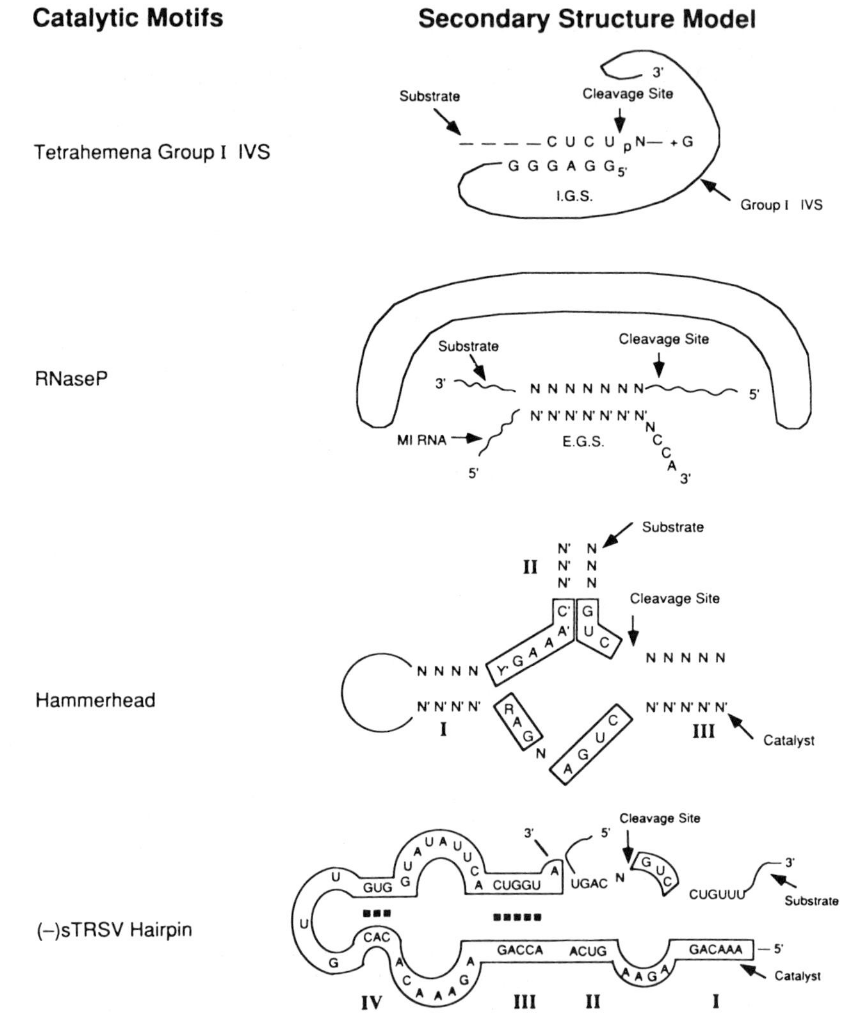

FIGURE 1. Diversity of catalytic RNA motifs. Examples of catalytic RNAs and their respective secondary structure are shown including: (1) *Tetrahymena* group I intervening sequence (IVS) with internal guide sequence (IGS), (2) RNAseP shown as RNA component (M1 RNA) with external guide sequence (EGS) and 14 kD protein component, (3) hammerhead ribozyme with stems I, II, and III, and (4) hairpin ribozyme of negative strand of satellite RNA of tobacco ringspot virus [(-)sTRSV]. All four ribozymes are depicted in the *trans* configuration with separate substrate and catalyst molecules, and the respective cleavage site on the target RNA is indicated by the *arrow.* (Reproduced from Sarver *et al.*,[6] with permission.)

and IV are formed by folding of the ribozyme strand itself. The cleavage initiated by both hammerhead and hairpin ribozymes generates a 2′, 3′-cyclic phosphate and a 5′ hydroxyl group at the cleavage site.

In addition to these four ribozyme groups, additional catalytic motifs that perform a cleavage function exist in other RNA molecules. Hepatitis delta virus, which contains ribozyme activity, is the first natural ribozyme discovered to be active in higher forms of animal life.[13–15] In addition, ribosomal RNAs are found in fungal mitochondria.[16] Additional RNA molecules that mediate catalysis will likely be discovered and will have important application to biologic systems.

UNDERSTANDING HAMMERHEAD RIBOZYMES

Sequence analysis of hammerhead ribozymes indicates that a consensus nucleotide sequence of 14 nucleotides comprises a catalytic core. This represents a predictable secondary structure in which single base alterations destroy the catalytic function of the molecule.[17] Adjacent to these are flanking regions that by base-pairing mechanisms bring the critical core sequences into apposition. Mechanisms by which this specific structure causes cleavage are not known, but the cleavage is always at a specific site that is on the 3′ side of the XUX′, where X is any ACUG and X′ is any ACU. In the natural environment, the ribozyme and its substrate are part of the same RNA molecule, and cleavage occurs in a process by which one portion of the molecule creates a loop and results in site-specific cleavage. The conserved domains of the catalytic center define three stems of the hammerhead ribozyme (FIG. 1) that stabilize the substrate-catalyst complex and complete the three-dimensional topological motif required for cleavage. Thus, natural ribozymes exist as *cis* operational structures relative to the cleavage site. However, these components of the active center can be brought together in *trans* configurations to target specific XUX′ sites. Independently, Uhlenbeck[17] and coworkers, Koizumi *et al.,*[18] and Jeffries and Symons[19] demonstrated that hammerhead ribozymes can be targeted to RNA substrates which are separate from the active RNA molecule. In this situation, the catalytic center consists of two separate strands, one serving as the catalyst and the other as a substrate; the annealing of the two strands forms the hammerhead structure and results in the formation of the catalytic center. Because the target strand needs only to contain the XUX′, the transacting ribozyme, containing the other elements of the active center, can be configured to target many genetic sequences.

In addition to its structure, the ribozyme-substrate complex also requires the presence of magnesium for cleavage to occur.[20] Once the ribozyme and the cleaved substrate disassociate, the ribozyme remains intact and is free to interact with other substrate molecules. This is the theoretic advantage of ribozymes over antisense.

The similarities in terms of both advantages and disadvantages between ribozymes and antisense are as follows: dependence on Watson/Crick base-pairing, requirement for target access, theoretical high specificity/low toxicity, degradation by cellular enzymes, limitations regarding exogenous delivery, and requirement for efficient expression. Both possess a specificity that is based on DNA/RNA sequence complementarity. The targeting by these two molecules to site-specific sequences accounts for the potential relative lack of toxicity for these antisense molecules. Similarly, both antisense and ribozyme have the same limitations in terms of delivery, restrictions on expression, and target access. This will be discussed in more detail, but, clearly, current development and application of antisense and ribozyme to clinical medicine are focused on methods to address these limitations. It is possible that with short flanking regions adjacent to the active center, ribozyme could target native RNA, but by increasing the size of the flanking regions, this becomes less probable. However, because these molecules can be targeted specifically to host

RNA, if desired, it is possible that both antisense and ribozyme will have application in many different diseases. In this report, we describe only applications in human immunodeficiency virus type 1 (HIV-1) infection; nevertheless, it is possible to apply these to other infectious diseases, to diseases mediated by gene rearrangements, such as chronic myelogenous leukemia,[21] and to diseases mediated by abnormal control of normal growth factors, such as solid tumors.[5] Finally, because ribozyme relies on the flanking sequences to target a specific RNA site, it is possible that by mutation this site could be lost and resistance could occur. It is possible, for example, in HIV-1 that a single site could undergo mutation during viral replication and eventually select for resistant virus. Again, by appropriate selection of flanking regions or by inclusion of multiple ribozymes, it is possible to minimize the likelihood of such mutational loss.

CONSIDERATIONS REGARDING RIBOZYME EFFICIENCY

Role of Secondary Structure of Ribozyme Transcript on Enzyme Function of Ribozyme

To determine the effect of secondary RNA structure on enzyme function, we made a ribozyme that was stable and was produced in *Escherichia coli* using the stable 4.5S RNA message for a signal recognition protein.[22] An anti-HIV *gag* ribozyme of 47 bases was cloned into an internal *Mlu*I restriction site of the 4.5S RNA under control of the pTac promotor with RRNA E gene T1 and T2 transcriptional terminators in an ampicillin resistance-encoding plasmid. The resulting molecule was 199 nucleotides in length and could form 25 antisense base-pairs with its *gag* sequence substrate (FIG. 2). Computer-assisted RNA fold-modeling programs predicted that the structure of the 4.5S RNA would be maintained except for the addition of the ribozyme itself. The ribozyme was stable enough to accumulate at adequate concentrations to be visualized and harvested from an ethidium bromide-stained electrophoretic gel of total bacterial RNA. Specific HIV *gag* cleavage was achieved using this embedded ribozyme (rz.GH4.5S).

To test the cleavage rate of the 4.5S ribozyme, we compared its initial velocity of cleavage with two other ribozymes of the same sequence, rz.GH and rz.GHpBK. rz.GH was the same 47-nucleotide ribozyme as that cloned into the 4.5S RNA, and rz.GHpBK was the same core sequence cloned into a plasmid and created by *in vitro* transcription with 50 vector-derived, nonstructural bases preceding the ribozyme at the 5′ end and 25 nucleotides at the 3′ end. Kinetic analyses of the three ribozymes show that the nonstructural flanking bases of rz.GHpBK have little effect on substrate binding (K_m) but slow catalysis (K_{cat}) by half. The structured sequences surrounding the rz.GH4.5S ribozyme, however, lead to more significant impairment. rz.GH4.5S has twice the K_m of the other ribozymes, signifying poor binding to the substrate, and a drop to one seventh the (rz.GH) rate of catalysis of the more simple structures. These data signify that flanking sequences appended to ribozymes, either to impart stability or as a by-product of synthesis, do have an effect on the activity of the ribozyme they surround. Although the stability of an embedded ribozyme may compensate for the lower cleavage rate *in vivo,* it is at least useful to assess the effects of flanking regions on catalysis rates before *in vivo* cleavage is attempted.

Role of Secondary Structure of Target on Enzyme Function of Ribozyme

To shed light on the optimal reaction sites for hammerhead ribozymes, we selected three sites within the target RNA molecule, a 291-nucleotide portion of

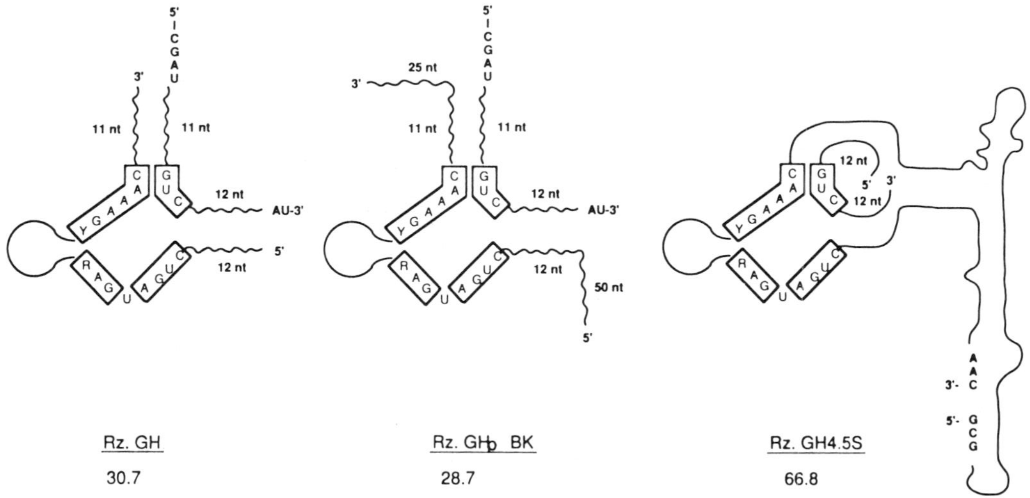

Ribozyme:	Rz. GH	Rz. GHpBK	Rz. GH4.5S
Km (nM):	30.7	28.7	66.8
Vmax (pmol min^{-1} ml^{-1}):	2.54	1.15	0.355
kcat (min^{-1}):	0.254	0.115	0.0355
$\frac{kcal}{Km} \times 10^{-3}$:	8.3	4.0	0.53

FIGURE 2. Effect of ribozyme RNA structure on function. Three hammerhead ribozymes, targeted to the same substrate, were constructed with short flanking sequences (Rz.GH), with intermediate length flanking sequences (Rz.GHpBK), and with a complex flanking sequence of *E. coli* 4.5S RNA (Rz.GH4.5S). Kinetic parameters for each ribozyme depicted are derived from reference 22. (Reproduced from Sarver,[6] with permission.)

HIV-1 *gag* RNA, transcribed *in vitro,* and designed T7 templates for transcription of unique hammerhead ribozymes directed against them (FIG. 3). A computer-generated prediction of the secondary structure of the target RNA displays these regions, A, B, and C, which were targeted by the ribozymes, *rbz*A, *rbz*B, and *rbz*C, respectively. Assays performed with these three ribozymes revealed that they are more active in cleaving the target RNA at temperatures below 37°C; hence, we

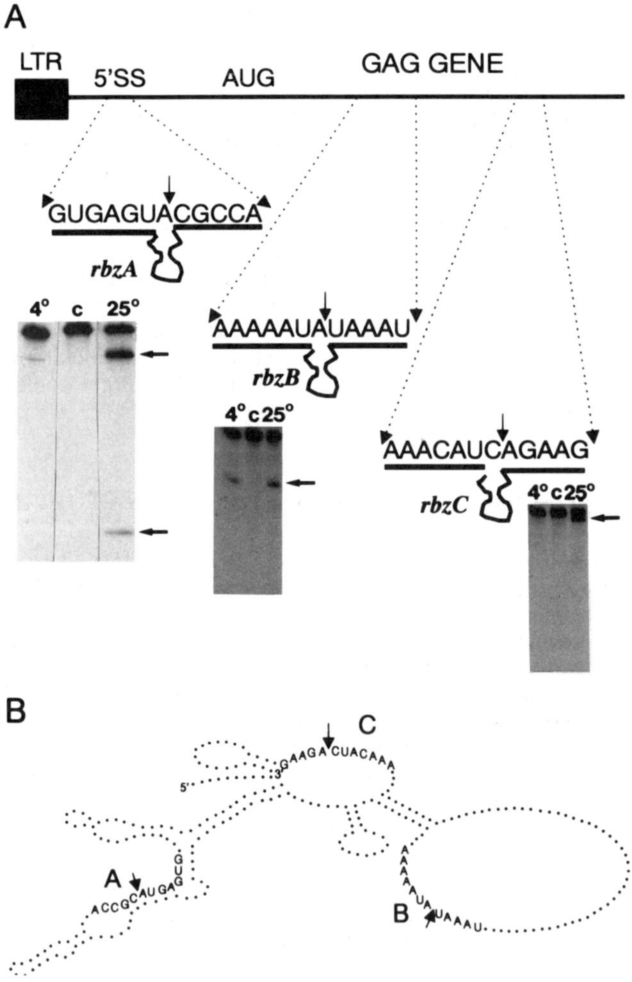

FIGURE 3. Effect of substrate structure on ribozyme function. (**A**) Three hammerhead ribozymes were synthesized and targeted to the 5′ splice site of HIV-1 RNA (*rbz*A) or to sites (*rbz*B and *rbz*C) downstream from the AUG site in the *gag* gene. Electrophoretic patterns are shown for each cleavage reaction performed at 4°C, at 25°C, and without cleavage (c). The *solid arrows* indicate either cleavage site or cleavage product. (**B**) RNA modeling structure of targeted *gag* RNA in which regions A, B, and C correspond to the cleavage site (*arrow*) for *rbz*A, and *rbz*B, and *rbz*C, respectively.

compared their activity at temperatures between 4 and 25°C. Assays using these very small ribozymes did not show cleavage at true "catalytic" ratios of ribozyme to target, and reactions had to be carried out at ribozyme:substrate ratios of 10:1 or higher. We attribute this limitation, at least in part, to the shortness of the flanking regions that mediate base-pairing of the ribozyme to the target. FIGURE 3A illustrates additional characteristics of the individual ribozymes: *rbz*A targets a GUA cleavage site and exhibits optimal cleavage activity at 25°C, but it will also cleave at 15°C (not shown) and displays much reduced activity at 4°C. *Rbz*B cleaves an AUA site and, interestingly, shows activity at 4°C. *Rbz*C is directed against an AUC target and is most active at 25°C, slightly active at 15°C (not shown), and inactive at 4°C. Thus, the cleavage activity of these ribozymes varied significantly with temperature, and the temperature optima for individual species were different. For example, moderate activity at 4°C was seen with *rbz*B, but no activity at 4°C occurred with *rbz*C. Factors giving rise to these differences probably include the secondary and tertiary structure of the target RNA region and the nucleotide composition of the base-pairing flanks. For example, the cleavage site targeted by *rbz*B lies within a large predicted loop (FIG. 3B), which would render it quite open to binding with and cleavage by a ribozyme molecule. The activity of *rbz*B toward this site indicates that this loop may be an attractive candidate region for eventual ribozyme intervention *in vivo.* It is noteworthy that all 11 flanking base pairs of *rbz*B are A-Us, which consist of only two hydrogen bonds and therefore contribute less stability than do G-C base pairs. This composition may account for the greater activity of the ribozyme at lower temperatures, as the base-pairing interactions between *rbz*B and its target would be stabilized at lower temperatures.

ANTISENSE AND RIBOZYME INHIBITION OF HIV-1

Antisense DNA and RNA have been shown to inhibit HIV-1 as either synthetic DNA[23–38] or expressed RNA.[39–47] The potential of antisense molecules as an inhibitor of HIV-1 remains open to study because of the variability in results. Like ribozymes, it is likely that appropriate target selection, efficiency of expression, and effective delivery are required for antisense activity. Ribozymes represent a "second generation" of antisense, and expression of ribozymes that cleave HIV-1 RNA *in vitro* and *in vivo* has been described.[48–52] It remains to be determined if antisense or ribozyme can be developed for consistent HIV-1 inhibition. Here, the development of improved expression vectors will undoubtedly be essential, and it is likely that the efficient inhibition of HIV-1 with antisense-based elements will undergo significant application when this problem is surmounted.

Adeno-associated Virus (AAV) Antisense Vector-Mediated Induction of Resistance to Immunodeficiency Viruses

The capacity to control HIV-1 infection by the constitutive intracellular expression of molecules designed to inhibit HIV-1 replication represents a significant potential for AIDS therapy, and this has been termed intracellular immunization.[53] In this regard, efficient systems are currently being developed for inserting into the host cell genome the DNA sequences that could potentially confer resistance to HIV replication. Certain of us (S. C. and K. K. W.) have modified a vector system that is based on the nonpathogenic, replication-defective parvovirus, adeno-associated virus (AAV).[54,55] Upon infection of permissive cells, AAV directs the stable integration of foreign DNA encoding an HIV-1–specific antisense RNA. Specifically, the target for the antisense RNA transcript is the TAR sequence, a *cis*-acting RNA stem-loop structure in the HIV-1 long terminal repeat (LTR). The interaction of this RNA sequence with the HIV-1 *tat* protein is essential for virus transcription and replication.[56] AAV vectors have high transduction frequencies,[57] and in the absence of a helper virus, they integrate via the inverted terminal repeat sequences in a site-specific manner into the telomeric region of human chromosome 19.[58] AAV vectors appear to be suitable vehicles for the introduction of foreign DNA sequences into host cells.

The vector used in this study was constructed by removing all endogenous promoters and protein-coding sequences from an infectious molecular clone of AAV. These sequences were replaced by the Rous sarcoma virus (RSV) LTR, a multiple cloning site, and a cassette containing the neomycin resistance gene under the control of the SV40 early promoter. Complementary synthetic oligodeoxynucleotides spanning bases +13 to +75 (+1 being the RNA cap site) of the HIV-1 LTR were cloned into the polylinker immediately downstream of the RSV LTR. The predicted RNA transcript would be complementary to the 5′-untranslated leader sequence including TAR, common to all HIV transcripts.

A human embryonic kidney cell line (293 cells) transformed by the adenovirus type 5 E1A and E1B genes[59] was transfected with the vector construct just described, and G418-resistant colonies were harvested and propagated. 293 cells do not possess surface CD4 molecules and therefore are not infectable by HIV-1 or simian immunodeficiency virus. However, high titers of retrovirus can be generated following transfection of an infectious proviral molecular clone. Thus, in this system, viral inhibitory function can be analyzed during a single round of virus replication. Antisense RNA expression in these clones was demonstrated by Northern blot analysis of total cellular RNA. Several clones were assayed for the production of virus as recorded by the presence of reverse transcriptase activity in culture supernatants. Transfection efficiencies of the cell lines were controlled by cotransfection of an indicator plasmid (pE2-CAT) encoding the chloramphenicol acetyltransferase gene.

Because the targeted region, TAR, is highly conserved between HIV-1 and SIV, antisense-expressing clonal cell lines were tested for their ability to restrict both HIV-1 and SIV replication. As shown in FIGURE 4, both HIV-1 and SIV production was clearly inhibited in the antisense clones, SIV to a slightly lesser degree than HIV-1. There was 85% inhibition of SIV replication in clone IIC4 versus 92% inhibition of HIV on day 3 posttransfection. The significant inhibition of SIV production was likely due to the sequence similarity in SIV TAR and HIV-1 TAR.

Potential Gene Therapy for AIDS Using AAV

The *in vivo* efficacy of conferring intracellular resistance to HIV-1 remains unproven. For the treatment of human AIDS, it may be necessary to confer intracellular resistance to progenitor bone marrow cells that give rise to renewable target cells such as T lymphocytes, monocytes, and macrophages for HIV-1 infection.[60] Clearly, retroviral vectors are already established in human gene therapy,[61] and *in vitro* data suggest that inhibition of HIV-1 can result from several strategies of intracellular immunization, including antisense, transdominant negative mutant proteins such as rev[62] and *tat*,[63] and decoy systems based on protein-RNA recognition sequences such as TAR[64,65] and kappa B.[66] It is likely that additional transdominant mutant and decoy strategies will be added as more is understood about the regulation of HIV-1 infection. Eventually perhaps, these will be used singly or in combination with antisense-based strategies in clinical studies.

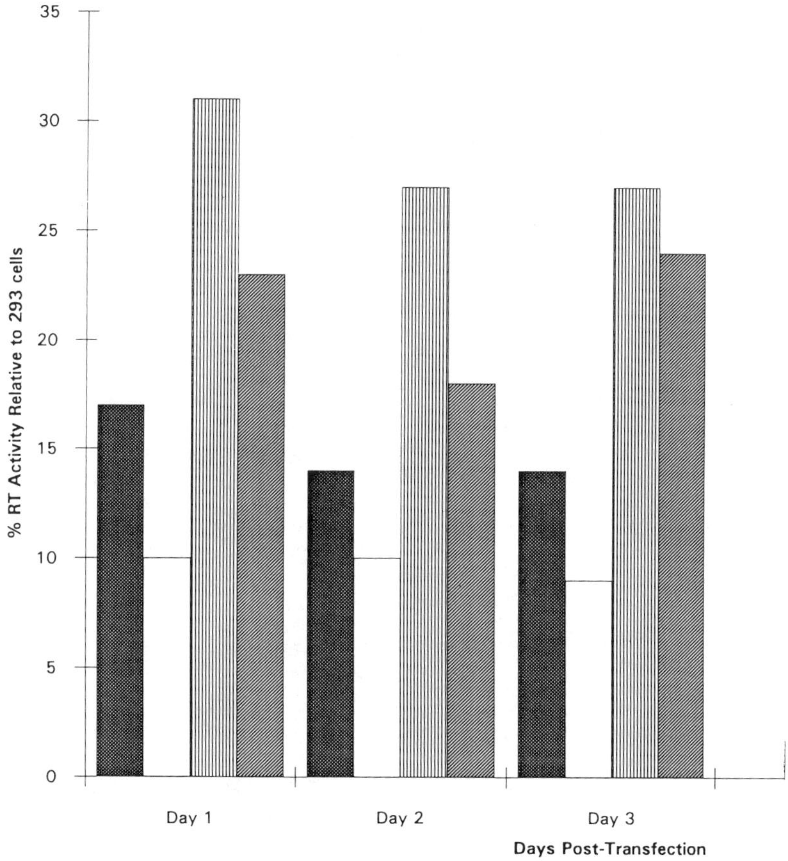

FIGURE 4. A comparison of HIV-1 and SIV replication in antisense-expressing clones. Reverse transcriptase activity from the culture supernatants of antisense clone IIC4 transfected with 100 ng proviral pSIV$_{sm}$H-4 (SIV$_{sm}$) DNA (▩) or 100 ng pHXB-2 (HIV-1) DNA (□) or supernatants from clone alpha 1 transfected in the same ways with proviral SIV$_{sm}$ DNA (▥) or with proviral HIV-1 (▨). The reverse transcriptase (RT) activity is expressed as a percentage of the same in 293 cells following identical transfection with proviral SIV$_{sm}$ or HIV-1 DNA into 1.5×10^6 cells with 3 μg E2-CAT as carrier.

AAV was chosen as a vector for the delivery of the antisense DNA for several reasons: (1) AAV is a nonpathogenic virus that can transduce cells of many types, including cells of human, simian, and rodent origin[67]; (2) multiple copies of the vector integrate into cellular DNA in a tandem fashion following intracellular amplification[68]; (3) AAV does not prevent superinfection,[39] thus allowing for infection with several different vectors or multiple rounds of infection with the same vector; (4) AAV integrates in a site-specific manner into the q arm of human chromosome 19,[58] thereby reducing the possibility of insertional mutagenesis associated with many other forms of gene therapy; and (5) as AAV and HIV-1 are genetically distinct, the likelihood of recombination resulting in new retroviral pseudotypes, as recently described for HIV-1 and other retroviruses,[69] seems remote. Therefore, AAV appears to be a candidate vector for use in antiviral gene therapy, and studies of safety and efficacy will be needed to validate this.

ACKNOWLEDGMENT

We are grateful for the collaborative work of Dr. Nava Sarver which generated many of the results summarized here and to Joan Cass for technical assistance.

REFERENCES

1. Kruger, K., P. J. Grabowski, A. J. Zaug, J. Sands, D. E. Gottschling & T. R. Cech. 1982. Cell **31:** 147.
2. Zamecnik, P. C. & M. L. Stephenson. 1978. Proc. Natl. Acad. Sci. USA **75:** 280–294.
3. Blake, K. R., A. Murakami & P. S. Miller. 1985. Biochemistry **24:** 6132–6138.
4. Walder, R. Y. & J. A. Walder. 1988. Proc. Natl. Acad. Sci. USA **85:** 5011–5015.
5. Scanlon, K. K., L. Giao, T. Funato, W. P. Wang, T. Ton, J. J. Rossi & M. Kashani-Sabat. 1991. Proc. Natl. Acad. Sci. USA **88:** 10591.
6. Sarver, N., J. A. Zaia & J. J. Rossi. 1992. *In* AIDS Research Reviews, 2:259–285. W. C. Koff, F. Wong-Staal & R. C. Kennedy, eds. Marcel Dekker, Inc. New York.
7. Zaug, A. J. & T. R. Cech. 1986. Science **231:** 470.
8. Guerrier-Takada, C., K. Gardiner, T. Marsh, N. Pace & S. Altman. 1983. Cell **35:** 849.
9. Forster, A. C. & R. H. Symons. 1987. Cell **49:** 211.
10. Keese, P. & R. H. Symons. 1987. *In* Viroids and Viroid-like Pathogens. J. S. Semancik, ed. 1–47. CRC Press. Boca Raton, FL.
11. Hampel, A. & R. Tritz. 1989. Biochemistry **28:** 4929.
12. Hample, A., R. Tritz, M. Hicks & P. Cruz. 1990. Nucleic Acids Res. **18:** 299.
13. Sharmeen, L., M. Kuo, G. Dinter-Gottlieb & J. Taylor. 1988. J. Virol. **62:** 2674.
14. Kuo, M., L. Sharmeen, G. Dinter-Gotlieb & J. Taylor. 1988. J. Virol. **62:** 4439.
15. Wu, H., Y. Lin, F. Lin, S. Makino, M. Chang & M. Lai. 1989. Proc. Natl. Acad. Sci. USA **86:** 1831.
16. Saville, B. J. & R. A. Collins. 1990. Cell **61:** 685.
17. Uhlenbeck, O. C. 1987. Nature **328:** 596.
18. Koizumi, M., S. Iwai & E. Ohtsuka. 1988. FEBS Lett. **239:** 285.
19. Jeffries, A. C. & R. H. Symons. 1989. Nucleic Acid Res. **17:** 1371.
20. Heus, H. A. & A. Pardi. 1991. J. Mol. Biol. **217:** 113.
21. Szczylik, C., T. Skorski, N. C. Nicolaides, *et al.* 1991. Science **253:** 562–565.
22. Taylor, N. R. & J. J. Rossi. 1991. Antisense Res. Dev. **1:** 173.
23. Zamecnik, P. C., J. Goodchild, Y. Taguchi & P. S. Sarin. 1986. Proc. Natl. Acad. Sci. USA **83:** 4143–4146.
24. Zaia, J. A., J. J. Rossi, G. J. Murakawa, P. A. Spallone, D. A. Stephens, B. Kaplan, R. Eritja, R. B. Wallace & E. M. Cantin. 1988. J. Virol. **62:** 3914.
25. Agrawal, S., T. Ikeuchi, D. Sun, P. S. Sarin, A. Konopka, J. Maizel & P. C. Zamecnik. 1989. Proc. Natl. Acad. Sci. USA **86:** 7790–7794.
26. Buck, H. M., L. H. Koole, M. H. van Genderen, L. Smit, J. L. Geelen, S. Jurriaans & J. Goudsmit. 1990. Science **248:** 208–212.

27. Ebright, Y., K. Raska, Jr., S. Gaur, L. Frenkel, J. Tsao & S. Stein. 1990. N. J. Med. **87:** 1011–1015.
28. Germany, J. & D. J. Ecker. 1991. Nucleic Acids Res. **19:** 3359–3368.
29. Goodchild, J. & V. Kohli. 1991. Arch. Biochem. Biophys. **284:** 386–391.
30. Goudsmit, J., J. Geelen, W. Keulen, D. Notermans, C. Kuiken, C. Ramautarsing, L. Smit, L. Koole, M. van Genderen & H. Buck. 1990. AIDS **4:** 559–564.
31. Kinchington, D., S. Galpin, J. W. Jaroszewski, K. Ghosh, C. Subasinghe, & J. S. Cohen. 1992. Antiviral Res. **17:** 53–62.
32. Laurence, J., S. K. Sikder, J. Kulkosky, P. Miller & P. O. Ts'o. 1991. J. Virol. **65:** 213–219.
33. Letsinger, R. L., G. R. Zhang, D. K. Sun, T. Ikeuchi & P. S. Sarin. 1989. Proc. Natl. Acad. Sci. USA **86:** 6553–6556.
34. Matsukura, M., G. Zon, K. Shinozuka, M. Robert-Guroff, T. Shimada, C. A. Stein, H. Mitsuya, F. Wong-Staal, J. S. Cohen & S. Broder. 1989. Proc. Natl. Acad. Sci. USA **86:** 4244–4248.
35. Mori, K., C. Boiziau, C. Cazenave, M. Matsukura, C. Subasinghe, J. S. Cohen, S. Broder, J. J. Toulm:e & C. A. Stein. 1989. Nucleic Acids Res. **17:** 8207–8209.
36. Renneisen, K., L. Leserman, E. Matthes, H. C. Schroder & W. E. Muller. 1990. J. Biol. Chem. **265:** 16337–16342.
37. Stein, C. A., M. Matsukura, C. Subasinghe, S. Broder & J. S. Cohen. 1989. AIDS Res. Hum. Retroviruses **5:** 639–646.
38. Stevenson, M. & P. L. Iversen. 1989. J. Gen. Virol. **70:** 2673–2682.
39. Gyotoku, J., M. A. el-Farrash, S. Fujimoto, W. T. Germeraad, Y. Watanabe, K. Teshigawara, S. Harada & Y. Katsura. 1991. Virus Genes **5:** 189–202.
40. Hasler, J. M., T. F. Weighous, T. W. Pitts, D. B. Evans, S. K. Sharma & W. G. Tarpley. 1989. AIDS Res. Hum. Retroviruses **5:** 507–516.
41. Joshi, S., A. Van Brunschot, S. Asad, I. van der Elst, S. E. Read & A. Bernstein. 1991. J. Virol. **65:** 5524–5530.
42. Rhodes, A. & W. James. 1991. AIDS **5:** 145–151.
43. Rhodes, A. & W. James. 1990. J. Gen. Virol. **71:** 1965–1974.
44. Sczakiel, G. 1991. Nucleic Acids Res. **19:** 1421–1426.
45. Sczakiel, G., K. Rittner, M. Homann & M. Pawlita. 1991. Biochem. Soc. Trans. **19:** 437S.
46. Sczakiel, G. & M. Pawlita. 1991. J. Virol. **65:** 468–472.
47. Sczakiel, G., M. Pawlita & A. Kleinheinz. 1990. Biochem. Biophys. Res. Commun. **169:** 643–651.
48. Chang, P. S., E. M. Cantin, J. A. Zaia, *et al.* 1990. Clin. Biotech. **2:** 1–9.
49. Sarver, N., E. M. Cantin, P. S. Chang, *et al.* 1990. Science **247:** 1222–1225.
50. Dropulic, B., N. H. Lin, M. A. Martin & K. T. Jeang. 1992. J. Virol. **66:** 1432–1441.
51. Weerasinghe, M., S. E. Liem, S. Asad, S. E. Read & S. Joshi. 1991. J. Virol. **65:** 5531–5534.
52. Sioud, M. & K. Drlica. 1991. Proc. Natl. Acad. Sci. USA **88:** 7303–7307.
53. Baltimore, D. 1988. Nature **335:** 395.
54. Berns, K. I. & R. A. Bohenzky. 1987. **32:** 343.
55. Samulski, R. J., M. Berns, M. Tan & N. Myzyczka. 1982. Proc. Natl. Acad. Sci. USA **79:** 2077.
56. Cullen, B. R. & W. C. Gren. 1989. Cell **58:** 423.
57. Hermonat, P. & N. Muzycka. 1984. Proc. Natl. Acad. Sci. USA **81:** 6466.
58. Kotin, R., M. Siniscalo, J. Samulski, X. Shu, L. Hunter, C. Laughlin, S. McLaughlin, N. Muzycka, M. Rocchi & K. I. Berns. 1990. Proc. Natl. Acad. Sci. USA **87:** 2211.
59. Graham, F. L., J. Smiley, W. C. Russell & R. Narin. 1977. Virology **52:** 456.
60. Laface, D., P. Hermonat, E. Wakeland & A. Peck. 1988. Virology **162:** 483.
61. Miller, A. D. 1992. Nature **357:** 455–460.
62. Malim, M. H., S. Bohnlein, J. Hauber & B. R. Cullen. 1989. Cell **58:** 205–214.
63. Pearson, L., J. Garcia, F. Wu, N. Modesti, J. Nelson & R. Gaynor. 1990. Proc. Natl. Acad. Sci. USA **87:** 5079–5083.
64. Graham, G. J. & J. J. Maio. 1990. Proc. Natl. Acad. Sci. USA **87:** 5817–5821.

65. Sullenger, B. A., H. F. Gallardo, G. E. Ungers & E. Gilboa. 1990. Cell **63:** 601–608.
66. Logeat, F., N. Israel, R. Ten, V. Blank, O. Le Bail, P. Kourilsky & A. Israel. 1991. EMBO J. **10:** 1827–1832.
67. McLaughlin, S. K., P. Collis, P. L. Hermonat & N. Myzycka. 1988. J. Virol. **62:** 1963.
68. Lusso, P., F. M. Veronese, B. Ensolli, G. Franchini, C. Jemma, S. DeRocco, V. S. Kalyanraman & R. C. Gallo. 1990. Science **247:** 848.
69. Srivastava, A., E. W. Lusby & K. I. Berns. 1983. J. Virol. **45:** 555.

Epstein-Barr Viral Latency and Cell Immortalization as Targets for Antisense Oligomers[a]

JOSEPH S. PAGANO,[b,c,d] GUSTAVO JIMENEZ,[d]
NANCY S. SUNG,[c,d] NANCY RAAB-TRAUB,[c,d]
AND JUNG-CHUNG LIN[d,e]

Departments of Medicine,[b]
Microbiology and Immunology,[c]
and UNC Lineberger Comprehensive Cancer Center[d]
University of North Carolina at Chapel Hill
Chapel Hill, North Carolina 27599–7295

Replication of the Epstein-Barr virus (EBV), one of the seven human herpesviruses, can be effectively inhibited by available antiviral drugs, but latent EBV infection is unaffected by any of these conventional antiviral agents regardless of the mode of action.[1] The difficulty in treating latent forms of viral infection is a general one that includes not only the herpesviruses but also all other viruses that produce latent infection. This work is directed toward establishing a conceptual approach for treatment of latent EBV infection and its consequences.[2]

EBV infection is the causative agent of infectious mononucleosis and hairy leukoplakia of the tongue. It also causes polyclonal B-cell lymphoproliferative conditions in immunosuppressed patients who are recipients of organ transplants or have AIDS. EBV infection is universally linked to carcinoma of the posterior nasopharynx and is also associated with parotid carcinoma.[3] The virus predisposes to the development of Burkitt's lymphoma in endemic regions. In each one of these conditions except perhaps hairy leukoplakia of the tongue, latent as well as, at times, productive infection is present.[4] Latent infection is particularly important in the lymphoproliferative conditions that are stimulated by the nonproductive EBV infectious state.[5,6]

The molecular and cellular pathogenesis of EBV infection helps to explain these diverse virus-disease associations. EBV initially infects epithelial cells in the oropharynx and ductal cells of the parotid gland where the virus causes productive infection and cytolysis. The virus also infects B lymphocytes in the tonsil, and a small number of EBV-infected B lymphocytes circulate in the peripheral blood during primary infection. In B lymphocytes the infection is latent in the normal nonimmunosuppressed host, and these lymphocytes are stimulated to proliferate rather than lyse. After the acute phase of lymphoproliferation has subsided, a very small subset of B cells that carries the EBV genome persists in the circulation for life and can be recovered in culture. Infected B cells are capable of dividing normally and of

[a]This work was supported by a grant from the National Institute of Allergy and Infectious Diseases (RO1 AI17205-13). Nancy Sung is supported by a USPHS training grant (5T32 AIO7001). Gustavo Jiménez is a recipient of a fellowship of Ministerio de Educación y Ciencia de España.

[e]Current address: Division of Immunologic, Oncologic and Hematologic Diseases, Center for Disease Control, Atlanta, Georgia 30333.

maintaining the latent form of the EBV genome without virus production in hosts with intact immune mechanisms. If the host becomes immunosuppressed, EBV-infected lymphocytes sometimes resume unchecked proliferation, and some cells may also begin to produce virus.[5,6]

EBV has two well-established modes of replication, one using virus-encoded DNA polymerase to produce the linear double-stranded ~180 kbp DNA genome that is encapsidated in virions and the other using host DNA polymerase to replicate the circular form of the same genome, the EBV episome, which is found only intracellularly within the chromatin. The episome is the molecular basis for latent EBV infection. Integrated forms of the EBV genome have also been detected in cell lines but not in EBV-infected tumors or other tissues that contain episomes.[7,8]

The episome, which is present in low, constant copy number in each cell, replicates once every cell cycle. Its replication is the primary target for inhibition by antisense oligonucleotides. The strategy is to interfere with the synthesis of a *trans*-acting viral protein synthesized in latently infected cells and needed to initiate and maintain replication of the episome. Although nine EBV genes are expressed in latently infected cells and in EBV-immortalized B lymphocytes—EBNA1, EBNA2, 2a, and 2b, EBNA3, LP, LMP1, and LMP2a and b plus 2 species of EBERS[9]—the strategy focuses on blocking the synthesis of a single protein, the EBNA1 protein, to block replication of EBV episomes and, secondarily, synthesis of other EBV latent products.[2]

The EBV genome has two distinct origins of replication, one, *ori-lyt,* used by the viral DNA polymerase in productive infection, and the other, *ori-P,* which is the plasmid or episomal origin of replication.[10,11] The two *cis*-acting elements that comprise the EBV *ori-P* are shown in FIGURE 1. The episome of the simian EBV-like agent, *Herpesvirus papio,* has a similar topology.[12] For both viruses the motif of repeated nucleotides is a transcriptional enhancer, and the downstream element containing 4–5 repeat units in a dyad array is the site at which episomal DNA replication is initiated.[13] The EBNA1 protein enables episomal replication by binding to the repeat units.[14,15] The EBNA1 coding region is located in the *Bam*HI fragment K of the EBV genome as part of a gene whose upstream leader sequence is extensively spliced.[9]

In this work, both natural and phosphorothioated oligodeoxynucleotides antisense to several sites within the EBNA1 ORF were tested. All results presented were obtained with two sequential 18 mers commencing immediately after the AUG of the EBNA1 ORF; the sequence of these oligomers is given in FIGURE 2. The oligomers, both modified and unmodified, were synthesized by the UNC-LCCC

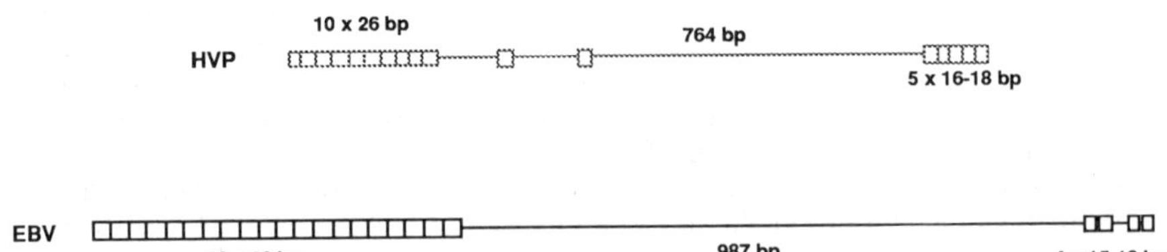

FIGURE 1. Comparison of the Epstein-Barr virus and Herpesvirus Papio (HVP) origins of plasmid (episomal) replication (*ori-P*). The *ori-P* region of each virus has the general structure of an array of monomer units, followed by an intervening sequence, followed by four or five partial copies of the monomer unit which has the potential to form a large dyad. (Reprinted with permission from Loeb *et al.*[12])

SENSE: 5' --AUG **TCTGACGAGGGGCCAGGT ACAGGACCTGGAAATGGC**--- 3'

ANTISENSE: 3'--**AGACTGCTCCCCGGTCCA TGTCCTGGACCTTTACCG**---5'

FIGURE 2. Sequence of antisense and sense oligodeoxynucleotides used for treatment of latently EBV-infected Raji Burkitt lymphoma cells and EBV-immortalized cord-blood lymphocytes.

Nucleic Acids Core Facility or by ISIS Pharmaceutical Company and subjected to double ethanol purification.

DISRUPTION OF EBV EPISOMAL MAINTENANCE IN LATENTLY INFECTED CELLS

Raji cells are a clonal Burkitt's lymphoma cell line that contains EBV episomes in a constant copy number of 60 per cell.[16,17] In Raji cells EBNA1 mRNA is expressed in low abundance, and a small constant amount of EBNA1 protein is found in the nucleus.[18] The cells were grown as stationary suspension cultures, and for each condition 10^7 cells were exposed to 40 μM of unmodified oligomers in serum-free medium for 1 hour. Cells were then incubated in 20 ml of RPMI medium containing 10% fetal bovine serum. Every 3 days cells were pelleted, and exposure to the oligomers was repeated with 10^7 cells. The remaining cells were harvested and assayed for expression of EBNA1 (FIG. 3) and the presence of the viral episome (FIG. 4). The Western immunoblot (FIG. 3), prepared with human EBV-positive antiserum, disclosed the EBNA1 67-kD protein before treatment which was unaffected in the control cells throughout treatment with the sense oligomer. Cells treated with the antisense oligomer had progressive reduction in EBNA1 which was linear with respect to time, so that by day 23 EBNA1 protein was no longer detected.[19] Thus, EBNA1 antisense oligomer treatment had a striking effect on synthesis of the *ori-P trans*-activator.

Episomes were detected by Southern blot hybridization analysis of the treated cells in comparison with control hybridizations showing 50, 5, and 1 copy of the EBV genome (FIG. 4). In Raji cells treated with the sense oligomer there was no effect on episomal copy number. In cells treated with the antisense oligomer, episomal copy number was significantly reduced at least 90%; this result could be discerned despite the unequal loading of channels in the Southern blot. Thus, concomitant with elimination of synthesis of the EBNA1 protein was a highly significant reduction of EBV episomal copy number by at least 90%, when the experiment was terminated because of contamination.[19] This is the first example of suppression of expression of a latent gene product in EBV-infected cells and, further, represents the first direct proof that EBNA1 is required for maintenance of authentic episomes as against artificial *ori-P*-containing constructs transfected into cells. The result indicates that a single viral gene product is necessary for episomal maintenance.

However, this experimental result has not been reproduced. Likely variables including oligonucleotide uptake and stability, serum DNAse activity, and Raji cell strain variability may contribute to outcome and have been examined. The results of these tests showed that oligodeoxynucleotides are rapidly degraded in fetal as well as newborn calf serum and that this degradation varies widely with different lots of serum. Studies of cellular uptake of oligonucleotides indicated that incubation of end-labeled oligomers with a cationic lipid such as Lipofectin™[20] drastically increased the amount of cell-associated radioactivity and that uptake occurred very

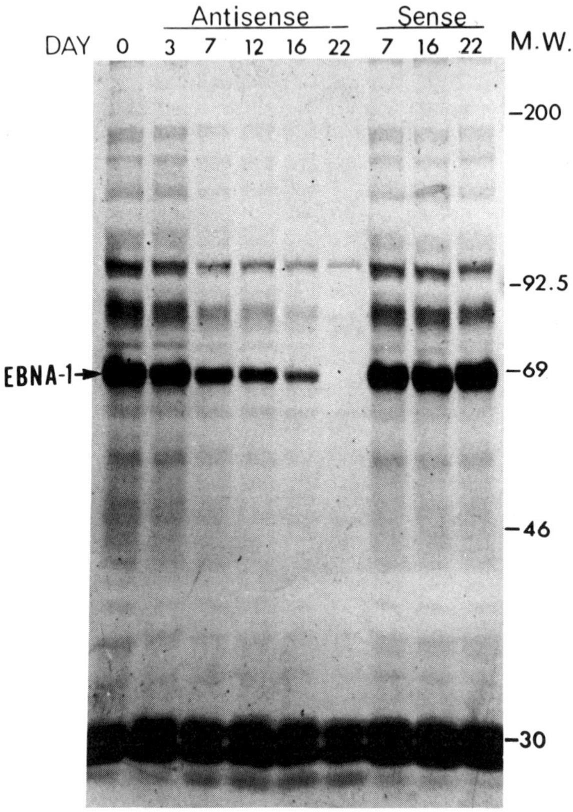

FIGURE 3. Effect of antisense oligodeoxynucleotide treatment on expression of EBNA1 in Raji cells as shown by Western immunoblotting. Cells were treated every 3 days with 40 μM of unmodified sense or antisense oligonucleotides (shown in FIG. 2). At the indicated times, total cellular proteins were prepared for Western blotting and reacted with EBV-positive human antisera.

rapidly (FIG. 5A). Lipofectin also appeared to increase oligomer stability, as shown by a higher percentage of cell-associated radioactivity bound to DE-81 paper (FIG. 5B). Despite the positive effects on oligomer uptake, we were unable to show that the use of Lipofectin produced an increase in inhibition of expression of the target gene.

Phosphorothioate-modified oligonucleotides were used in a similar experimental scheme, again without consistent results. The lack of reproducibility might be due to the fact that many cell divisions are required before all the episomes, which each replicate once every cell cycle, are successfully removed from each cell in the culture. Moreover, the cells that still carry EBV episomes might have a slight selective growth advantage. Finally cells that take up the synthetic oligonucleotides, especially the phosphorothioate compounds, might be subject to a slight toxic effect that retards their growth. Consequently the cells that have taken up a sufficient amount of the oligonucleotide to eliminate episomes might grow slightly less well. In any case none of the same components of the first successful experiment (e.g., cells, batch of oligonucleotide, serum in the culture media, etc.) was available to be used in any of the subsequent experiments.

EBV-IMMORTALIZED CELLS AS A TARGET FOR ANTISENSE TREATMENT

Studies of EBV-immortalized cell lines were initiated, because these are an important biological target provided by EBV infection and because effects on such cells might provide a better, short-term assay. Transforming strains of EBV, both wild-type and laboratory strains, produce immortalization of cord-blood B lymphocytes within 4–6 weeks. Uninfected cells remain static and eventually die; they never spontaneously immortalize.[21]

Although both Raji cells and EBV-immortalized cord-blood lymphocytes contain actively transcribed EBV episomes and the EBV oncoproteins LMP1 and EBNA2, only Raji cells contain activated *c-myc* expression brought about by chromosomal translocation. Freshly immortalized cord-cell lines are often polyclonal but rarely have chromosomal translocations or significant known oncogene activation.[22] Therefore, removal of the episomes contained in EBV-immortalized cell lines might be sufficient to cause cessation of cell growth. The kinetics of cell growth of these lines are such that growth-curve assays can be conducted conveniently over a 5- to 10-day period in 96-well microtiter plates with a percentage reduction in cell numbers relative to untreated cells used as the end-point.

Three cord-blood B-lymphocyte lines that were generated by exposure to EBV-containing throat washings (SFC-1, 3, and 4) and had been in culture for 8–12 weeks

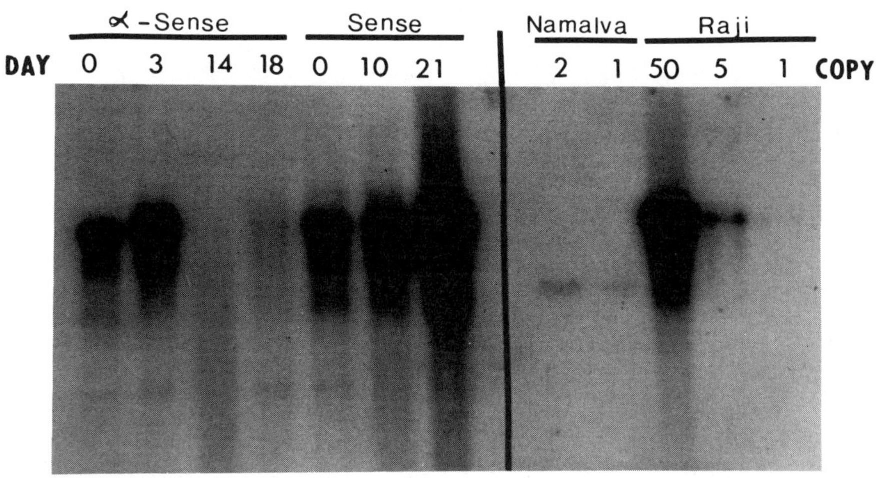

FIGURE 4. Effect of antisense oligodeoxynucleotide treatment on EBV episomal copy number as shown by Southern blot hybridization. Raji cells were treated every 3 days with 40 μM of unmodified sense or antisense oligomers (shown in FIG. 2). At the indicated timepoints, DNA was prepared by guanidinium thiocyanate extraction and purified by centrifugation over a 5.7 M CsCl cushion and digestion with proteinase K. DNA samples were then digested with *Bam*HI, electrophoresed through an agarose gel, blotted onto nitrocellulose membranes, and probed with a 1.9-kb *Xho*I fragment from the EBV genome.[15,19]

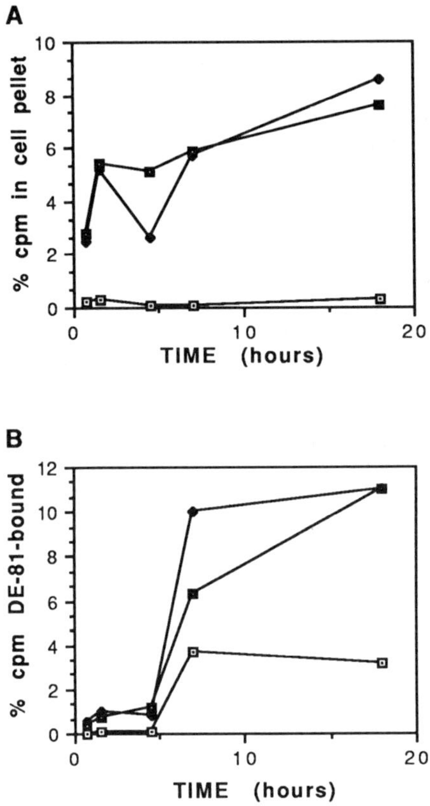

FIGURE 5. Cellular uptake of oligonucleotides in the presence of Lipofectin. End-labeled oligodeoxynucleotides (FIG. 2) were incubated in the presence of 5 μg/ml (■), 10 μg/ml (◆), or no Lipofectin (⊡) with Raji cells in serum-free medium. At the given timepoints, cells were pelleted and washed three times with phosphate-buffered saline solution. **(A)** Cell-associated radioactivity was determined by the percentage of cpm remaining in the cell pellet after three washings. **(B)** Oligomer stability was determined by the percentage of cell-associated cpm that bound to DE-81 paper.

were tested for presence and copy number of EBV episomes by Southern blot hybridization with the use of terminal genomic probe analysis to distinguish episomes from linear genomes.[23] EBV-immortalized cord-blood-cell lines may produce small amounts of virus initially, but usually evolve rapidly into non-virus-producing cell lines. As shown in FIGURE 6, SFC-1 contained two fused genomic terminal fragments, SFC-3 a single fused terminal fragment and SFC-4 two fused terminal fragments; an alternate explanation for the smaller fragment in SFC-4 is that it may represent a single terminus of a linear genome, either integrated or replicating. SFC-3 (approximately 20 genomes per cell compared with the Raji control) and SFC-4 (approximately 15 genomes per cell), both of which grew well, were selected for assay. In a first screening experiment, cells in the amount of 2×10^5 per milliliter (0.2 ml per well) were exposed to a range of both unmodified and phosphorothioated oligonucleotides, targeted at different sites within the EBNA1 ORF. Cell growth was monitored by Coulter counting cells in aliquots removed from the cultures at different times. Of three different targets, only the tandem antisense oligomers shown in FIGURE 2 were found to have an inhibitory effect when used as phosphorothioates at a 10-μM concentration. Hence, we focused on this target in further experiments.

Representative results with the SFC-4 cell line are shown in FIGURE 7. As shown in FIGURE 7A, after an initial lag phase the cells entered logarithmic growth. By day 7

(168 hours) the scrambled sense phosphorothioated (P = S) oligomer produced a 55% reduction in cell numbers, the sense (P = S) oligomer produced 86% reduction, whereas the antisense (P = S) oligomer produced complete cytostasis with essentially 110% inhibition of cell growth, all at 20-μM concentrations. Dose-response curves were constructed as shown in FIGURE 7B: 20 μM antisense oligomer produced 110% inhibition, 10 μM produced 87%, and 5 μM produced 63%. Thus, there was a linear inhibitory response. Testing of the SFC-3 cell line gave essentially similar results to those with the antisense compound.

In initial experiments the sense compounds were not inhibitory for the SFC-4 cell line, but this result changed in subsequent experiments, and the sense compound (20 μM) produced significant inhibition, as shown in FIGURE 7A. In general, concentrations of 10–20 μM of the antisense oligomers were consistently strong inhibitors of cell growth in both cell lines, but in later experiments the sense and scrambled sense oligonucleotides began to have larger inhibitory effects, perhaps because of selection

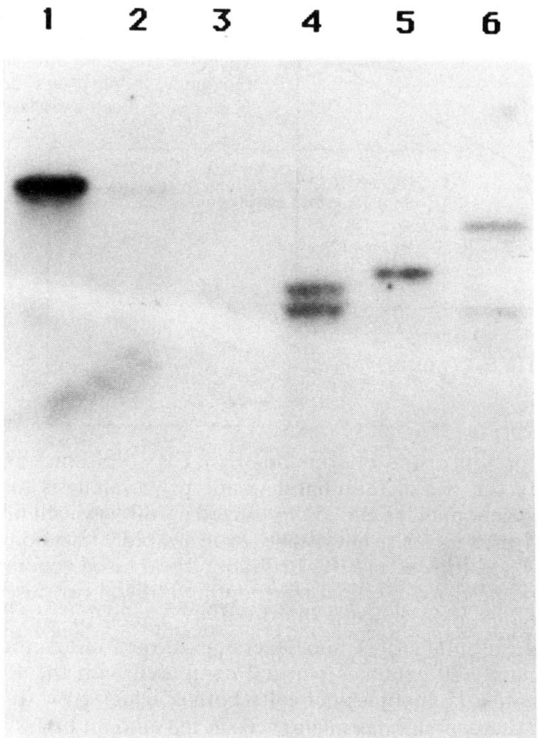

FIGURE 6. Southern blot hybridization of EBV-immortalized B-lymphocyte DNA. The DNA of 50 × 10^6 cells was extracted by the guanidinium thiocyanate method and purified by CsCl-cushion centrifugation and digestion with proteinase K. For hybridization, the DNA was digested with *Bam*HI, electrophoresed, and blotted onto a nitrocellulose membrane, then hybridized with a probe specific for the right terminus of the EBV sequence (*Xho*I fragment a). *Lane 1,* Raji cell line, 10 μg DNA (equivalent to 50 copies of EBV episome); *lane 2,* Raji cell line, 1 μg (5 copies); *lane 3,* Raji cell line, 0.1 μg (0.5 copies); *lane 4,* SFC-1, 10 μg; *lane 5,* SFC-3, 10 μg; *lane 6,* SFC-4, 10 μg.

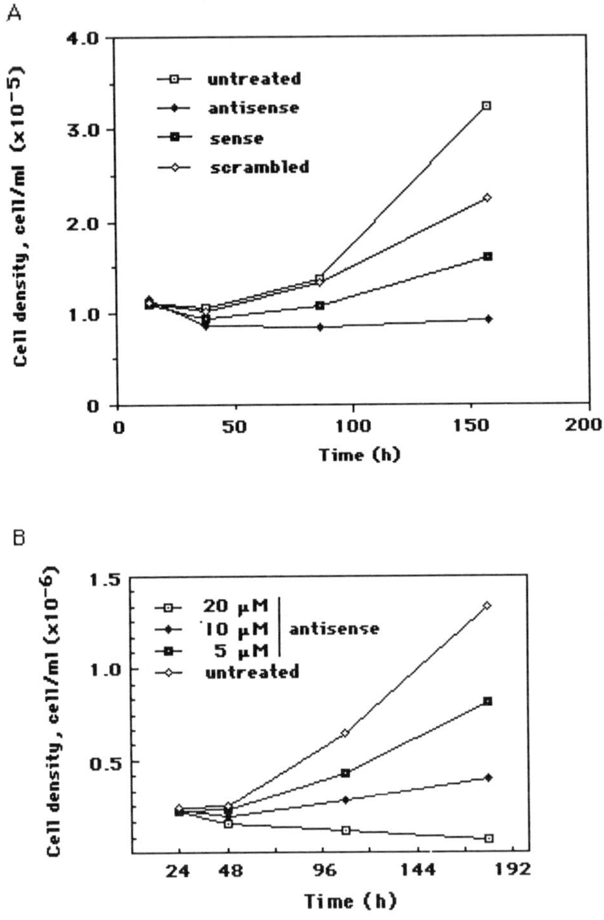

FIGURE 7. **(A)** Growth inhibition of EBV-immortalized cord lymphocytes by EBNA1-specific antisense oligonucleotides. EBV-immortalized human cord-blood lymphocytes (2×10^5) were incubated in 200 μl of RPM1 1640 (supplemented with 15% fetal calf serum, glutamine, penicillin, and streptomycin) in the presence of 20-μM concentrations of antisense, sense, and scrambled sense phosphorothioated oligodeoxynucleotides. Cells were incubated at 37°C over a period of 7 days. Proliferation was monitored by removing 100-μl volumes and counting cells in a Coulter counter, then replenishing with fresh medium with oligomer at the same concentration. **(B)** Dose response of inhibition of EBV-immortalized cord lymphocytes by EBNA1-specific antisense oligonucleotides. Cells (3×10^5) were treated as in **A** in the presence of increasing concentrations of the antisense and scrambled sense (not shown) oligonucleotides, and cell growth was monitored by periodic Coulter counting.

during initial propagation of frozen stock. However, the antisense compounds consistently produced greater inhibitory effects on cell growth than did either sense or scrambled sense compounds. These variable results make it essential to prove whether the inhibitory effects with the antisense compounds are specific or are caused by the nonspecific effects often seen with phosphorothioates. Western blot analysis for EBNA1 protein and Southern blot analysis to determine episomal copy number, which cannot be performed with the small number of cells remaining after treatment in the microtiter plate wells, are in progress in scaled-up experiments that will provide more cells for analysis.

SUMMARY

The approach of using an antisense oligonucleotide to oppose the synthesis of a single selected viral gene product needed to maintain the EBV episome in non-virus-producing cells appears to be promising not only for possible cure of latent viral infection, but also for reversal of EBV-driven cell proliferation. It is also possible that targeting multiple latent viral genes might mount a synergistic effect that would prove to be more efficient at curing latent infection. However, further work is needed to identify the reasons for the variable results that are observed as well as to prove specificity of the inhibitory effects. Finally, we are also working on other assays and methods to screen compounds rapidly.

ACKNOWLEDGMENTS

We thank Dr. Susan Fiscus for establishing the immortalized cord lymphocyte lines and Francois Besencon and Kathy Flynn for technical assistance.

REFERENCES

1. Lin, J. C. & J. S. Pagano. 1985. Cellular transformation by the herpesviruses and antiviral drugs. Pharmac. Ther. **28:** 135–161.
2. Pagano, J. S. 1991. Gertrude and Werner Henle Lecture on Viral Oncology. From Latency to Replication: Recent Studies of the Epstein-Barr Virus. *In* Epstein-Barr Virus and Human Disease 1990. D. V. Ablashi, A. T. Huang, J. S. Pagano, G. R. Pearson & C. S. Yang, eds.: 19–32. The Humana Press. Totowa, NJ.
3. Raab-Traub, N., P. Rajadurai, K. Flynn & A. P. Lanier. 1991. Epstein-Barr virus infection in carcinoma of the salivary gland. J. Virol. **65:** 7032–7036.
4. Gilligan, K., P. Rajadurai, L. Resnick & N. Raab-Traub. 1990. Epstein-Barr virus small nuclear RNAs are not expressed in permissively infected cells in AIDS-associated leukoplakia. Proc. Natl. Acad. Sci. USA **87:** 8790–8794.
5. Pagano, J. S. 1991. Epstein-Barr Virus. Textbook of Internal Medicine. W. N. Kelley, ed. 2nd Ed., chapt. 307: 1499–1503. J. B. Lippincott Co. Philadelphia, PA.
6. Raab-Traub, N. & J. S. Pagano. 1989. The Epstein-Barr Virus and Its Antigens. *In* Human Immunogenetics. S. D. Litwin, ed. Vol. **43:** 477–498. Marcel Dekker, Inc. New York, NY.
7. Hurley, E. A., S. Agger, J. A. McNeil, J. B. Lawrence, A. Calendar, G. Lenoir & D. A. Thorley-Lawson. 1991. When Epstein-Barr virus persistently infects B-cell lines, it frequently integrates. J. Virol. **65:** 1245–1254.
8. Gulley, M., M. Raphael, C. T. Lutz, D. W. Ross & N. Raab-Traub. 1992. Epstein-Barr virus integration in human lymphomas and lymphoid cell lines. Cancer, in press.
9. Raab-Traub, N. & K. Gilligan. 1991. Comparison of EBV transcription in lymphoid and epithelial cells. *In* Herpesvirus Transcription and Its Regulation. E. K. Wagner, ed.: 339–362. CRC Press, Inc. Boca Raton, FL.
10. Reisman, D., J. Yates & B. Sugden. 1985. A putative origin of replication of plasmids derived from Epstein-Barr virus is composed of two *cis*-acting components. Mol. Cell Biol. **5:** 1822–1832.

11. HAMMERSCHMIDT, W. & B. SUGDEN. 1988. Identification and characterization of *ori-lyt,* a lytic origin of DNA replication of Epstein-Barr virus. Cell **55:** 427–433.
12. LOEB, D., N. S. SUNG, R. L. PESANO, C. J. SEXTON, C. A. HUTCHISON, III & J. S. PAGANO. 1990. Plasmid origin of replication of herpesvirus papio: DNA sequence and enhancer function. J. Virol. **64:** 2876–2883.
13. GAHN, T. A. & C. L. SCHILDKRAUT. 1989. The Epstein-Barr virus origin of plasmid replication, *ori-P,* contains both the initiation and termination sites of DNA replication. Cell **58:** 527–535.
14. YATES, J. L., N. WARREN & B. L. SUGDEN. 1985. Stable replication of plasmids derived from Epstein-Barr virus in various mammalian cells. Nature **313:** 812–815.
15. RAWLINS, D. R., G. MILMAN, S. D. HAYWARD & G. S. HAYWARD. 1985. Sequence-specific DNA-binding of the Epstein-Barr virus nuclear antigen (EBNA-1) to clustered sites in the plasmid maintenance region. Cell **42:** 859–868.
16. NONOYAMA, M. & J. S. PAGANO. 1971. Complementary RNA specific to the DNA of the Epstein-Barr virus: Detection of EB viral genome in nonproductive cells. Nature New Biol. **233:** 103–106.
17. NONOYAMA, M. & J. S. PAGANO. 1972. Separation of Epstein-Barr virus DNA from large chromosomal DNA in nonvirus-producing cells. Nature New Biol. **238:** 169–171.
18. STERNÅS, L., T. MIDDLETON & B. SUGDEN. 1990. The average number of molecules of Epstein-Barr nuclear antigen 1 per cell does not correlate with the average number of Epstein-Barr virus (EBV) DNA molecules per cell among different clones of EBV-immortalized cells. J. Virol. **64:** 2407–2410.
19. LIN, J. C., N. RAAB-TRAUB & J. S. PAGANO. 1990. Disruption of Epstein-Barr virus episomal DNA maintenance by a specific oligodeoxyribonucleotide. Antiviral Res. **70:** 76.
20. FELGNER, P. & M. HOLM. 1989. Cationic liposome-mediated transfection. Focus **11:** 21–25.
21. MILLER, G. 1990. Epstein-Barr Virus. *In* Virology, 2nd Ed. B. N. Fields & D. M. Knipe, eds. Vol. **2:** 1921–1958. Raven Press. New York, NY.
22. KLEIN, C. & N. RAAB-TRAUB. 1987. Human neonatal lymphocytes immortalized after microinjection of Epstein-Barr virus DNA. J. Virol. **61:** 1552–1558.
23. RAAB-TRAUB, N. & K. FLYNN. 1986. The structure of the termini of the Epstein-Barr virus as a marker of clonal cellular proliferation. Cell **47:** 883–889.

Protooncogenes in the Regulation of Normal Hematopoiesis

An Antisense Approach[a]

BRUNO CALABRETTA,[b] THOMAS SKORSKI, CEZARY SZCZYLIK, DANIELE CARACCIOLO, MAURO VALTIERI, AND DONATELLA VENTURELLI

Jefferson Cancer Institute
Thomas Jefferson University
Philadelphia, Pennsylvania 19107

Most normal circulating blood cells are highly specialized end cells that are derived from a small number of marrow progenitor cells.[1,2] Progenitor cells, in turn, are derived from common pluripotent hematopoietic stem cells (PHSCs). PHSCs are capable of self-renewal, whereas committed cells have lost this potential but divide extensively during the process of differentiation.[1,2] The cloning of genes for several hematopoietic growth factors and the resulting production of the encoded recombinant proteins have led to major conceptual advances in understanding the regulation of normal hematopoiesis; however, intermediary metabolic events occurring within cells as steps distal to the interaction between growth factors and their own receptors are poorly understood, as are the nuclear processes that specifically regulate proliferation and differentiation of progenitor cells. In theory, activation of protooncogenes could provide a functional link between events occurring at the cellular membrane and the ensuing proliferative and maturative responses that occur in the pool of hematopoietic progenitors. The study of the regulation of hematopoietic proliferation and differentiation has primarily been hampered by difficulties in purifying a sufficient number of pure progenitor cells. In the last several years, it was shown that specific gene functions can be inhibited by exposing target cells to synthetic oligomers complementary to their mRNAs.[3–6] This approach eliminates the need to purify a large number of progenitor cells and provides a powerful tool for the study of the molecular basis of normal human hematopoiesis.

In the following paragraphs we discuss the role of N-ras, c-abl, and c-myb (three protooncogenes encoding proteins with distinct functions) in the regulation of normal hematopoiesis.

N-ras

N-ras together with H-ras and K-ras is a member of the RAS family of cellular protooncogenes. These genes encode for the p21 ras protein, a member of a family of

[a]This work was supported in part by National Institutes of Health (NIH) grant CA–46782 and grants CH-492 and CH-455A from the American Cancer Society. Dr. Calabretta is a scholar of the Leukemia Society of America.

[b]Address for correspondence: Bruno Calabretta, MD, Room 630, Bluemle Life Sciences Building, Jefferson Cancer Institute, Thomas Jefferson University, 10th and Locust Streets, Philadelphia, PA 19107.

signal-transducing GTP-binding proteins that appear to be involved in transduction of growth signals inside the cell via increased amounts of the active GTP-bound protein.[7] Increased levels of activated p21 ras were observed after treatment of several cell lines with IL-2, IL-3, and GM-CSF, suggesting that p21 ras activation is an essential feature in the response of different cell types to growth factor stimulation.[8] To determine if N-ras is directly involved in the regulation of normal hematopoiesis stimulated by selected growth factors, we studied the effects of inhibiting N-ras expression in normal progenitor cells with a specific oligomer complementary to 18 nucleotides of N-ras exon 1 beginning with the initiation codon.[9] Normal marrow mononuclear cells, T-lymphocyte and monocyte-macrophage depleted, were exposed to N-ras oligodeoxynucleotides and plated in culture in the presence of single hematopoietic growth factors. Treatment with the N-ras antisense oligomer resulted in 48.7, 36.0, and 67.5% inhibition of colony formation in the presence of IL-3, GM-CSF, or M-CSF, respectively, but it did not affect colony formation in the presence of G-CSF or IL-5 compared with sense-treated groups. To analyze the effect of N-ras antisense oligomer on the growth of a more homogeneous, less mature population of hematopoietic progenitors, we selected from marrow mononuclear cells the fraction expressing the CD34 antigen. This subpopulation represents as little as 0.5–2.0% of marrow mononuclear cells.[10] Treatment with the antisense oligomer inhibited colony formation in the presence of IL-3 (73.6%), GM-CSF (53.3%), and M-CSF (61.5%). Colony formation stimulated by G-CSF was not affected.

The influence of N-ras antisense oligomers on erythroid colony formation stimulated by erythropoietin was also evaluated. Neither sense nor antisense oligomers affected the number or size of erythroid colonies growing from marrow mononuclear cells depleted of T lymphocytes and monocyte-macrophages.

c-abl

The c-abl protooncogene encodes a protein with tyrosine kinase activity localized to the inner cellular membrane.[11] Although c-abl encoded protein is not a growth factor receptor, when constitutively expressed, it can abrogate the requirement for growth factors, suggesting that is coupled to the signal transduction pathway activated by growth factors.[12] However, the role of this protein in cellular functioning is not certain. More than 10% of patients with chronic myelogenous leukemia (CML) have structural alteration of the c-abl gene in DNA of hematopoietic cells.[13] These structural alterations are caused by the translocation of c-abl from chromosome 9 to the breakpoint cluster region (bcr) on chromosome 22, which results in the formation of bcr-abl hybrid genes.[13] Expression of the bcr-abl encoded protein is thought to play an important etiologic role in the pathogenesis of CML, a hypothesis that is strongly supported by recent experiments in which mice infected with a retrovirus encoding the p^{210} bcr-abl gene developed a CML-like disease.[14,15] These observations raise the possibility that c-abl function is required during normal human hematopoiesis. Partially purified and more highly purified bone marrow and peripheral blood progenitor cells were exposed to c-abl sense and antisense oligomers before culturing those cells in semisolid medium. The oligomers employed were 18 bases in length, corresponded to codons 2–7, and were complementary to either the 7.0- or the 6.0-kb c-abl transcripts; in addition, a third c-abl antisense oligomer complementary to the common second exon of c-abl was also examined. The colony-forming ability of treated cells was assayed under lineage-selective growth conditions. Exposure of bone marrow mononuclear cells to c-abl antisense oligomers did not affect erythroid colony formation deriving from BFU-E and CFU-E progenitors, but markedly inhibited (10–20% of residual growth in comparison to controls) myeloid colony formation deriving from CFU-G and CFU-GM progenitors. In addition, the residual myeloid colonies were much smaller than those formed in the presence of c-abl sense oligomers.[10] Similar effects were observed on the growth of bone marrow cells selected on the basis of the expression of the My10 antigen (CD34+ cells). Such cells constitute a more homogeneous and more primitive population of progenitor cells. In these cells, the c-abl antisense oligomers inhibited the formation of granulocyte-macrophage colonies (CFU-GM growth) but did not affect the growth of primitive erythroid colonies (BFU-E growth). Of note, a bcr antisense oligomer did not have any effect on colony number or size.[16] To determine if the lack of inhibitory effects of c-abl antisense oligomers on erythroid colony formation was due to the absence of c-abl expression in those specific colonies, single colonies obtained after stimulation with IL-3 + GM-CSF (GFU-GM colonies) or IL-3 + erythropoietin (BFU-E colony) were analyzed for c-abl expression. Detection of c-abl mRNA by RT-PCR in each colony type (FIG. 1) indicates that the effect of c-abl antisense oligodeoxynucleotides on the different colonies is not associated with a lineage-dependent pattern of c-abl expression.

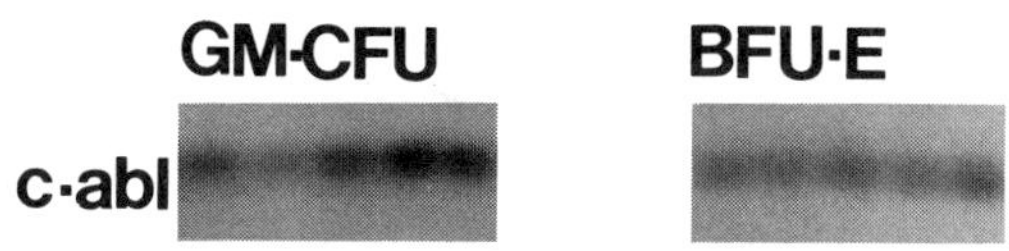

FIGURE 1. Levels of N-ras mRNA in individual myeloid and erythroid colonies. Five individual erythroid (BFU-E) and myeloid (CFU-GM) colonies were isolated after 10 days of culture in methylcellulose in the presence of hematopoietic growth factors (IL-3 and erythropoietin for BFU-E colony growth and IL-3 + GM-CSF for CFU-GM colony growth). Cytoplasmic RNA was then extracted and c-abl transcripts were detected by RT-PCR. (**A**) BFU-E; (**B**) CFU-GM colonies.

c-myb

The protooncogene c-myb is the normal cellular homolog of the avian myeloblastosis virus (AMV) transforming gene, v-myb, that causes myeloblastic leukemia in chickens and transforms myelomonocytic hematopoietic cells in culture.[17] The avian leukemia virus E26 also contains the v-myb oncogene in addition to a second oncogene, ets. The v-myb containing E26 virus, like the AMV, can transform hematopoietic cells with a distinct myeloid phenotype and, in association with the ets oncogene, can induce erythroleukemia in chicken erythroid precursor cells.[17] The myb protein localizes to the cell nucleus,[18] where it binds to DNA, specifically recognizing the nucleotide sequence pyAACG/TG.[19] Concatemers of this consensus sequence confer myb-dependent inducibility on otherwise unresponsive promoters, suggesting that myb protein acts as a sequence-specific DNA binding transcription factor.[20,21]

In the hematopoietic system, the c-myb protooncogene is thought to play a particularly important role in regulating cell proliferation and perhaps some differentiation-specific functions. Evidence to support these hypotheses includes a number of observations: (1) c-myb is preferentially expressed in immature hematopoietic cells[22]; (2) levels of myb expression decline during terminal differentiation[22]; (3)

constitutive expression of exogenously introduced c-myb inhibits the ability of a murine erythroleukemia cell line (MEL) to undergo erythroid differentiation in response to known inducing agents.[23,24] Finally, it was also recently shown that v-myb directly regulates the expression of a cellular gene in chicken myeloblasts infected with a temperature-sensitive AMV mutant.[20] This cellular gene, designated mim-1, encodes a secretable protein contained in the granule of both normal and v-myb–transformed promyelocytes.[20] The promoter of the mim-1 gene contains three closely spaced binding sites of c-myb protein and is strongly transactivated by v-myb in cotransfection assays.[20]

To directly determine the importance of c-myb in regulating normal human hematopoiesis, we inhibited myb protein synthesis in bone marrow progenitors cultured in the presence of myb antisense oligodeoxynucleotides and we determined formation of hematopoietic colonies in a semisolid medium. A substantial decrease in both colony number and size for each hematopoietic lineage was observed in the presence of c-myb antisense, but not sense, oligodeoxynucleotides.[6] These studies provided direct evidence that the c-myb gene product plays an important role in regulating the proliferation of hematopoietic cells; however, it remained unclear whether c-myb function was most relevant for the proliferation of distinct progenitor subsets or was independent to the stage of differentiation of hematopoietic progenitors. To address these questions, the growth of late and early progenitors, separated on the basis of cell surface marker expression, was analyzed *in vitro* after exposure to c-myb antisense oligodeoxynucleotides. The number of colonies derived from late progenitors was significantly diminished (75–80% reduction) by c-myb deprivation. On the contrary, the number of colonies derived from early progenitors was not affected, although the number of cells within each colony was greatly diminished (75–80% reduction). These findings suggest that c-myb gene function is preferentially required for the proliferation of intermediate-late hematopoietic progenitors and is less critical for the proliferation of early progenitors.[25] As the compartment of late progenitors is composed of a high number of cells in S phase as compared to that in the compartment of early progenitors,[26] the differential requirement of c-myb during normal hematopoiesis might be explained by the differential cycling activity of early and late progenitors. To determine if c-myb mRNA levels were correlated with progenitor cycling activity, highly purified progenitors (CD34+ cells) were cultured in the presence of IL-3 for 12 days and c-myb mRNA expression was analyzed at 2-day intervals (FIG. 2); expression of c-myb reached a peak after 6–8 days in culture, the length of time needed for the early progenitors to expand and acquire the properties (surface marker expression, cycling activity) of late progenitors. Thus, c-myb requirement during normal hematopoiesis appears tightly linked to cycling activity of progenitor subsets.

DISCUSSION

The molecular regulation of hematopoiesis proliferation and differentiation is still poorly understood. Present theories regarding the control of these processes suggest that protooncogenes encode a variety of proteins that are critical for balanced regulation of those events by their interaction with distinct targets in different subcellular locations. To investigate the functional significance of protooncogenes in normal hematopoietic progenitors we took advantage of the ability of synthetic oligodeoxynucleotides to inhibit gene expression, and we investigated the consequences of this inhibition in culture systems designed to study the proliferation of distinct progenitor subsets in response to hematopoietic growth factors. We focused our attention on these genes, c-abl, N-ras, and c-myb, because they have long been implicated in the regulation of normal hematopoiesis, and their aberrant expression, resulting from either point mutations or rearrangements or amplification, has been associated with the development or maintenance of the leukemic phenotype. c-abl appears to be required in a lineage-specific manner in that its function is needed for the formation of myeloid colonies. Also, c-abl functional requirements appear to be independent of proliferative activity and the differentiation stage of myeloid progenitor cells. Thus, c-abl gene function is important for carrying out normal proliferation for early and late myeloid progenitor cells. The selective requirement of c-abl during myelopoiesis does not appear to be due to restricted expression of the gene in myeloid progenitors compared to erythroid

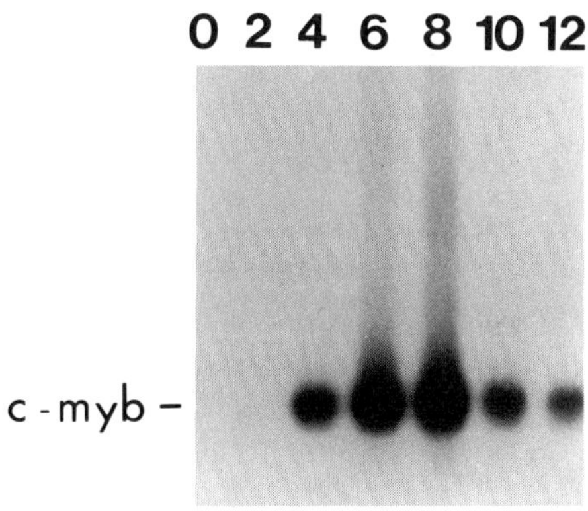

FIGURE 2. Levels of c-myb expression in CD34+ cells incubated in the presence of IL-3 for different days. CD34+ cells were obtained from normal marrow mononuclear cells as previously described.[25] 4×10^4 cells per each timepoint were plated into 96-well Falcon plates in 0.1 ml of Iscove-modified Dulbecco's medium (IMDM) supplemented with 2% heat-inactivated human AB serum, 1 mM L-glutamine, and IL-3 (500/ml). Cells were collected at different timepoints and total RNA was extracted as described.[25] c-myb mRNA was amplified by the RT-PCR technique with 22-mer specific primers corresponding to c-myb cDNA sequence[27] at nucleotides 2,258–2,279 (5′ primer) and 2,466–2,487 (3′ primer). c-myb mRNA expression was then detected by hybridizing the products of cDNA amplification with a diagnostic ^{32}P-end-labeled 50-base oligomer corresponding to a region within the amplified c-myb mRNA from nucleotides 2,351–2,400.[27]

progenitors (FIG. 2). As c-abl encodes a tyrosine kinase activity that might be coupled to that of growth factor receptors, it is likely that this gene is required in a lineage-specific signal transduction pathway.

N-ras appears to be required by normal progenitor cells in response to stimulation with distinct growth factors, so that formation of granulocytic colonies in response to G-CSF is not affected by N-ras deprivation, whereas that elicited by IL-3 is significantly suppressed. Similarly, erythroid colony formation induced by erythropoietin is not affected by down-regulation of N-ras expression, whereas that induced by IL-3 is diminished by N-ras antisense treatment.

On the basis of the observation that growth factor-induced colony formation has differential requirements for N-ras, at least two distinct pathways appear to be associated with proliferation of hematopoietic progenitor cells, one associated with IL-B, GM-CSF, and M-CSF stimulation and the other with erythropoietin, G-CSF, and IL-3 activity.

At variance with those of c-abl and N-ras, c-myb requirements are not lineage specific or growth-factor dependent; rather, it appears that c-myb function is needed for the proliferation of hematopoietic progenitors of each lineage, although the subsets containing progenitor cells with the highest proliferating activity are those relying most on c-myb function.

Each step in the process of generating terminally differentiated cells from multipotent and unipotent hematopoietic progenitors is probably under the control of several genes encoding products with distinct functions. Products of at least three protooncogenes (c-abl, N-ras, and c-myb) have definite roles at different stages of blood cell formation or in response to stimulation with specific growth-factors. Additional studies of the type illustrated here will undoubtedly reveal complex requirements for many other protooncogenes underscoring the complexity of the regulatory mechanisms involved in hematopoiesis. Finally, such studies will provide a rational basis to explain the pathological effects that may result from aberrant activation of protooncogenes in leukemic cells.

REFERENCES

1. Till, J. E. & E. A. McCulloch. 1989. Hemopoietic stem cell differentiation. Biochem. Biophys. Acta **605:** 431–459.
2. Metcalf, D. 1985. The granulocyte-macrophage colony-stimulating factors. Science (Wash., DC) **229:** 16–22.
3. Zamecnik, P. C., J. Goodchild, Y. Taguchi & P. S. Sarin. 1986. Inhibition of replication and expression of human T-cell lymphotropic virus type III in culture cells by exogenous synthetic oligonucleotides complementary to viral RNA. Proc. Natl. Acad. Sci. USA **83:** 4143–4146.
4. Heikkila, R., G. Schwab, E. Wickstrom, S. L. Loke, D. H. Pluznik, R. Watt & L. M. Neckers. 1987. A c-myc antisense oligodeoxynucleotide inhibits entry into S phase but not progress from G_0 to G_1. Nature (Lond) **328:** 445–449.
5. Holt, J. T., R. L. Redner & A. W. Nienhuis. 1988. An oligomer complementary to c-myc messenger RNA inhibits proliferation of HL-60 promyelocytic cells and induces differentiation. Mol. Cell. Biol. **8:** 963–973.
6. Gewirtz, A. M. & B. Calabretta. 1988. A c-myb antisense oligodeoxynucleotides inhibits normal hematopoiesis *in vitro.* Science **242:** 1303–1306.
7. Bourne, H. R., D. A. Sanders & F. McCormick. 1990. The GTPase superfamily: A conserved switch for diverse cell functions. Nature (Lond) **348:** 125–130.
8. Satoh, T., M. Nakafuku, A. Miyajima & Y. Kaziro. 1991. Involvement of ras p21 protein in signal-transduction pathways from interleukin 2, interleukin 3 and granulocyte/macrophage colony-stimulating factor, but not from interleukin 4. Proc. Natl. Acad. Sci. USA **88:** 3314–3318.
9. Hall, A. & R. Brown. 1985. Human N-ras: cDNA cloning and gene structure. Nucleic Acids Res. **13:** 5255–5268.
10. Brandt, J., L. Baird, L. Lu, E. Srour & R. Hoffman. 1988. Characterization of a human hematopoietic progenitor cell capable of forming blast cell containing colonies *in vitro.* J. Clin. Invest. **82:** 1017–1027.
11. Konopka, J. B. & O. N. Witte. 1985. Detection of c-abl tyrosine kinase activity *in vitro* permits direct comparison of normal and altered abl gene products. Mol. Cell. Biol. **5:** 3116–3123.
12. Cook, D. W., D. Metcalf, N. A. Nicole, A. W. Burgess & F. Walker. 1985. Malignant transformation of a growth factor-dependent myeloid cell line by Abelson virus without evidence of an autocrine mechanism. Cell **41:** 677–683.

13. KURZROCK, R., J. U. GUTTERMAN & M. TALPAZ. 1988. The molecular genetics of Philadelphia chromosome-positive leukemias. N. Engl. J. Med. **319:** 990–998.
14. DALEY, G. Q., R. A. VAN ETTEN & D. BALTIMORE. 1990. Induction of chronic myelogenous leukemia in mice by the P210 bcr/abl gene of the Philadelphia chromosome. Science **247:** 824–830.
15. HEISTERKAMP, N., G. JENSTER, J. TEN HOEVE, D. ZOVICH, P. K. PATTENGALE & J. GROFFEN. 1990. Acute leukemia in bcr/abl transgenic mice. Nature **344:** 251–253.
16. CARACCIOLO, D., M. VALTIERI, D. VENTURELLI, C. PESCHLE, A. M. GEWIRTZ & B. CALABRETTA. 1989. Lineage specific requirement of c-abl function in normal hematopoiesis. Science **245:** 1107–1109.
17. MOSCOVICI, C. & L. GAZZOLO. 1982. Transformation of hemopoietic cells and avian leukemia viruses. *In* Advances in Viral Oncology, vol. 1. Oncogene Studies. :83–106. Raven Press. New York.
18. KLEMPNAUER, K. H., G. SYMONDS, G. I. EVAN & J. M. BISHOP. 1984. Subcellular localization of protein encoded by oncogenes of avian myeloblastosis virus and avian leukemia virus E26 and by the chicken c-myb gene. Cell **37:** 537–547.
19. BEIDENKAPP, H., U. BORGMEYER, A. E. SIPPEL & K. H. KLEMPNAUER. 1988. Virus myb oncogene encodes a sequence-specific DNA-binding activity. Nature **335:** 835–837.
20. NESS, S. A., A. MARKNELL & T. GRAF. 1989. The v-myb oncogene product binds to and activates the promyelocyte-specific MIM-1 gene. Cell **59:** 1115–1125.
21. WESTON, K. & J. M. BISHOP. 1989. Transcriptional activation by the v-myb oncogene and its cellular progenitor c-myb. Cell **58:** 85–93.
22. WESTIN, E. H., R. C. GALLO, S. K. ARYA, A. EVA, L. M. SOUZA, M. A. BALUDA, S. A. AARONSON & F. WONG-STAAL. 1982. Differential expression of the AMV gene in human hematopoietic cells. Proc. Natl. Acad. Sci. USA **79:** 2194–2198.
23. CLARKE, M. F., J. F. KUKOWSKA-LATALLO, E. H. WESTIN, M. SMITH & E. PROCHOWNICK. 1988. Constitutive expression of a c-myb cDNA blocks Friend murine erythroleukemia cell differentiation. Mol. Cell. Biol. **8:** 884–892.
24. TODOKORO, K., R. J. WATSON, H. HIGO, H. AMANUMA, S. KURAMUCHI, H. YANAGISAWA & Y. IKAWA. 1988. Down-regulation of c-myb gene expression is a requisite for erythropoietin-induced erythroid differentiation. Proc. Natl. Acad. Sci. USA **85:** 8900–8904.
25. CARACCIOLO, D., D. VENTURELLI, M. VALTIERI, C. PESCHLE, A. M. GEWIRTZ & B. CALABRETTA. 1990. Stage-related proliferative activity determines c-myb functional requirements during normal human hematopoiesis. J. Clin. Invest. **85:** 55–61.
26. ISCOVE, N. N. 1977. Role of erythropoietin in regulation of population size and cell cycle of early and late erythroid precursors in mouse marrow. Cell Tissue Kinet. **10:** 323–350.
27. MAJELLO, B., L. C. KENYON & R. DALLA-FAVERA. 1986. Human c-myb proto-oncogene: Nucleotide sequence of cDNA and organization of the genomic locus. Proc. Natl. Acad. Sci. USA **83:** 9636–9640.

Reversal of Transformed Phenotypes by Antisense *fos*

MATTHEWS O. BRADLEY, SUJATA MANAM,[a]
ANDREW R. KRAYNAK,[a] WARREN W. NICHOLS,[a]
AND BRIAN J. LEDWITH[a]

Genetic MediSyn Corporation
Rockville, Maryland 20850

The impact of therapeutic agents that are based on the genetic code will revolutionize pharmacology as we know it today. The genetic code evolved to transmit biological information in an absolutely precise form. The therapeutic potential of antisense pharmacology derives from turning the genetic code to the absolutely precise regulation of protein(s) related to a given disease. Antisense drugs contain much more information than do traditional drugs of low molecular weight. Thus, a reduction in nonspecific toxicity and a huge increase in the precision of therapeutic targeting are the result. As a model system for developing antisense therapeutics targeted against cancer, we sought to interrupt signal transduction in H-*ras* transformed cells to attempt to reverse their oncogenic phenotype and thus gain a therapeutic advantage.

Protooncogenes are thought to regulate normal cellular proliferation and differentiation by functioning within complex signal transduction pathways that translate the extracellular stimulus from growth and differentiation factors into a specific program of *de novo* gene expression in the nucleus. Perturbation of these regulatory pathways by mutation or overexpression of a protooncogene can result in uncontrolled growth and cellular transformation. The *ras* genes encode GTP-binding proteins that are believed to transduce mitogenic signals across the plasma membrane by coupling growth factor receptors to enzyme systems capable of generating second-messenger molecules.[1] Activating point mutations within the *ras* genes have been detected in a wide variety of animal tumors[2] and are transforming *in vitro.*[1–3] These mutations inhibit the deactivation of *ras*-regulated signal transduction pathways which induces cell proliferation.

A major consequence of activating signal transduction pathways is the rapid induction of immediate-early genes[4,5] which include transcription factors that regulate the expression of other genes that are critical for the proliferative response. The induction of c-*fos* gene expression is one of the earliest nuclear responses to a wide variety of growth and differentiation factors. The c-*fos* protein interacts with the c-*jun* protein and transcription factor AP-1 to regulate the transcription of AP-1 responsive genes.[6,7]

Stacey *et al.*[8] demonstrated that microinjection of *ras* proteins into NIH 3T3 cells causes a rapid induction of c-*fos* expression, suggesting that stimulation of c-*fos* expression may be one of the downstream effects of *ras*-regulated signal transduction. Recent evidence suggests that this enhanced c-*fos* expression may be required for the transforming effects of the *ras* oncogene.

In this study, we investigated if inhibition of c-*fos* expression would reverse

[a] Merck, Sharp & Dohme Research Laboratories, West Point, Pennsylvania 19486.

neoplastic transformation by *ras*. To do so, we used antisense RNA to inhibit c-*fos* expression in EJ c-Ha-*ras*-transformed NIH 3T3 cells and examined the effects of diminished c-*fos* expression on several EJ cell phenotypes, including morphological transformation, multilayer growth *in vitro,* anchorage-independent growth, and tumorigenicity in nude mice. We found[9] that each of these *ras*-induced phenotypes was significantly inhibited by the antisense-*fos* RNA, supporting the role of c-*fos* as an important downstream mediator of *ras*-regulated signal transduction. These results suggest that the model developed here is useful in searching for effective antisense therapeutic agents and, furthermore, that *ras*-induced tumors may be controlled by antisense agents that inhibit signal transduction through c-*fos* or other pathways of signal transduction.

RESULTS

Transfection of EJ cells with an antisense-fos RNA-producing plasmid. EJ *ras*-transformed NIH 3T3 (EJ) cells were transfected with a plasmid, pM(84)fos^{AS},[10] containing an 84-base-pair sequence from the 5′ end of the mouse c-*fos* gene downstream from the dexamethasone-inducible MMTV promoter. The c-*fos* sequence was in an antisense orientation relative to the promoter, so that activation of transcription by dexamethasone resulted in the production of antisense-*fos* RNA that could bind to and thereby inhibit translation of c-*fos* mRNA.[10] EJ cells were also transfected with a similar construct that contained the c-*fos* sequence in the sense orientation pM(84)fos^{S}. As an additional control, pSV2neo[11] was also used. All three plasmids contained an integral copy of the *neo* gene and, therefore, stable transfectants were selected in geneticin (G418). Twelve G418-resistant colonies were isolated for each transfection.

Antisense-*fos* clones as well as the sense-*fos* and *neo* control clones were initially screened to determine if dexamethasone inhibited *in vitro* cell growth. Two of the antisense-*fos* clones, designated EJ/AS-1 and EJ/AS-10, exhibited density-dependent inhibition of growth relative to EJ cells. All of the sense *fos* and *neo* control lines grew identically to the parental EJ cells.

Southern analysis confirmed the presence of intact plasmid inserts in EJ/S-1, EJ/AS-1, and EJ/AS-10 cells. EJ/AS-10 cells contain approximately 20-fold more copies of pM(84)fos^{AS} than do EJ/AS-1 cells.

*Antisense-*fos *RNA restores density-dependent growth arrest to EJ cells.* Antisense-*fos* RNA did not inhibit the ability of EJ cells to proliferate and be continually passaged *in vitro.* However, a marked reduction in the saturating cell density was observed in the cell lines expressing antisense-*fos* RNA. Growth curves for EJ/AS-1 and EJ/AS-10 cells are presented in FIGURE 1A and B, respectively. The growth curves of control EJ/n and NIH 3T3 cells (not affected by dexamethasone) are superimposed on each figure. EJ/AS-1 cells exhibited growth inhibition relative to EJ/n cells even in the absence of dexamethasone (FIG. 1A), probably because of deregulation of the MMTV promoter upon insertion into the EJ cell genome (so that leaky transcription occurred in the absence of dexamethasone). EJ/AS-1 cells did not exhibit further growth inhibition when grown in the presence of dexamethasone. EJ/AS-10 cells were also partially inhibited (relative to control cells) in the absence of dexamethasone (FIG. 1B), again probably because of leaky transcription from the MMTV promoter. However, EJ/AS-10 cells exhibited a further, titratable inhibition of growth in response to increasing concentrations of dexamethasone. Growth inhibition of EJ/AS-10 cells was observed at 50 nM dexamethasone, with near-

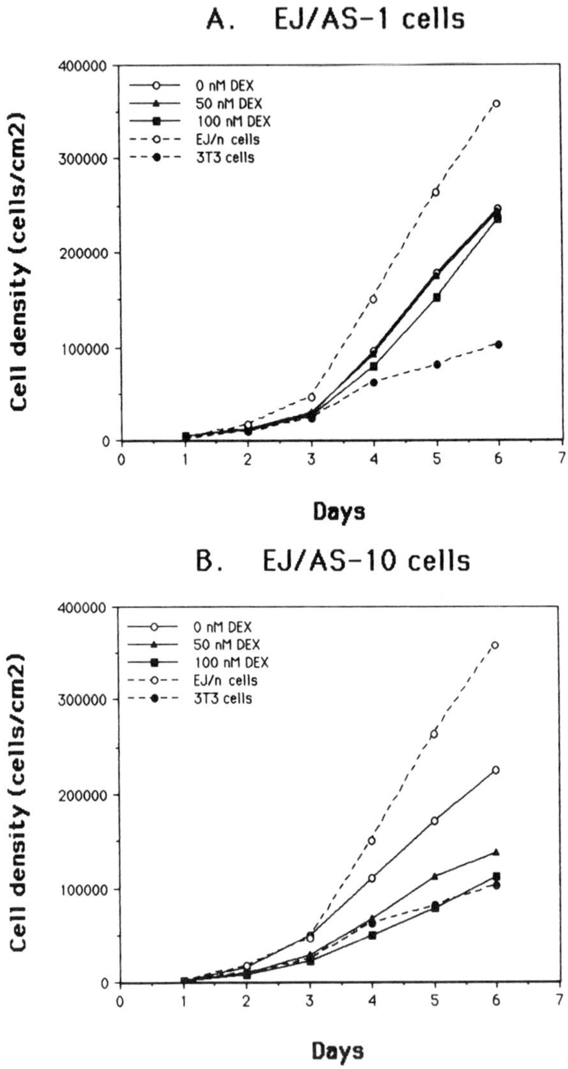

FIGURE 1. *In vitro* growth curves. Cells were seeded at 10¾-cm^2 well. (**A**) EJ/AS-1 cells; (**B**) EJ/AS-10 cells. Cells were grown in 0, 50, and 100 nM dexamethasone (DEX). Growth curves for EJ/n and NIH 3T3 cells, neither of which is affected by 100 nM dexamethasone (not shown), are superimposed on each figure. (Reprinted with permission from Ledwith *et al.*[9])

maximal inhibition at 100 nM. The greater inhibition of cell growth in EJ/AS-10 cells relative to EJ/AS-1 cells may be due to the larger number of antisense-*fos* plasmids stably inserted into EJ/AS-10 cell DNA and thus to a greater potential for producing antisense RNA capable of inhibiting c-*fos* expression.

*Reversion to a flat morphology in antisense-*fos *EJ cells.* An immediately obvious phenotypic conversion in the antisense-*fos* EJ cells was the reversion to a flat morphology characteristic of NIH 3T3 cells. EJ/n cells, like the parental EJ cells, had a refractile, spindle-shaped morphology and grew as a multilayered sheet. EJ/n cell morphology was unaffected by the presence of dexamethasone. In contrast, EJ/AS-1 cells were remarkably flat, even in the absence of dexamethasone, again suggesting promoter leakage. EJ/AS-10 cells, in the absence of dexamethasone, resembled EJ cells. However, they reverted to a flat, monolayer morphology in the presence of dexamethasone. EJ/AS-10 cells again responded in a titratable fashion to increasing concentrations of dexamethasone.

Maintenance of the cells in normal growth medium for several days after the cells became confluent dramatically demonstrated the reversion of uncontrolled growth and morphological transformation caused by antisense-*fos* RNA in EJ cells. EJ/n cells, and EJ/AS-10 cells in the absence of dexamethasone, continued growing as a multilayered sheet that, within 2 days postconfluence, began peeling off the plastic substratum, eventually resulting in complete cell death. However, EJ/AS-10 cells in the presence of dexamethasone and EJ/AS-1 cells with or without the inducer maintained a monolayer and remained firmly attached to their substratum for several days postconfluence, after which they could be routinely passaged.

*Antisense-*fos *RNA facilitates quiescence in serum-deprived EJ cells.* We investigated whether antisense-*fos* RNA would prevent the stimulation of quiescent EJ cells into the S phase by 10% serum. Cells were incubated for 3 days in 0.1% serum in the presence of various concentrations of dexamethasone and then stimulated with 10% serum and assayed for (^{3}H)thymidine incorporation. Both EJ/AS-1 and EJ/AS-10 cells, like the control EJ/n and NIH 3T3 cells, were stimulated into a new round of DNA synthesis by 10% serum regardless of the presence of dexamethasone. Thus, antisense-*fos* RNA did not inhibit the progression of quiescent EJ cells into the S phase. However, antisense-*fos* RNA did appear to restore the ability of serum deprivation to induce quiescence in EJ cells. After 3 days in 0.1% serum (without dexamethasone), 30–50% of EJ/n, EJ/AS-1, and EJ/AS-10 cells were still labeled with (^{3}H)thymidine without serum stimulation, whereas virtually all NIH 3T3 cells were quiescent. Although EJ/n cells were unaffected by the presence of dexamethasone during serum deprivation, both EJ/AS-1 and EJ/AS-10 cells were facilitated into quiescence by the addition of dexamethasone, so that fewer than 5% of their nuclei were labeled. Thus, antisense-*fos* RNA again caused a reversion to the phenotype associated with nontransformed NIH 3T3 cells.

*Soft-agar growth is inhibited by antisense-*fos *RNA in EJ cells.* EJ/n cells grew extremely well in soft agar and were not greatly affected by the presence of dexamethasone. EJ/AS-1 and EJ/AS-10 cells also grew well in soft agar, but when they were grown in the presence of dexamethasone, nearly complete inhibition of colony formation was observed. Maintenance of the cells in dexamethasone for 3 days prior to seeding in soft agar was required to observe the inhibitory effect of the antisense-*fos* RNA. Reversion of EJ/AS-10 cells to a flat morphology also required 2–3 days after the addition of dexamethasone.

*Tumorigenicity in nude mice is inhibited by antisense-*fos *RNA in EJ cells.* EJ/n, EJ/AS-1, and EJ/AS-10 cells were grown *in vitro* for 3 days, in the presence or absence of dexamethasone, prior to subcutaneous injection into nude mice that had previously been implanted with slow-release dexamethasone or placebo pellets, respectively. EJ/n cells rapidly formed aggressively growing tumors, and dexamethasone actually enhanced the rate of EJ/n tumor growth (FIG. 2). In the absence of dexamethasone, EJ/AS-10 cells formed tumors at a rate close to that of the EJ/n cells. However, when EJ/AS-10 cells were grown in the presence of dexamethasone, marked inhibition of tumor growth rate was observed. EJ/AS-1 cells had an inhibited rate of tumor growth relative to EJ/n cells, even in the absence of dexamethasone,

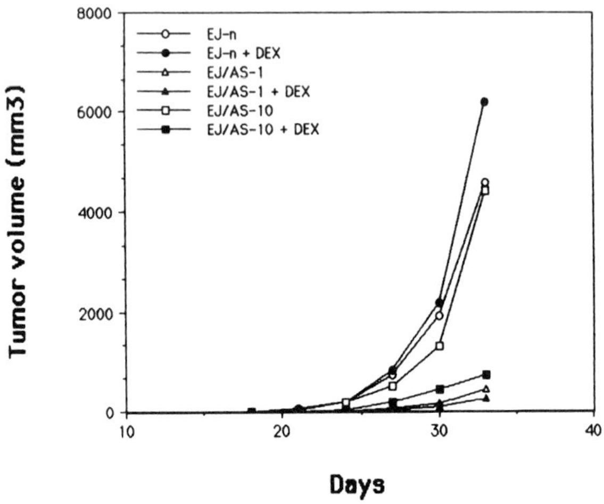

FIGURE 2. Tumor growth in nude mice. Cell lines (see figure key) were grown *in vitro* in the presence (*solid symbols*) or absence (*open symbols*) of 500 nM dexamethasone (DEX) for 3 days and then subcutaneously injected (10^3 cells per mouse) into the shoulders of nude mice that had previously been implanted with slow-release 0.5-mg dexamethasone or placebo pellets, respectively. Cross-sectional tumor diameters were externally measured, and the approximate tumor volume was determined as described in Materials and Methods. (Reprinted with permission from Ledwith *et al.*[9])

and exhibited a further decrease in tumor growth rate in the presence of the inducer. Thus, both anchorage-independent growth *in vitro* and tumor growth *in vivo* were inhibited by antisense-*fos* RNA in EJ cells.

ras *remains overexpressed in the presence of antisense-*fos *RNA.* Northern analysis of c-HA-*ras* gene expression was performed to rule out the possibility that the reversion of transformation observed in EJ/AS-1 and EJ/AS-10 cells was due to inhibited expression of the *ras* oncogene. No expression was observed in NIH 3T3 cells, which do not contain the pEJ plasmid, whereas *ras* was strongly expressed in the parental EJ cells and in the control transfectants EJ/n and EJ/S-1. The *ras* oncogene was also overexpressed in EJ/AS-1 and EJ/AS-10 cells regardless of the presence of dexamethasone. Thus, inhibited *ras* expression was not responsible for the reversion of transformation in the antisense-*fos* cell lines.

*Antisense-*fos *RNA causes diminished* c-fos *expression.* Although antisense RNA inhibition of gene expression is posttranscriptional, it is often associated with diminished levels of the target mRNA. We carried out Northern analysis to determine if our antisense-*fos* cell lines had decreased c-*fos* expression at the mRNA level. Serum-deprived, quiescent NIH 3T3 cells exhibited a transient burst of c-*fos* expression 1 hour after stimulation with 10% serum (FIG. 3, lanes a−). The presence of dexamethasone during serum stimulation enhanced the c-*fos* response in NIH 3T3 cells (lanes a+). Likewise, all of the EJ control lines, including EJ cells (lanes b), EJ/n cells (lanes c), and EJ/S-1 cells (lanes d), exhibited enhanced serum induction of c-*fos* expression in the presence of dexamethasone. EJ/AS-1 cells (lanes e) also exhibited the augmented, and thus uninhibited, c-*fos* response in the presence of dexamethasone. However, EJ/AS-10 cells, which require dexamethasone for maximal reversion of the transformed phenotypes, did not show enhanced c-*fos* mRNA expression in the presence of dexamethasone. Thus, c-*fos* mRNA expression in EJ/AS-10 cells was qualitatively inhibited by dexamethasone, relative to the effects of dexamethasone in control cells.

The inhibition of c-*fos* expression by antisense-*fos* RNA was much more apparent at the protein level. FIGURE 4 shows immunofluorescent staining for c-*fos* protein in serum-deprived cells and in cells that were stimulated with either 10% serum or 10% serum plus dexamethasone. The leakiness of antisense-*fos* expression in EJ/AS-1 cells is apparent in this experiment. In the absence of dexamethasone, EJ/AS-1 cells (FIG. 4E) exhibited a less intense induction of c-*fos* protein expression following serum stimulation than did either EJ/n cells (FIG. 4B) or EJ/AS-10 cells (FIG. 4H). However, antisense-*fos* expression was also inducible in EJ/AS-1 cells in that dexamethasone caused a further decrease in serum-stimulated c-*fos* protein levels (FIG. 4F). EJ/AS-10 cells exhibited strong serum induction of c-*fos* protein expression (FIG. 4H) which was nearly completely inhibited by dexamethasone (FIG. 4I). In contrast, control EJ/n cells were unaffected by dexamethasone (FIG. 4C). These results are consistent with the effects of dexamethasone on the neoplastic phenotypes described here and suggest that the reversion of transformation that we observed was a specific consequence of down-modulation of c-*fos* protein expression.

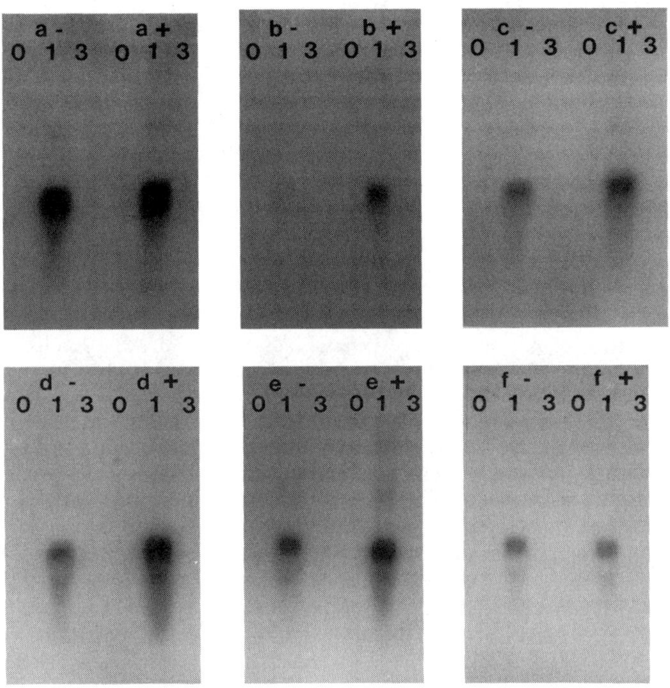

FIGURE 3. c-*fos* mRNA expression. **(a)** NIH 3T3 cells; **(b)** EJ cells; **(c)** EJ/n cells; **(d)** EJ/S-1 cells; **(e)** EJ/AS-1 cells; (f) EJ/AS-10 cells. Cells were maintained in 0.5% serum for 3 days prior to stimulation with 10% serum for either 0, 1, or 3 hours (as indicated). Dexamethasone was either absent (−) or present at 500 nM (+) during both serum deprivation and serum stimulation. A 10-μg sample of total RNA was analyzed by Northern blot, using the intragenic *Pvu*II-*Sst*I fragment of the mouse c- *fos* gene as the probe. (Reprinted with permission from Ledwith *et al.*[9])

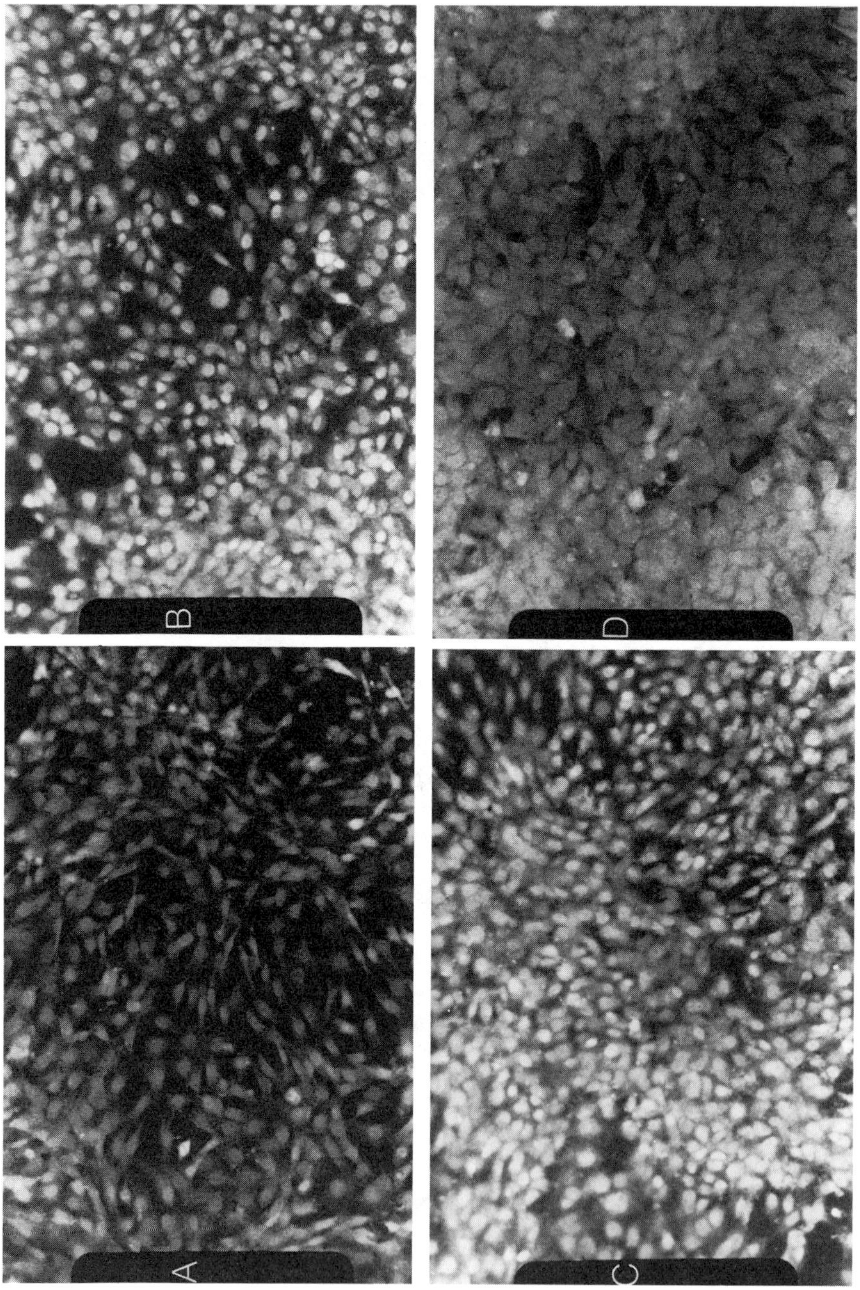

FIGURE 4A-D. Legend on page 132.

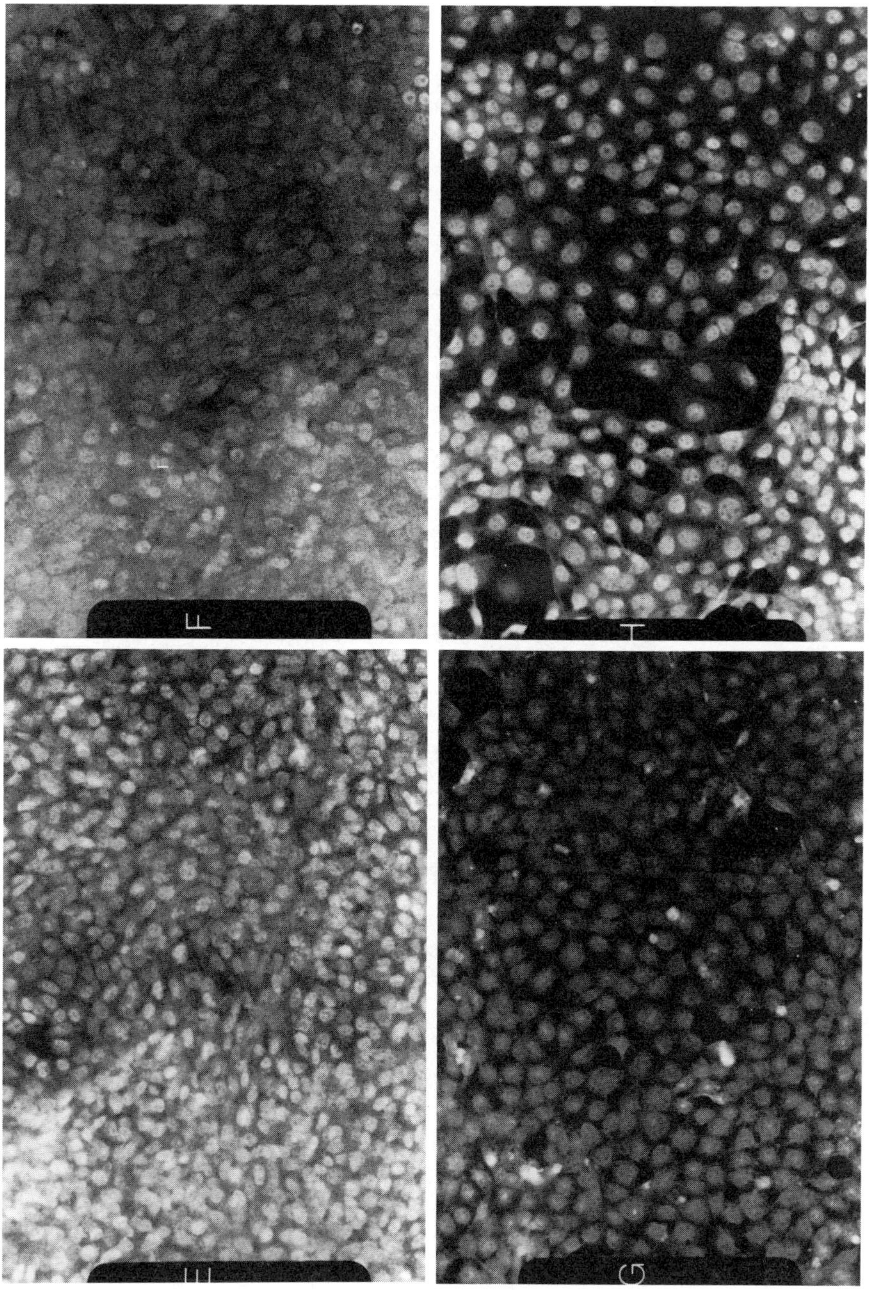

FIGURE 4E–H. Legend on page 132.

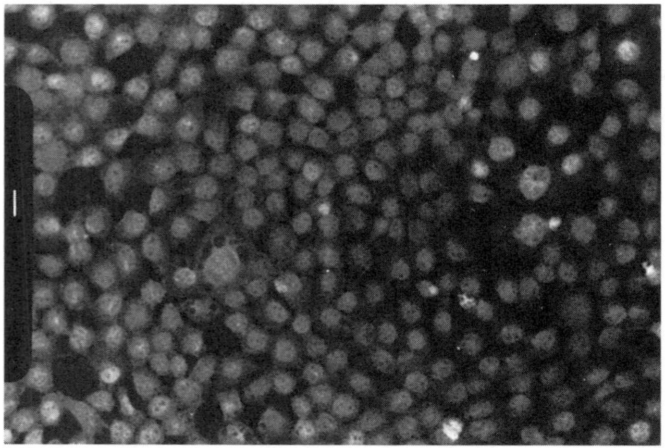

FIGURE 4I.

FIGURE 4. Immunofluorescent staining of c-*fos* protein (**A** to **C**) EJ/n cells; (**D** to **F**) EJ/AS-1 cells; and (**G** to **I**) EJ/AS-10 cells. All cells were incubated for 3 days in 0.5% serum and then were either left untreated (panels **A, D,** and **G**), stimulated in 10% serum for 1 hour (panels **B, E,** and **H**), or stimulated in 10% serum plus 500 nM dexamethasone for 1 hour (panels **C, F,** and **I**). After fixation and permeabilization, the cells were first incubated with rabbit anti-*fos* antiserum and then incubated with fluorescein-conjugated goat anti-rabbit immunoglobulin G antibody. Fluorescein-labeled cells were observed under a 20× objective and photographed as described in Materials and Methods. All photographs were taken with the same exposure time. Although cellular background staining is apparent in all samples, strong nuclear staining is observed only in serum-stimulated cells, and this nuclear staining is inhibited by antisense-*fos* RNA (from panels **F** and **I**). (Reprinted with permission from Ledwith *et al.*[9])

DISCUSSION

Several observations in this study suggest that the inhibitory effect of antisense-*fos* RNA on *ras*-induced neoplasia is a direct consequence of the inhibition of c-*fos* gene expression. First, immunofluorescent staining demonstrated that c-*fos* protein expression is inhibited by dexamethasone in both EJ/AS-1 and EJ/AS-10 cells. Second, EJ/AS-10 cells revert from their transformed phenotype only in the presence of dexamethasone, and they respond to dexamethasone in a titratable manner at concentrations previously found to activate the MMTV promoter in NIH 3T3 cells. EJ/AS-1 cells also require dexamethasone for maximal reversion of transformation in some, but not all, assays (to be described). Third, the *ras* oncogene remains overexpressed in both EJ/AS-1 and EJ/AS-10 cells. Finally, no flat revertants were observed in control transfections with pSV2neo or with a sense-*fos* RNA-producing plasmid.

EJ/AS-1 cells apparently have a considerable degree of promoter leakage (antisense-*fos* RNA synthesis in the absence of dexamethasone), so that serum stimulation of c-*fos* protein expression is not as intense as it is in either EJ/n or EJ/AS-10 cells. However, EJ/AS-1 cells are still inducible in the sense that dexamethasone causes a further decrease in c-*fos* protein levels. In some assays, including *in vitro* growth rate and morphology, the basal antisense-*fos* RNA level in EJ/AS-1 cells is apparently sufficient for maximal reversion of transformation. However, in other assays, including anchorage-independent growth, tumorigenicity in nude mice, and quiescence following serum deprivation, EJ/AS-1 cells require dexamethasone for maximal reversion. Different growth conditions and different assay end points probably account for the inconsistency in dexamethasone requirements between the various assays.

On the basis of this and other evidence, it seems likely that c-*fos* acts downstream from *ras* in a signal transduction pathway. However, our results do not rule out the possible involvement of c-*fos* in a cooperative pathway that has been postulated to allow NIH 3T3 cells to be dominantly transformed by *ras*.[1]

Although c-*fos* expression can be induced by the microinjection of *ras* proteins, the continuous expression of the *ras* oncogene in transfected cells does not cause elevated levels of c-*fos* expression.[8] In fact, the induction of c-*fos* expression by growth factors has been shown to be desensitized in *ras*-transformed cells.[12,13] Likewise, we observed a decreased induction of c-*fos* mRNA after serum stimulation in EJ cells relative to NIH 3T3 cells (FIG. 3, lanes a and b). We also observed this desensitization at the protein level (about sixfold) by Western immunoblot analysis. Thus, c-*fos* induction by *ras*-regulated signal transduction pathways appears to be desensitized in cells that continuously overexpress *ras* oncogenes. Desensitization of phospholipase C[14,15] and protein kinase C[17] activation has also been observed in *ras*-transformed cells, further implicating the involvement of these activities in signal transduction pathways that are regulated by *ras* proteins and that result in c-*fos* gene expression.

Several days of c-*fos* inhibition by antisense RNA were required for reversion of morphological transformation and inhibition of anchorage-independent growth. These observations are consistent with the hypothesis that c-*fos* acts as a transcriptional regulator that converts short-term stimuli into long-term phenotypic changes within the cell.[17] Thus, the c-Ha-*ras* oncogene protein may act through c-*fos* to begin a program of gene expression that ultimately produces the neoplastic phenotype, and several days may be required for the reversal of those phenotypic changes after the c-*fos* pathway has been inhibited.

The therapeutic implications of these results are clear: inhibition of key junctions in oncogenic signal transduction pathways by antisense therapeutic compounds could reverse the effects of prior oncogenic transformation so that malignant cells behave normally.

SUMMARY

The therapeutic use of antisense DNA has started a revolution in pharmacology. As a model system for demonstrating the therapeutic power of the antisense concept, we sought to interrupt signal transduction in H-*ras* transformed cells to attempt to down-regulate their oncogenic phenotype. We hypothesized that down-regulation of c-*fos* translation by antisense-*fos* expression would decrease oncogenic signal transduction through the *fos* pathway and thus reverse the tumorigenic phenotype of these cells. To test this hypothesis, we transfected H-*ras* cells with a plasmid containing an 84-base sequence antisense to the 5′ end of the mouse c-*fos* gene. The antisense-*fos* was under the transcriptional control of the MMTV promoter and inducible by dexamethasone.

Two of the antisense-*fos* clones grew in a density-dependent manner, exhibiting both a flat morphology and a quiescence in low serum medium unlike the sense-*fos* controls. Antisense-*fos* also inhibited soft agar growth to 1% of control values and dramatically reduced tumor growth in nude mice. Antisense-*fos* had no effect on *ras* expression but greatly reduced c-*fos* protein levels as assayed by immunofluorescence. These findings suggest that down-regulation of signal transduction pathways by antisense therapeutic compounds might have major therapeutic benefits against malignant cells transformed by *ras* or other oncogenes.

REFERENCES

1. Barbacid, M. 1987. *ras* genes. Annu. Rev. Biochem. **56:** 779–827.
2. Feramisco, J. R., M. Gross, T. Kamata, M. Rosenberg & R. W. Sweet. 1984. Microinjection of the oncogene form of the human Ha-*ras* (T-24) protein results in rapid proliferation of quiescent cells. Cell **38:** 109–117.
3. Bradley, M. O., A. R. Kraynak, R. D. Storer & J. B. Gibbs. 1986. Experimental metastasis in nude mice of NIH 3T3 cells containing various *ras* genes. Proc. Natl. Acad. Sci. USA **83:** 5277–5281.
4. Cohen, D. R. & T. Curran. 1988. *fra*-1: Serum-inducible, cellular immediate-early gene that encodes a *Fos*-related antigen. Mol. Cell. Biol. **8:** 2063–2069.
5. Lau, L. F. & D. Nathans. 1987. Expression of a set of growth-regulated immediate-early genes in BALB/c 3T3 cells: Coordinate regulation with c-*fos* or c-*myc*. Proc. Natl. Acad. Sci. USA **84:** 1182–1186.
6. Ciu, R., W. J. Boyle, J. Meek, T. Smeal, T. Hunter & M. Karin. 1988. The c-*fos* protein interacts with c-*jun*/AP-1 to stimulate transcription of AP-1 responsive genes. Cell **54:** 541–552.
7. Angel, P., E. A. Allegretto, S. Okino, K. Hattori, W. J. Boyle, T. Hunter & M. Karin. 1988. The *jun* oncogene encodes a sequence-specific *trans*-activator similar to AP-1. Nature (Lond.) **332:** 166–171.
8. Stacey, D. W., T. Watson, H.-F. Kung & T. Curran. 1987. Microinjection of transforming *ras* protein induces c-*fos* expression. Mol. Cell. Biol. **7:** 523–527.
9. Ledwith, B. J., S. Manam, A. R. Kraynak, W. W. Nichols & M. O. Bradley. 1990. Antisense-*fos* RNA causes partial reversion of the transformed phenotypes induced by the c-Ha-*ras* oncogene. Mol. Cell. Biol. **10:** 1545–1555.

10. HOLT, J. T., T. V. GOPAL, A. D. MOULTON & A. W. NIENHUIS. 1986. Inducible production of c-*fos* antisense RNA inhibits 3T3 cell proliferation. Proc. Natl. Acad. Sci. USA **83:** 4794–4798.
11. SOUTHERN, P. J. & P. BERG. 1982. Transformation of mammalian cells to antibiotic resistance with a bacterial gene under the control of the SV40 early region promoter. J. Mol. Appl. Genet. **1:** 327–341.
12. COLLETTA, G., A. M. CIRAFICI, E. CONSIGLIO & G. VECCHIO. 1987. Forskolin and tumor promoter are able to induce c-*fos* and c-*myc* expression in normal, but not in a v-*ras*-transformed rat thyroid cell line. Oncogene Res. **1:** 459–466.
13. LIN, A. H., V. E. GROPPI & R. R. GORMAN. 1988. Platelet-derived growth factor does not induce c-*fos* in NIH 3T3 cells expressing the EJ-*ras* oncogene. Mol. Cell. Biol. **8:** 5052–5055.
14. BENJAMIN, C. W., J. A. CONNER, W. G. TARPLEY & R. R. GORMAN. 1988. NIH-3T3 cells transformed by the EJ-*ras* oncogene exhibit reduced platelet-derived growth factor-mediated Ca^{2+} mobilization. Proc. Natl. Acad. Sci. USA **85:** 4345–4349.
15. BENJAMIN, C. W., W. G. TARPLEY & R. R. GORMAN. 1987. Loss of platelet-derived growth factor-stimulated phospholipase activity in NIH-3T3 cells expressing the EJ-*ras* oncogene. Proc. Natl. Acad. Sci. USA **84:** 546–550.
16. WEYMAN, C. M., E. J. TAPAROWSKY, M. WOLFSON & C. L. ASHENDEL. 1988. Partial down-modulation of protein kinase C in C3H10T½ mouse fibroblasts transfected with the human Ha-*ras* oncogene. Cancer Res. **48:** 6535–6541.
17. CURRAN, T. 1988. The *fos* Oncogene. *In* The Oncogene Handbook. E. P. Reddy, A. M. Skalka & T. Curran, eds.: 307–325. Elsevier Biomedical Press. Amsterdam.

Transferrinfection: A Highly Efficient Way to Express Gene Constructs in Eukaryotic Cells

KURT ZATLOUKAL, ERNST WAGNER,
MATT COTTEN, STEPHEN PHILLIPS,
CHRISTIAN PLANK, PETER STEINLEIN,
DAVID T. CURIEL,[a] AND MAX L. BIRNSTIEL

Research Institute of Molecular Pathology
Dr. Bohr-Gasse 7
A-1030 Vienna, Austria

[a] *University of North Carolina at Chapel Hill*
CB#7020, 724 Burnett-Womack Building
Chapel Hill, North Carolina 27599-7020

The expression of antisense RNA is a powerful means of extinguishing unwanted gene activity in the eukaryotic cell (see this volume). This approach has worked particularly well in plants[1–4] and animals[5] in which antisense-producing gene constructs can be inserted into the germ cell. In man, germ cell manipulation is not yet possible. The problem arises as to how somatic cells can be targeted with genes generating antisense RNAs.

Apart from retroviral vectors, a multitude of techniques exist for introducing DNA into eukaryotic somatic cells (ref. 6 and references therein). During the last few years, we and others[7,8] have been interested in adapting the cellular mechanisms of receptor-mediated endocytosis of macromolecules for the importation of DNA molecules into cells.[9–11] The principle of the method is to covalently connect a polycation, usually polylysine, to a ligand for a receptor on the cell surface. DNA, usually in the form of cloned genes, is then bound ionically to the polylysine. During importation of the DNA by receptor-mediated endocytosis the ligand binds to the receptor, and during internalization of both of these entities the DNA is thought to become colocalized, first in the coated pits and later in the endosome. The further fate of the DNA can be particularly well studied when reporter genes such as the firefly luciferase gene or the bacterial β-galactosidase gene, driven from a viral enhancer/promoter, are employed.

It can be predicted that DNA in the endosome is directed to the lysosomal compartment, where it can be expected to be destroyed by resident nucleases. Despite this, apparently a small amount of DNA does escape destruction, so that reporter gene activity is observed in a few cells of a cell population (see below). Because the genes are expressed, it is probably safe to assume that some of the DNA has reached the cell nucleus. When transferrin is used as a ligand, the transfection procedure is called "transferrinfection."

Consistent with the postulated chain of events during transferrinfection, we find that the addition of the lysosomatropic agent chloroquine, which prevents acidification and concomitant activation of lysosomal degradative enzymes, greatly enhances expression of the chosen reporter gene.[11] This enhancement can be spectacular, as for instance in the human erythroleukemic cell line K562 in which subsequent to the addition of chloroquine, virtually 100% of the cells express the reporter gene for up to 10 days at a very high level.[6] The reason for the great ease with which K562 cells can be transfected can be correlated with a particular physiological feature of K562 cells. The very high level of gene expression in these cells is probably due to the absence of the Na^+, K^+-ATPase regulation of endosomal acidification in these cells. In K562 cells the pH of the endosomal compartment reaches unusually low values,[12] and consequently the endosomal compartment is expected to accumulate exceptionally high amounts of chloroquine which can be expected to act osmotically, vacuolarizing and finally disrupting the endosome (R. F. Murphy, personal communication) and thus releasing the DNA from endosomal confinement. These findings and hypotheses suggest that the rate-limiting step for reporter gene expression during transferrinfection lies at the level of the endosome and that to obtain high levels of gene expression, strategies need to be developed that allow exit of the DNA from the endosome into the cytoplasmic compartment.

Many viruses enter cells via the endosomal route but escape destruction in the lysosome owing to their molecular design which allows them to exit from the endosomes by disrupting the endosomal membranes at low pH.[13] This endosomolytic property of the virus is a feature of the proteinacious capsid and is not dependent on viral gene activity during endosomolysis. From this it can be predicted that after inclusion of such viruses, for instance adenovirus, during receptor-mediated endocytosis, the viruses will bind to viral receptors and will become colocalized in the endosome together with the DNA/polylysine-ligand conjugates (FIG. 1). Because of the disruption of the endosome by the virus, the DNA complexes may be released into the cytoplasm and obtain access to the nuclear compartment. Indeed, it is found experimentally that the inclusion of virus during transferrinfection greatly enhances gene expression, by a factor of as much as 1,000, for cells that have both virus and transferrin receptors.[14] As the endosomolytic function of the virus resides in the capsid proteins, the use of genetically[15] or biochemically[16] inactivated virus minimizes the danger of viral infection while providing the membrane-disrupting function of the virus.

Although the addition of adenovirus can enhance transferrinfection by a large factor,[14] the strategy still suffers from the drawback that relatively high levels of virus are required to ensure colocalization of virus and DNA/polylysine-transferrin binary complexes in the same endosome. Furthermore, viral enhancement is only possible for those cells that express adenoviral receptors on their surface. These difficulties can be circumvented by a physical linkage between the adenovirus and the polylysine-transferrin/DNA complex. Such a link is afforded, for instance, by covalently linking polylysine with an antibody (Ab) that is directed against an epitope of the viral adeno capsid proteins.[17]

We now report on two new kinds of ternary complexes in which the link between adenovirus and polylysine is established by covalently linking the two entities by means of transglutaminase.[18,19] Transglutaminase catalyzes the formation of epsilon-(gamma-glutamyl)lysine bonds between glutamines of the viral capsid proteins and the epsilon-amino groups of the polylysine. In an alternative protocol we biotinylated adenovirus and established a link with streptavidinylated polylysine. A "combi" complex can be generated by adding, first, a limited amount of virus-polylysine to a given amount of DNA, neutralizing only a fraction of the negative charges of the DNA, and then adding polylysine-transferrin conjugate to the complex, neutralizing the remainder of the charges. Such a ternary complex containing adenovirus-polylysine/polylysine-transferrin and DNA has in theory the capacity to enter the eukaryotic cell by means of *either* the adenovirus receptor or the transferrin receptor (FIG. 2). Once internalized within the endosome, the endosomolytic property of the

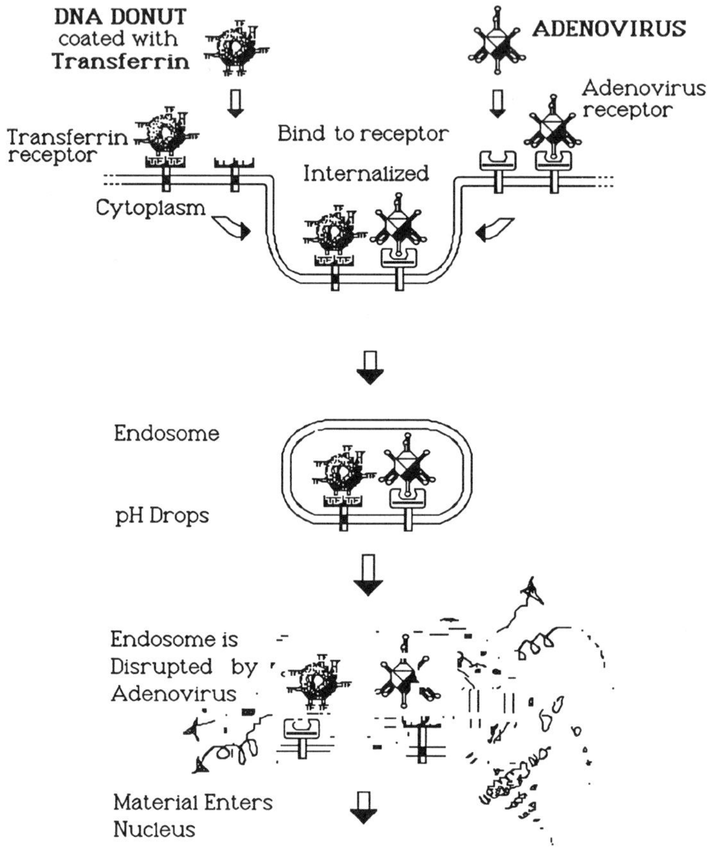

FIGURE 1. Cartoon of adenovirus-enhanced transferrinfection. The adenovirus in many instances is thought to become colocalized with the DNA conjugate which interacts with the transferrin receptor. As adenovirus disrupts the endosome, the DNA exits to the cytoplasm.

conjugated virus will be exploited, and the DNA will gain access to the cytoplasm. A short report of these results has been published previously.[18]

MATERIALS AND METHODS

Human transferrin-polylysine conjugates were obtained from Serva (Heidelberg, Germany).

Preparation of Fusogenic Peptide-Polylysine Conjugate

The influenza peptide ("Influ") of the sequence Gly-Leu-Phe-Gly-Ala-Ile-Ala-Gly-Phe-Ile-Glu-Asn-Gly-Trp-Glu-Gly-Met-Ile-Asp-Gl y-Gly-Gly-Cys (compare ref. 20) was synthesized by the Fmoc-procedure[22] on SASRIN resin (Bachem) using an

Applied Biosystems 431A peptide synthesizer. The side-chain protecting groups were *t*-butyl- for Cys, Glu, and Asp and trityl- for Asn. Coupling was monitored by a ninhydrin test which showed > 98% coupling efficiency for each step. Beginning with amino acid 19, double couplings were carried out. Cleavage from solid support and removal of most protective groups were performed with trifluoroacetic acid and various scavengers according to standard procedures.[23] The crude product obtained, still containing the cysteine in *t*-butyl protected form, was purified by reverse phase HPLC (BIORAD Hi-Pore RP-304 column; buffer A: 20 mM aqueous NH_4HCO_3 + 300 μl 25% NH_3/l; buffer B: A in 98% methanol, gradient elution

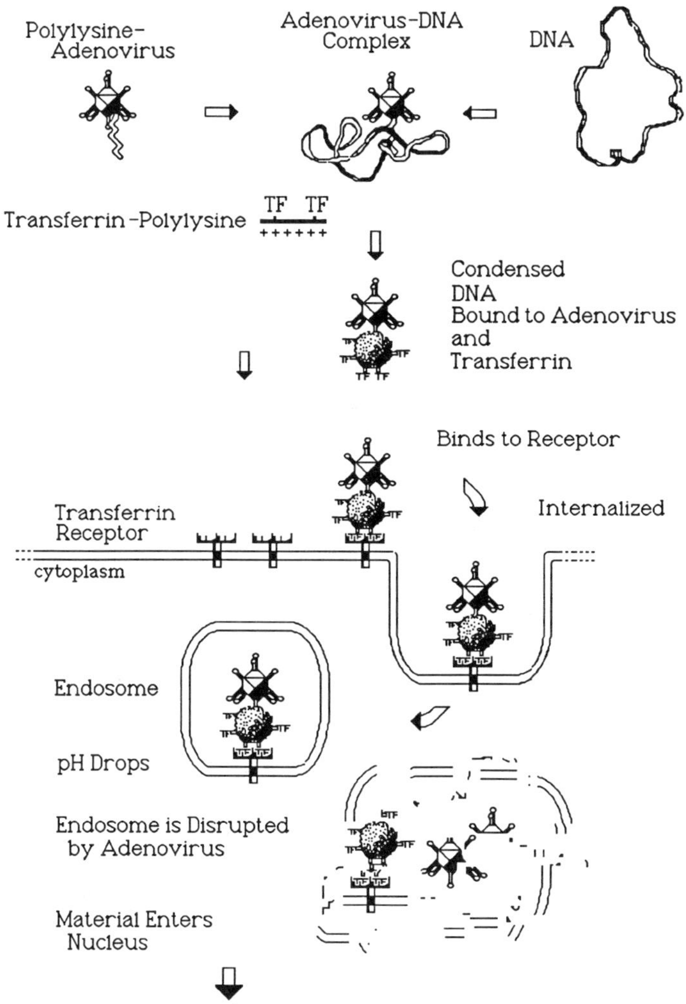

FIGURE 2. Cartoon of linked virus-enhanced transferrinfection. Colocalization of the endosomolytic virus and the transferrin receptor complex is assured even at low input of adenovirus. As only the capsid function of the virus is exploited, the viral genome can be inactivated without loss of endosomolytic activity of the virus.

50–100% buffer B). The pure, *t*-butyl protected peptide was eluted at 80% B and converted to the free mercapto form by treatment with thioanisol/ethanedithiol/trifluoroacetic acid/trifluoro-methanesulfonic acid (2/1/40/3). The deprotected peptide was isolated by precipitation with ether and subsequent gel filtration (Sephadex G-25) under an argon atmosphere. This influenza peptide was conjugated to pyridyldithiopropionate-modified polylysine $pLys_{300}$ (prepared as described in ref. 9) via disulfide bond formation. Conjugates were isolated by cation-exchange chromatography on a Mono S HR/5/column (Pharmacia) using a gradient from 20% to 100% B; (A: 20 mM HEPES, pH 7.3; B: 20 mM HEPES, pH 7.3, 3 M sodium chloride). The main product fraction, eluting with 1.5 M sodium chloride, was pooled and dialyzed twice against HBS (150 mM NaCl, 20 mM HEPES, pH 7.3) to yield, according to a ninhydrin assay and absorption at 280 nm, conjugates ("InflupL") with a molar ratio of peptide:polylysine of 4:1.

Adenovirus Preparation

The adenovirus strain dl312 described by Jones and Shenk[15] with a deletion in the E1a region was used. Replication of the virus was carried out in E1a-*trans*-complementing cell line 293, and purification was carried out on a large scale as described by Davidson and Hassell.[24] The purified virus was diluted with an equal volume of glycerol, and aliquots were stored at −70°C. The virion concentration was determined by UV-spectrophotometric analysis of the extracted viral genomic DNA (conversion factor: one optical density unit [OD, A_{260}] corresponds to 10^{12} viral particles/ml).[25]

Biotinylation of Adenovirus dl312

Two milliliters of a gel-filtered (Sephadex G-25 PD10, Pharmacia) solution of adenovirus dl312 (about 10^{11} particles per ml) in 150 mM NaCl, 5 mM HEPES, pH 7.9/10% glycerol, was mixed with 10 μl (10 nmol) of a 1 mM solution of NHS-LC biotin (Pierce 21335). After 3 hours at ambient temperature the biotin-modified virus was separated from the excess reagent by gel filtration on Sephadex G-25 in 150 mM NaCl, 5 mM HEPES, pH 7.9/40% glycerol (total volume 2.2 ml) and stored at −70°C. Biotinylation of the virus was demonstrated by qualitative detection after the dropwise addition of various dilutions onto a nitrocellulose membrane. After drying at 80°C for 2 hours in a vacuum dryer, blocking with BSA, incubating with streptavidin-conjugated alkaline phosphatase (BRL), washing, and incubating for 1 hour with the developing solution NBT/X-phosphate (nitro-blue-tetrazolium salt/5-bromo-4-chloro-3-indolylphosphate, toluidine salt; Boehringer Mannheim), a positive color reaction showed that it had been biotinylated.

Preparation of Streptavidin-Polylysine Conjugates

The coupling of streptavidin to polylysine was effected using the method described by Wagner *et al.*[9]

Seventy-nine nmol (4.7 mg) of streptavidin in 1 ml of 200 mM HEPES, pH 7.9, and 300 mM NaCl were treated with a 15-mM ethanolic solution of SPDP (236 nmol). After 1.5 hours at ambient temperature the modified protein was gel filtered over a Sephadex G-25 column, yielding 75 nmol of streptavidin modified with 196 nmol of dithiopyridine linker. The modified protein was reacted under an argon atmosphere with 3-mercaptopropionate-modified polylysine (75 nmol, average chain-length 290 lysine monomers, modified with 190 nmol mercaptopropionate linker) in 2.6 ml of 100 mM HEPES, pH 7.9, and 150 mM NaCl. Conjugates were isolated by cation exchange chromatography on a Mono S HR/5/column (Pharmacia) (gradient: 20–100% buffer B, buffer (A and B, see above)). The product fraction eluted at a salt concentration of between 1.2 and 1.7 M. Dialysis against HBS (20 mM HEPES, pH 7.3, 150 mM NaCl) resulted in a conjugate consisting of 45 nmol of streptavidin and 53 nmol of polylysine (Strept-pL).

Enzymatic Linkage of Adenovirus dl312 and Polylysine

Two milliliters of the adenovirus preparation (strain dl312; 5×10^{10} PFU/ml) were applied to a Sephadex G-25 gel filtration column (Pharmacia) equilibrated with 25 ml of reaction buffer (0.1 M Tris-HCl; pH 8.0, 2 mM DTT, 30% glycerol). Elution was carried out with 3.5 ml of reaction buffer. The reaction mixture for enzymatic coupling consisted of 1,150 μl of the virus elution fraction, 0.5 nmol guinea pig liver transglutaminase (Sigma), 2 or 20 nmol of polylysine 290, 10 mM $CaCl_2$ and reaction buffer in a final volume of 1,500 μl. The reaction was carried out at 37°C for 1 hour and then stopped by the addition of 30 μl of 0.5 M EDTA. To monitor the specificity of the coupling, reaction mixtures were also prepared without transglutaminase. Nonincorporated polylysine was separated from the viruses by centrifugation in a CsCl gradient (density 1.33 g/ml; $170{,}000 \times g$, 2 hours). The fraction containing the viruses was collected, mixed with an equal volume of glycerol, frozen in liquid nitrogen, and stored at −70°C.

To demonstrate that polylysine and adenovirus were covalently coupled, the reaction was carried out as described with polylysine that had been labeled with ^{125}I by means of Bolton-Hunter reagent (Amersham). After CsCl-gradient centrifugation the virus fraction was drawn off and separated by means of another CsCl gradient. The gradient was then fractionated, and the radioactivity in every fraction was determined using a scintillation counter. It became apparent that radioactive polylysine had accumulated in the virus fraction. In the control mixture without transglutaminase there was no accumulation of radioactive polylysine in the virus fraction.

Formation of the Binary Complexes of Virus-Polylysine/DNA

Forty μl of the biotin-modified virus fraction (ca. 3×10^9 viral particles) were mixed with 800 ng Strept-pL in 160 μl HBS. After 30 minutes, 6 μg of the DNA plasmid pCMV-L (reporter gene construct containing the *Photinus pyralis* luciferase gene under the control of the cytomegalovirus promoter) in 160 μl HBS were added and incubated for 30 minutes at ambient temperature. For full condensation of the DNA, 3 μg polylysine 290 in 160 μl HBS was added, and after further incubation at ambient temperature, the binary complexes were applied to tissue culture cells.

Formation of Ternary Complexes of Virus-Polylysine/Polylysine-Transferrin/DNA

Forty μl of biotinylated viruses were mixed with Strept-pL and DNA as above. Then DNA was condensed by the addition of 9 μg human transferrin-polylysine (hTfpL) to the mixture, incubation was continued for a further 30 minutes, and finally the ternary complexes were applied to tissue culture cells. Alternatively, 50 μl of the enzymatically polylysine-modified viruses (ca. 4 × 10^8 viral particles) were complexed to 6 μg of DNA in 30 μl HBS by incubation for 30 minutes. Full condensation of DNA was achieved by the addition of 8 μg of murine transferrin-polylysine 290 (mTfpL 290).

Transfection of HeLa Cells and Mouse Hepatocytes with a β-Galactosidase Reporter Gene-Construct and in situ *Demonstration of β-Galactosidase Expression*

HeLa cells and mouse hepatocytes (BNL Cl.2, ATCC No.: TIB73) were grown in DMEM containing 10% fetal calf serum (FCS), 100 U/ml penicillin, 10 μg/ml streptomycin, and 2 mM glutamine in 3 cm culture dishes on coverslips (5 × 10^4 cells/dish).

For transfection, 6 μg pCMV-β-gal in 160 μl HBS were complexed with 12 μg of hTfpL (for HeLa cells) or 8 μg mTfpL 290 (for mouse hepatocytes) in 160 μl HBS and incubated for 30 minutes at room temperature. In another experiment 6 μg pCMV-β-gal were complexed with 6 μg of TfpL in a volume of 200 μl HBS for 30 minutes at room temperature. Thereafter, 10 μg of the fusogenic peptide InflupL in 100 μl HBS were added to the mixture and incubated for a further 30 minutes. Ternary complexes consisting of transglutaminase-mediated polylysine-modified adenovirus were formed as described above.

These polycation-DNA complexes were then mixed with 1 ml DMEM containing 2% FCS, antibiotics, and glutamine and added to the cells. To demonstrate the effect of chloroquine and free adenovirus on transfection efficiency, chloroquine to a final concentration of 100 μM or 50 μl of virus stock solution dl312 was added. After an incubation period of 4 hours at 37°C the medium was removed and the cells were cultured for 48 hours in 3 ml of fresh DMEM containing 10% FCS, antibiotics, and glutamine.

For histochemical detection of β-galactosidase expression, the cells were washed once with phosphate-buffered saline solution (PBS) and fixed with 0.5% glutaraldehyde in PBS for 5 minutes at room temperature. Then the fixative was removed and the cells were washed once with PBS followed by incubation with the staining solution containing 10 mM phosphate buffer pH 7.0, 150 mM NaCl, 1 mM $MgCl_2$, 3.3 mM $K_4Fe(CN)_6 3H_2 0$, 3.3 mM $K_3Fe(CN)_6$, and 0.2% 5-bromo-4-chloro-3-indolyl-beta-galactopyranosid at 37°C for 20 minutes to 3 hours.[26] Thereafter, the coverslips were rinsed in PBS, water, and 96% ethanol and mounted in Mowiol® on slides.

RESULTS AND DISCUSSION

From Transferrinfection to Transfection with Transferrin-Containing Ternary DNA Complexes

Eukaryotic cells contain many different kinds of receptors for the transport of macromolecules across the cell membrane. These receptors not only are numerous (there are, for instance, 10^5 to 5 × 10^5 asialoglycoprotein receptors[29] on hepatocytes), but also have exceedingly high internalization rates. Asialoglycoprotein receptors are calculated to internalize as many as 200 ligands each[27] per day (or ca. 500 ligands per second per cell). This compares to an internalization rate for transferrin of about 300 ligands per second per cell.[28] If we succeed in pirating these mechanisms for the introduction of DNA into cells, we can expect to obtain a procedure that is not only highly efficient, but also physiological. Wu and Wu were first to show that DNA can be combined with a receptor ligand by the expedient means of covalently linking a polylysine tail to the ligand.[8,9] The polylysine binds DNA ionically. Unexpectedly, we found that complexes formed from transferrin-polylysine conjugates and circular plasmid DNA form distinct molecular condensates in the shape of virus-like "donuts" with a diameter of about 80 nm (FIG. 3A), whereas linear DNA also yields some donuts, but in the main cigar-shaped structures such as those indicated by the arrow in FIGURE 3A. The number of DNA molecules per donut is not known, but we find that the overall shape and size do not vary noticeably when circular DNA in the range of 5,000–50,000 base pairs is used. Such donuts can be calculated to contain as many as 100 transferrin molecules attached per DNA molecule via the polylysine moiety. Despite the structural modifications that must ensue from the conjugation with polylysine and the binding of DNA, the association of transferrin with the transferrin receptor is little impaired[10] and the physiology of the cell appears to be unaffected as far as the capacity to transport iron is concerned.[9] Indeed, cells can be subjected to repeated transferrinfection without loss of viability.[10]

To visualize the importance of transferrin-polylysine/DNA conjugates, chicken HD3 cells were incubated with transferrin (conalbumin)-free differentiation medium[10] at 37°C for 18 hours. After the addition of transferrin-polylysine DNA conjugate (labeled with fluoresceinisothiocyanate [FITC] at the polylysine moiety), cells were incubated for 2 hours. FIGURE 3B shows the fluorescence of FITC. In FIGURE 3C, the DNA contents of the cells were fluorescently labeled with DAPI. FITC labeling is distributed over the vesicular system of the cell and the cell membrane, whereas DAPI staining reveals the DNA in both the cell nucleus as well as the vesicular system.

In the normal course of events one would expect DNA complexes to become targeted to the lysosomal compartment and to be degraded therein. The inclusion of chloroquine is a possible remedy because DNA is released into the cytoplasm as a consequence of vacuolarization and disruption of the endosome (see introductory paragraphs), but the behavior of the various cell types is idiosyncratic in that some cell types do not respond to chloroquine. A more general means to liberate the DNA from the vesicular system is the inclusion of adenovirus during transferrinfection[14] with an additional improvement arising from physically linking the virus to the transferrin-polylysine conjugate.[17,18]

Linkage between DNA Condensates and Adenovirus Greatly Enhances Luciferase Reporter Gene Expression

Physical linkage between adenovirus strain dl312 and polylysine can be brought about by either incubating the two components with transglutaminase or biotinylating the adenovirus and streptavidinylation of the polylysine. The effect of linkage on transfection efficiency is clearly demonstrated in FIGURE 4 where hepatocytes were incubated with transferrin-polylysine/DNA complexes (TfpL) in the presence of chloroquine or in the presence of adenovirus (AdenoV + TfpL). Transfection in the presence of adenovirus is elevated, showing the typical enhancement of the transport of transferrin-polylysine DNA complexes into the cells.[14] In pLAdenoV/TfpL, polylysine was attached to adenovirus by means of transglutaminase. The complex was reacted with DNA, neutralizing part of the negative charges of the DNA, and later transferrin-polylysine was added to neutralize the remainder of the charges. In

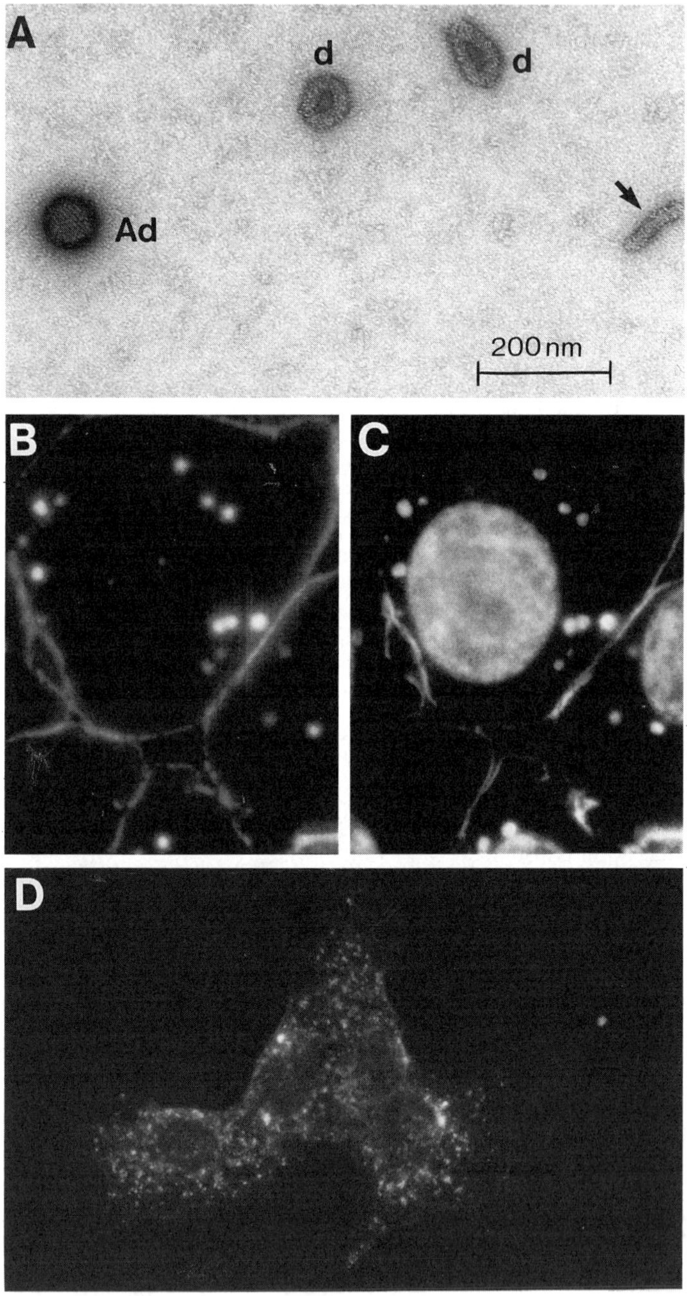

FIGURE 3.

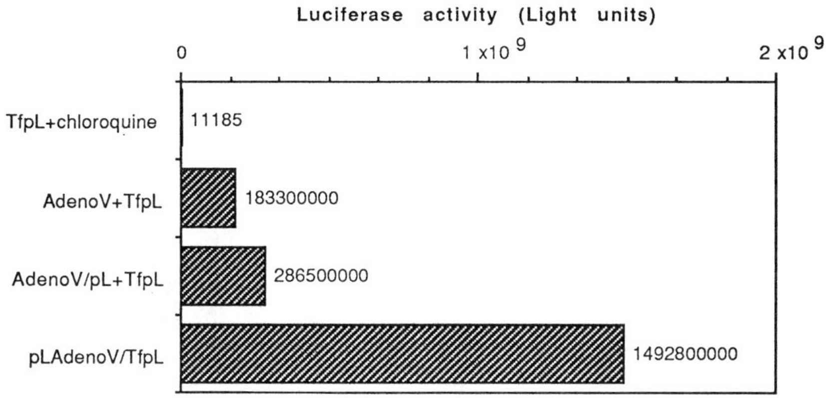

FIGURE 4. Effect of enzymatically polylysine-modified adenovirus on transfection efficiency in mouse hepatocytes. Six μg of pCMV-L was complexed to adenovirus that had been covalently modified with polylysine by transglutaminase (dl312/TG-pL). Full condensation of the DNA was achieved by the addition of mTfpL 290. To demonstrate the specific effect of transglutaminase, transfection was performed in parallel with an adenovirus preparation in which the enzyme was omitted during the coupling reaction (dl312/pL). For comparison, DNA was directly complexed to mTfpL 290, and transfections were performed either with the addition of free adenovirus (dl312) or in the presence of 100 μM chloroquine.

this way a ternary complex of adenovirus-polylysine/transferrin-polylysine DNA was synthesized. As can be seen, an extraordinarily high value of 1.5×10^9 luciferase light units was obtained. In slot AdenoV + pL + TfpL, adenovirus and pL were mixed as for pLAdenoV/TfpL, but transglutaminase treatment was omitted. TfpL was then added. Transfection is of the order of AdenoV + TfpL, because colocalization of virus and TfpL/DNA complexes in the endosome is a stochastic process in contrast to pLAdenoV/TfpL where, through linkage of pL and AdenoV, colocalization is assured, yielding a very high level of transferrinfection (5,000 light units per cell).

A similar point is also made in FIGURE 5 where biotinylated virus and streptavidinylated polylysine was used to link the two components. The figure further demonstrates that complexes with physically linked viruses also work in a highly diluted state, whereas experiments with free, unlinked virus require high concentrations of both virus and transferrin-polylysine/DNA complexes. In the experiments reported in FIGURE 5, 6 μg of covalently circular plasmid DNA CMV-L were combined with the proteinacious moieties indicated in the figure legend. A series of 10-fold dilutions

←

FIGURE 3. (**A**) Electromicrograph of transferrin-polylysine/DNA complexes and of added adenovirus. Negative stain. *Arrow* points to a cigar-shaped complex that is typical for the polylysine-transferrin conjugated with linear DNA. The "donut" structure of covalantly circular DNA complexes is clearly seen (d). Ad denotes adenovirus. (**B** and **C**) Fluorescent micrographs of HD3 cells. (**B**) FITC-labeled polylysine in polylysine-transferrin/DNA complex. (**C**) DAPI fluorescent stain of DNA (see text). Magnification × 1250. (**D**) Demonstration of uptake of biotinylated DNA complexed to TfpL in hepatocytes (Hep G2). Transfection was performed in the presence of 100 μM chloroquine for 4 hours. Endocytosed DNA was visualized with phycoerythrin-coupled streptavidin. Note the uniform uptake of DNA/TfpL complexes over the whole cell population. Magnification × 510.

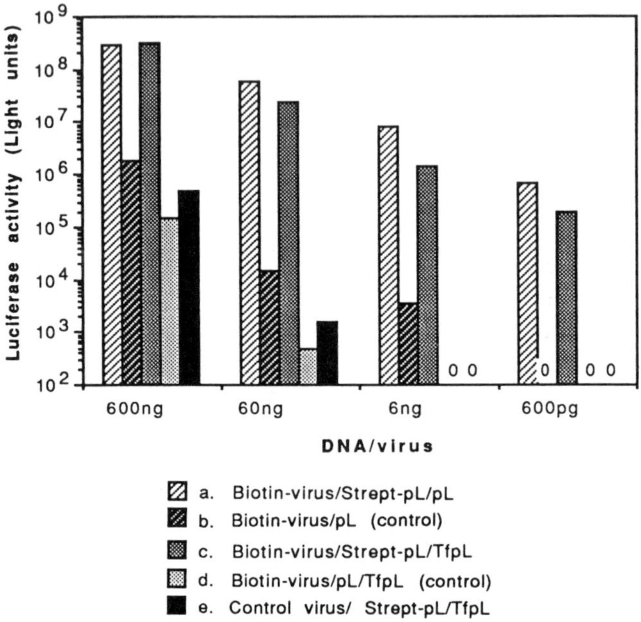

FIGURE 5. Impact of linking adenovirus with polylysine-transferrin/DNA complexes on transfection efficiency. Six μg of covalently circular CMV-L plasmid were combined with various conjugates and diluted to yield the final DNA amounts denoted at the bottom of the graph. (**a**) Biotin-virus/Strept-pL/pL is a binary complex in which 800 ng streptavidinylated polylysine were mixed with 40 μl biotinylated virus and 6 μg plasmid DNA and then 3 μg polylysine (pL) were added to saturate the charges of the DNA. (**b**) Biotin-virus/pL (control) is the control for the **a** construct in which 4 μg polylysine were added to 6 μg plasmid DNA and biotinylated virus. Because polylysine is not streptavidinylated in this case, polylysine forms "donuts" with DNA that are not linked to the virus. (**c**) Biotin-virus/Strept-pL/TfpL is the same as the biotin-virus/Strept-pL/pL except that 9 μg polylysine-transferrin conjugates were added to the initial binary complex consisting of 6 μg plasmid and 800 ng streptavidinylated polylysine and biotinylated virus. This yields a combi-complex capable of binding to both transferrin and viral receptors on the cell, revealing the capacity of linked virus to enhance transferrinfection. (**d**) Biotin-virus/pL/TfpL (control) is the control for the **c** construct in which transferrin-polylysine conjugates are not linked to virus, because polylysine has not been streptavidinylated. This is an example of enhancement of transferrinfection with free virus. (**e**) Control virus/Strept-pL/TfpL is another control for construct **c**. Eight hundred ng Strept-pL were incubated with unmodified virus followed by the addition of polylysine-transferrin conjugate. Formation of polylysine-transferrin "donuts" is expected, whereas because of the unmodified nature of the virus, there is no formation of a ternary complex.

were then made down to 600 pg DNA per 6 cm diameter tissue culture dish (ca. 3×10^5 HeLa cells).

From inspection of the figure it becomes clear that where the adenovirus is not physically linked with the DNA condensate as in slot biotin-virus/pL (control), biotin-virus/pL/TfpL, or control virus/Strept-pL/TfpL, luciferase expression is low even at high DNA input, and these compositions were all very highly sensitive to dilution. In these experiments the adenovirus can only enhance transfection if by chance it colocalizes in the same endosome, and this is only guaranteed at high concentration of both virus and DNA condensate.

For biotin-virus/Strept-pL/pL and biotin-virus/Strept-pL/TfpL, which represent binary and ternary DNA condensate, respectively, in which the virus is physically linked with the DNA via a biotin-streptavidin bridge, very high levels of gene expression are obtained (4×10^8 light units per 3×10^5 HeLa cells at 600 ng DNA per dish) and the expression remains high even after a 1,000-fold dilution of the complexes. For the last point (600 pg DNA per dish) there were only 300 DNA copies and 1 virus per cell and still 8×10^5 light units were obtained. This contrasts with the calcium phosphate precipitation transfection procedure in which modest numbers of luciferase light units are obtained at an input of 5×10^5 DNA molecules per cell.[29] Clearly, the linked virus constructs are a most efficient means for introducing DNA into cells, because colocalization of DNA condensate, virus and/or transferrin is guaranteed.

One striking feature of these HeLa cell transfection experiments is that both binary complexes consisting solely of virus and linked polylysine (supplemented with free polylysine to reach electroneutrality) and the ternary complex containing transferrin-polylysine transfect cells at a similar level and that there is no synergism between transferrin- and adenovirus-receptors. We suggest that this may be a consequence of HeLa cells having a relatively large number of adenovirus receptors and relatively few transferrin receptors, so that in both binary and ternary complexes it is actually the adenovirus receptor and not so much the transferrin receptor that transports DNA.

This situation contrasts with that of K562 cells. Figure 6 demonstrates, first, that a high level of luciferase activity was obtained in K562 cells in the presence of chloroquine (but in the absence of adenovirus) for reasons that were already discussed. The addition solely of polylysine-transferrin/DNA, in the absence of chloroquine, yielded background levels of luciferase activity, whereas the simultaneous addition of both transferrin-polylysine conjugate and free virus yielded

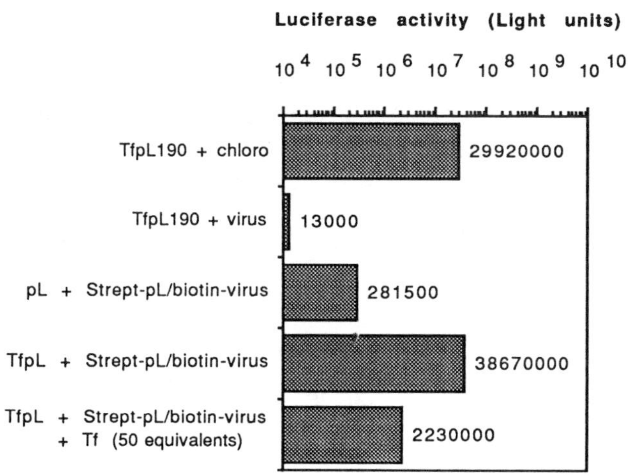

FIGURE 6. Influence of chloroquine and adenovirus on transferrinfection of K562 cells (see text).

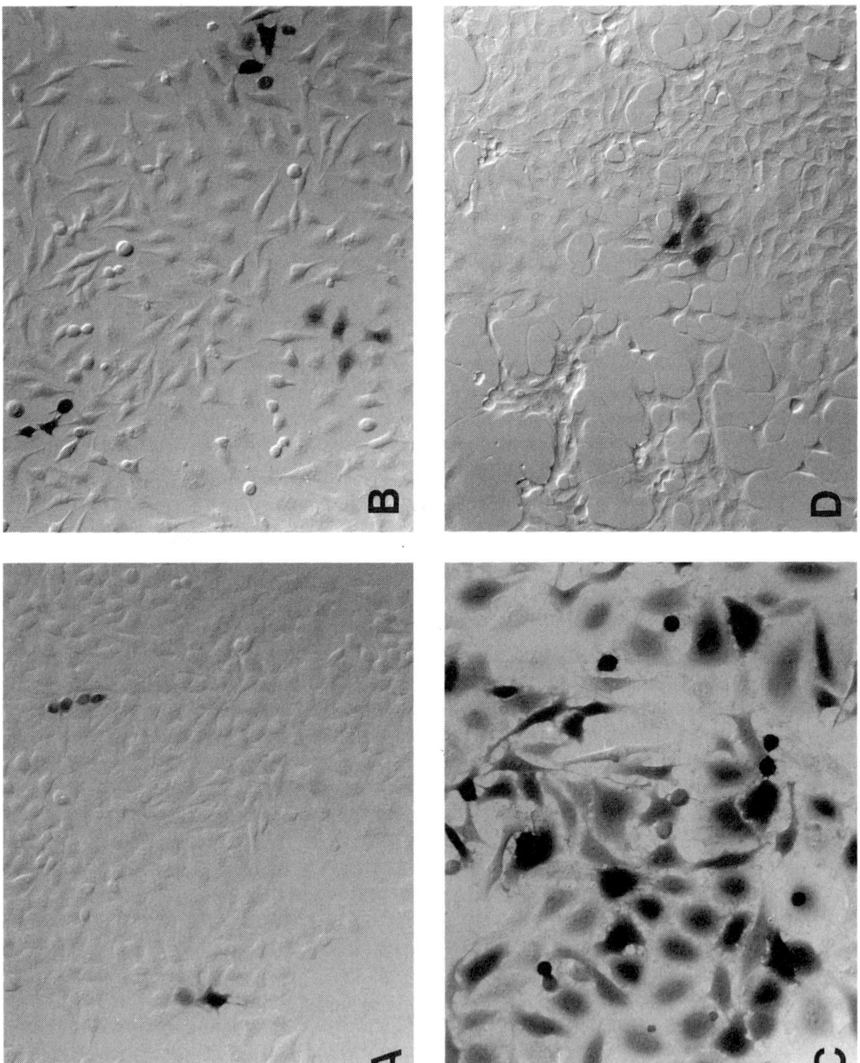

FIGURE 7A–D. Legend on facing page.

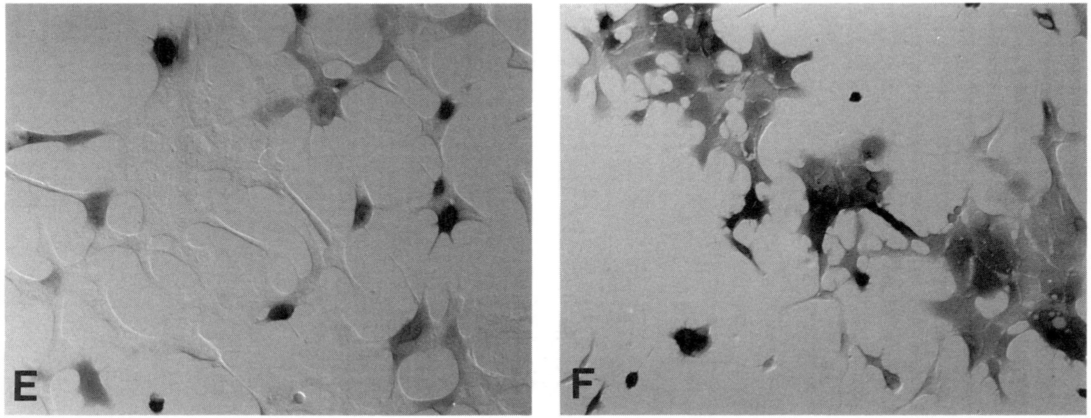

FIGURE 7E–F.

FIGURE 7. Comparison of the frequency with which cells can be transfected using different transfection protocols. (**A**) HeLa cells were transfected with pCMV-β-gal complexed to TfpL in the presence of 100 μM chloroquine (less than 1 in 1,000 cells produced β-galactosidase). (**B**) HeLa cells were transfected with transferrin-polylysine/DNA complexes containing the fusogenic peptide InflupL (about 2% of the cells were positively stained). (**C**) HeLa cells were transfected with transferrin-polylysine/DNA complexes with the addition of free adenovirus. Note that almost all cells express the delivered reporter gene. (**D**) Mouse hepatocytes were transfected with pCMV-β-gal complexed to mTfpL 290 in the presence of 100 μM chloroquine (less than 1 in 1,000 cells produced β-galactosidase). (**E**) Mouse hepatocytes were transfected with transferrin-polylysine/DNA complexes with the addition of free adenovirus (10% of the cells were positive for β-galactosidase). (**F**) Mouse hepatocytes were transfected with ternary complexes consisting of enzymatically modified adenovirus-polylysine, pCMV-β-gal, and mTfpL 290. Note that the ternary complexes result in a strong expression of β-galactosidase in more than 90% of the hepatocytes.

relatively few light units; even when the virus was linked to polylysine in a binary complex, only moderate levels of luciferase activity were obtained. By contrast, the ternary complexes containing transferrin increased luciferase expression to nearly 4×10^7 light units, presumably because here the transferrin receptor, which abounds on K562 cells, comes into play. This contention is supported by our finding that the addition of 50 mass equivalents of transferrin lowers transferrinfection by ternary complexes by a factor of about 18 due to the competition between free transferrin and ternary complexes for the same receptors.

The Fusogenic Peptide of Influenza Virus Can Partially Replace the Function of the Linked Adenovirus

The fusogenic peptide domain(s) of the viruses which enter via endocytosis are thought to change their conformation upon acidification of the endosomal milieu and to expose hydrophobic domains that initiate lysis of the endosomal membrane[13] and allow entry of the virus into the cytoplasm. If the function of the physically linked or free virus during transferrinfection is to disrupt the endosome, then it should be possible to obtain a similar endosomolytic property for a fusogenic peptide when such a peptide is included in a complex containing DNA, transferrin, and polylysine.

Hemagglutinin HA2 of the influenza virus contains such a fusogenic peptide.[20] The peptide from the NH_2-terminus of HA2, 23 amino acids long, was synthesized and coupled to polylysine (see Materials and Methods). As we shall show, combi-complexes consisting of DNA, transferrin-polylysine, and fusogenic peptide-polylysine conjugates facilitate transferrinfection (FIG. 7B), although not at the level of free adenoviruses in excess (FIG. 7C) or linked adenovirus (FIG. 7F).

Efficiency of Reporter Gene Expression at the Cellular Level

In FIGURE 7 the frequency with which cells in a cell population became transferrinfected is shown as a consequence of using different transfection protocols. FIGURE 7A–C depicts expression of the β-galactosidase reporter gene in HeLa cells, FIGURE 7D–F that in mouse hepatocytes.

Transferrinfection with human transferrin-polylysine/DNA complex plus chloroquine yielded less than 1:1,000 of treated HeLa cells that expressed the reporter gene. Positive cells occurred in clusters of two or four cells, presumably because galactosidase-positive cells have divided once or twice. In FIGURE 7B HeLa cells were transfected with DNA to which transferrin-polylysine had been added to neutralize a fraction of the DNA charges. The remainder of the negative charges were then neutralized by the addition of influenza virus fusogenic peptide-polylysine conjugate to yield a ternary complex consisting of transferrin-polylysine/fusogenic peptide-polylysine/DNA. This ternary complex allowed expression of the reporter gene in about 2% of cells. Note the typical grouping of positive cells. In FIGURE 7C HeLa cells were transfected with the classical transferrin-polylysine/DNA complexes in the presence of free, replication-defective dl312 adenovirus. Virtually all the cells respond to this transfection protocol.

FIGURE 7D–F shows that mouse hepatocytes are somewhat more refractory to transferrinfection. The reporter gene complexed with murine transferrin-polylysine was applied in the presence of chloroquine. Less than 1:1,000 cells became positive for β-galactosidase (FIG. 7D). The low probability of obtaining gene expression has to be compared with the high efficiency with which cells in a culture take up the transferrin-polylysine/DNA complexes (FIG. 3D). The same protocol, but in the presence of free adenovirus, yielded about 10% positive cells (FIG. 7E). This number can be increased to nearly 100% by using adenovirus at high multiplicity, which, however, is rather toxic for the cells (unpublished results). Virtually all cells expressed β-galactosidase when the ternary combi-complex consisting of transferrin-polylysine/adenovirus-polylysine/DNA (containing linked adenovirus) was used (FIG. 7F).

Stable Transfection of Tissue Culture Cells

Although adenovirus-enhanced transferrinfection excels by the high level of gene expression that is obtained transiently (see below), so far the technique has been less

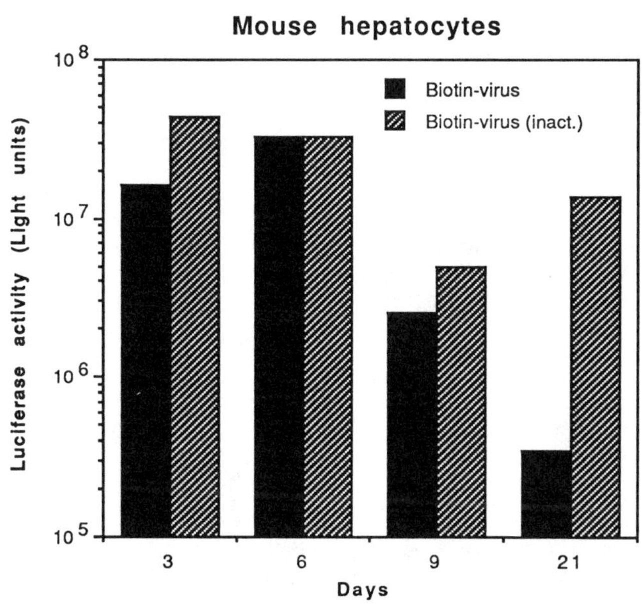

FIGURE 8. Time course of luciferase reporter gene expression in a transferrinfected embryonic hepatocyte cell line. The use of psoralene inactivated adenovirus[16] prolongs a high level of reported gene expression (see text).

useful for the production of stably transformed cell lines. The reason for this is that many cells that express the inserted gene (having experienced adenovirus-enhanced transferrinfection) stopped dividing and often round up and detach from the support, making clonal selection more difficult. Such behavior is not seen when the fusogenic peptide is used. At the moment it is unclear if these deleterious effects seen when using adenovirus are caused by residual viral gene activity in adeno strain dl312 or if they are ascribable to the intense endosomolytic property of the virus.

DNA Reporter Genes Are Expressed Transiently but for a Protracted Period

K562 cells, as discussed, can be transfected at a very high level by inclusion of chloroquine, and these cells were shown to express the β-galactosidase gene transiently for at least 10 days.[6] We wanted to know if adenovirus-enhanced transferrinfected cells also express the transfected reporter gene (luciferase) transiently but for a relatively long time. For this we used the adherent embryonic hepatocyte cell line BNL Cl.2 and transferrinfected it with ternary complexes amounting to 200 viruses per cell. We know from parallel experiments that at high viral input (2,000 viruses per cell) $<95\%$ of the hepatocytes become transfected. As can be seen in FIGURE 8, hepatocytes express luciferase to levels of 4×10^8 light units. High expression prevails at least 21 days, especially when the virus is inactivated by psoralene treatment.[16]

Efficient Transfer of Cosmids into Tissue Culture Cells

When viral vectors, such as retroviruses or adenoviruses, are used to infect cells, there is a severe limitation as to the size of the DNA that can be inserted into the viral genome. When the virus is used as an endosomolytic principle, the DNA is deposited on the outside of the intact virus, as for instance in the ternary complexes consisting of transferrin-polylysine/virus-polylysine/DNA or in the binary complexes consisting of virus-polylysine/polylysine/DNA, and here limitations as to the size of the DNA to be transferred are less obvious. Thus, we find that we can transfer 48-kb cosmids with similar efficiency as plasmids of 6 and 15 kb using binary and ternary complexes.[16] We are presently investigating whether yeast artificial chromosomes (YACs) can be inserted into cells using transferrinfection protocols.

REFERENCES

1. VAN DER KROL, A. R., J. N. M. MOL & A. R. STUITJE. 1988. Antisense genes in plants: An overview. Gene **72:** 45–50.
2. POWELL, P. A., D. M. STARK, P. R. SANDERS & R. N. BEACHY. 1989. Protection against tobacco mosaic virus in transgenic plants that express tobacco mosaic virus antisense RNA. Proc. Natl. Acad. Sci. USA **86:** 6949–6952.
3. MOL, J. N. M., A. R. VAN DER KROL, A. J. VAN TUNEN, R. VAN BLOKLAND, P. DE LANGE & A. R. STUITJE. 1990. Regulation of plant gene expression by antisense RNA. FEBS **268:** 427–430.
4. DAY, A. G., E. R. BEJARANO, K. W. BUCK, M. BURRELL & C. P. LICHTENSTEIN. 1991. Expression of an antisense viral gene in transgenic tobacco confers resistance to the DNA virus tomato golden mosaic virus. Proc. Natl. Acad. Sci. USA **88:** 6721–6725.
5. STOUT, J. T. & C. T. CASKEY. 1990. Antisense RNA Inhibition of HPRT synthesis. Somatic Cell Mol. Gen. **16:** 369–382.
6. COTTEN, M., E. WAGNER & M. L. BIRNSTIEL. 1992. Receptor mediated transport of DNA into eukaryotic cells. *In* Methods in Enzymology. R. Wu, ed. Academic Press. New York. In press.
7. WU, G. Y. & C. H. WU. 1987. Receptor-mediated *in vitro* gene transformation by a soluble DNA carrier system. J. Biol. Chem. **262:** 4429–4432.
8. WU, G. Y. & C. H. WU. 1988. Receptor-mediated gene delivery and expression *in vivo.* J. Biol. Chem. **263:** 14621–14624.
9. WAGNER, E., M. ZENKE, M. COTTEN, H. BEUG & M. L. BIRNSTIEL. 1990. Transferrin-polycation conjugates as carriers for DNA uptake into cell. Proc. Natl. Acad. Sci. USA **87:** 3410–3414.

10. ZENKE, M., P. STEINLEIN, E. WAGNER, M. COTTEN, H. BEUG & M. L. BIRNSTIEL. 1990. Receptor-mediated endocytosis of transferrin-polycation conjugates: An efficient way to introduce DNA into hematopoietic cells. Proc. Natl. Acad. Sci. USA **87:** 3655–3659.
11. COTTEN, M., F. LÄNGLE-ROUAULT, H. KIRLAPPOS, E. WAGNER, K. MECHTLER, M. ZENKE, H. BEUG & M. L. BIRNSTIEL. 1990. Transferrin-polycation-mediated introduction of DNA into human leukemic cells: Stimulation by agents that affect the survival of transfected DNA or modulate transferrin receptor levels. Proc. Natl. Acad. Sci. USA **87:** 4033–4037.
12. SIPE, D. M., A. JESURUM & R. F. MURPHY. 1991. Absence of Na^+, K^+-ATPase regulation of endosomal acidification in K562 erythroleukemia cells. J. Biol. Chem. **266:** 3469–3474.
13. PASTAN, I., P. SETH, D. FITZGERALD & M. WILLINGHAM. 1986. *In* Virus Attachment and Entry into Cells. eds.: 141–146. R. L. Crowell & K. Lonberg-Holm, eds. Am. Soc. Microbiol., Washington.
14. CURIEL, D. T., S. AGARWAL, E. WAGNER & M. COTTEN. 1991. Adenovirus enhancement of transferrin-polylysine-mediated gene delivery. Proc. Natl. Acad. Sci. USA **88:** 8850–8854.
15. JONES, N. & T. SHENK. 1979. An adenovirus type 5 early gene function regulates expression of other early viral genes. Proc. Natl. Acad. Sci. USA **76:** 3665–3669.
16. COTTEN, M., E. WAGNER, K. ZATLOUKAL, S. PHILIPPS, D. T. CURIEL & M. L. BIRNSTIEL. 1992. High efficiency receptor-mediated delivery of small and large (48kb) gene constructs using the endosome disruption activity of defective or chemically-inactivated adenovirus particles. Proc. Natl. Acad. Sci. USA **89:** 6094–6098.
17. CURIEL, D. T., E. WAGNER, M. COTTEN, M. L. BIRNSTIEL, CH. LI, ST. LOECHEL, S. AGARWAL & P. HU. 1992. High efficiency gene transfer by adenovirus coupled to DNA-polylysine complexes via an antibody bridge. Human Gene Therapy **3:** 147–154.
18. WAGNER, E., K. ZATLOUKAL, M. COTTEN, H. KIRLAPPOS, K. MECHTLER, D. T. CURIEL & M. L. BIRNSTIEL. 1992. Coupling of adenovirus to polylysine-DNA complexes greatly enhances receptor mediated gene delivery and expression of transfected genes. Proc. Natl. Acad. Sci. USA **89:** 6099–6113.
19. FOLK, J. E. & S. I. CHUNG. 1985. Transglutaminases. Methods Enzymol. **113:** 358–375.
20. WAGNER, E., M. COTTEN, K. ZATLOUKAL & M. L. BIRNSTIEL. 1992. Influenza virus hemagglutinin HA-2 NH_2-terminal fusogenic peptides augment gene transfer by transferrin-polylysine/DNA complexes: Towards a synthetic virus-like gene transfer vehicle. Proc. Natl. Acad. Sci. USA, in press.
21. WHARTON, S. A., S. R. MARTIN, R. W. H. RUIGROK, J. J. SKEHEL & D. C. WILEY. 1988. Membrane fusion by peptide analogues of influenza virus haemagglutinin. J. Gen. Virol. **69:** 1847–1857.
22. ATHERTON, E., M. J. GAIT, R. C. SHEPPARD & B. J. WILLIAMS. 1979. The polyamide method of solid phase peptide and oligonucleotide synthesis. Bioorganic Chem. **8:** 351–370.
23. ATHERTON, E. & R. C. SHEPPARD. 1989. Solid phase peptide synthesis. The Practical Approach Series. IRL Press. Oxford, England.
24. DAVIDSON, D. & J. A. HASSELL. 1987. Overproduction of polyoma middle T antigen in mammalian cells through the use of an adenovirus vector. J. Virol. **61:** 1226–1239.
25. CHARDONNET, Y. & S. DALES. 1970. Early events in the interaction of adenoviruses with HeLa cells. I. Penetration of type 5 and intracellular release of the DNA genome. J. Virol. **40:** 462–477.
26. LIM, K. & CH.-B. CHAE. 1989. A simple assay for DNA transfection by incubation of the cells in culture dishes with substrates for β-galactosidase. BioTechniques **7:** 576–579.
27. SPIESS, M. 1990. The asialoglycoprotein receptor. A model for endocytic transport receptors. Biochemistry **29:** 10009–10017.
28. STRYER, L. 1988. *In* Biochemistry, 3rd ed., W. H. Freeman. New York, NY.
29. SAMBROOK, J., E. F. FRITSCH & T. MANIATIS. 1989. Molecular Cloning III. Cold Spring Harbor Laboratory Press. New York.

Antisense and Differentiation

C. THOMAS CASKEY[a]

Institute for Molecular Genetics,
Human Genome Center, and
Howard Hughes Medical Institute
Baylor College of Medicine
Houston, Texas 77030

The aryl hydrocarbon hydroxylase (AHH) enzyme is one of a large family of drug-metabolizing enzymes encoded by the P450 superfamily of genes. Expression from the *Cyp1a-1* gene encoding this enzyme is inducible by administration of polycyclic aromatic hydrocarbons such as 3-methylcholanthrene (3-MC) or 2,3,7,8-tetrachlorodibenzo-*p*-dioxin (TCDD). An intracellular aromatic hydrocarbon (Ah) receptor that binds 3-MC and TCDD is necessary for induction of the *Cyp1a-1* gene. The mechanism of transcription activation involves binding of the polycyclic aromatic hydrocarbon to the Ah receptor, translocation of the receptor-ligand complex to the nucleus, and binding of the complex to multiple enhancer regions 5′ to the *Cyp1a-1* gene.

A 2.6-kb fragment from the promoter region of the *Cyp1a-1* gene contains multiple dioxin-responsive enhancer motifs and has been used to drive expression from the chloramphenicol acetyltransferase (CAT) reporter gene in transgenic mice. The response of this reporter construct in male and female transgenic mice was investigated with regard to inducibility and tissue specificity.

HPRT catalyzes the metabolic salvage of the preformed purine bases hypoxanthine and guanine, and is constitutively expressed at a low level in all mammalian tissues; HPRT mRNA accounts for less than 0.01% of all polyadenylated messages in most tissues. In cultured cells it is possible to select for both the presence and the absence of HPRT activity, making this a useful system for studying mutation and reversion within the HPRT gene.

Experiments over the last few years have demonstrated that intracellular nucleic acids complementary to endogenous RNA transcripts can suppress gene expression in a targeted manner. Such antisense inhibition of gene expression provides the potential for "turning off" selected genes and studying the effects of their deficiency. Because the HPRT system is so well characterized, this gene was considered a good candidate for experiments to determine the extent of sequences necessary for antisense inhibition.

A variety of constructs was assembled, comprising portions of the mouse *Hprt* gene under transcriptional control of the chicken β-actin promoter. These constructs were transfected into cultured mammalian cells, and their effect on endogenous HPRT mRNA and enzyme activity was measured.

RESULTS

A construct was engineered to contain the CAT gene linked to simian virus 40 (SV40) large T-antigen splice donor/acceptor sites and polyadenylation signals,

[a] Howard Hughes Medical Institute Investigator.

under the transcriptional control of the promoter/enhancer region of the *Cyp1a-1* gene. This construct was then used to generate three independent strains of transgenic mice (named Fo#2, Fo#6, and Fo#7).

Matched male and female F2-generation siblings from all three transgenic lines were induced by interperitoneal injection of 3-MC in corn oil or mock-induced with corn oil alone; 24 hours later the mice were anesthetized and sacrificed. No physiological effect was observed following injection of 3-MC or corn oil alone. Tissues harvested from the sacrificed mice were assayed for protein concentration and CAT activity.

Basal levels of CAT activity were detected in the spleen and small bowel (and to a lesser extent in the liver, kidney, and lung) of mock-induced mice. These levels were

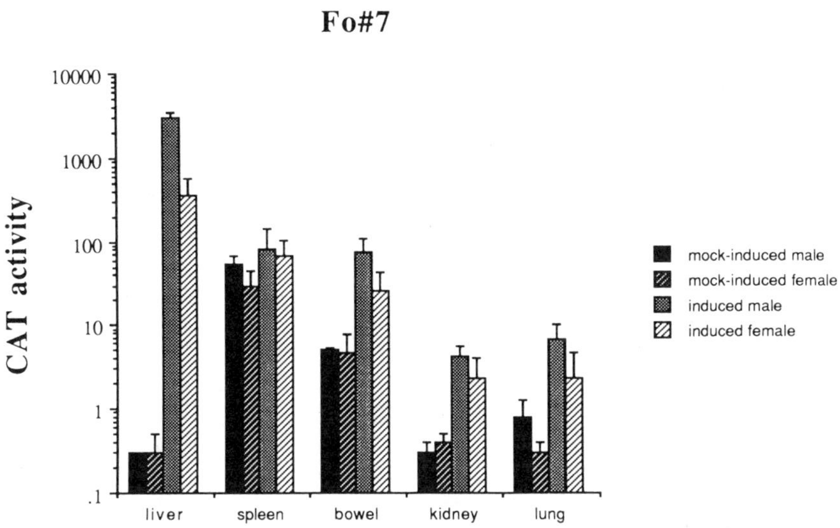

FIGURE 1. Induction of *Cyp1a-1*-CAT expression in Fo#7 transgenic mice. Four male and four female sibling mice from transgenic line Fo#7 were induced with 3-MC in corn oil or mock-induced with corn oil alone and sacrificed 24 hours later. Tissue extracts were assayed for CAT activity, and the results are expressed as a percentage conversion of chloramphenicol substrate to acetylated products per hour per 0.1 mg extract, on a logarithmic scale. Reproduced with permission from Jones *et al.*[5]

increased in induced animals, to a greater extent in male mice. Dramatic induction was observed in liver tissue, in which induced levels of CAT activity were up to 10,000-fold higher than mock-induced levels. FIGURE 1 shows an example of CAT activities of four Fo#7 mice from each sex, in response to 3-MC or mock induction; because the differences between basal and induced levels are so great, it has been necessary to plot CAT activity on a logarithmic scale. Neither basal nor induced transgene expression was detected in heart, muscle, brain, or testis/ovary tissue from the three transgenic mouse lines. Thus, these investigations indicate that the *Cyp1a-1* promoter/enhancer region functions as an inducible promoter in the spleen, small bowel, kidney, and lung of mice and as a strongly inducible promoter in the liver.

In one series of experiments, constructs were engineered to contain the full-length (1.3 kb) mouse HPRT cDNA or portions of the 5′ end of the mouse *Hprt* gene, in sense and antisense orientations, under the control of the chicken β-actin promoter. The larger gene fragment was 1.4 kb containing the first exon, 400 bp of the first intron, and approximately 850 bp of 5′ flanking sequence; the smaller gene fragment lacked the promoter region 5′ to the exon. When the antisense constructs were transfected into NIH-3T3, COS, or HeLa cells, numerous colonies resistant to 6-thioguanine (6-TG) were isolated, with <1.5% normal HPRT activity and abundant production of antisense HPRT RNA. The genomic antisense constructs were as effective as the cDNA construct in generating 6-TG resistance. Transfection of sense constructs did not result in the development of 6-TG–resistant colonies.

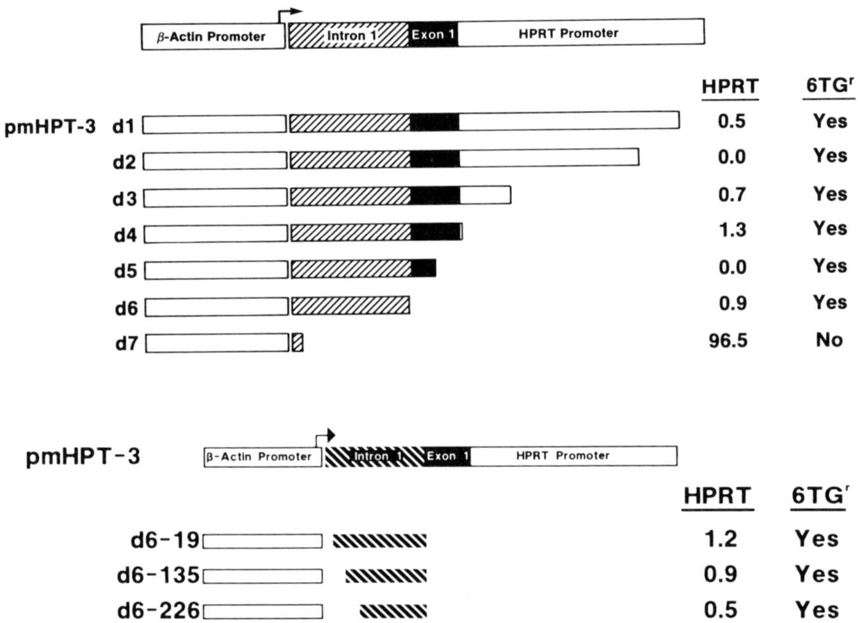

FIGURE 2. Deletion analysis of HPRT antisense gene constructs. pmHPT-3 consists of a 1.4-kilobase fragment from the 5′ end of the mouse *Hprt* gene, including the first exon, 400 base pairs of the first intron, and approximately 850 base pairs of the 5′ flanking sequence. Deletions were made in this construct as shown, with the resulting effect on HPRT activity and 6-TG resistance after transfection into NIH-3T3 cells. HPRT activity is expressed as a percentage of the specific activity seen in the parental cells. Reproduced with permission from Stout and Caskey.[3]

To define more accurately the region of the *Hprt* gene that could suppress gene expression by antisense inhibition, truncated versions of the original larger genomic fragment were tested. A diagram of these constructs is shown in FIGURE 2, together with the resulting cellular levels of HPRT activity compared to normal and an indication of 6-TG-resistance. The constructs named pmHTP-3d6, pmHTP-3d6-19, pmHTP-3d6-135, and pmHTP-3d6-226 all lack any coding sequence and yet are able to confer 6-TG resistance to transfected cells. Further deletion variants of pmHPT-3d6 were generated, removing DNA from the end of the fragment adjacent to the exon 1/intron 1 boundary. Constructs with termini 2 and 10 nucleotides from the exon/intron boundary were able to confer 6-TG resistance in the antisense orientation, but a fragment ending 37 nucleotides from the exon/intron boundary did not result in reduced HPRT levels.

NIH-3T3 or HeLa cells were cotransfected with pmHPT-3 and another plasmid containing the G418-resistance gene; G418-resistant cells were then assayed for the amount of antisense HPRT RNA and 6-TG resistance. These experiments indicated that a 65- to 86-fold excess of antisense RNA over normal HPRT mRNA levels was necessary to confer 6-TG resistance.

DISCUSSION

Expression of a CAT transgene driven by the *Cyp1a-1* promoter is enhanced up to 10,000-fold by administration of the inducer 3-MC. This would suggest a potential use for this promoter as an inducible regulator of hepatic gene expression in transgenic animals.

Antisense RNA complementary to 172 base pairs of the first intron of the mouse *Hprt* gene was as effective as RNA complementary to all nine exons in reducing cellular HPRT levels. This would suggest that one mechanism of antisense inhibition of HPRT gene expression is interference with RNA splicing in the nucleus.

It has been possible to achieve antisense inhibition of *Drosophila* (Krüppel) and mouse (shiverer) genes, resulting in mutant phenocopies.[1,2] We found this dominant transgene strategy an exciting concept for readily generating mouse mutants. In this report we provide evidence for splice junction sequences at the HPRT locus, serving to inhibit completely HPRT mRNA and protein synthesis in cultured cells.[3] We reported elsewhere our failure to inhibit HPRT in transgenic animals carrying and expressing the same antisense constructions.[4] Although one can never be certain of the reasons for a failed experiment, it was clear in this instance that the 60:1 ratio of antisense RNA: sense RNA achieved in cultured cells was never achieved in the transgenic animals. Furthermore, the ability to select 6-TG–resistant cells offered an advantage over transgenic mouse studies because the latter lacked selection power. In this report we identified a very powerful drug-inducible promoter that has functioned well in transgenic mice. This advancement together with the rapid advance of embryonic stem cell technology for mouse genetics may now provide a powerful tool for the creation of transgenic mice with inducible inhibition of developmental genes. Such a tool should provide a new genetic approach toward the study of developmental genes and lineage of expression.

SUMMARY

The use of antisense technology in animals has great potential for studying individual genes and for generating animal models of human disorders. Further control can be obtained by the use of inducible promoters to regulate the expression of antisense constructs, so that the timing and degree of antisense inhibition can be manipulated. In addition, tissue-specific promoters and enhancers provide the potential for targeting antisense inhibition to specific organs.

Transgene expression in mice directed by the *Cyp1a-1* promoter and enhancer elements can be increased in the liver up to 10,000-fold by administration to the animals of the inducer, 3-methylcolanthrene.[5] Transfected antisense constructs containing splice site regions of the hypoxanthine guanine phosphoribosyltransferase

(HPRT) gene can reduce HPRT activity to less than 1% of levels in parental NIH-3T3, COS, or HeLa cells.[3] Antisense constructs including splice site regions and driven by the inducible *Cyp1a-1* promoter should prove to be powerful tools in generating mouse models of human disorders.

ACKNOWLEDGMENTS

Most of the described work was performed by Drs. Stephen N. Jones[5] and J. Timothy Stout.[3] Dr. Belinda J. F. Rossiter assisted in preparation of the manuscript.

REFERENCES

1. ROSENBERG, U. B., A. PREISS, E. SEIFERT, H. JÄCKLE & D. C. KNIPPLE. 1985. Production of phenocopies by *Krüppel* antisense RNA injection into *Drosophila* embryos. Nature **313:** 703–706.
2. KATSUKI, M., M. SATO, M. KIMURA, M. YOKOYAMA, K. KOBAYASHI & T. NOMURA. 1988. Conversion of normal behavior to shiverer by myelin basic protein antisense cDNA in transgenic mice. Science **241:** 593–595.
3. STOUT, J. T. & C. T. CASKEY. 1990. Antisense RNA inhibition of HPRT synthesis. Somat. Cell Mol. Genet. **16:** 369–382.
4. MUNIR, M. I., B. J. F. ROSSITER & C. T. CASKEY. 1990. Antisense RNA production in transgenic mice. Somat. Cell Mol. Genet. **16:** 383–394.
5. JONES, S. N., P. G. JONES, H. IBARGUEN, C. T. CASKEY & W. J. CRAIGEN. 1991. Induction of the *Cyp1a-1* dioxin-responsive enhancer in transgenic mice. Nucleic Acids Res. **19:** 6547–6551.

Nonionic Oligonucleotide Analogs (Matagen™) as Anticodic Agents in Duplex and Triplex Formation

P. O. P. TS'O,[a] L. AURELIAN,[b] E. CHANG,[c]
AND P. S. MILLER[a]

[a] *The Johns Hopkins University School of Hygiene and Public Health*
615 North Wolfe Street
Baltimore, Maryland 21205

[b] *University of Maryland School of Medicine*
10 South Pine Street
Baltimore, Maryland 21201

[c] *Uniformed Services University*
4301 Jones Bridge Road
Bethesda, Maryland 20814

The promise of and enthusiasm for the antisense strategy is amply self-evident in this volume. The antisense strategy is no longer an avant-garde approach; rather, it has become a reliable tool for the identification of gene function in complex, living biological systems/hosts and has great potential for development of therapeutic agents. However, the contributions to this volume also reflect two fundamentally different approaches to this antisense strategy. A clarification at the onset will be of value.

The first approach is *biological,* in which a vector is constructed with the DNA of the target gene inserted with *reversed* polarity. After successful transfection, an "antisense-messenger" RNA (or any RNA transcribed by design) is produced inside the cell, based on controlled transcription of the vector. As in current molecular biological nomenclature the "transcribing" strand of the DNA duplex is termed the "sense" strand and the "nontranscribing" strand in the DNA duplex is termed the "non-sense" strand, naturally the transcribed RNA with reversed polarity is termed "antisense".[1,2] The antisense RNA is expected to form a double-stranded RNA duplex with the complementary target/"sense" RNA in the cell. This expected event may not always take place in transfected cells, because both the antisense RNA and the sense RNA have extensive secondary and tertiary structures. Therefore, the ability to form duplexes depends heavily on the linearly exposed portions of the two RNAs. Although certain genetic engineering steps can be taken to promote this process, the approach and the outcome are not under strict rational design and cannot be predicted precisely.

The second approach, pioneered in our laboratory in the late 1960s and early 1970s,[3–8] involves *chemical synthesis* of oligonucleotide or oligonucleotide analogs based on the sequence information of the target RNA. These oligomers can first be tested in cell-free biochemical systems for hybridization-arrest of protein synthesis or in biophysical systems for binding between the target RNA (or target RNA sequence) and the oligomers. The *a priori* requirement of this approach is precise sequence information on the target RNA or DNA. This approach can be adopted for single-stranded targets (mostly RNA), which involves duplex formation, or for double-stranded targets (mostly DNA), which involves triplex formation. The oligomers can be further modified by the attachment of chemical groups that promote covalent linkage to the target or cleavage of the target. This approach, which is based on an understanding of the chemical and physical properties of the oligomers and which uses molecular genetics/biology information, is totally under rational design.

To make a clear distinction between these two approaches, we suggested at several meetings[9,10] that the biological approach should retain the term "antisense approach," whereas the chemical approach should be termed the "anticode approach." The targets of the anticode approach can be at either the single-stranded RNA level or the double-stranded DNA level. The anticode approach appears to be much more amenable to efficient development and rational design of therapeutic agents. Our experiences over the last 20 years in this area of anticode approach are briefly described.

THE ANTICODE STRATEGY

Cornerstones

The rational design for the construction of active oligomers using the anticode strategy is based on the following two cornerstones.

1. To design oligomers that bypass or overcome the barriers and defense systems of living cells and tissues. The principal barriers are the membranes and protective layers of cells and tissues. The principal defense systems are nucleases.

2. To design oligomers that have absolute base sequence specificity coupled with as high an affinity for the target as possible or acceptable.

After careful consideration, one realizes that high sequence specificity and high affinity are contradictory demands. To gain high sequence specificity, the oligomer-nucleic acid interaction must have a high on-rate *and high off-rate* in order for the oligomer to search for the precise sequence. As a first approximation, the on-rate is primarily dependent upon diffusion, and the affinity or association constant can only be substantially increased by reducing the off-rate. Thus, any factor, such as the addition of positive charges or intercalators not related to sequence specificity, that significantly reduces the off-rate will increase affinity but at the sacrifice of sequence selectivity.

Similarly, for a very long oligomer, the off-rate of the mismatched duplex and the off-rates of shorter duplexes formed between the long oligomer and other targets may be sufficiently similar to that of the perfectly matched, full-length duplex to generate many pseudo-equilibrium states. For example, if "M" is the number of monomeric units in an oligomer for binding, and if "n" is the number of *contiguous* monomeric units in the oligomer that upon duplex formation have a sufficiently long off-rate to produce biological effects, then the number of degenerate contiguous binding sequences equals $\Sigma(\text{m-n})_{n \rightarrow m} + 1$. Thus, if it is assumed that a duplex formed between a target and a 12 mer has a sufficiently long off-rate to produce a biological effect at a given concentration, then for an 18-nucleotide oligomer the theoretical degeneracy in sequence selectivity will be 28. In other words, in addition to binding to the intended target as an 18 mer, there are theoretically 27 other shorter contiguous sequences (ranging from 12–17) within the oligomer that could also have sufficiently stable binding constants for targets in the cell. Obviously to avoid this situation, the concentration of the oligomer should be reduced so that the binding of the shorter sequences within the oligomers will be sufficiently reduced. However, the free energy of binding (assuming a linear, additive situation) of a 10 mer *vs* a 9 mer will be reduced about 10%, whereas that of a 20 mer *vs* a 19 mer will be reduced only 5%. Thus, the difference in selectivity between an (m) oligomer and an (m-1) oligomer will be much greater if m is small (~10–12) and will be smaller if m is large (~18–20). On the other hand, the oligomer should be sufficiently long to provide a statistically unique selection among all the possible nucleic acid sequences available inside the cell.

A corollary to the second cornerstone is not only that the oligomer (probe) should be absolutely specific for the target nucleic acid, but also that it should be inert or noninteractive with any other biological components of the cell or tissue. In other words, the oligomer should be "noble" and should interact *only* with targeted nucleic acid. Any strong, adventitious interaction of the oligomer with proteins, enzymes, lipids, or membranes will reduce the predictability of the effectiveness of the oligomer in the anticode approach and possibly lead to undesirable side effects. Thus, interaction of negatively charged oligomers with nucleic acid enzymes would be of great concern.

Backbone Construction

In considering the two cornerstones described above, the Johns Hopkins group designed and studied anticode oligonucleotide analogs with a nonionic backbone. Of the many possible types of constructs, three are described extensively in publications beginning in 1970.

The first nonionic backbone studied by us was the nonionic ethylphosphotriester linkage first reported in 1971.[3] We were the first to note that the ethylphosphotriester linkages of deoxyoligonucleotides, contrary to the prevailing notion at that time, are stable in aqueous solution. We further characterized these oligomers by NMR and studied their hybridization properties in solution.[10] In 1977, a comprehensive and comparative investigation of the physical chemistry of nucleic acid interaction involving this nonionic backbone was published.[11]

We also studied the anticode effects of the ethylphosphotriester deoxyoligomer analogs or their 2′-O-methylribosyl derivatives.[12] In the early 1970s, the only naturally occurring nucleic acid sequences known were those of tRNAs (particularly phenylalanine tRNA). Thus, we constructed anticode oligomers targeted toward specific tRNAs for various amino acids and demonstrated sequence specificity. Furthermore, these compounds were shown to inhibit protein synthesis and cell growth in mammalian cells in culture.[8] However, we also observed a moderately rapid degradation of the phosphotriester analogs in the cells. Studies on ethylphosphotriester oligomers from 1969–1977 showed us that although the anticode approach was workable, a more biologically and chemically stable backbone linkage was required.

During this time, the Johns Hopkins group also investigated oligomers with hydrocarbon backbones. A paper on poly (1-vinyl uracil) was published in 1970.[3] The interaction of poly (1-vinyl uracil) with poly A was observed and characterized. Other types of oligomers with hydrocarbon backbones were also constructed with particular attention to the distances between the base pairs in a nucleic acid helix. It soon became evident that with many freely rotatable bonds in this type of backbone, the favorable entropic factors for interaction that are intrinsic to the nucleic backbone are lost. The lesson we learned from this set of compounds was that we should preserve as much of the naturally occurring nucleic acid backbone as possible.

On the basis of these observations, the Johns Hopkins group went on to develop the nonionic methylphosphonate backbone. Both deoxyoligomers[13–16] and the more difficult to synthesize 2′-O-methylribosyl oligomers[17] were built with the methylphosphonate linkage.

The chemical structure of a deoxyribosyl oligonucleoside methylphosphonate is shown in FIGURE 1. This family of compounds has the following general properties:

1. As shown, the 5′ terminus of this oligomer contains a phosphodiester linkage. The negatively charged phosphodiester group allows easy purification of the intact and full-length oligomer by DEAE-ionic exchange chromatography and allows the

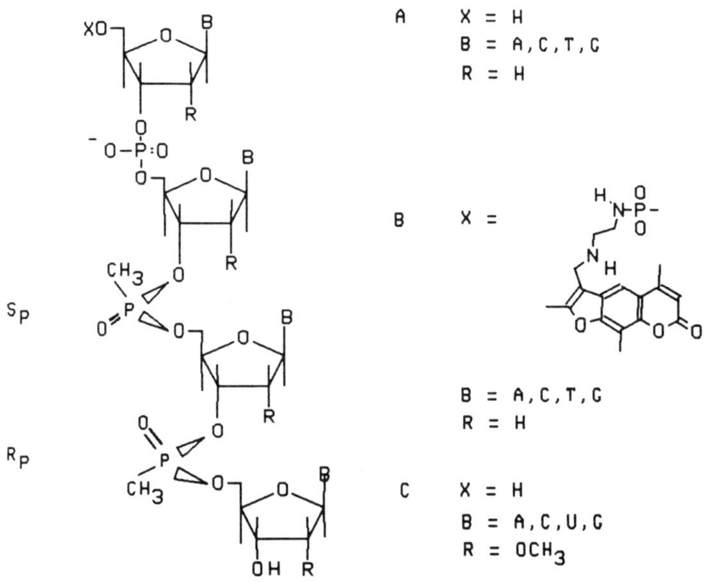

FIGURE 1. Structures of **(A)** an oligodeoxyribonucleoside methylphosphonate, **(B)** an oligodeoxyribonucleoside methylphosphonate derivatized with 4′-N-(aminoethyl)amino-methyl-4,5′,8-trimethylpsoralen, and **(C)** an oligo-2′-O-methylribonucleoside methylphosphonate.

oligomer to be enzymatically phosphorylated by polynucleotide kinase. With more advanced technology in the separation of oligomers by size, the presence of the charged terminus may not be required for the preparation of a full-length nonionic molecule.

2. As shown, each methylphosphonate linkage may have an R or an S configuration. The stereochemically pure R and S diastereoisomers of dApA (the symbol p indicating a methylphosphonate linkage) have been obtained, and their ability to undergo triplex formation with polypyrimidines nucleotides with (2U or T):(1A) stoichiometry has been studied. The dApA(S):2 poly U has a higher Tm (19.8°C) than does dApA(R):2 poly U (15.4°C). However, the Tms of dApA(S):2 poly dT and dApA(R):2 poly dT are nearly the same (18.4°C *vs* 18.7°C, respectively).[18] In addition, comparative studies on Tm(s) and the sharpness of the transition (d

Absorbance/dT) indicates the following series: $d\underline{A}_{16} \cdot dT_{16} > dA_{16} \cdot dT_{16} \cdot \gg dA_{16} \cdot d\underline{T}_{16}$.[19] This observation suggests that the methyl groups in $d\underline{T}_{16}$ methylphosphonate may provide some specific steric hindrance that is not found at least in dA or dG methylphosphonates. Molecular modeling studies suggest that this steric hindrance may not occur with RNA targets, suggesting that A type duplexes may be less sensitive to this steric hindrance effect than are B form duplexes (M. Reynolds, private communication). Despite the possible steric hindrance of the phosphonate methyl group to duplex or triplex formation, the configurations of the diastereoisomers produced during oligomer synthesis must be sufficiently random that the major portion (over 98%) if not the entire population of molecules in the preparation do participate in interactions with the target as indicated by experimental observations with gel analyses, NMR studies, sharpness of the thermal transition, and the like.

3. The solubility of this family of compounds in aqueous buffer depends on the size, composition, and possibly even sequence of the oligomer. A high percentage of guanine or, worse, an aggregate of contiguous guanine residues sharply reduces solubility. For example, $d(C\underline{p}T)_8$ is soluble up to millimolar concentrations, whereas $d(A\underline{p}G)_8$ has a solubility of less than 0.1 mM. The latter compound is more soluble at lower rather than higher temperatures, possibly as a result of hydrophobic factors. In general, for 12–15 mers, the solubilities will be between 0.1 and 0.3 mM. The oligomers are highly soluble in mixed aqueous organic solvents such as 50% acetonitrile/water or 50% ethanol/water.

4. The methylphosphonate oligomers are very stable chemically in aqueous solution at neutral pH. The purine-containing deoxyoligomers are hydrolyzed at pH 1 or less because of depurination, but the 2′-O-methylribosyl oligomers appear to be much more resistant to acid. The methylphosphonate linkage is hydrolyzed under alkaline conditions and in the presence of certain amines, such as aqueous piperidine. Thus, these oligomers cannot be deprotected using the concentrated ammonium hydroxide reagent normally used for deprotecting oligodeoxyribonucleotides.

Biochemically, these compounds are totally resistant to nucleases found in serum or in cell extracts. The oligomers are not bound to serum albumin, but they can be phosphorylated by polynucleotide kinase if the 5′ terminus contains a phosphodiester linkage. This feature can be used conveniently for labeling. In addition, oligonucleoside methylphosphonates can serve as primers for reverse transcriptase and for deoxynucleotidyl terminal transferase.

5. The physicochemical properties of these compounds have been extensively characterized by NMR, mass spectroscopy, circular dichromism, UV absorption spectroscopy, and the like.[20] The interaction of these compounds with the corresponding complementary oligomers, either phosphodiester linkages or methylphosphonate linkages, has also been characterized extensively by the aforementioned physicochemical approach.

6. The pharmacokinetics of an 11-mer methylphosphonate injected into the tail veins of mice have been investigated.[21] In brief, the compound was found to distribute readily to all organs of the body except the brain. In addition, the oligomer was excreted rapidly into the urine. The oligomer was recovered intact or minus the 5′ terminal nucleotide unit. There was no evidence of hydrolysis of the methylphosphonate linkages. However, in this study as in all other similar investigations, it is unclear if the compound was in the vascular system of the tissue or inside the cells.

7. Finally, this family of oligomers with methylphosphonate linkages has been termed "Matagen," an acronym for Masking tape for gene expression. This term indicates the biological function of this family of compounds in an anticode strategy. Matagen is a collective term for the methylphosphonate oligomers targeted against either RNA or DNA.

Antiviral Research

Application of anticode strategy to the study of viral gene function and to the development of antiviral therapeutic agents appears to be naturally appealing. The entire nucleic acid sequence of many viruses is now known, and the molecular biology of many viruses has also been investigated extensively. However, for a large virus such as herpes simplex virus (HSV), which has a molecular weight close to 100 million and consists of nearly 100 genes, the study of the function of each individual viral gene and development of an effective therapeutic agent remains a formidable challenge.

Only the salient features[23,24] of the extensive research performed with HSV at the University of Maryland are described here. The first phase of this investigation used

A

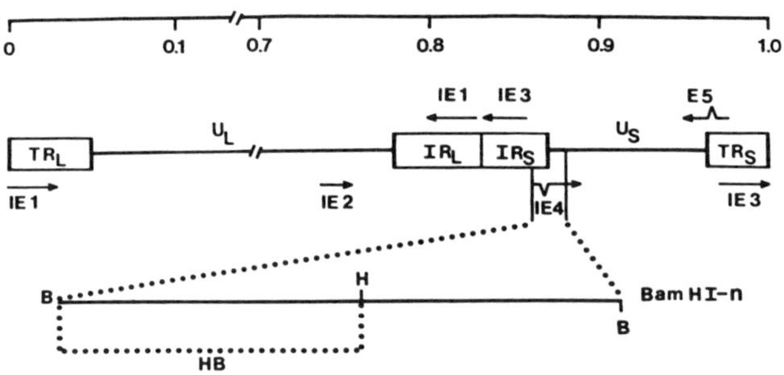

B

```
                      INTRON        EXON
              5'                              3'
HSV-1 IE mRNA 4:  CGTGCCTTCCCGCAG]GAGGAACGTCCTCGT
                           3'GGCGTC CTCCTT 5'

HSV-2 IE mRNA 4:  CGCGCTTTCTCGCAG]GCCGGGCGCCGCCTT
```

FIGURE 2. **(A)** Approximate map location of 1E mRNAs 4 and 5. The restriction fragment (HB) labeled with ^{32}P at the unique 5′ end *Hind*III site was used in S1 nuclease analysis of the 5′ portion of 1E mRNA 4. TR_L and TR_S, terminal repeats; IR_L and IR_S, inverted repeats; U_L and U_S, unique long and short DNA sequences. **(B)** Partial nucleotide sequence of HSV-1 and HSV-2 IE pre-mRNA 4 at the intron/exon splice acceptor site (]). The sequence of the dodecamer complementary to the acceptor junction is shown below the HSV-1 sequence. (Reproduced with permission from Kulka *et al.*[23])

sequence information on the splice junctions of immediate early pre-mRNA 4 and 5 of HSV-1 and HSV-2 (FIG. 2). The strategy was to mask the splice junction by an 8-mer Matagen, dTCCTCCTG, or by an 11-mer Matagen, dTTCCTCCTGCGG. The main assay was to measure viral titer as determined by a plaque assay, usually 24 hours postinfection. Additional biochemical assays of protein synthesis and viral mRNA synthesis were also performed.

The major thrust of this first phase of the investigation focused on three issues: (1) specificity, (2) mechanism; and (3) animal experimentation.

1. Specificity. The specificity issue was examined in four ways.

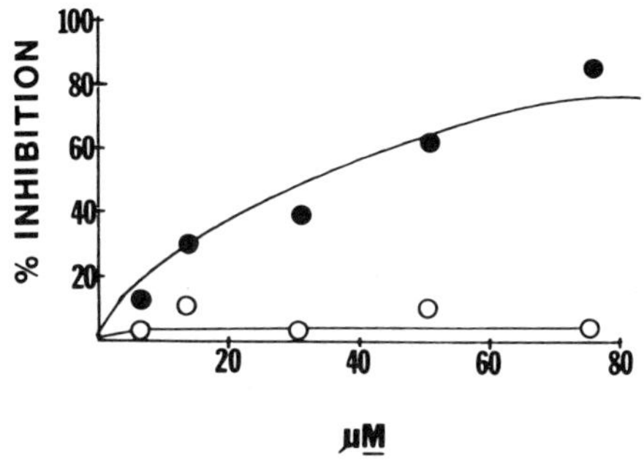

FIGURE 3. Inhibition of virus production. HSV-1 (F)-infected Vero cells were exposed to d(TpCCTCCTG) (●) or d(TpCCCTCTG) (○) at 0 hpi. (Reproduced with permission from Smith *et al.*[24])

A. Sequence specificity. Reversal of the two center bases of dTpCCTCCTG, an oligomer complementary to the acceptor splice junction of HSV IE mRNA 4 and 5, or the change of sequence to d(TpCCCTCTG), was sufficient to remove all the inhibitory activity of d-TpCCTCCTG (FIG. 3).

B. Strain specificity. As shown in TABLE 1, slight changes in the lengths or sequences of Matagens complementary to the split junction results in distinct strain-specific inhibitory effects on HSV-1 and HSV-2 replication.

C. Timing specificity. In two separate and different experiments, the data clearly indicate that the addition of Matagen to the cell culture is most effective when it takes place prior to or simultaneously with the onset of viral infection. The addition of Matagen 1 to 4 hours after infection is largely ineffective. This observation supports the notion that the added Matagen indeed inhibits immediate early gene expression and that expression of these IE genes is important for viral replication and pathogenesis.

D. Virus versus host specificity. The dodecamer d(TpTCCTCCTGCGG) caused 50% and 98% decreases in HSV-1 titers at concentrations of 15 and 100 μM,

TABLE 1. Effect of the Oligo(deoxynucleoside methylphosphonate) Sequence on HSV Growth[a]

Oligomer		% Inhibition	
Exon	Intron	HSV-1	HSV-2
(3′ AAGGAG	GACGCC 5′)		
TCCTC	CTG	85	40
TTCCTC	CT	14	ND
GCGTTCCTC	CTG	30	0
GTTCCTC	CTGCGG	98	0
TC	CTGCGG	0	0
TC	CTGCGGGAAG	9	12
	GCGGGAAGGCAC	0	0
TCCCT	CTG	0	0

[a]HSV-1- or HSV-2-infected Vero cells were treated with the respective oligomers (100 μM) at 0 hour postinfection. Infectious virus was determined by plaque assay at 24 hours postinfection. Results are the average of three to six experiments for each oligomer. They are expressed as the percentage of inhibition of virus titers compared to the appropriate control (no oligomer). HSV-1 IE mRNA 4 acceptor splice site is shown in parentheses. The HSV-2 IE mRNA 4 acceptor splice site is GGGCCG GACGCT. (Reproduced with permission from Kulka *et al.*[23])

respectively. At these concentrations of Matagens, both the dodecamer and the octomer d(TpCCTCCTG) caused only inhibition of viral protein and viral DNA synthesis, but not cellular protein synthesis, DNA synthesis, and cell growth.

2. Mechanism. The Matagens were designed to inhibit splicing of HSV IE mRNA 4 and 5. This proposed mechanism of inhibition is strongly supported by the following sets of experimental observations.

A. Matagens complementary to either the 5′ or the 3′ side of the splice junction have similar inhibitory effects on viral titers (Kulka *et al.,* submitted). These observations strongly suggest that these two very different but sequence-specific Matagens work by the same mechanism, that is, prevention of the splicing of the IE-4,5 gene.

B. Because specific antibodies against the gene products of IE-4 and IE-5 genes are not available, we are unable to show the corresponding decrease in the IE-4 or IE-5 protein after treatment with Matagen. However, we were able to show the presence of the *un*spliced IE-4,5 mRNA, which again indicates that inhibition of splicing took place in the presence of Matagen in cell culture.

C. Additional support for the foregoing conclusions derives from studies of Matagens complementary to the initiation codon region of either the IE-4 or the IE-5 protein mRNA. Individually, these two Matagens will presumably inhibit translation of IE-4 or IE-5 mRNA. However, the Matagen targeted against IE-4 mRNA had an inhibitory effect on viral titer similar to that of the Matagen targeted against the splice junction, whereas the Matagen targeted against the IE-5 mRNA had little inhibitory effect on viral yield. The most logical and simplest interpretation of this interesting observation is that inhibition of IE-4 protein synthesis is vital to viral replication which can be accomplished by either masking the splice junction level or interfering with protein synthesis. On the other hand, reduction of IE-5 protein caused by either masking event may have little effect on viral replication (Kulka *et al.,* submitted).

3. Animal experimentation (Kulka *et al.,* submitted). Success in cellular experiments with anti-HSV matagens obviously leads to the challenge of whether this approach is applicable to animal or human, at least at the level of HSV skin infection and topical application of Matagen. The experimental system is that of intradermal injection of HSV in the mouse ear, followed by daily application of Matagen emulsified in polyethylene glycol cream. The experimental observation is based on monitoring the lesion outbreak or the increase in viral titer in the infected ear. Three salient features can be summarized as follows:

A. Outbreak of herpetic lesions can be significantly suppressed.

B. Inhibition of viral titers in infected ears can be positively correlated with an increase in Matagen concentration and can be negatively correlated with an increase in the titer (PFU) of the viral inoculum.

C. Strain specificity can be demonstrated, that is, the Matagen-specific HSV-1 is less effective against an HSV-2 infection than is the respective HSV-2 Matagen.

These results indicate that indeed the nonionic Matagen has certain transdermal properties that have been confirmed by others (B. Tseng, unpublished results). Taken together, all these observations underscore the potential for the development of topical therapeutic agents against HSV-1 and HSV-2 infections. These Matagens do *not* attack the latent viral DNA or the presumptive messages of a few genes needed to maintain latency. However, they can prevent reinfection of healthy, uninfected cells if applied at an early stage during HSV recurrence, to prevent the outbreak of the lesion, or, if applied prophylactically, to prevent epidermal reinfection by the viruses migrating out of the neuronal cells.

Although the search for the most effective Matagen against HSV continues, we have preliminary evidence that when two pertinent viral genes are suppressed by the Matagens at the same time, there is a large synergism in inhibition of virus growth. This is another major principle in design of therapeutic treatment that should be kept in mind.

Suppression of Mutated Oncogene

The extensive research conducted in the Uniformed Services University of the Health Sciences[25–28] has two main characteristics: (1) a single goal and (2) a systematic and exhaustive investigation. This single goal is to investigate if any anticode agent can be sufficiently selective to be able to suppress expression of a single-point (single base) mutated gene (such as an oncogene). This investigation was reported in a series of papers, including a comparison of oligomers constructed with different types of backbones,[27] particularly at the onset. This investigation compared methylphosphonate linkage and phosphorothioate linkage as the anticode agent in a cell-free translational system. Anticode agents with methylphosphonate were selected owing to its superb selectivity in base sequence and its inertness in the interaction with the translation machinery. Current progress in this exciting investigation is summarized below.

1. The experimental system is based on the single base mutation (A to T transition) in the second base of codon 61 of Ha-*ras* gene which encodes the p21 protein. This A to T transition leads to replacement of glutamine by leucine in the protein, resulting in a slight increase in the gel electrophoretic mobility of the mutated *ras* p21 relative to that of its normal counterpart. Thus, an assay based on mobility in gel electrophoresis can readily be established in order to distinguish between selective suppression of the mutated *ras* p21 or the normal *ras* p21 by the added Matagens. The *ras* p21 is identified by the appropriate monoclonal antibody.

Two NIH-3T3 cell lines were used. RS504 cells contain multiple copies of a cloned human c-Ha-*ras* gene that has a mutation in the 12th amino acid codon while the 61st amino acid codon is normal. The other 453 cells contain multiple copies of a human c-Ha-*ras* gene that has a normal 12th amino acid codon, but contains a point-mutated 61st amino acid codon that is *not* of concern in these experiments. Both cell lines express elevated levels of p21, form colonies in soft agar, and grow well in nude mice. Thus, the investigation can be carried on four levels of biological organizations: (1) cell-free protein synthesis system; (2) intracellular labeling/synthesis of *ras* p21; (3) colony formation in soft agar; and (4) tumor growth in nude mice.

The comparative studies involve two anti-*ras* dodecameric matagens, that is, 5′ Tp CCTCCTGGCC (*ras* T) and 5′ TpCCTCCAGGCC (*ras* T) that differ in only one base in the middle of the oligomers.

2. In the cell-free protein synthesis system for inhibition of translation with a perfectly matched dodecameric matagen, inhibition at the 59% level required only 25 μM, at 90% 50 μM, and at 98% level 100–200 μM concentration. A single mismatched base caused a three to fivefold loss of activity and two mismatched bases caused a more than 10-fold loss of inhibitory activity. The striking results shown in specificity experiments (FIG. 4) involving mixed cell cultures indicate that indeed intracellular synthesis of *ras* p21 was suppressed significantly only by the perfectly matched Matagen and not by the singly mismatched Matagen. Furthermore, colony formation of these transformed cells in soft agar and the growth of the injected transformed cells in nude mice were suppressed significantly by the perfectly matched Matagen and demonstrably more than the singly mismatched Matagen

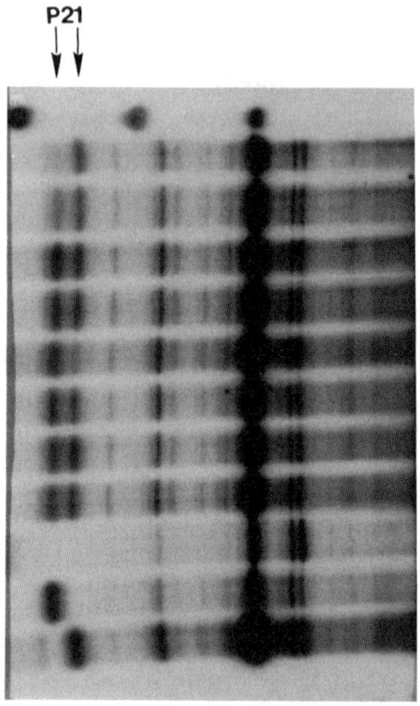

Cells	Oligo	Conc. (μM)	Ab.
Mix	RAS-A	150	+
Mix	RAS-A	75	+
Mix	RAS-A	37.5	+
Mix	RAS-A	0	+
Mix	RAS-T	150	+
Mix	RAS-T	75	+
Mix	RAS-T	37.5	+
Mix	RAS-T	0	+
Mix	——	0	-
453	——	0	+
504	——	0	+

FIGURE 4. Inhibition of *ras* p21 expression in 453 and RS504 cells by sequence-specific anti-*ras* ONMPs. Twenty-four hours prior to treatment, 2×10^5 cells (453:RS504 = 3:1) were seeded in each well of a six-well tissue culture dish. Cells were treated with the indicated concentrations for *ras*-A or *ras*-T for 48 hours and then incubated in the presence of 150 μCi/mL [^{35}S]methionine in low methionine medium. The cell lysate was preincubated with protein A-Sepharose beads coated with goat anti-rat IgG. After removal of the beads, 2×10^7 cpm of TCA-precipitable counts was immunoprecipitated with monoclonal antibody (Ab.) Y13-259 directed against p21 and analyzed by SDS-13% polyacrylamide gel electrophoresis followed by autoradiography. (Reproduced with permission from Chang *et al.*[28])

(manuscripts in preparation). These results strongly suggest that the abnormal growth of these mouse cells, transformed by transfection of Ha-*ras* gene suffered from a single base mutation, can be suppressed with a high degree of selectivity.

3. As described in a subsequent section, the psoralen-derivatized Matagen combined with laser irradiation can have an activity 10-fold higher than that of the underivatized Matagen while maintaining the *same* level of high selectivity, that is, sensitive to a single-base mismatch (FIG. 5).

The foregoing observations provide the foundation for future investigation in animal systems. It would be exciting to know if carcinogenesis can be prevented by prophylactic treatment of Matagen or even if tumor growth can be suppressed or reduced by treatment of Matagen. The major question is will Matagen 11–13 mers in length exhibit the ultimate selectivity that can detect and distinguish a single-base mismatched target versus a perfectly matched target in an animal system.

Potency Enhancement

1. *The binding isotherm.* As shown in FIGURE 6, the calculated binding isotherm using a dissociation constant of 10^{-5} M shows that the binding process is very efficient

at low concentrations of the Matagen where the association process dominates. At higher concentrations of Matagen and at binding levels higher than 50–60%, the dissociation process begins to become dominant. From this thermodynamic consideration, 100% binding level is approached asymptotically. At high concentrations of oligomer, not only is binding inefficient, but also the specificity decreases because binding of less energetically favorable states is now driven by the high oligomer concentration. This thermodynamic factor is basic to the consideration of specificity and the extent of binding. The biochemical and biological data described in sections on "antiviral" activity and "oncogene suppression" support this theoretical consideration. At 15–25 μM concentrations, about 50% inhibitory activity is observed. To reach 95–99% inhibition, concentrations between 100 and 200 μM are often required in cell-free experiments.

2. *The appropriate length of the Matagen.* Obviously, a decrease in the dissociation constant can change the steepness of the binding isotherm and reduce the concentration needed for a high level of binding. However, as described above, an increase in

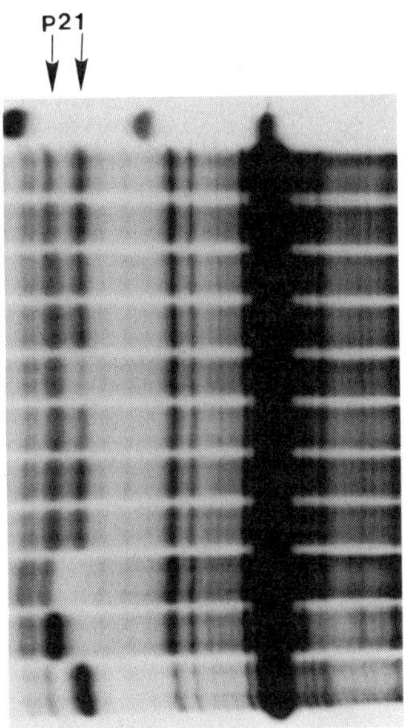

Cells	Oligo	Conc. (μM)	Ab.
Mix	psRAS-A	15	+
Mix	psRAS-A	7.5	+
Mix	psRAS-A	3.75	+
Mix	psRAS-A	0	+
Mix	psRAS-T	15	+
Mix	psRAS-T	7.5	+
Mix	psRAS-T	3.75	+
Mix	psRAS-T	0	+
Mix	———	0	−
453	———	0	+
504	———	0	+

FIGURE 5. Inhibition of *ras*-p21 expression in 453 and RS504 cells by psoralen-derivatized anti-*ras* ONMPs. Twenty-four hours prior to treatment, 2 × 10^5 cells (453:RS504 = 3.1) were seeded in each well of a six-well tissue culture dish. The psoralen-modified anti-*ras* ONMP ps*ras*-A or ps*ras*-T was then added at the indicated concentrations to duplicate wells. After incubation for 48 hours, one member of each pair of oligomer-treated cells was irradiated with a XeF laser (Questek 2000 excimer laser) at 40 Hz and 23 mJ/pulse for 110 seconds. This irradiation step was followed by metabolic labeling with 150 μCi of [^{35}S]methionine/well in 0.8 ml of low-methionine medium/well for 6 hours. The labeling medium was not supplemented with fresh oligomer. Cells were lysed and the lysates were immunoprecipitated with monoclonal antibody Y13-259 directed against p21 and then with protein A-Sepharose beads coated with goat anti-rat IgG. The immunoprecipitates were analyzed by SDS-13% polyacrylamide gel electrophoresis followed by autoradiography. (Reproduced with permission from Chang *et al.*[28])

oligomer chain length will also create a certain degree of sequence-specific degeneracy, which has to be kept in mind. In addition, from the standpoint of duplex formation, the target mRNA is unlikely to have very long stretches of open, single-stranded regions for duplex formation with the Matagen.

Nevertheless, the length of Matagen should be sufficient so that the target region would be unique for a potential binding. A discussion of this issue was published in the Annals of the New York Academy of Sciences in 1987.[29] This discussion is quoted verbatim below.

> 4. A discussion on the statistical basis of the chain length of Matagen versus its sequence specificity would be instructive. The average size of the mammalian haploid genome is about 3×10^9 base pairs and the average gene expression in any given cell is only 1–3% of the total genome or $3 \times 10^7 - 1 \times 10^8$ base pairs. It is estimated that only 10% of the mRNA sequences or $3 \times 10^6 - 1 \times 10^7$ nucleotides have fully exposed regions of 8–12 nucleotides in length. The number of unique sequences is a combination of 4 bases, equivalent to 4^n, where n is the chain length of the sequence. Thus, 4^{10} equals 1.05×10^6, 4^{12} equals 1.68×10^7, 4^{14} equals 2.68×10^8 and 4^{16} equals 4.16×10^9. According to these considerations, a 12-mer could be a unique sequence in a genome size of approximately 1.5×10^7 base pairs. Thus, on a statistical basis, a 12-mer is sufficiently long to mask a specific mRNA species. In order to mask a specific gene in DNA, a 14-mer, which would give 3×10^8 sequence specificity, may be more appropriate. Since approximately 50% of the DNA is in the form of repetitive or mid-repetitive sequences, a chain length of 14 may be sufficient to uniquely define the sequence specificity (1.5×10^9 b.p. in a single copy sequence). In any event, a chain length of 16 is sufficiently unique on a statistical basis for the entire mammalian genome and may be used in solution hybridization experiments.

3. Changing the "off-rate" of the binding process. As described above, decreasing the off-rate is really the only practical method of decreasing the dissociation constant. However, decreasing the off-rate without including the contribution of the

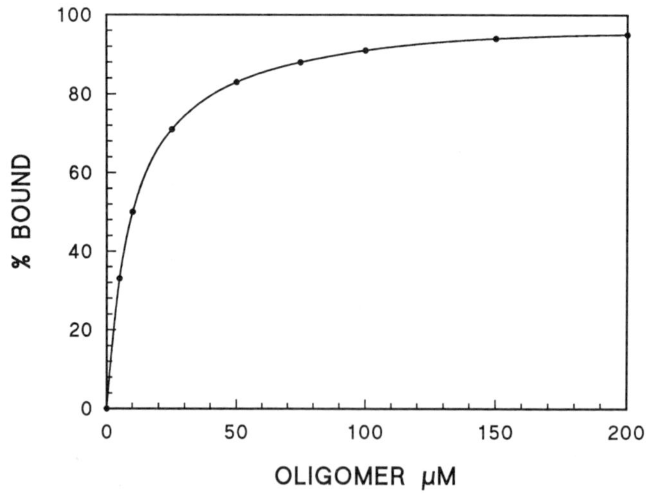

FIGURE 6. The binding isotherm between an oligomer and a target with a dissociation constant of 10^{-5} M.

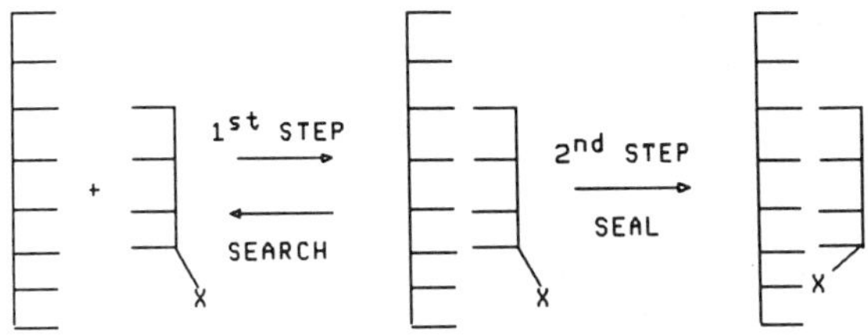

FIGURE 7. The diagram of a two-step *search* and *seal* approach between a target and an oligomer probe. X denotes a chemical linkage.

base-pairing mechanism, which occurs, for example, when an intercalator is conjugated to the oligomer, would decrease the binding selectivity. Lengthening the Matagen so that reduction of the off-rate is due to the increase in the base-pairing mechanism may cause a sufficient reduction in the rate of the searching process to allow the Matagen to bind and stay on the wrong target long enough to generate undesirable biological side effects.

What is the theoretical solution to this limitation? Obviously, the only way to have high affinity and high selectivity together is *to reduce the off-rate after a perfectly matched duplex is formed.* This consideration demands a two-step approach, that is, a "*search* and *seal*" approach (FIG. 7).

The first experimental adaptation of this "search and seal" approach is shown in FIGURE 8. The "sealer" used is a photoreactive psoralen group derivatized to the 5′ end of the Matagen.[30–33] Sealing is achieved by irradiation with long wavelength (365 nm or greater) light.

The following five points about this experimental approach are worth noting.

a. Psoralen nucleic acid photochemistry is well known. Photocyclization between psoralen and nucleic acid bases is specific for pyrimidines and selective for uracil residues in RNA. Thus, the sealing reaction only occurs when the psoralen is properly oriented with respect to a pyrimidine in the target RNA. Because irradiation is at wavelengths above those at which nucleic acid bases absorb, a very specific linkage can be induced *after* the physical pairing between the Matagen and the target region of RNA. The dependence of this photochemical reaction on duplex formation has been characterized extensively. This process not only reduces the off-rate but

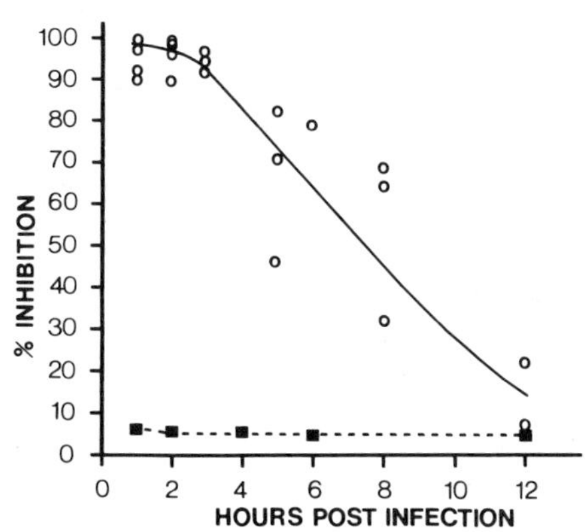

FIGURE 8. Inhibition of HSV-1 growth by psoralen-derivatized oligo(deoxynucleoside methylphosphonate). HSV-1-infected Vero cells were treated (○) or not treated (■) with 5 μM aeAMT-d(pTpTCCTCCT-GCGG) at 0 hours postinfection. At selected times (1–12 hours) after infection, the cells were exposed (10 minutes) to near UV light (365 nm), and virus titers were determined 24 hours later. Results are expressed as the percentage of inhibition of virus titer compared to untreated (no oligomer), nonirradiated HSV-1-infected Vero cells. (Reproduced with permission from Kulka *et al.*[23])

also gains a certain level of selectivity because of the base specificity of the cross-linking reaction.

b. Psoralen has been used in medicine for a long time. The amount of psoralen attached to Matagen for therapeutic purposes will be much smaller than that of psoralen used alone as a drug.

c. Photochemical linkage of a psoralen-derivatized Matagen to a target nucleic acid provides a unique and excellent tool to study the specificity of duplex formation. This approach has been used to investigate the sequence specificity of binding of psoralen-derivatized Matagens to various mRNAs of the vesicular stomatitis virus (VSV). Highly selective binding and cross-linking to M or N protein mRNAs was observed at 20°C. This binding specificity decreased when irradiation was carried out at 0°C. The photocross-linking reaction can also be used to unambiguously demonstrate triplex formation between Matagen and double-stranded DNA.

d. Obviously, this approach is limited to target sites that are accessible to irradiation. However, in addition to topical application and irradiation, many cavities inside the body can be accessed by photoirradiation by laser fiberoptics. It should also be noted that in addition to cross-linking, the psoralen group undergoes photodegradation. The photodegraded psoralen cannot participate in further cross-linking. Thus, Matagen molecules derivatized with photodegraded psoralen can act as competitive inhibitors for the photochemically induced linking reaction, because these molecules can bind but cannot cross-link. To increase the cross-linking yield, fresh Matagen must be added before further irradiation.

e. Finally, as shown in FIGURES 5 and 8 in the anti-HSV and the anti-ras systems, psoralen-derivatized Matagens are 10- to 20-fold more active than are the parent underivatized Matagens. Preliminary experiments also show that psoralen-derivatized Matagen is approximately 10 times more effective as an inhibitor of herpetic lesions than is the parent underivatized Matagen in the mouse ear infection model.

Triplex Formation

Formation of triple-stranded helices of nucleic acid was observed over three decades ago and was regarded as a novelty by nucleic acid physical chemists at that time.[34–37] Current interest in these structures derives from the possibility of using anticode oligonucleotides to target double-stranded DNA in the nucleus of cells.[38–41] Methylphosphonate oligomers have two unique properties that make them suitable for targeting double-stranded DNA: (1) As shown below, the nonionic backbone of methylphosphonate oligomers reduces the electrostatic repulsion normally encountered in triplex formation. This electrostatic repulsion between the three negatively charged strands of nucleic acid is a major deterence to triplex formation as indicated by the requirement of high concentrations of salt, divalent metal ions, or polyamines in the medium: (2) As described in the section on "antiviral research," Matagen inhibits splicing of HSV mRNA in a highly specific manner in living cells. Because mRNA splicing takes place only in the nucleus, this result demonstrates that Matagen can readily penetrate this organelle.

The following summarizes the two major areas of progress in our laboratories in the research on triplex formation. At the outset, it should be noted that only the purine bases (adenine and guanine) of double-stranded DNA or RNA have acceptor or donor sites suitable for formation of two additional hydrogen bonds with the bases of an incoming third strand. The pyrimidines (cytosine, thymine, or uracil) in the original duplex possess only one hydrogen bonding site for interaction with the bases of a third strand. This fact implies that from the standpoint of both stability and selectivity, only the following types of motifs are expected for triplex formation: pyrimidine:purine-pyrimidine or purine:purine-pyrimidine, where: ": symbol" denotes Hoogsteen type or Hoogsteen-like base pairing and "- symbol" denotes Watson-Crick type base pairing.

Triplex formation having a pyrimidine:purine-pyrmidine motif has long been observed.[34–37] In this motif the base triads are T:A-T and C^+:G-C. Whereas the T:A-T triad can exist at neutral pH, the C^+:G-C triad can only exist at acidic pH, because protonation of cytosine is required for formation of a second hydrogen bond with G and for reduction of electrostatic repulsion with the negatively charged phosphate groups. The substitution of cytosine by 5-methyl cytosine increases the stability (presumably as a result of greater hydrophobic-stacking interactions) of the triplex, thereby allowing the formation at higher pH, such as pH 6.5. However, protonation of 5-methyl cytosine is still required. If a cluster of 5-methyl cytosines exists in the oligomer to match the cluster of G-C sequence in the duplex, then these positively charged 5-methyl cytosines in the cluster will tend to repel each other and to destabilize the triplex. The third pyrimidine strand is located in the major groove of the duplex and the polarity of the sugar-phosphate backbone of the third strand is the same as that of the purine strand.

To address the foregoing problem, our laboratory has synthesized 2′-O-methylribopseudo-isocytosine (PiC) as a deoxycytidine substitute in the third strand.[42,43] As shown in FIGURE 9, this analog contains an N-H group at position 3 that can serve as a hydrogen bond donor to the N7 of guanine in the duplex. The formation of PiC:G-C triplex was demonstrated employing UV absorption and stoichiometric mixing curve, CD studies, and gel electrophoresis. Furthermore, we studied a system composed of a duplex (duplex 22)

5′d (ACCAAAAGAAAAGGGGGACCA) 3′
3′d (TGGTTTTCTTTTCCCCCCTGGT) 5′

and a third strand, 5′TTTT1TTTT111111T 3′. Nucleoside "1" in the third strand can be 2′-O-methylribopseudo-isocytidine or 5-methyl-2′-deoxycytidine. We observed that a triplex is formed at 0.01 M Na cacodylate, 0.5 M NaCl, 5 mM $MgCl_2$, pH 7.2, with a $Tm = 20°C$ for the third strand containing 2′-O-methylribopseudo-isocytidine but not with the third strand containing 5-methyl-2′-deoxycytidine ($Tm < 0°C$). It should be noted that so far, all the studies in this series employ oligomers with a phosphodiester backbone.

The success in formation and characterization of triplexes having the pur:pur-pyr motif is far less extensive. Such triplexes appear to require the presence of specific cations and a large proportion of guanine in the third strand in order to form. We have studied triplex formation between $d(AG)_8$, or its nonionic methylphosphonate analog $d(\underline{AG})_8$, and the complementary homopyrimidine 16 mer, $d(CT)_8$.[44] In this model system, the interactions of these oligomers at pH 8.0 were studied by UV and CD spectroscopy and by polyacrylamide gel electrophoresis (PAGE) under native conditions. The 16 mer, $d(AG)_8$, tends to self-associate, particularly in the presence of Mg^{2+} and at low temperature, whereas the nonionic $d(\underline{AG})_8$ does not. As expected, both $d(AG)_8$ and the nonionic $d(\underline{AG})_8$ form duplexes with the complementary 16 mer, $d(CT)_8$. However, although the duplexes have similar Tms, the CD spectra suggest that the conformations of $d(AG)_8 - d(CT)_8$ and of $d(\underline{AG})_8 - d(CT)_8$ are dissimilar. Triplex formation between two $d(AG)_8$ and $d(CT)_8$ is *not* found under any conditions, including low temperature and high Mg^{2+} concentration. In contrast, a 2:1 $d(\underline{AG})_8 \cdot d(CT)_8$ triplex readily forms with a Tm of 50°C in 0.1 M sodium chloride, 0.01 M Tris-Cl, 0.01 mM EDTA, pH 8.0, in the absence of Mg^{2+}. Furthermore, the native PAGE studies also indicate the presence of a $d(\underline{AG})_8 \cdot d(AG)_8 - d(CT)_8$ triplex.

In view of the strong self-association of guanine-containing oligomers and polymers, we were eager to investigate the purine · purine-pyrimidine triplex

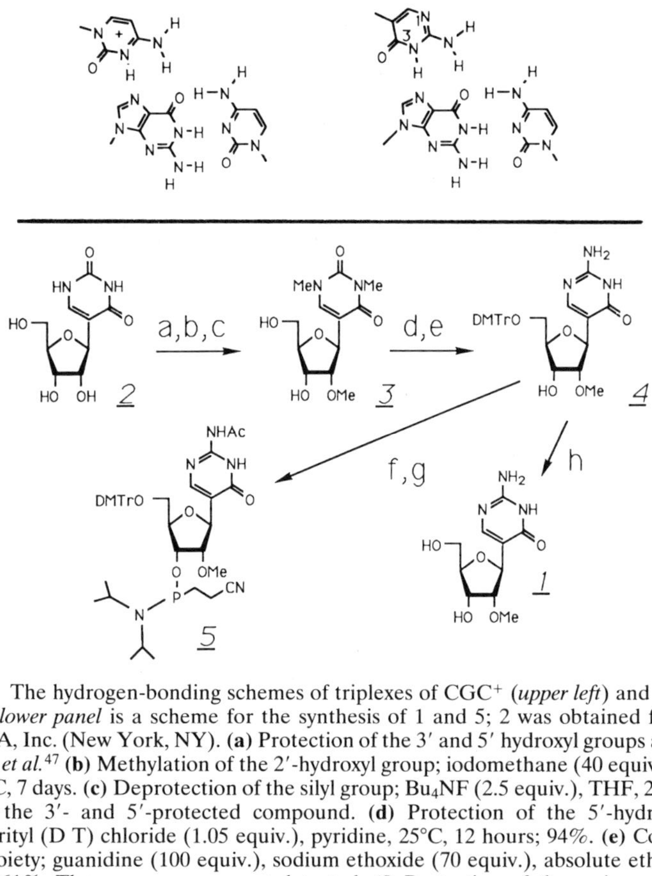

FIGURE 9. The hydrogen-bonding schemes of triplexes of CGC^+ (*upper left*) and CG1 (*upper right*). The *lower panel* is a scheme for the synthesis of 1 and 5; 2 was obtained from Kyowa Hakko, USA, Inc. (New York, NY). **(a)** Protection of the 3′ and 5′ hydroxyl groups according to Pankiewicz *et al.*[47] **(b)** Methylation of the 2′-hydroxyl group; iodomethane (40 equiv), Ag_2O (10 equiv), 25°C, 7 days. **(c)** Deprotection of the silyl group; Bu_4NF (2.5 equiv.), THF, 25°C, 1 hour; 58% from the 3′- and 5′-protected compound. **(d)** Protection of the 5′-hydroxyl group; dimethoxytrityl (D T) chloride (1.05 equiv.), pyridine, 25°C, 12 hours; 94%. **(e)** Conversion of the base moiety; guanidine (100 equiv.), sodium ethoxide (70 equiv.), absolute ethanol, reflux for 1 hour 61%. The α-anomer was not detected. **(f)** Protection of the amino group; acetic anhydride, DMF, 25°C, 12 hours, 78%. **(g)** (2-Cyanoethyl)-*N,N*-diisopropylphosphonamidous chloride (1.5 equiv.), $EtiPr_2N$ (2 equiv.), CH_2Cl_2, 25°C, 30 minutes; 71%. **(h)** Acetic acid (80%), 25°C, 12 hours; 78%. (Reproduced with permission from Ono *et al.*[42])

formation in the absence of guanine bases. Thus, we have investigated the systems involving $d(A)_{16}$, nonionic $d(\underline{A})_{16}$, the $d(T)_{16}$, and nonionic $d(\underline{T})_{16}$, again by UV and CD spectroscopy and by PAGE at pH 7.0.[19] The data clearly show the formation of duplexes $d(\underline{A})_{16} - d(T)_{16}$ and $d(A)_{16} - d(\underline{T})_{16}$ as well as the formation of triplexes $d(A)_{16} : d(\underline{A})_{16} - d(\underline{T})_{16}$; $d(\underline{T})_{16} : d(\underline{A})_{16} - d(T)_{16}$; and $d(T)_{16} : d(\underline{A})_{16} - d(T)_{16}$, in 0.1 M NaCl, pH 7.0, with Tm about 45°–55°C. The triplex $d(A)_{16}$: $d(A)_{16} - d(T)_{16}$ was *not* observed under any conditions. Thus, the advantage of the nonionic methylphosphonate linkage as the backbone of the third strand in the triplex formation is clearly demonstrated in these comparative studies on the stable $d(\underline{AG})_8 : d(\underline{AG})_8 - d(CT)_8$ and the stable $d(\underline{A})_{16} : d(\underline{A})_{16} - d(T)_{16}$ versus the <u>un</u>stable $d(AG)_8 : d(AG)_8 - d(CT)_8$ and the <u>un</u>stable $d(A)_{16} : d(A)_{16} - d(T)_{16}$ triplexes.

Returning to the issue of targeting genomic DNA in the nucleus through triplex formation, despite the successes just cited, the challenge is still very formidable. We need to consider that an anticode oligomer 16–20 nucleotides in length must bind to a uniquely designated target sequence in a genome comprising 3×10^9 base pairs and not to any other target sequence. It is not obvious that a chemical and a biological assay system can readily be established to study such specific interactions. However, we could attempt to form triplexes with a single-stranded target, such as mRNA in the cell. In this situation, triplex formation is much more readily measured in view of the experiment described thus far. Furthermore, the ease of escape of the single-stranded target nucleic acid from the triplex is likely to be significantly lower than that of the third strand from the duplex. The use of two Matagens to capture a single-stranded target RNA is likely to afford a more stable complex than is the duplex formation using only one Matagen. In addition, because triplex formation is specified by base pairing (Hoogsteen-like),[45,46] capture of single-stranded RNA can be achieved with higher affinity and without the loss of specificity.

CONCLUDING REMARKS

We have briefly defined the difference between an antisense biological approach based on biology and an anticode chemical approach to control gene expession. The two cornerstones of the anticode approach are the bypass of the cellular barrier and defense system, and high selectivity coupled with specificity. The basic concepts concerning the requirements of both high specificity and selectivity are delineated. One successful solution in meeting both requirements is reported, that is, physical binding coupled with photoinduced covalent linkage.

Two applications from our laboratories are described. The first approach involves antiviral research with HSV as the target virus. This approach is based on the masking of an intron-exon junction to prevent splicing of precursor in RNA. The studies in cellular and animal model systems show that this approach has a very high degree of selectivity in preventing viral infection and replication. The second approach involves suppression of oncogene, with H-ras p21 protein as the target. The results show that a single base mutation can be distinguished by an 11-mer Matagen, so that synthesis of a point-mutated ras p21 protein is inhibited with minimal effects on the synthesis of normal ras p21 protein. This specific inhibition of mutated p21 synthesis leads to suppression of tumor cell growth in soft agar and nude mice.

Studies on triplex formation yield significant progress in two areas. The substitution of deoxycytidine by the 2′-O-methylribopseudo-isocytidine allows a pyrimidine · purine-pyrimidine triplex to form at neutral pH, without pronation of the cytosine. The nonionic methylphosphonate backbone promotes triplex formation, allowing for ready formation of purine · purine-pyrmidine triplexes, a process that only occurs with great difficulty with third-strand oligomers containing phosphodiester backbones. Although targeting DNA in the genome remains a formidable task, it appears that single-stranded nucleic acids such as mRNA could be captured with high affinity and high selectivity by the use of two Matagens designed to form triplexes with the target.

REFERENCES

1. MURRAY, J. A. H. (ed). 1990. Antisense RNA and DNA. *In* Antisense Nucleic Acids and Proteins. Fundamentals and Applications. Wiley-Liss Publishing Co. J. N. M. Mol & A. R. van der Krol, eds. Wiley-Liss Publishing Co. New York.
2. (a) IZANT, J. G. & H. WEINTRAUB. 1984. Cell **36:** 1007–1015.
   (b) IZANT, J. G. & H. WEINTRAUB. 1985. Science **229:** 345–352.
3. PITHA, J., P. PITHA & P. TS'O. 1970. Biochem. Biophys. Acta **204:** 39–48.
4. FRIEDMAN, S. & P. TS'O. 1971. Biochemistry **30:** 3099–3104.
5. KAN, L.-S., J. BARRETT, P. MILLER & P. TS'O. 1973. Biopolymers **12:** 2226–2240.
6. BARRETT, J., P. MILLER & P. TS'O. 1974. Biochemistry **13:** 4898–4906.
7. MILLER, P., J. BARRETT & P. TS'O. 1974. Biochemistry **13:** 4887–4896.
8. MILLER, P., L. BRAITERMAN & P. TS'O. 1977. Biochemistry **16:** 1988–1996.
9. TS'O, P. 1991. Antisense Res. Dev. **1:** 273–276.
10. TS'O, P. 1992. Clin. Chem. In press.
11. PLESS, R. & P. TS'O. 1977. Biochemistry **16:** 1239–1250.
12. TS'O, P., J. BARRETT, L.-S. KAN & P. MILLER. 1973. Ann. N.Y. Acad. Sci. **222:** 290–306.
13. MILLER, P., J. YANO, E. YANO, C. CARROLL, K. JAYARAMAN & P. TS'O. 1979. Biochemistry **18:** 5134–5143.
14. KAN, L.-S., D. CHENG, P. MILLER, J. YANO & P. TS'O. 1980. Biochemistry **19:** 2122–2132.
15. JAYARAMAN, K., P. MCPARLAND, P. MILLER & P. TS'O. 1981. Proc. Natl. Acad. Sci. **78:** 1537–1541.
16. MILLER, P., C. CUSHMAN & J. LEVIS. 1991. *In* Oligonucleotides and Analogues. A Practical Approach. F. Eckstein, ed. :137–154. IRL Press at Oxford. University Press. Oxford.
17. MILLER, P., P. BAHN, C. CUSHMAN, J. KEAN & J. LEVIS. 1991. Nucleosides and Nucleotides **10(1–3):** 37–46.
18. MILLER, P., C. AGRIS, L. AURELIAN, K. BLAKE, S.-B. LIN, A. MURAKAMI, M. REDDY, C. SMITH & P. TS'O. 1985. *In* Interrelationship Among Aging, Cancer and Differentiation. B. Pullman *et al.,* eds. :207–219. Reidel Publishing Co.
19. TRAPANE, T. & P. TS'O. 1992. Biophys. J. **61:** A423. (Abstract)
20. CALLAHAN, D., T. TRAPANE, P. MILLER, P. TS'O & L.-S. KAN. 1991. Biochemistry **30:** 1650–1655.
21. CHEN, T.-L., P. MILLER, P. TS'O & M. O. COLVIN. 1990. Drug Metabolism & Disposition **18:** 815–818.
22. MILLER, P., C. AGRIS, L. AURELIAN, K. BLAKE, S. GLAVE, S.-B. LIN, A. MURAKAMI, M. REDDY, C. SMITH, S. SPITZ & P. TS'O. 1986. *In* Working Group on Molecular Mechanisms of Carcinogenic and Antitumor Activity. C. Chagas & B. Pullman, eds. :169–203. Vatican Workshop, Oct. 21–25.
23. KULKA, M., C. SMITH, L. AURELIAN, R. FISHELEVICH, K. MEADE, P. MILLER & P. TS'O. 1989. Proc. Natl. Acad. Sci. USA **86:** 6868–6872.
24. SMITH, C., L. AURELIAN, M. REDDY, P. MILLER & P. TS'O. 1986. Proc. Natl. Acad. Sci. USA **83:** 2787–2791.
25. BROWN, D., Z. YU, P. MILLER, K. BLAKE, C. WEI, H. KUNG, R. BLACK, P. TS'O & E. CHANG. 1989. Oncogene Res. **4:** 243–252.
26. YU, Z., D. CHEN, R. BLACK, K. BLAKE, P. TS'O, P. MILLER & E. CHANG. 1989. J. Exp. Pathol. **4:** 97–108.
27. CHANG, E., Z. YU, K. SHIMIZUKA, W. WILSON, A. STREKOSKA & G. ZON. 1989. Anticancer Drug Des. **4:** 221.
28. CHANG, E., P. MILLER, C. CUSHMAN, K. DEVADAS, K. PIROLLO. P. TS'O & Z. YU. 1991. Biochemistry **30:** 8283–8286.
29. TS'O, P., P. MILLER, L. AURELIAN, A. MURAKAMI, C. AGRIS, K. BLAKE, S.-B. LIN, B. LEE & C. SMITH. 1987. Ann. N. Y. Acad. Sci. **507:** 220–241.
30. LEE, B., A. MURAKAMI, K. BLAKE, S.-B. LIN & P. MILLER. 1988. Biochemistry **27:** 3197–3203.
31. LEE, B., K. BLAKE & P. MILLER. 1988. Nucleic Acids Res. **16:** 10681–10697.
32. KEAN, J., A. MURAKAMI, K. BLAKE, C. CUSHMAN & P. MILLER. 1988. Biochemistry **27:** 9113–9121.

33. BHAN, P. & P. MILLER. 1990. Bioconjugate Chem. **1:** 82–88.
34. FELSENFELD, G., D. DAVIES & A. RICH. 1957. J. Am. Chem. Soc. **79:** 2023–2024.
35. LIPSETT, M. 1963. J. Biol. Chem. **239:** 1256–1260.
36. MORGAN, A. & R. WELLS. 1968. J. Mol. Biol. **37:** 63–80.
37. LEE, J., D. JOHNSON & A. MORGAN. 1979. Nucleic Acids Res. **6:** 3073–3091.
38. MILLER, J. & H. SOBEL. 1966. Proc. Natl. Acad. Sci. USA **55:** 1201–1205.
39. COONEY, M., G. CZERNUSZEWICZ, E. POSTEL, S. FLINT & M. HOGAN. Science **241:** 456–459.
40. MAHER, L., B. WOLD & P. DERVAN. 1989. Science **245:** 725–730.
41. FRANCOIS, J.-C., T. SAISON-BEHMOARAS, N. THUONG & C. HELENE. 1989. Biochemistry **28:** 9617–9619.
42. ONO, A., P. TS'O & L.-S. KAN. 1991. J. Am. Chem. Soc. **113:** 4032–4033.
43. ONO, A., P. TS'O & L.-S. KAN. 1992. J. Org. Chem. In press.
44. TRAPANE, T., R. HOGREFE, M. REYNOLDS, L.-S. KAN & P. TS'O. 1992. Proc. Natl. Acad. Sci. USA. In press.
45. ROBERTS, R. & D. CROTHERS. 1991. Proc. Natl. Acad. Sci. USA **88:** 9397–9401.
46. SUN, J., J.-L. MERGNY, R. LAVERY, T. MONTENAY-GARESTIER & C. HELENE. 1991. J. Biomolecular Structure & Dynamics **9:** 411–424.

Therapeutic Applications of Antisense DNA in the Treatment of Human Leukemia[a]

ALAN M. GEWIRTZ

Hematology Sections
Departments of Pathology and Internal Medicine
University of Pennsylvania School of Medicine
Philadelphia, Pennsylvania 19104

Though great strides have been made in the treatment of hematological malignancies, therapy of these diseases remains problematic. The acute leukemias are illustrative in this regard. Initial remission induction rates are high, using now standard chemotherapy, but unfortunately at least half to two thirds of the responding patients are doomed to a relapse.[1,2] These individuals, and patients who are refractory to initial therapy, are destined to die of their disease.

Bone marrow transplantation is an effective, but highly morbid and expensive strategy for saving some individuals who would otherwise succumb.[1,2] Nevertheless, this approach too has significant problems as most patients do not have a histocompatible donor, and many are too old to undergo the procedure. Autologous transplantation with bone marrow purged of leukemic cells surmounts these issues and is therefore an ideal alternative, but relapse rates remain high among patients treated in this manner.[3] Increasing the efficiency of the purging process might be expected to improve this situation,[4] but novel purging agents may be required to accomplish this goal.

A class of compounds that might be particularly suitable for bone marrow purging are the antisense oligodeoxynucleotides. These molecules can be targeted to a specific gene's mRNA and prevent its translation by forming duplex structures and/or by accelerating the targeted mRNA's destruction.[5] If a leukemic cell were more dependent on the targeted gene's function for proliferation and/or cell maintenance than its normal counterpart, targeted disruption with antisense DNA might have therapeutic utility. To test this hypothesis, we targeted several genes whose function is postulated to be of critical importance for hematopoietic cell growth, in particular the c-myb and c-kit protooncogenes.[6,7] In the case of c-myb we also demonstrated that leukemic colony-forming cells (CFC) are more sensitive to c-myb deprivation than are normal CFC.[8] Some of these data are presented here. They suggest that perturbation of protooncogene function with antisense oligodeoxynucleotides might eventually form the basis for a novel molecular approach to antileukemic therapy, perhaps initially in the form of *ex vivo* bone marrow purging agents.

[a]This work was supported in part by grants CA 36896, CA 01324, and CA 54384 from the National Cancer Institute and a grant from the W. W. Smith Charitable Trust. Dr. Gewirtz is the recipient of a Research Career Development Award from the National Cancer Institute, NIH.

MATERIALS AND METHODS

Cells

Normal bone marrow mononuclear cells (MNC) were aspirated from consenting volunteers and enriched for hematopoietic progenitors, as previously reported.[6–8] Human leukemia cells utilized consisted of established cell lines[8,9] and primary material isolated from patients.[8,10]

Oligodeoxynucleotides: Synthesis, Purification, and Sequences Employed

Unmodified, 18-base oligodeoxynucleotides were synthesized on an Applied Biosystems 380B DNA synthesizer and purified as reported.[6–10] Oligomer sequences utilized were based on the human c-myb[11] and c-kit cDNA sequence.[12] Oligomers were synthesized in a sense, antisense, and scrambled sequence orientation.

Cell Culture and Oligomer Exposure

Adherent cell and T lymphocyte depleted normal mononuclear cells (0.5 to 2×10^5/ml) and leukemic blast cells (0.5 to 2×10^5/ml) were placed in liquid suspension cultures (RPMI 1640 with 20% fetal bovine serum) alone or in a 1:1 mix. Treated cultures received varying amounts of oligomers at the start of the culture period. A second dose (25–50% of the initial dose) was added to the medium after ~18 hours. Cells were either left in suspension culture for 4–9 days during which time cell counts and viability were periodically determined or were immediately seeded into duplicate methylcellulose or plasma clot cultures containing recombinant human interleukin-3 (IL-3, 20 U/ml) and granulocyte-macrophage colony-stimulating factor (GM-CSF, 5 ng/ml). Cells were cultured 10–12 days. Total numbers of CFU-GM derived colonies and clusters were then counted with the aid of an inverted microscope. Control cultures were handled in an identical manner but did not contain oligomers.[6–10]

Cell Phenotyping

Leukemic cells were morphologically or immunochemically identified after histochemical or immunochemical staining.[8]

Detection of bcr-abl and c-kit Gene Expression

Previously published RT-PCR based analysis was employed.[7,8,10] Philadelphia chromosome-positive chronic myelogenous leukemia blasts were detected in liquified methylcellulose cultures. Amplified DNA was electrophoresed, transferred to nitrocellulose, and then detected by Southern hybridization with a 40-nt, [γ-^{32}P]ATP end-labeled probe that recognized a 325-nt sequence within the amplified bcr-abl junction cDNA. c-kit was detected in a similar manner using primer pairs specific for this mRNA.

Statistical Analysis

Statistical significance of differences in numbers of colonies arising in the duplicate control, sense, or antisense plates was determined using the Student *t* test for unpaired samples. *p* values less than 0.05 were judged to be of statistical significance.

RESULTS

Role of c-myb and c-kit Encoded Protein in Normal Human Hematopoiesis

Exposing normal bone marrow mononuclear cells to c-myb antisense oligomers resulted in a decrease in cloning efficiency and progenitor cell proliferation.[6] The effect was lineage indifferent as c-myb antisense DNA inhibited granulocyte-macrophage colony-forming units (CFU-GM), CFU-E (erythroid), and CFU-Meg (megakaryocyte). In contrast, c-myb sense oligomers had no consistent effect on hematopoietic colony formation when compared to growth in control cultures ($p = 0.778$, $p = 0.796$, and $p = 0.375$ for CFU-GM, CFU-E, and CFU-Meg, respectively). Inhibition of colony formation was also dose related and sequence specific. If progenitor cells were exposed to 0.3, 1.4, 7, and 14 μM of c-myb oligodeoxynucleotides, the sense oligomers did not affect colony formation significantly, whereas the antisense oligomers at these concentrations led to the formation of 21 ± 3, 17 ± 1, 10 ± 2, and 4 ± 2 megakaryocyte colonies, respectively.[6] Exposure to 1.4, 7, and 14 μM concentrations of the antisense oligodeoxynucleotide led to the formation of 775 ± 127, 515 ± 9, and 244 ± 25 myeloid colonies.[6] Finally, for progenitor cells cloned in culture, requirements for MYB protein seem related to the cell's stage of development.[13] The protein appears to be most important for progenitor cells at the stage of development where they are actively cycling and less important for the processes of lineage commitment and early progenitor cell amplification.

We also investigated potential roles for c-kit expression in normal mononuclear cells using a similar strategy.[7] We first demonstrated that c-kit expression could be abrogated in a sequence-specific manner (FIG. 1). Mononclear cells were exposed to

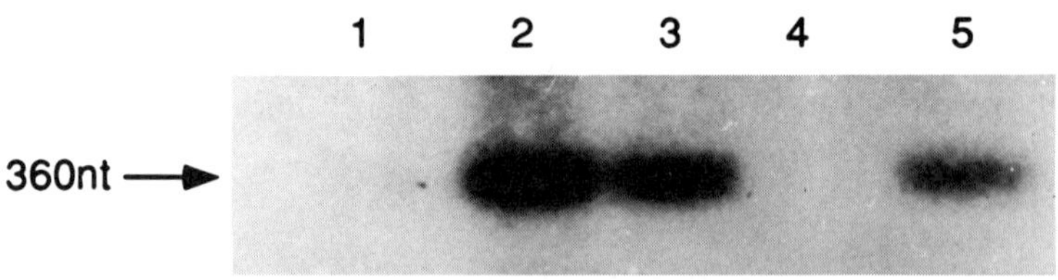

FIGURE 1. Expression of c-*kit* mRNAs in normal MNC exposed to c-kit oligomers. Cells were exposed to oligomers as previously described.[6] 2×10^6 A-T-MNC were incubated in 5 ml polypropylene tubes (Fisher Scientific, Pittsburgh, PA) in a total volume of 0.4 ml of Iscove's modified Dulbecco's medium (IMDM) containing 2% human AB serum and 10 mM HEPES buffer. Oligomers were added at time 0 (100 μg/ml), and 50% of the initial dose was added again 18 hours later. At time 0 and 36 hours after the first addition of oligomers, total RNA was extracted for RT-PCR analysis of c-kit expression.[10] *Lane 1,* time 0 control; *lane 2,* 36-hour control; *lane 3,* cells exposed to c-kit sense oligomers; *lane 4,* cells exposed to c-kit antisense oligomers; *lane 5,* cells exposed to scramble sequence c-kit oligomers. (From Ratajczak *et al.*[7])

TABLE 1. Effect of c-kit Oligomers on A-T-Colony Formation[a]

Progenitor Cell Type	Study	Control	Sense	Scrambled Sequence	Antisense
CFU-E	1	182, 209	153, 142	119, 128	33, 59
	2	1943, 543	1635, 1135	627, 649	243, 213
	3	148, 110	129, 176	149, 206	97, 107
		Mean (± SE) =	Mean (± SE) =	Mean (± SE) =	Mean (± SE) =
		522 ± 291	562 ± 268	313 ± 103	125 ± 34
BFU-E	1	133, 152	117, 106	94, 64	60, 149
	2	534, 392	601, 249	273, 246	126, 113
	3	206, 172	215, 258	162, 246	59, 51
		Mean (± SE) =	Mean (± SE) =	Mean (± SE) =	Mean (± SE) =
		265 ± 66	258 ± 74	181 ± 36	76 ± 14
CFU-GM	1	212, 189	231, 179	282, 193	195, 220
	2	412, 408	395, 421	457, 384	407, 471
	3	217, 241	230, 237	201, 199	293, 187
		209, 246	Mean (± SE) =	Mean (± SE) =	Mean (± SE) =
		Mean (± SE) =	282 ± 41	286 ± 46	296 ± 49
		280 ± 42			
CFU-MEG	1	114, 107	133, 117	154, 113	127, 112
	2	93, 100	58, 52	53, 40	47, 54
		Mean (± SE) =	Mean (± SE) =	Mean (± SE) =	Mean (± SE) =
		104 ± 5	90 ± 20	90 ± 27	85 ± 20

[a] 2×10^5 A-T-MNC were incubated in 5 ml polypropylene tubes (Fisher Scientific, Pittsburgh, PA) in a total volume of 0.4 ml of Hepes buffer. Oligomers were added at time 0 (2.5–100 μg/ml), and 50% of the initial dose was added again 18 hours later (final total concentration 3.7–150 μg/ml; ~1–52.5 μM). Twenty-four hours after the first addition of oligomers, cells were prepared for plating in plasma clot or methylcellulose cultures.[6] Cells were not washed before plating. Control cell cultures were untreated. Values are actual colony counts from two or three individual studies at the highest concentration tested. (From Ratajczak *et al.*[7])

c-kit antisense oligodeoxynucleotides. A-T-MNC (2×10^6) were incubated at 4°C for 24 hours in a low serum (2%)-containing medium. Cells were rewarmed to 37°C and exposed to growth factors as already described in the presence of c-kit oligomers. After 36 hours, total RNA was extracted for RT-PCR detection of c-kit message. C-kit mRNA was expressed at very low levels after 24 hours at 4°C. Thirty-six hours after warming, c-kit mRNA was detected in control cells and in cells exposed to sense and scrambled sequence oligomers. In contrast, antisense treated cells had no detectable c-kit mRNA. Accordingly, inhibition of c-kit expression was highly efficient and sequence specific. The effect of c-kit antisense, and control sequence oligomers on normal hematopoietic progenitor cell growth was then investigated by assessing the oligomer's effect on CFU-E, BFU-E, CFU-GM, and CFU-Meg colony growth. c-kit antisense oligomers inhibited erythroid colony formation in a dose-dependent and sequence-specific manner. Antisense oligomers inhibited growth of CFU-E and BFU-E of ~75% and ~71%, respectively. Residual colonies were much smaller than those of the untreated controls. In contrast to the erythroid colony results, CFU-GM and CFU-Meg derived colony formation was unaffected by exposure to any of the oligomers, at any of the doses employed (TABLE 1).

Role of c-myb and c-kit Encoded Protein in Leukemic Hematopoiesis

As the c-myb antisense oligomers inhibited normal cell growth, it was also of interest to determine their effect on the growth of leukemic cells. This was done using myeloid and lymphoid leukemic cell lines[8,9] and primary patient material.[7–10]

The effect of myb sense and myb antisense oligomers on the growth of four myeloid leukemia cell lines, HL-60, K562, KG-1, AND KG-1a, has been determined.[9] The antisense oligomer inhibited the proliferation of each leukemia cell line, although the effect was most pronounced on HL-60 cells. Specificity of this inhibition was demonstrated by the fact that the sense oligomer had no effect on cell proliferation and neither did "antisense" sequences with 2 or 4 nucleotide mismatches. To determine if the treatment with c-myb antisense modified the cell cycle distribution of HL-60 cells, we measured the DNA content in exponentially growing cells exposed to either sense or antisense myb oligomers. Control cells and cells treated with c-myb sense oligomers had twice the DNA content of HL-60 cells exposed to the antisense oligomers. The majority of these cells either appeared to reside in the G1 compartment or were blocked at the G1/S boundary.

To examine the effect of the c-myb oligomers on lymphoid cell growth we employed a lymphoid leukemia cell line designated CCRF-CEM.[8] As noted earlier in normal lymphocytes,[14] the CCRF-CEM cells were extremely sensitive to the antiproliferative effects of the c-myb antisense DNA. When exposed to the sense oligomers, the effects on CEM cell growth in short-term suspension cultures were negligible. In contrast, exposure to c-myb antisense DNA resulted in a daily decline in cell numbers. Compared to untreated controls, antisense DNA inhibited growth ~2 logs. Growth reduction was not a cytostatic effect, as cell viability was reduced only ~70% after exposure to the antisense oligomers and CEM cell growth did not recover when cells were left in culture for an additional 9 days. The c-myb antisense oligomers also appeared effective in inhibiting cell growth of primary patient material when derived from patients with both acute and chronic myeloid leukemias.[10] Colony inhibition was ~58–95% (FIG. 2), and as determined by elimination of bcr-abl expressing CML clones, it apparently was quite efficient.

We then studied patients with acute leukemia, chronic myelogenous leukemia, and polycythemia vera for sensitivity to c-kit antisense oligomers.[7] Sequences encompassing c-kit codons 1–6 were employed in a sense (S), antisense (AS), or scrambled (ScSq) configuration. In 2 of 7 patients with acute leukemia, 5 of 10 with chronic myelogenous leukemia, and 3 of 3 with polycythemia vera, granulocyte colony formation decreased in the AS-treated cultures in a dose-dependent manner. In comparison to control cultures, a mean (± SE) decrease in CFU-GM colonies of 65 (± 3)% in acute leukemia, 65 (± 23)% in chronic myelogenous leukemia, and 89 (± 8)% in polycythemia vera was observed at the highest AS doses employed. These results are in marked contrast to those observed in normal cells where c-kit AS oligomers had no detectable effect on CFU-GM colony formation. Erythroid colony formation (CFU-E, BFU-E) was also assessed in patients with polycythemia vera. Dose-dependent inhibition was again observed. At the highest AS concentrations employed, a mean (± SE) decrease of 65 (± 23)% in CFU-E colonies (n = 3) and 84% in BFU-E (n = 1) was noted in comparison with controls. These results appear to correlate with our finding that c-kit mRNA was easily detectable by RT-PCR in patients with polycythemia vera. Finally, although inhibition of colony formation was occasionally noted in control S and ScSq treated dishes, no consistent or dose-dependent effect was observed. These results suggest that the protein encoded by the c-kit protooncogene plays a role in the maintenance and/or growth of some malignant hematopoietic cells and that this role may depend on where in the cell's developmental hierarchy maturation is blocked. These results also suggest that the lineage-related functions of c-kit in malignant hematopoiesis may differ from those in normal hematopoietic cell development. Use of AS oligomers to exploit these differences may also prove to be of therapeutic value in these diseases.

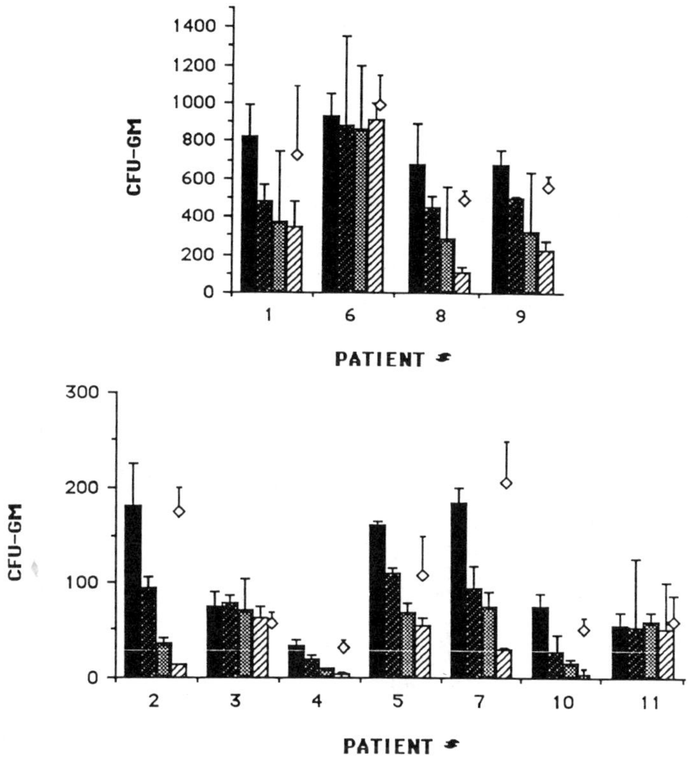

FIGURE 2. Effect of c-myb oligomers on chronic myeloid leukemia cell colony formation by cells with "high" (**A**) and "low" (**B**) cloning efficiency. Colony-forming cells were enriched from patient peripheral blood or bone marrow (TABLE 1) and exposed to oligomers as detailed in the text. Control cells were not exposed to oligomers. Oligomers were added to suspended mononuclear cells at culture inception (Time 0) and after 18 hours of incubation. At 24 hours, cells were plated, and resulting colonies were enumerated in untreated control (■), antisense (■, 20 μg/ml, then 10 μg/ml; ▩, 40 μg/ml, then 20 μg/ml; ▨, 100 μg/ml, then 50 μg/ml), and sense (◇, 100 μg/ml, then 50 μg/ml) containing plates. Values plotted are mean ± SD of actual colony counts compared to growth in control cultures that contained no oligomers. (From Ratajczak *et al.*[7])

Evidence That Normal and Leukemic Progenitor Cells Rely Differentially on c-myb Function

If leukemic cells depended more on MYB protein than did their normal counterparts, inhibition of c-myb function might prove therapeutically useful. To test this possibility, we incubated phagocyte and T-cell-depleted normal human marrow mononuclear cells (MNC), human T-lymphocyte leukemia cell line blasts (CCRF-CEM), or 1:1 mixtures of these cells with sense or antisense oligodeoxynucleotides to codons 2–7 of human c-myb mRNA (TABLE 2). Oligomers were added to liquid suspension cultures at time 0 and at time +18 hours (25% 0 hour dose). Control cultures were untreated. In controls or in high-dose sense oligomers, CCRF-CEM proliferated rapidly, whereas mononuclear cell numbers and viability decreased < 10%. In contrast, when CCRF-CEM were incubated for 4 days in c-myb antisense DNA, cultures contained $4.7 \pm 0.8 \times 10^4$ cells/ml (mean ± SD; $n = 4$) compared to $285 \pm 17 \times 10^4$/ml in controls. At the effective antisense dose, mononuclear cells were largely unaffected. After 4 days in culture, remaining cells were transferred to methylcellulose supplemented with recombinant hematopoietic growth factors. Myeloid colonies/clusters were enumerated at day 10 of culture inception. Depending on cell number plated, control mononuclear cells formed from 31 ± 4 to 274 ± 18 colonies. In dishes containing equivalent numbers of untreated or sense oligomers exposed CCRF-CEM, colonies were too numerous to count (TNTC). When mononuclear cells were mixed 1:1 with CCRF-CEM in antisense oligomer concentrations ≤ 5 μg/ml, only leukemic colonies could be identified by morphologic, histochemical, and immunochemical analysis. However, when antisense oligomer exposure was intensified, normal myeloid colonies could now be found in the culture whereas leukemic colonies could no longer be identified with certainty using the same analytic methods (TABLE 2).

TABLE 2. Effect of c-myb Oligomer Exposure on Colony/Cluster Formation by T-Cell Leukemia and Normal Bone Marrow Progenitor Cells (MNC)[a]

Cells Plated	No. Added	Oligomer/Amt. Added[b]	Colony/Cluster
MNC	5×10^4/ml	None	24 ± 4
		MYB S (20; 5.0)	31 ± 4
		MYB AS (20; 5.0)	30 ± 6
T Leukemia	5×10^4/ml	None	TNTC[c]
		MYB S (20; 5.0)	TNTC
		MYB AS (20; 5.0)	1 ± 1
MNC + Leukemia	5×10^4/ml of each	None	TNTC
		MYB S (20; 5.0)	TNTC
		MYB AS (2; 0.5)	TNTC
		MYB AS (5; 1.0)	TNTC
		MYB AS (10; 2.5)	41 ± 5
		MYB AS (20; 5.0)	34 ± 1

[a]Cells were exposed to oligomers for 4 days in suspension cultures and then transferred to semisolid media as described in the Methods section. After 12 days in culture, colonies and clusters were counted in paired dishes with an inverted microscope. Colony/cluster counts are presented as mean ± SD. (From Calabretta *et al.*[8])

[b]μg/ml added to the culture medium at time 0 and +18 hours, respectively.

[c]Too numerous to count (> 1,000 colonies). (From Calabretta *et al.*[8])

Finally, at antisense DNA doses used in the foregoing studies, acute myelogenous leukemia blasts from 18 of 23 patients exhibited ~75% decrease in colony and cluster formation compared to untreated or sense oligomer treated controls. When 1:1 mixing experiments were carried out with primary acute myelogenous leukemia blasts and normal mononuclear cells, we were again able to preferentially eliminate acute myelogenous leukemia blast colony formation while normal myeloid colonies continued to form.[8]

Use of c-myb Oligomers as Bone Marrow Purging Agents

The experiments described here strongly suggested that leukemic cell growth could be preferentially inhibited after exposure to c-myb antisense oligodeoxynucleotides. We therefore determined if the antisense oligomers could be used as *ex vivo* bone marrow purging agents. To test this hypothesis, normal mononuclear cells were mixed (1:1) with primary acute or chronic myelogenous leukemia blast cells and then exposed to the oligomers using a slightly modified protocol designed to test the feasibility of a more intensive antisense exposure. Towards this end, an additional oligomer dose (20 μg/ml) was given just prior to plating the cells in methylcellulose. In control growth factor-stimulated cultures, leukemic cells formed 25.5 ± 3.5 (mean ± SD) colonies and 157 ± 8.5 clusters (per 2×10^5 cells plated). Exposure to c-*myb* sense oligomers did not significantly alter these numbers (19.5 ± 0.7 colonies and 140.5 ± 7.8 clusters; $p > 0.1$). In contrast, equivalent concentrations of antisense oligomers totally inhibited colony and cluster formation by the leukemic blasts. Colony formation was also inhibited in the plates containing normal mononuclear cells, but only by ~50% in comparison with untreated control plates (control colony formation, 296 ± 40 per 2×10^5 cells plated; treated colony formation, 149 ± 15.5 per 2×10^5 cells).

Evaluation of the Efficiency of an Antisense Purge

In antisense treated cocultures of normal mononuclear cells and primary acute myelogenous leukemia blast cells (1:1 ratio), only normal colonies could be identified. To assess the potential effectiveness of an antisense purge we conducted similar coculture studies with chronic myelogenous leukemia cells obtained from patients in blast crisis and in the chronic phase of their disease. As these cells express the tumor-specific, and easily detected, bcr-abl fusion gene, it was an easy task to measure residual leukemic cells in oligonucleotide-treated cultures.[8,10]

To carry out these studies, RNA was extracted from cells cloned in methylcellulose cultures after exposure to the highest c-myb antisense oligomer dose. The RNA was then reverse transcribed and resulting cDNA amplified as described in the Materials and Methods section. For each patient studied, mRNA was also extracted from a comparable number of cells derived from untreated control colonies using the same technique. Eight cases were evaluated, and in each case bcr-abl expression as detected by RT-PCR correlated with colony growth in cell culture. In cases that were inhibited by exposure to c-myb antisense oligomers (7 of 11), bcr-abl expression was also greatly decreased or nondetectable.[10]

These results suggested that bcr-abl expressing CFU might be substantially or entirely eliminated from a population of blood or marrow mononuclear cells by exposure to the antisense oligodeoxynucleotides.[10] To explore this possibility further, re-plating experiments were carried out on samples from two patients. We hypothesized that if CFU belonging to the malignant clone were present at the end of the original 12-day culture period, but not detectable because of failure to express bcr-abl, they might reexpress the message upon regrowth in fresh cultures. Accordingly, cells from these patients were exposed to oligomers and then plated into methylcellulose cultures formulated to favor growth of either CFU-GM or CFU-GEMM. As was found with the original specimens, untreated control cells and cells exposed to sense oligomers had RT-PCR detectable bcr-abl transcripts. Those exposed to c-myb antisense oligomers had none. One of the paired dishes from these cultures was solubilized with fresh medium, and all cells contained therein were washed, disaggregated, and re-plated into fresh methylcellulose cultures *without* re-exposing the cells to oligomers. After 14 days, CFU-GM and CFU-GEMM colony cells were again probed for bcr-abl expression. Control and sense-treated cells had RT-PCR detectable mRNA, but none was found in the antisense-treated colonies. These results suggest that elimination of bcr-abl expressing cells and CFU was highly efficient and perhaps permanent.

DISCUSSION

The results described above are consistent with the hypothesis that many types of leukemic blast cells depend on the function of known protooncogenes. The relative importance of these genes in regulating leukemic versus normal cell growth is an active area of investigation, as is the mechanism whereby deprivation of the encoded protein leads to preferential inhibition of leukemic cell growth. Nevertheless, several explanations for our results are possible. For example, as MYB is required for development of normal hematopoietic progenitor cells at a specific stage of maturation, we could hypothesize that if leukemic blasts were blocked as a cohort at a susceptible maturation stage, it might lead to differential sensitivity to targeted disruption of c-myb expression. Alternatively, as c-myb is now known to be a transactivator of several developmentally regulated genes,[15,16] loss of c-myb regulated gene function could have a more serious impact on malignant cell development.

The ability to block gene function with antisense oligodeoxynucleotides is now an accepted and powerful research tool. As overexpression of protooncogenes may be an important mechanism in the development of many human malignancies,[17] the antisense strategy could also have significant therapeutic potential. Many technical problems relating to oligodeoxynucleotide targeting, stability, and cellular uptake remain to be solved before the *in vivo* use of antisense oligomers can be contemplated. However, results from our group suggest that in the near term, antisense DNA might find utility as an *ex vivo* bone marrow purging agent. Problems such as those just alluded to are minimized, and potential systemic toxicity is avoided. Although significant cell kill will have to be demonstrated before antisense oligomers can be used for purging, establishment of relevant *in vivo* models should allow this and other issues relevant to the therapeutic potential of antisense DNA to be explored.

ACKNOWLEDGMENT

The editorial assistance of Elizabeth R. Bien is gratefully acknowledged.

REFERENCES

1. Tricot, G. J. K. 1991. Treatment of acute nonlymphocytic leukemia. *In* Hematology. Basic Principles and Practice. R. Hoffman, E. J. Benz, Jr., S. J. Shattil, B. Furie & H. J. Cohen, eds.: 732–748. Churchill Livingstone Inc., New York, NY.
2. Hoelzer, D. 1991. Acute lymphoblastic leukemia in adults. *In* Hematology. Basic Principles and Practice. R. Hoffman, E. J. Benz, Jr., S. J. Shattil, B. Furie & H. J. Cohen, eds.: 793–804. Churchill Livingstone Inc. New York, NY.
3. Santos, G. W., A. M. Yeager & R. Jones. 1991. Autologous bone marrow transplantation. Ann. Rev. Med. **40:** 99–112.

4. ROWLEY, S. D., C. B. MILLER, S. PIANTADOSI, J. M. DAVIS, G. W. SANTOS & R. J. JONES. 1991. Phase I study of combination drug purging for autologous bone marrow transplantation. J. Clin. Oncol. **9:** 2210–2218.
5. VAN DER KROL, A. R., J. N. M. MOL & A. R. STUITJE. 1988. Modulation of eukaryotic gene expression by complementary RNA or DNA sequences. Biotechniques **6:** 958–976.
6. GEWIRTZ, A. M. & B. CALABRETTA. 1988. A c-myb antisense oligodeoxynucleotide inhibits normal human hematopoiesis *in vitro.* Science **242:** 1303–1306.
7. RATAJCZAK, M., S. LUGER, K. DERIEL, J. ABRAHM, B. CALABRETTA & A. M. GEWIRTZ. 1992. Role of the c-kit protooncogene in normal and malignant human hematopoiesis. Proc. Natl. Acad. Sci. USA **89:** 1710–1714.
8. CALABRETTA, B., R. B. SIMS, M. VALITERI, D. CARACCIOLO, C. SZCZYLIK, D. VENTURELLI, M. RATAJCZAK, M. BERAN & A. M. GEWIRTZ. 1991. Normal and leukemic hematopoietic cells manifest differential sensitivity to inhibitory effects of c-myb antisense oligodeoxynucleotides: An *in vitro* study relevant to bone marrow purging. Proc. Natl. Acad. Sci. USA **88:** 2351–2355.
9. ANFOSSI, G., A. M. GEWIRTZ & B. CALABRETTA. 1989. An oligomer complementary to c-myb encoded mRNA inhibits proliferation of human myeloid leukemia cell lines. Proc. Natl. Acad. Sci. USA **86:** 3379–3383.
10. RATAJCZAK, M. Z., N. HIJIYA, L. CATANI, K. DERIEL, S. L. LUGER, P. MCGLAVE & A. GEWIRTZ. 1992. Acute and chronic phase chronic myelogenous leukemia colony forming units are highly sensitive to the growth inhibitory effects of c-myb antisense oligodeoxynucleotides. Blood **79:** 1956–1961.
11. MAJELLO, B., L. C. KENYON & R. DALLA-FAVERA. 1986. Human c-myb proto-oncogene: Nucleotide sequence of cDNA and organization of the genomic locus. Proc. Natl. Acad. Sci. USA **83:** 9636–9640.
12. YARDEN, Y., W. J. KUANG, T. YANG-FENG, L. COUSSENS, S. MUNEMITSU, T. J. DULL, E. CHEN, J. SCHLESSINGER, U. FRANCKE & A. ULLRICH. 1987. Human protooncogene c-kit: A new cell surface ligand for an unidentified ligand. EMBO J. **6:** 3341–3351.
13. CARACCIOLO, D., D. VENTURELLI, M. VALITERI, C. PESCHLE, A. M. GEWIRTZ & B. CALABRETTA. 1990. Stage-related proliferative activity determines c-myb functional requirements during normal human hematopoiesis. J. Clin. Invest. **85:** 55–61.
14. GEWIRTZ, A., G. ANFOSSI, S. VALPREDA, R. SIMS & B. CALABRETTA. 1989. Normal human lymphocytes G1/S transition requires the nuclear protein encoded by the proto-oncogene c-myb. Science **245:** 180–183.
15. NESS, S. A., A. MARKNELL & T. GRAF. 1989. The v-myb oncogene product binds to and activates the promyelocyte specific mim-1 gene. Cell **59:** 1115–1125.
16. SIU, G., A. L. WURSTER, J. S. LIPSICK & S. M. HEDRICK. 1992. Expression of the CD4 gene requires a MYB transcription factor. Mol. Cell Biol. **12:** 1592–1604.
17. SLAMON, D. 1987. Protooncogenes and human cancers. N. Engl. J. Med. **317:** 955–957.

Myelin-Deficient Mutant Mice

An *in Vivo* Model for Inhibition of Gene Expression by Natural Antisense RNA[a]

JEAN-MARIE MATTHIEU,[b,c] MIRJANA TOSIC,[b] AND ARTHUR ROACH[d]

[b]*Service de Pédiatrie*
Centre hospitalier universitaire vaudois
CH-1011 Lausanne, Switzerland

[d]*Mount Sinai Hospital Research Institute*
Toronto, Ontario M5G 1X5, Canada

Myelin basic protein (MBP) is a functionally important constituent of myelin, the specialized membrane that surrounds and insulates axons. MBP is localized at apposing cytoplasmic faces of the myelin lamellae and plays a key role by keeping them in tight contact.[1-3] Myelin-deficient mutant mice (*mld*) are characterized by trembling of hind limbs, occasional seizures in adults, reduced fertility, and shortened life span. The mutation shows full recessiveness and it is transmitted in simple mendelian fashion, indicating a change in a single gene. Morphological investigations showed that the central nervous system (CNS) is severely deficient in myelin.[4,5] The few myelinated axons in *mld* CNS are surrounded by thin, uncompacted lamellae, suggesting that the primary defect lies in the impaired function or absence of MBP. (For review see ref. 6.) Indeed, only traces of MBP[4-7] were identified in *mld* CNS myelin as a consequence of drastically reduced MBP synthesis.[8-10]

In *mld* mice, reduced MBP synthesis is due to a low concentration of normal MBP mRNA which in the cytoplasm represents approximately 2% of that of wild-type mRNA.[11-15] MBP mRNA showed a normal size and was efficiently translated in an *in vitro* system.[11] The reduced mRNA levels, however, are not caused by a mutation within the promoter region.[16] The intermediate concentration of MBP mRNA in *mld* heterozygotes suggested that the intact Mbp allele in *mld* heterozygotes is normally regulated during development.[11,12]

The Mbp gene cloned in rat[17] and mouse[18-20] made possible direct analysis of the Mbp^{mld} allele. In mice, the Mbp gene is located on the distal part of chromosome 18q.[16] The *mld* Mbp gene is duplicated in tandem with some rearrangements.[13,15,16,21,22] A large portion of the upstream gene contains an inversion encompassing exons 3 to 7, whereas the downstream gene is normal[21] (FIG. 1). The overall transcription rate of the Mbp locus in *mld* mutants is normal,[21-23] and both Mbp genes in *mld* mice have functional promoters and regulatory elements within 1.3 kb upstream of exon 1.[16,22,25] The upstream gene transcript contains antisense RNA which is complementary to exons 3 and 7 of the downstream gene.[21,24,25] This antisense RNA was detected in nuclei but not in the cytoplasm.[25-27] The downstream gene gives rise to apparently normal mRNA and functional protein but in extremely low concentrations.

[a]This research was supported by the Swiss National Science Foundation (grant 31–9400.88), the Multiple Sclerosis Society of Switzerland, and the Jubiläumsstiftung der Rentenanstalt.

[c]Author for correspondence.

To explain the discrepancy between normal transcription rate and drastic reduction in MBP mRNA concentration in the cytoplasm, we measured the concentration of nuclear MBP-specific RNA. MBP-specific RNA concentration in *mld* nuclei is only slightly lower than that in control brains, in contrast to the 50-fold reduction found for cytoplasmic MBP mRNA.[25] Therefore, *mld* MBP gene transcripts appeared to be processed abnormally or transported inefficiently to the cytoplasm. One possible explanation for the arrest of MBP-specific RNA in the nuclei lies in the structure and size of the primary transcripts.

Transcription of the upstream gene produces antisense RNA complementary to exons 3 and 7.[21,24–26] The rate of transcription of the upstream gene was slightly higher than that of the downstream gene.[25] This difference was reflected in the concentration of sense and antisense RNA found in nuclei.

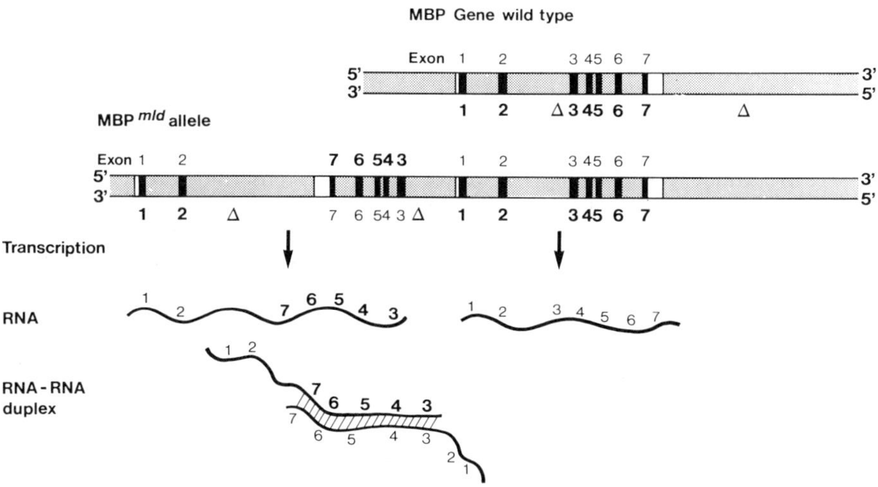

FIGURE 1. Organization of wild-type and *mld* allele of myelin basic protein gene in mice with possible formation of sense-antisense duplexes from two distinct transcripts. The inverted segment is marked by *open triangles.*

Transcriptional analysis of exon 1 and the 5′ flanking region indicated that transcription of the downstream gene starts from its own promoter, although very low levels of uninterrupted transcription could be detected.[25] A low level of transcription initiation upstream of the major initiation site has been reported[24,28] in normal mice and is probably responsible for this signal.

Independent transcription of the two Mbp genes indicates that the RNA detected in the *mld* nuclei is a mixture of two distinct transcripts, one of which contains the partial antisense sequence. To discriminate between these two types of RNAs, we measured concentrations of the nuclear RNAs using strand-specific probes. The antisense RNA was detected only in *mld* nuclei.[25,26] As the inverted 3′ region includes the polyadenylation signal, the antisense RNA is most probably not polyadenylated and not transported to the cytoplasm.

The proportion of sense and antisense RNA in nuclei reflects the observed transcription rates of upstream and downstream genes. The drastic decrease of sense

RNA concentrations between nucleus and cytoplasm indicates that the transcript from the downstream gene is abnormally processed or inefficiently transported to the cytoplasm. The large concentration of MBP antisense RNA may favor the formation of sense-antisense duplexes (FIG. 1). Some experimental work suggests that these duplexes might occur,[27] which could lead to a low concentration of MBP and hypomyelination by interfering with the processing of the sense transcript.[25,27] Furthermore, inhibition of Mbp gene expression by antisense RNA has been demonstrated in transgenic mice.[29] Introduction of antisense Mbp cDNA into normal murine embryos inhibits expression of the Mbp gene and correlates phenotypically with hypomyelination and tremor in adult transgenic mice.[29]

Uninterrupted transcription of the two adjacent Mbp genes in *mld* mice would give rise to extremely long RNA molecules with impaired function. Studies of steady-state levels of Mbp transcripts observed by *in situ* hybridization using a 5′ non-coding region of genomic Mbp have been interpreted to suggest that transcription of the downstream *mld* Mbp gene is inhibited by readthrough transcripts that originate at the upstream promoter[26] and not by posttranscriptional inhibition as proposed previously.[25,27] These results conflict with RNAse protection[24] and run-on experiments[25] which showed clearly that the amount of readthrough was negligible. Furthermore, taking into account that steady-state levels of primary transcript do not reflect transcription rates because of different stabilities for the different parts of the transcript, evidence supporting the hypothesis that transcriptional interference inhibits Mbp gene expression[26] is not definitive. Experimental evidence[24,25,27] supporting the hypothesis that the Mbp gene is repressed by posttranscriptional events can be summarized as follows: (1) Upstream and downstream Mbp genes are transcribed independently from their own promoters. (2) The promoter activity of the downstream gene is not severely reduced. (3) The proportion of sense and antisense RNA in nuclei reflects the observed transcription rates of upstream and downstream genes. (4) High levels of sense RNA are present in *mld* nuclei and protected from degradation by RNAse.

CONCLUSION

The *mld* mutation is an example of posttranscriptional inhibition affecting gene expression by synthesis of antisense RNA. Antisense RNA may form double-stranded RNA with sense primary transcript, which may interfere with processing or transport of mature RNA to cytoplasm. This mechanism occurs naturally and has become an experimental tool useful for the study of gene function.

REFERENCES

1. GOLDS, E. E. & P. E. BRAUN. 1976. Organization of membrane proteins in the intact myelin sheath: Pyridoxal phosphate and salicylaldehyde as probes of myelin structure. J. Biol. Chem. **251:** 4729–4735.
2. OMLIN, F. X., H. DEF. WEBSTER, C. G. PALKOVITS & S. R. COHEN. 1982. Immunocytochemical localization of basic protein in major dense line regions of central and peripheral myelin. J. Cell Biol. **95:** 242–248.
3. PODUSLO, J. F. & P. E. BRAUN. 1975. Topographical arrangement of membrane proteins in the intact myelin sheath. J. Biol. Chem. **250:** 1099–1105.
4. MATTHIEU, J.-M., H. GINALSKI, R. L. FRIEDE, S. R. COHEN & D. P. DOOLITTLE. 1980. Absence of myelin basic protein and major dense line in CNS myelin of the *mld* mutant mouse. Brain Res. **191:** 278–283.

5. MIKOSHIBA, K., H. OKANO, Y. INOUE, M. FUJISHIRO, K. TAKAMATSU, F. LACHAPELLE, N. BAUMANN & Y. TSUKADA. 1987. Immunohistochemical, biochemical and electron microscopic analysis of myelin formation in the central nervous system of myelin deficient (*mld*) mutant mice. Dev. Brain Res. **35:** 111–121.
6. TOSIC, M. & J.-M. MATTHIEU. 1990. From myelin basic protein to myelin deficient mice. *In* Cellular and Molecular Biology of Myelination. G. Jeserich *et al.,* eds. NATO ASI Series, Vol. H43: 489–503. Springer-Verlag. Berlin.
7. BOURRE, J. M., C. JACQUE, A. DELASSALLE, J. NGUYEN-LEGROS, O. DUMONT, F. LACHAPELLE, M. RAOUL, C. ALVAREZ & N. BAUMANN. 1980. Density profiles and basic protein measurements in the myelin range of particulate material from normal developing mouse brain and from neurological mutants (Jimpy; Quaking; Trembler; Shiverer and its *mld* allele) obtained by zonal centrifugation. J. Neurochem. **35:** 458–464.
8. CAMPAGNONI, A. T., C. W. CAMPAGNONI, J. M. BOURRE, C. JACQUE & N. BAUMANN. 1984. Cell-free synthesis of myelin basic proteins in normal and dysmyelinating mutant mice. J. Neurochem **42:** 733–739.
9. GINALSKI-WINKELMANN, A., G. ALMAZAN & J.-M. MATTHIEU. 1983. *In vitro* myelin basic protein synthesis in the PNS and CNS of myelin deficient (*mld*) mutant mice. Brain Res. **277:** 386–388.
10. MATTHIEU, J.-M., G. ALMAZAN & T. V. WAEHNELDT. 1984. Intrinsic myelin proteins are normally synthesized *in vitro* in the myelin-deficient (*mld*) mutant mice. Dev. Neurosci. **6:** 246–250.
11. ROCH, J. M., M. BROWN-LUEDI, B. J. COOPER & J.-M. MATTHIEU. 1986. Mice heterozygous for the *mld* mutation have intermediate levels of myelin basic protein mRNA and its translation products. Mol. Brain Res. **1:** 137–144.
12. ROCH, J.-M., B. J. COOPER, M. RAMIREZ & J.-M. MATTHIEU. 1987. Expression of only one myelin basic protein allele in mouse is compatible with normal myelination. Mol. Brain Res. **3:** 61–68.
13. AKOWITZ, A. A., E. BARBARESE, K. SCHELD & J. H. CARSON. 1987. Structure and expression of myelin basic protein gene sequences in the *mld* mutant mouse: Reiteration and rearrangement of the MBP gene. Genetics **116:** 447–464.
14. OKANO, H., M. MIURA, A. MORIGUCHI, K. IKENAKA, Y. TSUKADA & K. MIKOSHIBA. 1987. Inefficient transcription of the myelin basic protein gene possibly causes hypomyelination in myelin-deficient mutant mice. J. Neurochem. **48:** 470–476.
15. POPKO, B., C. PUCKETT, E. LAI, H. D. SHINE, C. READHEAD, N. TAKAHASHI, S. W. HUNT, III, R. L. SIDMAN & L. HOOD. 1987. Myelin-deficient mice: Expression of myelin basic protein and generation of mice with varying levels of myelin. Cell **48:** 713–721.
16. OKANO, H., T. TAMURA, M. MIURA, A. AOYAMA, K. IKENAKA, M. OSHIMURA & K. MIKOSHIBA. 1988. Gene organization and transcription of duplicated MBP gene of myelin-deficient (shi^{mld}) mutant mouse. EMBO J. **7:** 77–83.
17. ROACH, A., K. BOYLAN, S. HORVATH, S. B. PRUSINER & L. HOOD. 1983. Characterization of cloned cDNA representing rat myelin basic protein: Absence of expression in brain of *shiverer* mutant mice. Cell **34:** 799–806.
18. ZELLER, N. K., M. J. HUNKELER, A. T. CAMPAGNONI, J. SPRAGUE & R. A. LAZZARINI. 1984. Characterization of mouse myelin basic protein messenger RNAs with a myelin basic protein cDNA clone. Proc. Natl. Acad. Sci. USA **81:** 18–22.
19. KIMURA, M., H. INOKO, M. KATSUKI, A. ANDO, T. SATO, T. HIROSE, H. TAKASHIMA, S. INAYAMA, H. OKANO, K. TAKAMATSU, K. MIKOSHIBA, Y. TSUKADA & I. WATANABE. 1985. Molecular genetic analysis of myelin-deficient mice: *Shiverer* mutant mice show deletion of gene(s) coding for myelin basic protein. J. Neurochem. **44:** 692–696.
20. TAKAHASHI, N., A. ROACH, D. B. TEPLOW, S. B. PRUSINER & L. HOOD. 1985. Cloning and characterization of the myelin basic protein gene from mouse: One gene can encode both 14 kd and 18.5 kd MBPs by alternate use of exons. Cell **42:** 139–148.
21. POPKO, B., C. PUCKETT & L. HOOD. 1988. A novel mutation in myelin-deficient mice results in unstable myelin basic protein gene transcripts. Neuron **1:** 221–225.
22. ROCH, J. M., M. TOSIC, A. ROACH & J.-M. MATTHIEU. 1989. The duplicated myelin basic protein gene in *mld* mutant mice does not impair transcription. Brain Res. **477:** 292–299.

23. TOSIC, M., J.-M. ROCH, A. ROACH & J.-M. MATTHIEU. 1988. MBP gene duplication in *mld* mice does not affect transcription initiation. Experientia **44:** A 84.
24. OKANO, H., K. IKENAKA & K. MIKOSHIBA. 1988. Recombination within the upstream gene of duplicated myelin basic protein genes of myelin deficient shi^{mld} mouse results in the production of antisense RNA. EMBO J. **7:** 3407–3412.
25. TOSIC, M., A. ROACH, J.-C. DE RIVAZ, M. DOLIVO & J.-M. MATTHIEU. 1990. Posttranscriptional events are responsible for low expression of myelin basic protein in myelin deficient mice: Role of natural antisense RNA. EMBO J. **9:** 401–406.
26. FREMEAU, R., JR. & B. POPKO. 1990. *In situ* analysis of myelin basic protein gene expression in myelin-deficient oligodendrocytes: Antisense hnRNA and readthrough transcription. EMBO J. **9:** 3533–3538.
27. OKANO, H., J. ARUGA, T. NAKAGAWA, C. SHIOTA & K. MIKOSHIBA. 1991. Myelin basic protein gene and the function of antisense RNA in its repression in myelin-deficient mutant mouse. J. Neurochem. **56:** 560–567.
28. KITAMURA, K., S. L. NEWMAN, C. W. CAMPAGNONI, J. M. VERDI, T. MOHANDAS, V. W. HANDLEY & A. T. CAMPAGNONI. 1990. Expression of a novel transcript of the myelin basic protein gene. J. Neurochem. **54:** 2032–2041.
29. KATSUKI, M., M. SATO, M. KIMURA, M. YOKOYAMA, K. KOBAYASHI & T. NOMURA. 1988. Conversion of normal behavior to shiverer by myelin basic protein antisense cDNA in transgenic mice. Science **241:** 593–595.

Antisense Oligodeoxynucleotides as Probes of T-Lymphocyte Gene Function[a]

DOUGLAS R. GREEN, HEGUANG ZHENG,
AND YUFANG SHI

The La Jolla Institute for Allergy and Immunology
11149 N. Torrey Pines Rd.
La Jolla, California 92037
and
Department of Immunology
University of Alberta
Edmonton, AB, Canada

Since the discovery that resting lymphocytes respond to mitogenic stimuli by entry into the cell cycle, T cells have been an excellent model system for the study of cellular activation. Progress in understanding signal transduction, second messenger systems, the activation and function of transactivator proteins, and the regulation of entry into G_1 and S phase has been made with some success in T lymphocytes. T cells play important roles in the generation and regulation of immune responses and in the pathogenesis of AIDS, and the study of T-cell activation therefore takes on added importance.

This brief discussion reviews the application of one technique, the use of antisense oligodeoxynucleotides, to the study of T-lymphocyte function. By targeting the expression of specific genes with such oligonucleotides, the roles of those gene products in T-cell function have been elucidated. For example, antisense oligodeoxynucleotides corresponding to *c-myc,*[1,2] *c-myb,*[3] and cdc2[4] have all been shown to interfere with mitogenic activation of T lymphocytes. Similarly, antisense oligonucleotides corresponding to the genes encoding interleukin-2 (IL2) and IL4 inhibited proliferation in T-cell clones producing one or the other of these cytokines for autocrine growth.[5] These important observations underscore the value of the antisense approach in elucidating gene function in T cells.

Here, two T-cell phenomena will be considered in detail: (1) the phenomenon of activation-induced apoptotic cell death in T-cell hybridomas and developing T cells, and the role of protooncogenes in this process, and (2) antigen-specific immunoregulatory T-cell activity and its relation to the molecules of the T-cell receptor for antigen.

ANALYSIS OF ACTIVATION-INDUCED APOPTOSIS IN T CELLS WITH ANTISENSE OLIGONUCLEOTIDES

During the development of T lymphocytes in the thymus, genes encoding the variable chains of the T-cell receptor (TCR), usually TCRα and β, are rearranged

[a]This research was supported by grant AI31591 from the National Institutes of Health and grants from the Medical Research Council of Canada, the National Cancer Institute of Canada, and from Gemini Science. This is publication number 42 from the La Jolla Institute for Allergy and Immunology.

and expressed on the cell surface. Due to this rearrangement process, each developing T cell bears a different TCR and is therefore potentially specific for a different antigenic ligand.[6] In some cases, this ligand will be present in the thymus, that is, it will represent a self-antigen, and therefore the maturation of such a T cell will carry a potential risk of autoimmunity. Several strategies exist to prevent this eventuality, one of which is the phenomenon of negative selection: the disappearance from the maturation pathway of developing T cells bearing autoreactive TCR.[7,8] An important mechanism of negative selection is activation-induced apoptosis.

Following induction through the T-cell receptor, immature T cells[9–11] and some T-cell hybridomas[12–14] undergo a program of active cell death, or apoptosis. Apoptosis is a form of cell death characterized by chromatin condensation, DNA fragmentation, and cytoplasmic blebbing, following which the dying cells fragment into small, membrane-bound apoptotic bodies to be taken up by resident phagocytic cells.[15] Frequently the process of apoptosis is dependent upon RNA and protein synthesis, and this is the case for activation-induced apoptosis. Despite extensive analysis of T-cell activation, little is known about those aspects of activation that lead to the induction of apoptosis.

Several genes known to be transcribed during T-cell activation were demonstrated to be regulated by antisense oligodeoxynucleotides.[1–3,5] We therefore prepared antisense oligonucleotides corresponding to several such genes and assessed their effects on activation-induced apoptosis in a T-cell hybridoma, A1.1. We had previously observed that this cell undergoes apoptosis within a few hours of activation via the TCR.[14]

Therefore, A1.1 cells were activated by culture in plastic wells that had previously been coated with a monoclonal antibody specific for murine CD3ϵ, a component of the TCR complex. This antibody is capable of activating T cells[16] and causes activation-induced apoptosis in T-cell hybridomas *in vitro*[12–14] and in developing T cells *in vivo*.[10,11] Conventional oligonucleotides were added at a concentration of 100 μM to some of the cultures. Following culture, viability of the cells was assessed by vital staining with tetrazolium (MTT). As shown in FIGURE 1, only the antisense oligodeoxynucleotides corresponding to *c-myc* prevented cell death after activation of A1.1. This result was confirmed with several preparations of the antisense oligodeoxynucleotide, including phosphorothioate-derivatized oligonucleotides, and was also reproduced using other T-cell hybridomas (results not shown). Furthermore, although *c-myc* antisense oligonucleotides consistently prevented cell death, oligonucleotides with the same base composition but different sequences (*c-myc* nonsense oligonucleotides) failed to have any effects (results not shown).

One hallmark of apoptotic cell death is a characteristic fragmentation of DNA into an oligonucleosomal "ladder."[15] In the experiment shown in FIGURE 2A, DNA from untreated (control) A1.1 cells was intact, whereas DNA from activated A1.1 cells (treated with anti-CD3 antibody) displayed the ladder pattern characteristic of apoptosis. A1.1 cells that were activated in the presence of *c-myc* antisense oligodeoxynucleotides showed dramatically less DNA fragmentation. Nonsense oligonucleotide or *c-fos* antisense oligonucleotide had no effect on DNA fragmentation. The ability of the *c-myc* antisense oligonucleotide to inhibit DNA fragmentation in this system was also observed using a quantitative assay of DNA fragmentation (FIG. 2B).

In addition to apoptosis, activation of T-cell hybridomas also induces the production of lymphokines, such as IL2. Whereas *c-myc* antisense oligonucleotides consistently inhibited the induction of apoptosis in T-cell hybridomas, none of the oligonucleotides (including the *c-myc* antisense) had any effect on the production of lymphokines by activated A1.1 cells (results not shown). Thus, the *c-myc* antisense oligonucleotides specifically inhibited activation-induced death without inhibiting activation-induced lymphokine production in T-cell hybridomas.

Although the *c-myc* antisense oligonucleotides were the only ones tested that were capable of interfering with the process of activation-induced apoptosis, both the *c-myc* and *c-fos* antisense oligonucleotides inhibited proliferation in murine L929 cells. As shown in FIGURE 3, L929 cells that were cultured in the presence of *c-myc* nonsense oligonucleotide grew normally, whereas those cultured with either *c-myc* or *c-fos* antisense oligonucleotides failed to grow. Thus, it is likely that the *c-fos* antisense oligonucleotide was functional, and therefore its failure to affect activation-induced apoptosis suggests that *c-fos* may not be a requirement for this process.

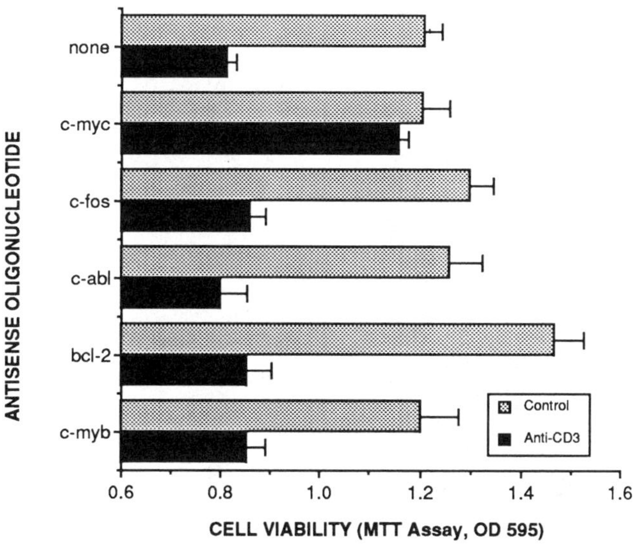

FIGURE 1. Effects of antisense oligonucleotides on activation-induced apoptosis. A1.1 cells were cultured in uncoated (*control-gray bars*) or anti-CD3-coated (*black bars*) 96-well plastic tissue culture plates for 20 hours, with or without 100 μM antisense oligodeoxynucleotides. Viability was assessed by MTT staining as described.[14] Conventional antisense oligonucleotides were provided by Regional DNA Synthesis (Calgary, AB) and sequences were as follows (all are based on murine protooncogene sequences): c-myc, GAAGTTCACGTTGAGGGGCAT; c-fos, GTTGAAACCCGAGAACATCAT; c-abl, CTTCAAACAGATCTCCAACAT; bcl-2, TGTTCTCCCGGCTTGCGCCAT; c-myb, GTGTCGGGGTCTCCGGGCCAT.

Ours is not the first observation to implicate the *c-myc* protooncogene in some forms of apoptosis. Wyllie and colleagues[17] observed that fibroblasts transformed with the myc oncogene displayed an increased incidence of apoptosis. Similarly, Wurm *et al.*[18] observed a high rate of cell death in CHO cells expressing an amplified *c-myc* gene under the control of a heat shock promoter. Recently, we determined that the death observed in these cells is due to apoptosis (Bissonnette *et al.*, submitted for publication). Thus, *c-myc* plays roles not only in cell growth and transformation,[19] but also in cell death as well. One interesting possibility is that *c-myc* is capable of driving some cells to a "decision" point, where an additional signal (e.g., one from a growth factor) will route the cell into the cell cycle. This is in

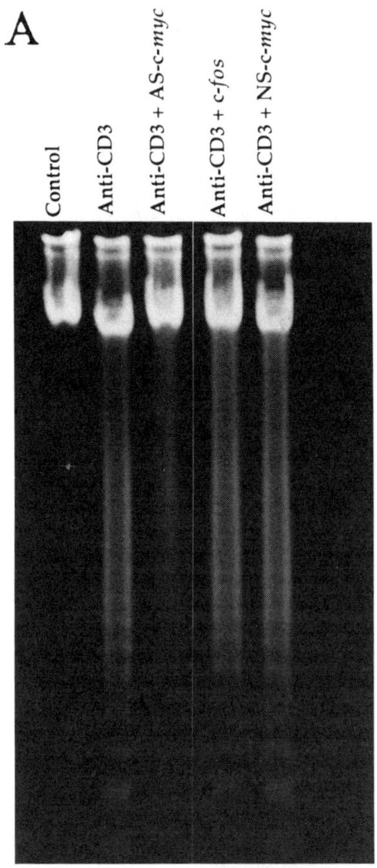

FIGURE 2. Inhibition by *c-myc* antisense oligonucleotides of DNA fragmentation associated with apoptosis. (**A**) A1.1 cells were cultured in uncoated (control) or anti-CD3-coated plastic tissue culture plates for 7 hours, with or without the indicated phosphorothioate-derivatized oligonucleotide (Regional DNA Synthesis, Calgary, AB). Oligonucleotide sequences were as follows: Antisense (AS)-c-myc, CACGTTGAGGGGCAT; AS-c-fos, ACCCGAGAACATCAT; nonsense (NS)-c-myc, AGTGGCGGAGACTCT. After culture, DNA was extracted and analyzed by agarose gel electrophoresis, stained with ethidium bromide, and photographed under UV light. (**B**) A1.1 cells were labeled by 1-hour pulse with ^{125}I-UdR, then cultured overnight. The labeled cells were then cultured on uncoated (control) or anti-CD3-coated plastic for 8 hours with or without the conventional c-myc antisense oligonucleotide described in the legend to FIGURE 1. Cells were lysed and the pellets centrifuged at 11,000 × *g* to separate soluble (fragmented) from intact DNA. The percentage fragmentation was calculated from the cpm of the separated samples.

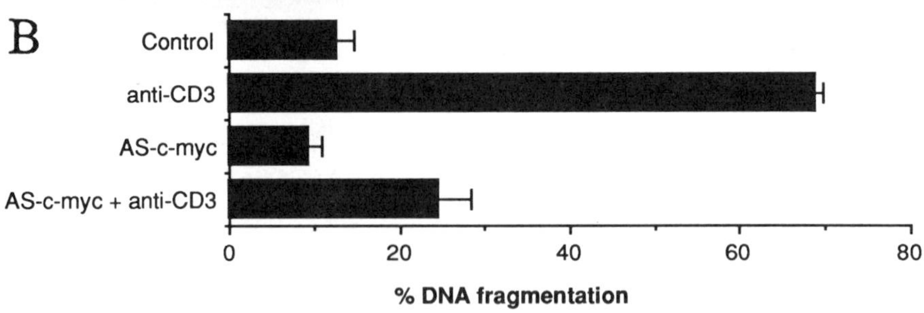

keeping with the observation that c-myc protein can act as a "competence factor."[20] In the absence of growth factor, the cell does not revert to resting, but rather it enters the apoptosis pathway.

Activation-induced apoptosis might follow a similar pattern. T-cell hybridomas constitutively express *c-myc* (unpublished observations) and presumably possess a constitutive signal analogous to a growth factor signal in normal cells. (These cells do not use T-cell growth factor, and therefore the autocrine growth signal is not the same as that of normal T cells.) The combination of *c-myc* and the growth factor-like signal results in constitutive growth. Stimulation via the TCR, then, might disrupt the growth factor-like signal, resulting in entry into the apoptosis pathway. This scenario is consistent with recent observations that T cells grown in growth factor IL2 become susceptible to induction of apoptosis upon activation via the TCR.[21] If the TCR signal interferes with the growth factor signal, then loss of the latter might result in apoptosis. This is possible, as withdrawal of IL2 from T cells maintained with this lymphokine can induce apoptosis.[22]

This brief discussion has outlined how the use of antisense oligodeoxynucleotides has given some insights into the paradoxical process of activation-induced apoptosis in T cells. We turn now to studies of a fundamentally different phenomenon, T-cell–derived antigen-specific immunoregulatory factors, and consider how the use of antisense oligonucleotides has shed light on this problem.

RELATION BETWEEN TCR AND A SOLUBLE FACTOR PROBED WITH ANTISENSE OLIGONUCLEOTIDES

In the 1970s and early 1980s, a great deal of immunological research was directed toward understanding regulatory T cells and their soluble products that control immune responses in an antigen-specific manner.[23] At least part of the interest in these antigen-specific T-cell factors was due to the belief that these represented soluble forms of the cell surface T-cell receptor for antigen, perhaps analogous to the soluble and membrane forms of immunoglobulin that had clearly been identified.

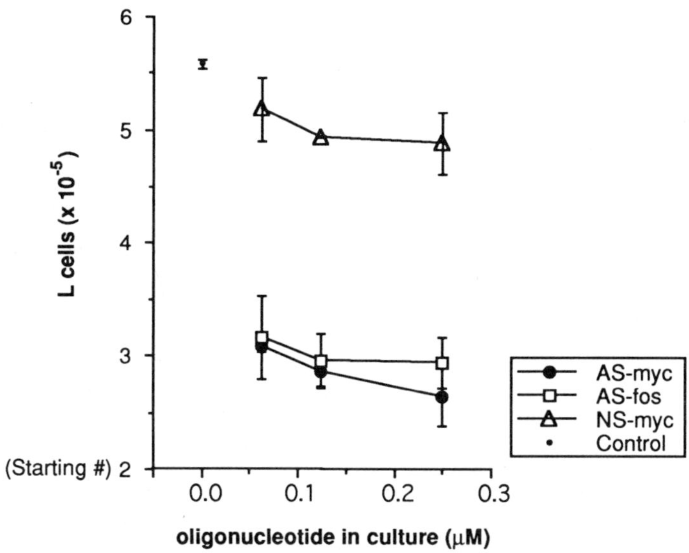

FIGURE 3. Effects of antisense oligonucleotides on L929 cell growth. L929 cells were grown with the phosphorothioate-derivatized oligodeoxynucleotides described in the legend to FIGURE 2A at the indicated concentrations. After 3 days, the cells were recovered and quantitated in a Coulter cell counter.

With the biochemical and molecular characterization of the chains of the TCR, and because of increasing frustration at the lack of characterization of the antigen-specific T-cell regulatory factors, interest in these soluble factors waned.

More recently, the study of antigen-specific, T-cell–derived immunoregulatory factors has undergone a resurgence, in part due to a renewed interest in their relation to the TCR. Before detailing the role of antisense oligonucleotides in the studies that contributed to this state of affairs, we will attempt to place the field into some perspective.

Although the TCRβ gene is sometimes lost from T-cell hybridomas producing antigen-specific immunoregulatory factors,[24] the TCRα gene is rearranged and expressed in such hybridomas.[25] Furthermore, Taniguchi and colleagues[26] found a strong correlation between the production of a factor specific for one antigen and expression of a particular VαJα in the T cells. It was at this point that our results with antisense oligonucleotides were reported, indicating a role for TCRα in the generation of such antigen-specific factors[27] (discussed in more detail below). At the same time, several groups independently showed that such factors were bound and eluted from a monoclonal antibody specific for murine TCRα.[28–30] The biologically active 46-kD molecule precipitated by this antibody showed the expected size of TCRα.[30] Together, these studies strongly supported a role for some TCRα molecules in antigen-specific immunoregulatory factors.

More recently, Kuchroo and colleagues[31] demonstrated that TCR^+ hybridomas producing antigen-specific factors can lose the expression of TCRβ without affecting the production of the factor, whereas loss of TCRα expression destroyed the factor production.[32] Transfection of the appropriate TCRα gene into such $TCR\alpha^-$ cells restored the ability to make the antigen-specific factor.[32] Similarly, we found that transfer of the TCRα gene from a factor-producing T-cell hybridoma to other lines transferred the ability to produce the antigen-specific activity.[33] These studies unambiguously demonstrated that the antigen-specific factor is composed, at least in part, of TCRα molecules.

Here, we briefly overview the use of antisense oligonucleotides in these studies and present previously unpublished observations supporting the idea that the biologically active TCRα molecule is secreted rather than shed from the cell surface.

A1.1, the T-cell hybridoma used in some of the studies just discussed, is specific for a synthetic polypeptide antigen, poly 18, presented by $I\text{-}A^d$ antigen-presenting cells.[34] We had previously found that supernatants of this cell contain poly 18-specific immunoregulatory activity detectable in an *in vitro* assay system, and that the antigenic fine-specificity of the soluble activity and that of the A1.1 TCR are the same.[30,35] This suggested a relation between TCR and soluble activity. Support for this idea came with the observation that the antigen-specific activity was bound by monospecific anti-TCR antiserum[36] and by a monoclonal anti-TCRα antibody.[30] Thus, it was possible that TCR genes encoded all or part of this soluble factor.

We tested this possibility through the use of antisense oligonucleotides. We took advantage of the fact that most T cells use different TCR V region gene segments in the TCRα and β genes.[6] Two cell lines were employed, A1.1 and B1.1, and antisense oligonucleotides corresponding to the Vα and Vβ translation start sequences were generated. The cells were treated with trypsin to remove cell surface TCR, and the cells were then cultured with the oligonucleotides for 48 hours. As cell surface expression of the TCR complex depends on the presence of most of the components (TCRα, TCRβ, CD3ε, CD3ζ, and either CD3δ or CD3γ) inhibition of one chain resulted in inhibition of reexpression of the entire complex (detected with anti-CD3ε antibody). We found that reexpression of the TCR complex was inhibited in A1.1 cells by the presence of antisense specific for the A1.1 Vα or Vβ, but not by antisense oligonucleotides specific for the B1.1 Vβ.[27] Conversely, B1.1 TCR reexpression was inhibited by the presence of antisense specific for B1.1 Vβ, but not by A1.1-specific oligonucleotides. These experiments clearly showed the effectiveness and specificity of these antisense oligonucleotides.

Related experiments are shown in FIGURE 4. Trypsin-treated A1.1 cells were cultured for 48 hours in the presence of antisense oligonucleotides specific for the

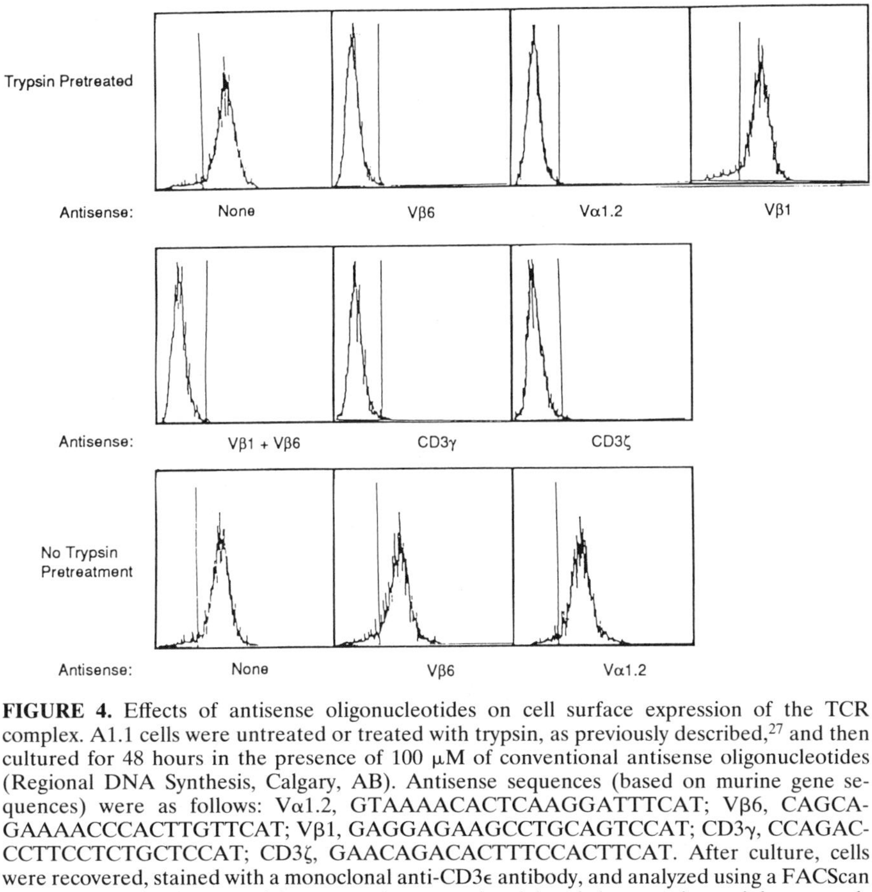

FIGURE 4. Effects of antisense oligonucleotides on cell surface expression of the TCR complex. A1.1 cells were untreated or treated with trypsin, as previously described,[27] and then cultured for 48 hours in the presence of 100 μM of conventional antisense oligonucleotides (Regional DNA Synthesis, Calgary, AB). Antisense sequences (based on murine gene sequences) were as follows: Vα1.2, GTAAAACACTCAAGGATTTCAT; Vβ6, CAGCAGAAAACCCACTTGTTCAT; Vβ1, GAGGAGAAGCCTGCAGTCCAT; CD3γ, CCAGACCCTTCCTCTGCTCCAT; CD3ζ, GAACAGACACTTTCCACTTCAT. After culture, cells were recovered, stained with a monoclonal anti-CD3ε antibody, and analyzed using a FACScan (Becton-Dickinson). *Vertical bars* were drawn to the right of the negative staining controls (controls are not shown) for comparison with the stained samples.

A1.1 Vβ (Vβ6), A1.1 Vα (Vα1.2), CD3γ, or CD3ζ. Reexpression of the TCR complex was assessed by staining with anti-CD3ε. As shown in FIGURE 4, reexpression was inhibited by the antisense oligonucleotides corresponding to the A1.1 Vα or Vβ (Vα1.2, Vβ6), CD3γ, or CD3δ. If the A1.1 cells were not pre-treated with trypsin, the presence of oligonucleotides during 48 hours of culture had little effect on the levels of cell surface TCR expression (indicating that turnover of TCR on the surface of these cells is relatively slow).

We had previously observed that antisense to Vβ2 had no effect on TCR reexpression in A1.1 cells.[27] In addition to expressing TCR Vβ6, A1.1 cells also express TCR Vβ1, derived from the BW5147 fusion partner (A. Fotedar, personal communication). Nevertheless, antisense to Vβ6 completely inhibited reexpression of TCR on A1.1, whereas antisense to Vβ1 had no effect (FIG. 4). Therefore, it is unlikely that Vβ1 contributes to the TCR of A1.1.

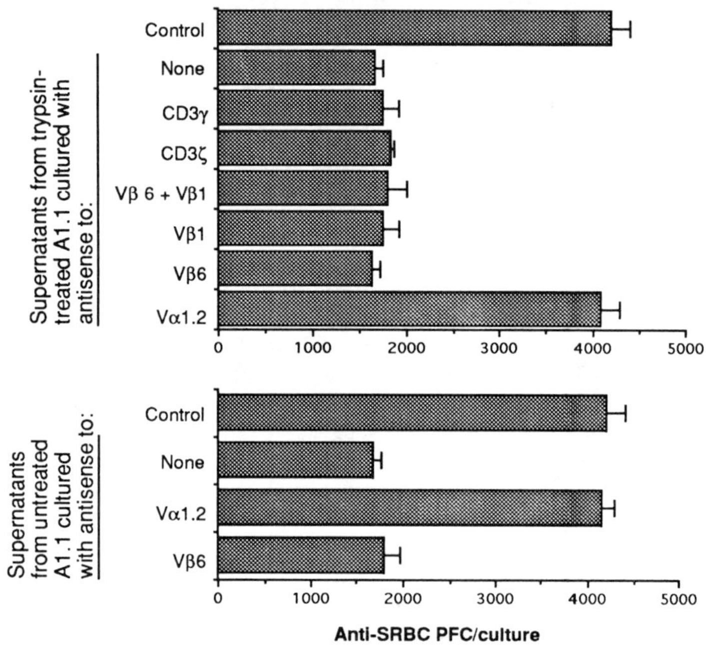

FIGURE 5. Effects of antisense oligonucleotides on production of an antigen-specific immunoregulatory activity. Supernatants from the 48-hour cultures described in the legend to FIGURE 4 were tested using a bioassay for the antigen-specific immunoregulatory activity produced from A1.1.[27,30,35] Briefly, cultures contained normal murine spleen cells, 10% supernatant from A1.1, 10% accessory supernatant (a supernatant from sheep red blood cell [SRBC] hyperimmune splenic T cells that was absorbed with SRBC; this contains an antigen-nonspecific activity required for the function of the A1.1 supernatant), and SRBC coupled with poly 18 or related peptides. After 5 days, the cells were recovered and anti-SRBC plaque-forming cells (PFC)/culture determined. The biological activity of the A1.1 supernatant is seen as an inhibition of anti-SRBC PFC. No effects were seen in the absence of either A1.1 supernatant or accessory supernatant or when inappropriate peptides were conjugated to the SRBC in the primary culture (not shown).

We next examined the biological activity present in the supernatants of treated A1.1 cells. As shown in FIGURE 5, the addition of supernatant from trypsin-pretreated (upper panel) or -untreated (lower panel) A1.1 cells significantly inhibited anti-SRBC PFC generation in the assay system, described in detail elsewhere.[27,30,35] The same inhibition was also observed in supernatants from A1.1 cells cultured with antisense corresponding to Vβ or CD3. On the other hand, no inhibitory activity was present in the supernatants of A1.1 cells cultured with antisense corresponding to the A1.1 Vα. This result was also obtained using untrypsinized cells cultured with antisense oligonucleotides (FIG. 5, lower panel). That is, despite surface expression of TCR, the presence of Vα1.2 antisense oligonucleotides prevented the production of the immunoregulatory activity by the A1.1 cells.

These results show that the production of immunoregulatory activity by A1.1 cells depends on expression of TCRα, as previously described.[27] Furthermore, cell surface TCR expression does not correlate with the production of the factor, either in cases of inhibition of cell surface TCR expression by Vβ or CD3 antisense or in cases of cell surface expression of TCR in untrypsinized A1.1 cells cultured with Vα antisense. This lack of correlation strongly suggests that the factor does not represent a shed form of the surface receptor; rather, a secreted form is suggested.

The immunoregulatory activity of TCRα and the conclusion that the soluble factor is not a shed form of this molecule are supported by gene transfer studies already discussed.[32,33] The fact that the same conclusions were drawn using antisense oligonucleotides underscores the power of the antisense approach.

CONCLUSIONS AND FUTURE PROSPECTS

Our studies employ antisense oligonucleotides to control gene function in T lymphocytes. Phosphorothioate-derivatized oligonucleotides, with their increased stability, may provide a method to temporarily control such gene function in intact animals. Some preliminary studies using *c-myc* antisense oligonucleotides to control activation-induced apoptosis *in vivo* suggest that this may be feasible. Long-term regulation of gene function, however, may require a modification of the approach.

Antigene strategies, discussed elsewhere in this volume, may represent such a modified approach. Furthermore, if organic compounds can be designed to have specific antigen effects,[37] these may provide an extremely powerful tool for further analysis of gene function *in vivo*. Studies that will attempt to modify T-cell function *in vivo* are currently underway.

SUMMARY

Conventional and thiophosphonate-derivatized oligonucleotides were employed to specifically regulate functional gene expression in murine T-cell hybridomas. For example, induction of apoptotic cell death following activation of T-cell hybridomas was examined using antisense oligonucleotides corresponding to several protooncogenes. We found that antisense oligodeoxynucleotides corresponding to *c-myc* inhibited both the characteristic DNA fragmentation and the loss of cell viability following activation without affecting production of lymphokines. Functional antisense oligonucleotides corresponding to *c-fos* had no effect in this system. These results demonstrate the use of antisense oligonucleotides to regulate function in T-cell hybridomas and provide valuable insights into the molecular bases of this biological phenomenon. Antisense oligonucleotides were also used to study another problem, the relation of T-cell–derived antigen-specific immunoregulatory factors to the T-cell receptor (TCR). Because the translation start of each TCR gene usually varies from one T cell to another, antisense oligonucleotides corresponding to the TCR Vα or Vβ of different cells were shown to act in a cell-specific manner.

Furthermore, this method was used to demonstrate that a soluble antigen-specific regulatory activity produced by one of the T-cell lines depends on expression of the specific TCRa, an observation that has since been confirmed by gene transfer experiments. Expression of the CD3-TCR complex on the cell surface was also blocked by antisense oligonucleotides corresponding to CD3γ and CD3ζ; however, neither these nor TCR Vβ antisense oligonucleotides had any effect on production of the soluble regulatory activity.

REFERENCES

1. HEIKKILA, R., G. SCHWAB, E. WICKSTROM, S. L. LOKE, D. H. PLUZNIK, R. WATT & L. M. NECKERS. 1987. Nature **328:** 445.
2. HAREL-BELLAN, A., D. K. FERRIS, M. VINOCOUR, J. T. HOLT & W. L. FARRAR. 1988. J. Immunol. **140:** 2431.
3. VENTURELLI, D., S. TRAVALI & B. CALABRETTA. 1990. Proc. Natl. Acad. Sci. USA **87:** 5963.
4. FURUKAWA, Y., H. PIWNICA-WORMS, T. J. ERNST, Y. KANAKURA & J. D. GRIFFIN. 1990. Science **250:** 805.
5. HAREL-BELLAN, A., S. DURUM, K. MUEGGE, A. K. ABBAS & W. L. FARRAR. 1988. J. Exp. Med. **168:** 2309.
6. HEDRICK, S. M. 1989. *In* Fundamental Immunology. W. E. Paul, ed.:291. Raven Press. New York.
7. PULLEN, A. M., J. W. KAPPLER & P. MARRACK. 1989. Immunol. Rev. **107:** 125.
8. VON BOEHMER, H., H. S. TEH & P. KISIELOW. 1989. Immunol. Today **10:** 57.
9. SMITH, C. A., G. T. WILLIAMS, R. KINGSTON, E. J. JENKINSON & J. T. T. OWEN. 1989. Nature **337:** 181.
10. SHI, Y., B. M. SAHAI & D. R. GREEN. 1989. Nature **339:** 625.
11. SHI, Y., R. P. BISSONNETTE, N. PARFREY, M. SZALAY, R. T. KUBO & D. R. GREEN. 1991. J. Immunol. **146:** 3340.
12. UCKER, D. S., J. D. ASHWELL & G. NICKAS. 1989. J. Immunol. **143:** 3461.
13. ODAKA, C., H. KIZAKI & T. TADAKUMA. 1990. J. Immunol. **144:** 2096.
14. SHI, Y., M. G. SZALAY, L. PASKAR, M. BOYER, B. SINGH & D. R. GREEN. 1990. J. Immunol. **144:** 3326.
15. WYLLIE, A. H., J. F. R. KERR & A. R. CURRIE. 1980. Int. Rev. Cytol. **68:** 251.
16. LEO, O., M. FOO, D. H. SACHS, L. E. SAMELSON & J. A. BLUESTONE. 1987. Proc. Natl. Acad. Sci. USA **84:** 1374.
17. WYLLIE, A. H., K. A. ROSE, R. G. MORRIS, C. M. STEEL, E. FOSTER & D. A. SPANDIDOS. 1987. Br. J. Cancer **56:** 251.
18. WURM, F. M., K. A. GWINN & R. E. KINGSTON. 1986. Proc. Natl. Acad. Sci. **83:** 5414.
19. COLE, M. D. 1986. Am. Rev. Genet. **20:** 361.
20. KACZMAREK, L., J. K. HYLAND, R. WATT, M. ROSENBERG & R. BASERGA. 1985. Nature **228:** 1313.
21. LENARDO, M. 1991. Nature **353:** 858.
22. DUKE, R. C. & J. J. COHEN. 1986. Lymphokine Res. **5:** 289.
23. GOLUB, E. S. & D. R. GREEN. 1991. Immunology: A Synthesis, 2nd ed.:494. Sinauer Press. Sunderland, MA.
24. HEDRICK, S. M., R. N. GERMAIN, M. J. BEVAN, M. DORF, I. ENGEL, P. FINK, N. GASCOIGNE, E. HEBER-KATZ, J. KAPP, Y. KAUFMAN, J. KAYE, F. MELCHERS, C. PIERCE, R. H. SCHWARTZ, C. SORENSEN, M. TANIGUCHI & M. M. DAVIS. 1985. Proc. Natl. Acad. Sci. USA **82:** 531.
25. IMAI, K., M. KANNO, H. KIMOTO, K. SHIGEMOTO, S. YAMAMOTO & M. TANIGUCHI. 1986. Proc. Natl. Acad. Sci. USA **83:** 8708.
26. KOSEKI, H., K. IMAI, T. ICHIKAWA, I. HAYATA & M. TANIGUCHI. 1989. Int. Immunol. **1:** 557.
27. ZHENG, H., M. BOYER, P. KILGANNON, A. FOTEDAR & D. R. GREEN. 1989. Proc. Natl. Acad. Sci. USA **86:** 3758.

28. FAIRCHILD, R. L., R. T. KUBO & J. W. MOORHEAD. 1990. J. Immunol. **145:** 2001.
29. IWATA, M., K. KATAMURA, R. T. KUBO, H. M. GREY & K. ISHIZAKA. 1989. J. Immunol. **143:** 3917.
30. BISSONNETTE, R., H. ZHENG, R. KUBO & D. R. GREEN. 1991. J. Immunol. **146:** 2898.
31. COLLINS, M., V. K. KUCHROO, M. J. WHITTERS, R. M. O'HARA, K. KELLEHER, R. T. KUBO & M. E. DORF. 1990. J. Immunol. **145:** 2809.
32. KUCHROO, V. K., M. C. BYRNE, Y. ATSUMI, E. GREENFIELD, J. B. CONNOLLY, M. J. WHITTERS, R. M. O'HARA, M. COLLINS & M. E. DORF. 1991. Proc. Natl. Acad. Sci. USA **88:** 8700.
33. GREEN, D. R., R. BISSONETTE, H. ZHENG, T. ONDA, F. ECHEVERRI, R. J. MOGIL, J. K. STEELE, M. VORALIA & A. FOTEDAR. 1991. Proc. Natl. Acad. Sci. USA **88:** 8475.
34. FOTEDAR, A., M. BOYER, W. SMART, J. WIDTMAN, E. FRAGA & B. SINGH. 1985. J. Immunol. **135:** 3028.
35. ZHENG, H., M. BOYER, A. FOTEDAR, B. SINGH & D. R. GREEN. 1988. J. Immunol. **140:** 1351.
36. ZHENG, H., A. FOTEDAR, B. SINGH & D. R. GREEN. 1989. *In* Cellular Basis of Immune Modulation. J. G. Kaplan, D. R. Green & R. C. Bleakley, eds.:379. Alan R. Liss, Inc. New York.
37. LOWN, J. W. 1992. Antiviral Res., in press.

Interference of Gene Expression in *Drosophila* by Antisense Ribosomal Protein Genes Integrated into the Germ Line

MARCELO JACOBS-LORENA

Department of Genetics
Case Western Reserve University
School of Medicine
10900 Euclid Ave.
Cleveland, Ohio 44106-4955

Antisense RNA interference experiments have been conducted in all major organisms, including bacteria, plants, and animals, to address different biological questions (reviewed in refs. 1–3). Despite the considerable volume of information accumulated, much remains to be learned about the mechanisms of antisense gene interference. A number of parameters can influence the effectiveness of antisense inhibition of gene expression. For example, the choice of target sequence, the concentration of antisense sequence relative to sense RNA, the stability of the antisense-sense hybrids, and the presence or absence of unwinding activities may all play a role.

Here I briefly review the *Drosophila* antisense literature and summarize experiments from this laboratory with antisense RNA expression in *Drosophila* whole organisms. The article ends with a discussion of the parameters that may influence the effectiveness of the antisense approach in *Drosophila*. A more detailed review of the subject can be found in reference 4.

EMBRYO INJECTIONS AND EXPRESSION IN CULTURED CELLS

Antisense interference experiments in *Drosophila* can be grouped into three categories: embryo injection, expression in cultured cells, and expression in whole organisms. By far the most successful approach has been the injection of *in vitro* synthesized antisense RNAs into early embryos. In general, the aim of these experiments was to verify that a cloned DNA corresponded to a gene previously characterized by classic genetic means. Genes identified in this way include *Krüppel*[5] *(Kr)*, *wingless*[6] *(wg)*, *snail*[7] *(sna)*, *knirps*[8] *(kni)*, *spalt*[9] *(sal)*, *pecanex*[10] *(pcx)*, and *teashirt*[11] *(tsh)*. An excess varying between 50-fold (*Kr*) and one-million-fold (*sal*) of antisense RNA was necessary to observe a weak phenotype. Complete gene inactivation was hardly ever observed, even when a very large excess of antisense RNA was injected.

A second approach for antisense inactivation is the transfection of *Drosophila* tissue culture cells with antisense genes driven by conditional promoters. McGarry and Lindquist[12] specifically interfered with the expression of the small heat shock protein hsp26, whereas Bunch and Goldstein[13] transfected cells with an antisense alcohol dehydrogenase (ADH) gene or antisense microtubule-associated protein (205-kD MAP) genes targeted to two different regions of the mRNA. Surprisingly, antisense ADH RNA sharply reduced ADH expression, whereas antisense MAP RNA was completely ineffective, despite the fact that ADH is the more abundant of the two messages and that the antisense/sense RNA ratios were much higher for MAP than for ADH. The reasons for these discrepancies are not understood.

Antisense RNA expression from a gene integrated into the germ line is a potentially powerful approach to investigate the function of genes at the organismal level. After an introduction to *Minute* mutations, we present the results of our experiments with antisense ribosomal protein (r-protein) genes integrated into the *Drosophila* genome.

DROSOPHILA MINUTE MUTATIONS

Minutes comprise a class of 50 or more mutations dispersed over the entire genome. Inactivation of one copy of the gene leads to a dominant phenotype typical of *Minutes* that includes delayed development, short and thin bristles, and reduced fertility. Homozygous *Minutes* are lethal. The similarity between *Minute* mutations and mutations that reduce expression of ribosomal RNAs has led to the hypothesis that *Minute* mutations interfere with some aspect of protein synthesis. It is likely that *Minute* mutations inactivate r-protein genes (reviewed in ref. 14). Conclusive evidence that a *Minute* corresponds to an r-protein gene is available only in one case, for *Minute(3)99D*. It has been shown that this mutation can be corrected by a transformed copy of the r-protein 49 (rp49) gene.[15] In some of the transformed lines, only partial rescue of the *Minute(3)99D* phenotype was observed, presumably because chromosomal position effects led to the underproduction of the rp49 gene product. This variability attests to the exquisite sensitivity of the *Minute* phenotype to rp49 gene dosage.

It has been apparent since the early days that target genes can rarely, if ever, be completely inactivated by antisense expression. *Minute* (and presumably r-protein genes) provided an ideal target for antisense experiments because of their characteristic haploinsufficiency: inactivation of one copy of the gene (that is, 50% reduction in gene expression) is in most cases sufficient to observe a strong phenotype. This was a key factor that led us to target r-protein genes for antisense experiments in whole *Drosophila.*

ANTISENSE EXPRESSION IN THE WHOLE ORGANISM

Antisense Ribosomal Protein A1 Expression Specifically Disrupts Oogenesis

The r-protein A1 (rpA1) gene has been extensively characterized in this laboratory (ref. 16 and references therein). It maps to the right arm of chromosome 2 on polytene chromosome band 53CD. Unfortunately, this region is genetically ill-defined. The only known *Minute* mutation in this general area [*M(2)S7*] appears not to correspond with the rpA1 gene (M. Kay and M. J.-L., unpublished data). Other as yet undiscovered *Minutes* may exist in the region. In the absence of a mutant in the rpA1 gene, we used an antisense gene integrated into the germ line to interfere with rpA1 expression in a whole organism setting.[17] We had two questions in mind: (1) Is it possible to interfere with *Drosophila* development by expression of an integrated antisense rpA1 gene? (2) If so, does antisense rpA1 expression induce the characteristic *Minute* phenotypes? An antisense rpA1 gene driven by the strongly inducible heat shock hsp70 promoter was introduced into the germ line by P-element–mediated transformation. Expression of antisense rpA1 severely disrupted oogenesis. About 2.5 days after a brief (1 hour) heat pulse to induce antisense rpA1 synthesis, up to 80% of the eggs laid were abnormally small. These eggs never developed and were probably unfertilized. Thereafter, egg production gradually reverted to pre-heat–treatment levels. Interestingly, the "small egg phenotype" appears to be caused by failure of nurse cells to transfer their contents to the oocyte at the end of oogenesis, presumably because of disruption of coordination between the developmental programs of the nurse and follicle cells. Further details can be found in reference 17.

An unexpected result of the antisense rpA1 experiments was that most of the phenotypes characteristic of *Minute* mutations were not observed, even after frequent induction of the antisense gene throughout the larval and pupal stages of development. As no rpA1 mutants are known, it was not determined if the divergence from the *Minute* phenotype is due to peculiarities of antisense interference experiments or if these are the only loss-of-function rpA1 phenotypes that can be observed. This question could best be addressed by targeting the rp49 gene, for which the *Minute*:r-protein relation had been firmly established (cf. previous section).

Phenocopying the Minute *Phenotype by Antisense Ribosomal Protein 49 Expression*

The following experiments were undertaken to compare the phenotype of antisense rp49 expression with the phenotype of a mutation [*Minute(3)99D*] in the same gene. The rp49 was placed in antisense orientation under the control of the strong heat-inducible hsp70 promoter and then introduced into the germ line by P-element–mediated transformation.[18] Activation of rp49 antisense expression by a 1-hour heat pulse (37°C) resulted in the accumulation of antisense RNAs to levels three- to fourfold higher than those of endogenous rp49 mRNA. As was the case for rpA1, antisense rp49 expression had its strongest effect on oogenesis. Egg production decreased to 20% of controls about 3 days after the heat pulse. However, for unknown reasons, anti-rp49 expression did not produce the "small egg phenotype."

We also measured the effect of antisense rp49 expression on the development of the organism. Anti-rp49 was induced in synchronized larvae or pupae carrying the transgene, by subjecting them to heat pulses every 12 hours. As controls, flies carrying a construct in the sense orientation were treated in the same way. About 25% of larvae homozygous for the antisense gene displayed a developmental delay of 12–24 hours, whereas no measurable delay was observed in controls. When the antisense gene was induced during pupal stages, thin and short bristles characteristic of *Minute* mutants were observed in about 15% of the flies. None of the control flies displayed this phenotype. Further details can be found in reference 18.

In summary, antisense rp49 expression specifically induced many of the characteristic *Minute* phenotypes that had also been observed in rp49 mutants. Thus, antisense rp49 expression appears to faithfully mimic the mutant phenotype. However, even after frequent antisense induction, the observed phenotypes never exceeded in severity those of weak *Minute* mutants.

CONCLUSIONS

Perhaps the most successful application of antisense technology to *Drosophila* so far has been the injection of *in vitro* synthesized antisense RNA into early embryos (cf. section on embryo injections above). This approach allows one to verify if a given DNA segment encodes the gene of interest immediately after cloning, even before any sequencing is performed. Antisense inhibition of gene expression in cultured cells has also been achieved, but the results have not always been predictable.

To our knowledge, the only experiments of antisense inactivation in whole *Drosophila* reported so far involve rpA1[17] and rp49,[18] as summarized in this article. There is anecdotal information of attempts of antisense interference in whole *Drosophila* by other research groups. Specifically, I am aware of attempts with three genes: the *white* eye color gene (V. Pirrotta, personal communication), the *ovarian tumor* (*otu*) gene (V. Schweickart and C. Laird, personal communication) and the *Drosophila* homolog of the *raf* oncogene (S. Qian, N. Perrimon, and M. Jacobs-Lorena, unpublished observations). In each case, antisense expression had no detectable effect on the phenotype of the flies. In the absence of a measurable effect, no firm conclusions can be derived from these experiments. Yet, it is worth pointing out that the *white* mRNA is quite rare and that the *white* antisense gene was driven by the same promoter (hsp70) that was used for the r-protein experiments described here. One might then ask if there is any way to evaluate *a priori* the chances of successful gene inactivation in whole *Drosophila.* The answer to this question has not yet been settled and will depend in part on the elucidation of the mechanisms of antisense gene interference in the different cell types (cf. Introduction). In any event, it appears that mRNA abundance is not a reliable predictor of effectiveness of antisense inactivation. Attempts at inactivating the rare *white* mRNA were unsuccessful, but a clear effect was observed with two abundant r-protein mRNAs. Similar observations were made in *Drosophila* cultured cells.[13] It appears that antisense gene inactivation in *Drosophila* may be inefficient. For instance, assuming that deletion of the rp49 gene is the only contributor to the phenotype in *Minute(3)99D,* a 50% reduction in gene dosage is sufficient to cause a strong *Minute* phenotype. The relatively weak phenotype of antisense rp49 flies implies that the reduction of gene activity in these flies is much less than 50%. Thus, we believe that the best targets for antisense interference experiments are genes for which there is some previous knowledge that a partial reduction in gene expression has a measurable phenotype.

In closing, I wish to touch on some practical aspects. The effectiveness of antisense interference can be expected to improve by increasing the sense-to-antisense RNA mass ratio. Thus, sequences that confer instability to the expressed antisense RNA should be avoided. For instance, when constructing the antisense r-protein genes, we added the hsp70 polyadenylation sequence at the 3′ end of the gene. It turns out that an element that confers RNA instability resides in this sequence,[19] which may reduce the effectiveness of antisense RNA action. Another element to consider is the promoter. In most cases the strongest possible promoter would be the most desirable one. It may be best to use a strong constitutive promoter even when targeting genes that are essential for the organism. In view of the relative ineffectiveness of antisense gene inactivation, we might expect to obtain only a partial loss-of-function phenotype. Finally, it is important to consider that a promoter that drives a gene encoding an abundant mRNA is not necessarily a strong promoter. A very stable and abundant mRNA might be driven by a relatively weak promoter.

REFERENCES

1. Walder, J. 1988. Antisense DNA and RNA: Progress and prospects. Genes Dev. **2:** 502–504.
2. Izant, J. G. 1989. Antisense "pseudogenetics." Cell Motil. Cytoskeleton **14:** 81–91.
3. Weintraub, H. M. 1990. Antisense RNA and DNA. Scientific Am. **262:** 40–46.

4. PATEL, R. C. & M. JACOBS-LORENA. 1992. Interference of gene expression by antisense RNAs in *Drosophila. In* Antisense RNA and DNA, Modern Biology Series. J. A. Murray, ed. John Wiley & Sons. New York, NY. In press.
5. ROSENBERG, U. B., A. PREISS, E. SEIFERT, H. JÄCKLE & D. C. KNIPPLE. 1985. Production of phenocopies by Krüppel antisense RNA injection into *Drosophila* embryos. Nature **313:** 703–706.
6. CABRERA, C. V., M. C. ALONSO, P. JOHNSTON, R. G. PHILLIPS & P. A. LAWRENCE. 1987. Phenocopies induced with antisense RNA identify the *wingless* gene. Cell **50:** 659–663.
7. BOULAY, J. L., C. DENNEFELD & A. ALBERGA. 1987. The *Drosophila* developmental gene *snail* encodes a protein with nucleic acid binding fingers. Nature **330:** 395–398.
8. NAUBER, U., M. J. PANKRATZ, A. KIENLIN, E. SEIFERT, U. KLEMM & H. JÄCKLE. 1988. Abdominal segmentation of the *Drosophila* embryo requires a hormone receptor-like protein encoded by the gap gene *knirps.* Nature **336:** 489–492.
9. SCHUH, R. & H. JÄCKLE. 1989. Probing *Drosophila* gene function by antisense RNA. Genome **31:** 422–425.
10. LABONNE, S. G., I. SUNITHA & A. P. MAHOWALD. 1989. Molecular genetics of *pecanex,* a maternal-effect neurogenic locus of *Drosophila melanogaster* that potentially encodes a large transmembrane protein. Dev. Biol. **136:** 1–16.
11. FASANO, L., L. RODER, N. CORE, E. ALEXANDRE, C. VOLA, B. JACQ & S. KERRIDGE. 1991. The gene *teashirt* is required for the development of *Drosophila* embryonic trunk segments and encodes a protein with widely spaced zinc finger motifs. Cell **64:** 63–79.
12. MCGARRY, T. J. & S. LINDQUIST. 1986. Inhibition of heat shock protein synthesis by heat-inducible antisense RNA. Proc. Natl. Acad. Sci. USA **83:** 399–403.
13. BUNCH, T. A. & L. S. B. GOLDSTEIN. 1989. The conditional inhibition of gene expression in cultured *Drosophila* cells by antisense RNA. Nucleic Acids Res. **17:** 9761–9782.
14. KAY, M. A. & M. JACOBS-LORENA. 1987. Developmental genetics of ribosome synthesis in *Drosophila.* Trends Genet. **3:** 347–351.
15. KONGSUWAN, K., Q. YU, A. VINCENT, M. C. FRISARDI, M. ROSBASH, J. A. LENGYEL & J. MERRIAM. 1985. A *Drosophila Minute* gene encodes a ribosomal protein. Nature **317:** 555–558.
16. QIAN, S., J.-Y. ZHANG, M. A. KAY & M. JACOBS-LORENA. 1987. Structural analysis of the *Drosophila* rpA1 gene, a member of the eucaryotic 'A' type ribosomal protein family. Nucleic Acids Res. **15:** 987–1003.
17. QIAN, S., S. HONGO & M. JACOBS-LORENA. 1988. Antisense ribosomal protein gene expression specifically disrupts oogenesis in *Drosophila melanogaster.* Proc. Natl. Acad. Sci. USA **85:** 9601–9605.
18. PATEL, R. & M. JACOBS-LORENA. 1992. Generation of *Minute* phenotypes by a transformed antisense ribosomal protein gene. Dev. Genetics, in press.
19. PETERSEN, R. B. & S. LINDQUIST. 1989. Regulation of HSP70 synthesis by messenger RNA degradation. Cell Regulation **1:** 135–149.

Cell Cycle Effects of Microinjected Antisense Oligodeoxynucleotides to p34^{cdc2} Kinase[a]

W. EDWARD MERCER,[b,c] STEPHEN J. ULLRICH,[d]
MICHELLE T. SHIELDS,[b] DAVID LIN,[b]
AND HANSJUERG ALDER[b]

[b]*Department of Microbiology and Immunology and*
Jefferson Cancer Institute
Thomas Jefferson University
Philadelphia, Pennsylvania 19107

[d]*Laboratory of Cell Biology*
National Cancer Institute
Bethesda, Maryland 20892

Antisense RNA or antisense oligodeoxynucleotide molecules (oligomers) within cells targeted toward the RNA transcript of a specific gene can inhibit the expression or promote the degradation of the RNA transcript, resulting in suppression of the function coded for by the gene. The addition of antisense oligomers to culture medium and their uptake by cells has been used most successfully to inhibit the expression of specific target genes in lymphoid or myeloid cell lines grown in suspension (reviewed by Calabretta[1]) and to a lesser degree for cells grown in monolayer cultures.[2–4] In some cases, when antisense oligomers known to be effective for cells grown in suspension are applied to cells grown in monolayer cultures, they are often much less active, resulting in little, if any, suppression of the targeted gene. The reason for the difference in sensitivity is unclear, but it could be due to differences in the efficiency of uptake of the oligomers by cells grown in monolayer cultures. In this study we used glass-capillary microinjection as an alternative method to introduce antisense oligomers directly into the cytoplasm of cells grown in monolayer culture.

As a target gene we chose to examine the effect of antisense oligomers directed against the RNA transcripts for the serine-threonine protein kinase p34^{cdc2}. The p34^{cdc2} protein is the catalytic subunit of a protein kinase activity that is involved in the regulation of the cell cycle in a wide range of eukaryotes (reviewed by Lewin[5]). For example, in the fission yeast *Schizosaccharomyces pombe,* p34^{cdc2} function is required at two checkpoints of the cell cycle, in the G_1 phase before the initiation of DNA replication and at the initiation of mitosis. The human homolog of p34^{cdc2} was cloned by marker rescue of an *S. pombe* mutation and is able to replace both the G_1 and the mitotic function of the yeast cdc2 gene.[6] In mammalian cells, p34^{cdc2} kinase appears to be most active during mitosis,[5] and microinjection of anti-cdc2 antibodies inhibits entry into mitosis.[7] Recent studies suggest that human p34^{cdc2} kinase may

[a]M. T. Shields was supported by training grant CA 09644.

[c]Address for correspondence: W. Edward Mercer, Jefferson Cancer Institute, Thomas Jefferson University, 609A Bluemle Life Sciences Building, 233 S. 10th Street, Philadelphia, PA 19107.

also be involved in cell cycle control of DNA replication through its association with a multiprotein replication complex (RF-S) present in S-phase cells.[8] The addition of $p34^{cdc2}$ antisense oligomers to the culture medium of mitogen-stimulated human T lymphocytes inhibits G_1 progression into S-phase.[9] This finding prompted us to examine the effects on G_1 progression into the S-phase of microinjected antisense oligomers to $p34^{cdc2}$ kinase.

In the present report, we present evidence that the abundance of the RNA transcripts encoding $p34^{cdc2}$ and the rate of p34 protein synthesis increase significantly before the initiation of DNA synthesis when growth-arrested cells are serum stimulated to proliferate. In addition, we show that microinjection of antisense oligomers to $p34^{cdc2}$ directly into the cytoplasm of cells grown in monolayer culture before or at early times after serum stimulation leads to depletion of nuclear p34 protein in the injected cells and inhibits G_1 progression into S-phase.

RESULTS AND DISCUSSION

$p34^{cdc2}$ Expression in Glioblastoma Tumor Cells

For these studies, the T98G cell line derived from a human glioblastoma tumor was used because these cells, although they are an established cell line, exhibit several aspects of normal cell cycle regulation similar to those of primary human diploid fibroblasts.[10,11] Moreover, they are resilient to the microinjection procedure. T98G cells were maintained in culture as previously described.[11–13] We first examined the expression of $p34^{cdc2}$ RNA transcripts in exponentially growing cells, in cells growth arrested by incubation in medium containing 0.1% calf serum (CS), and in cells stimulated from growth arrest by the addition of fresh medium containing 10% fetal calf serum (FCS). Total RNA was isolated from cells by the guanidinium isothiocyanate procedure.[14] RNA samples were normalized by A_{260}, and a 10-μg amount of each sample was size fractionated by formaldehyde-agarose gel electrophoresis. Northern blots were prepared by standard procedures,[15] and the blots were hybridized using a probe prepared from the cDNA fragment of plasmid pCDC2Ms that contains a full-length coding sequence for mouse $p34^{cdc2}$ protein.[16] The 1.1-kb cDNA fragment of pCDC2Ms was purified from vector sequences after excision with the restriction enzyme *Eco*RI by gel electrophoresis and radiolabeled using $[^{32}P]$dCTP to high specific activity by the oligo-labeling method.[17] The results are presented in FIGURE 1. Two RNA species of approximately 2.0 and 1.6 kb were detected in exponentially growing cells (lane 1). These RNA species were not detected in growth-arrested cells (lane 2); however, at 8 hours after serum stimulation the two transcripts were detected (lane 3). Similar results were reported for normal human diploid fibroblasts,[16] Hela cells,[18] and Balb/c3T3 mouse cells.[4] In contrast to mammalian cells, in *S. pombe* there does not appear to be any transcriptional control of cdc2 expression.[19]

The synthesis of p34 protein in T98G cells was examined by pulse labeling with $[^{35}S]$methionine at different times after serum stimulation of growth-arrested cells. p34 protein was immunoprecipitated from protein extracts by the addition of anti-p34 antibodies (K1 anti-p34). This anti-p34 antibody was raised in rabbits against the synthetic peptide K1 $(DNQIKKM)_2C$ conjugated via COOH-terminal cysteine to bovine serum albumin by standard procedures. The synthetic peptide corresponds to residues 291 to 297 of the human cdc2 protein. After incubation with the antibody the immunoprecipitates were reacted with protein A-Sepharose, and the immune complexes were collected by centrifugation and analyzed by sodium

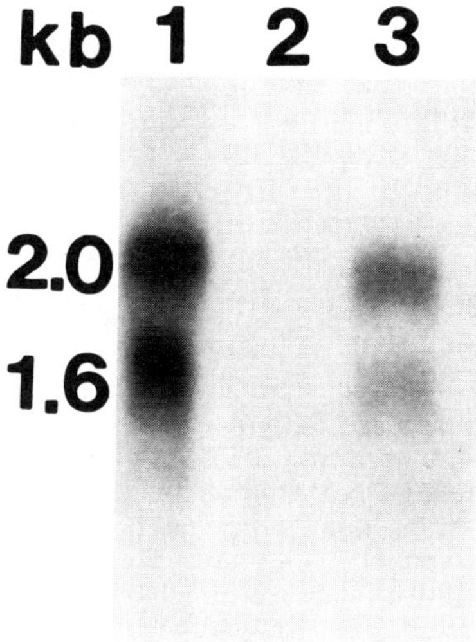

FIGURE 1. Northern blot analysis of p34[cdc2] RNA transcripts in T98G cells. Total RNA (10 μg) isolated from exponentially growing cells (*lane 1*), cells growth-arrested by incubation in medium containing 0.1% CS (*lane 2*), or cells stimulated from growth arrest for 8 hours by the addition of fresh medium containing 10% FCS (*lane 3*) was size fractionated on a 1.2% agarose gel containing 6.6% formaldehyde. Blotting of RNA to nitrocellulose filters followed standard procedures.[15] The blot was probed with a radiolabeled 1.1-kb cDNA fragment containing the entire coding region of mouse p34[cdc2] kinase.

dodecylsulfate/polyacrylamide gel electrophoresis using 10% gels.[20] Gels were fixed and processed for fluorography, as described previously.[11,13] The results are presented in FIGURE 2. In T98G cells, p34 proteins are not synthesized at high levels until 12 hours after serum stimulation. At this time, three proteins with apparent molecular weights of 34,000–36,000 are detected by the K1 anti-p34 antibodies in cells labeled with [^{35}S]methionine. On the contrary, metabolic labeling with [^{32}P]phosphate did not reveal the lower molecular weight (34,000) protein, indicating that this form was unphosphorylated; however, the two higher molecular weight forms were detected, indicating that they were the phosphorylated forms (data not shown).

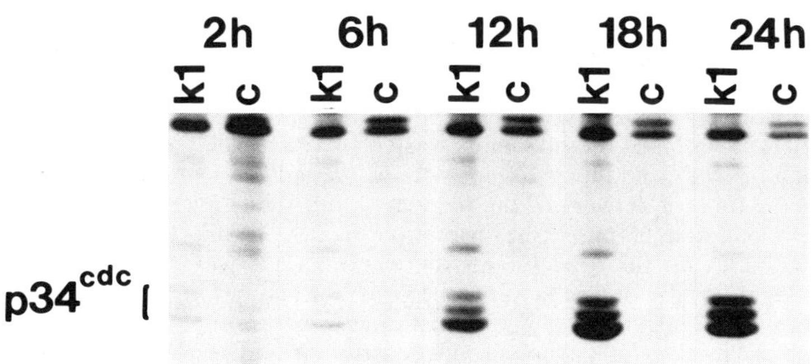

FIGURE 2. Synthesis of p34[cdc2] proteins in T98G cells during G_1 progression to S-phase. Cells were growth-arrested by incubation in medium containing 0.1% CS for 5 days and then stimulated to proliferate by replating at a 1:4 split ratio in medium containing 10% FCS. Cells were metabolically labeled with 75 μCi/ml of [^{35}S]methionine (SA 1100 Ci/mmol, Dupont/NEN) in methionine-free medium containing 2% dialyzed FCS at the times indicated in hours (h) after stimulation. Protein extracts were prepared and immunoprecipitated with either K1 anti p34 sera (k1) or preimmune sera (c). The immunoprecipitates were subjected to SDS-PAGE as described in the text.

These results are similar to what has been observed in other studies of p34^{cdc2} proteins in mammalian cells and are consistent with the finding that p34^{cdc2} protein is phosphorylated in a cell-cycle–dependent manner.[16,21] Thus, expression and phosphorylation of p34^{cdc2} protein appears to be regulated in T98G tumor cells in a manner similar to that observed in normal human diploid fibroblasts.[16]

Nuclear Localization of p34^{cdc2} Occurs before Entry into S-Phase

A significant portion of p34^{cdc2} protein shows nuclear localization by immunofluorescence staining during interphase in growing cells; however, in growth-arrested cells nuclear staining in barely detectable.[8] To examine more precisely the relation between p34 protein expression and the entry of T98G cells into S-phase, we simultaneously stained cells with affinity-purified K1 anti-p34 antibodies and monitored DNA replication by incorporation of [^{3}H]thymidine at different times after serum stimulation. The specificity of nuclear staining for p34 protein was confirmed by preincubating anti-p34 sera with the K1 peptide coupled to Sepharose beads, then testing the flow-through on cells by immunofluorescence. As expected, this led to a complete abolition of the nuclear staining pattern observed with the affinity-purified anti-p34 antibody (data not shown). As can be seen from the results presented in FIGURE 3, before nuclear incorporation of [^{3}H]thymidine is observed, the number of cells positive for nuclear p34^{cdc2} staining increases significantly. This result indicates that p34 protein accumulates in the nuclei of serum-stimulated cells in late G_1 phase before the onset of DNA synthesis.

p34^{cdc2} Antisense Oligomers Inhibit G_1 Progression into S-Phase

To determine if an increase in p34 protein was required for progression of serum-stimulated cells into S-phase, we examined if antisense oligodeoxynucleotides would inhibit S-phase entry. For this purpose, p34^{cdc2} oligomers (18 mers), which corresponded to antisense (5′-ATCTTCCATAGTTAGTCA-3′) sequences, nonsense (5-TGGCAAACTTATTCTACT-3′) sequences, and sense (5′TGACTAACTATGGAAGAT-3′) sequences, were synthesized and microinjected directly into the cytoplasm of growth-arrested cells before serum stimulation. Entry into S-phase was monitored by continuous labeling with [^{3}H]thymidine followed by autoradiography. Microinjection of the oligomers was employed instead of the more traditional approach of simply adding the oligomers to the culture medium to more precisely control the time of delivery of the oligomers into the cells. For these experiments, cells were plated in 60-mm dishes containing a 22 × 22 mm coverslip marked with a circle to delineate the area for microinjection. Cells were incubated in medium containing 0.1% CS for 5 days to induce growth arrest before microinjection. Under these conditions between 8 and 15% of the cells incorporated [^{3}H]thymidine during continuous labeling for 24 hours between days 5 and 6. Microinjection was performed using a Zeiss Axiovert inverted-microscope equipped with an Eppendorf microinjector and micromanipulator using commercially purchased glass-capillary femtotips (Eppendorf). All cells within a delineated area of the coverslip (usually 150–200) were microinjected in the cytoplasm with oligomers at concentrations ranging from 0.01–10 μg/μl in 10 mM Tris buffer (pH 7.6) containing 10 μg/μl FITC-conjugated dextran (MW 154,000) as a marker to establish the efficiency of microinjection.[22] Immediately following injection the cells were stimulated with fresh medium containing 10% fetal calf serum, and [^{3}H]thymidine (0.2 μCi/ml) was added. The cells were labeled continuously for 24 hours and then fixed in absolute methanol. The efficiency of microinjection for each experiment was determined by fluorescence microscopy before autoradiography and ranged typically for 50–80%. The autoradiographic results are shown in FIGURE 4. As can be seen from the data, microinjection of either sense or nonsense oligomers did not have a significant effect on the ability of cells to enter S-phase following serum stimulation. On the contrary,

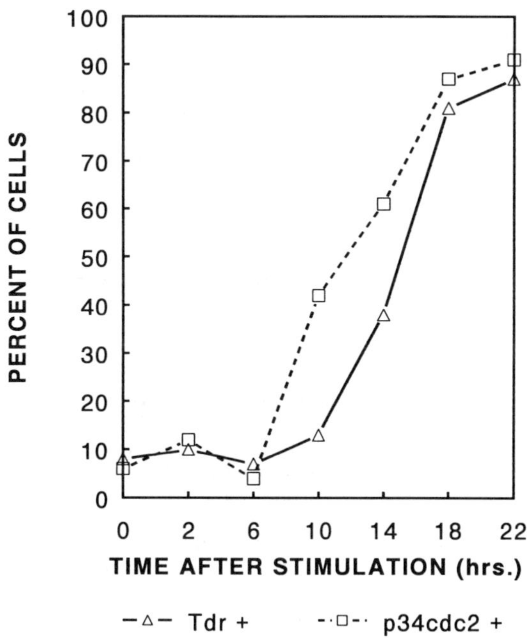

FIGURE 3. Nuclear localization of p34^{cdc2} protein during G_1 progression into S-phase. T98G cells were growth arrested and subsequently stimulated to proliferate as described in the text and the legend of FIGURE 2. [^{3}H]thymidine (0.2 μCi/ml) was added at the time of stimulation. Cells were fixed in methanol at −20°C for 5 minutes, air-dried, and subsequently reacted with affinity-purified K1 anti-p34 antibodies for 60 minutes in a humidified 37°C chamber. After rinsing in phosphate-buffered saline solution (PBS), cells were reacted with FITC-conjugated goat anti-rabbit IgG (Oncogene Science) as described above. After rinsing in PBS, the coverslips containing the cells were mounted in a solution of 90% glycerol and 10% PBS (pH 7.5) containing 3% (w/v) *N*-propyl gallate. The percentage of cells exhibiting intense nuclear staining (p34^{cdc2}+) was determined at each time point by fluorescence microscopy. The coverslips were subsequently removed from the mounting medium and mounted permanently on glass microscope slides using Protex and then processed for autoradiography. The percentage of cells labeled with [^{3}H]thymidine (Tdr+) was determined by light microscopy.

microinjection of antisense oligomers at a concentration of 10 μg/μl resulted in a 65% inhibition in the ability of cells to enter S-phase. The viability of cells microinjected with oligomers (determined at 4 hours postinjection by phase-contrast microscopy) ranged from 55–60%, which was identical to that observed for cells microinjected with 10 mM Tris and FITC-conjugated dextran alone (data not shown). This

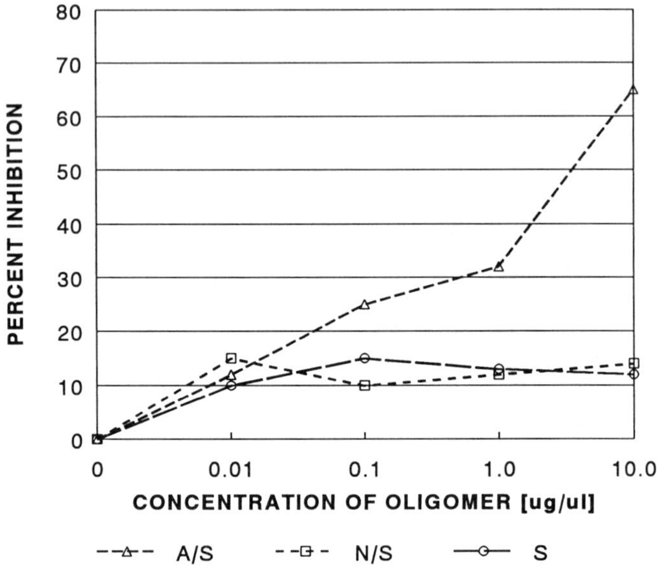

FIGURE 4. Effects of microinjected p34^{cdc2} oligomers on G_1 progression into S-phase. Cells were growth arrested as described in the legend for FIGURE 2. Growth-arrested cells were microinjected directly in the cytoplasm with antisense (A/S), nonsense (N/S), or sense (S) oligomers at concentrations ranging from 0.01–10 μg/μl. Immediately after microinjection, cells were stimulated with fresh medium containing 10% FCS and 0.2 μCi/ml [^{3}H]thymidine. The cells were labeled continuously for 24 hours and then fixed in methanol. The efficiency of microinjection (no. of fluorescent cells/total no. of cells in area of microinjection) was determined by fluorescence microscopy before processing for autoradiography. The percentage of inhibition was calculated as follows: [(no. of labeled fluorescent cells/no. of labeled background cells) − 1] × 100.

indicates that the introduction of oligomers into the cytoplasm of cells at a concentration up to 10 μg/μl by microinjection is not toxic to the cells.

Antisense Oligomers Decrease the Level of Nuclear p34^{cdc2} Protein

To determine if inhibition in the entry of cells into S-phase following microinjection of antisense oligomers was correlated with a decrease in the level of p34 protein, a second series of experiments was performed using antisense oligomers and control nonsense oligomers. In these experiments the steady-state level of nuclear p34 protein was quantitated in growth-arrested cells, serum-stimulated control cells, and cells microinjected with antisense oligomers or control nonsense oligomers at 24 hours after injection. Nuclear p34^{cdc2} protein was detected by immunofluorescence after staining with affinity-purified anti-p34 antibodies followed by FITC-conjugated goat anti-rabbit IgG. The relative intensity of p34^{cdc2} nuclear fluorescence was measured by fluorescence microscopy using a Zeiss Axiophot microscope equipped with an MPM 200 photometer and MSP 20 processor. The results are presented in FIGURE 5 as arbitrary units (au) of nuclear fluorescence. In cells growth-arrested by serum deprivation the mean fluorescence intensity per nucleus was 3 au and it increased to 25 au at 24 hours after serum stimulation. Similarly, in cells microinjected with nonsense oligomers the mean fluorescence intensity increased to 21 au at 24 hours after serum stimulation. On the contrary, in cells microinjected with antisense oligomers the mean fluorescence intensity increased to only 7 au at 24 hours after serum stimulation. Simultaneous labeling of cells with [^{3}H]thymidine continuously for 24 hours followed by autoradiography revealed that the labeling index was 11% for cells maintained in medium containing 0.1% CS, and upon stimulation with fresh medium containing 10% FCS the labeling index increased to 81%. In serum-stimulated cells microinjected with nonsense oligomers the labeling index was 75%; however, for cells microinjected with antisense oligomers the labeling index was only 32%. This difference in the labeling index translates into a 61% inhibition in the ability of cells microinjected with antisense oligomers to enter S-phase upon serum stimulation and is correlated with a threefold reduction in the steady-state level of nuclear p34^{cdc2} protein.

Time Course of Inhibition by Antisense Oligomers

We next examined the ability of growth-arrested cells to enter S-phase when antisense oligomers were microinjected at different times after serum stimulation.

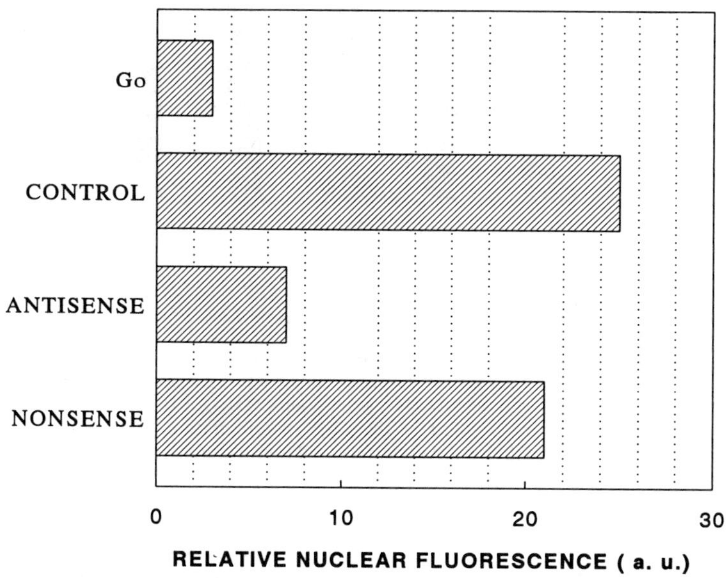

FIGURE 5. Effect of microinjected oligomers on the steady-state level of nuclear p34^{cdc2} protein. Growth-arrested cells were microinjected with antisense or nonsense oligomers (10 μg/μl) and subsequently stimulated with medium containing 10% FCS. At 24 hours after stimulation the cells were fixed and stained for p34^{cdc2} protein as described in the legend for FIGURE 3. The relative fluorescence intensity for growth-arrested (Go), serum-stimulated cells (control), and cells microinjected with antisense or nonsense oligomers was quantitated by fluorescence photomicroscopy as described in the text. The average of three independent experiments in which 100 cells per experiment were measured is expressed in arbitrary units (au).

For these experiments cells were plated for microinjection, growth arrested as described above, and then subsequently stimulated to reenter the cell cycle by the addition of fresh medium containing 10% FCS. At different times after stimulation the cells were microinjected with either antisense oligomers or control nonsense oligomers at a concentration of 15 μg/μl. [^{3}H]thymidine was added at the time of stimulation, and the cells were labeled continuously for 30 hours before fixation and processing for autoradiography. The autoradiographic results presented in FIGURE 6 represent the average of three independent microinjection experiments for each time point. A significant inhibition (approximately 80%) in the ability of cells to enter S-phase was observed when antisense oligomers were microinjected at 2 and 6 hours

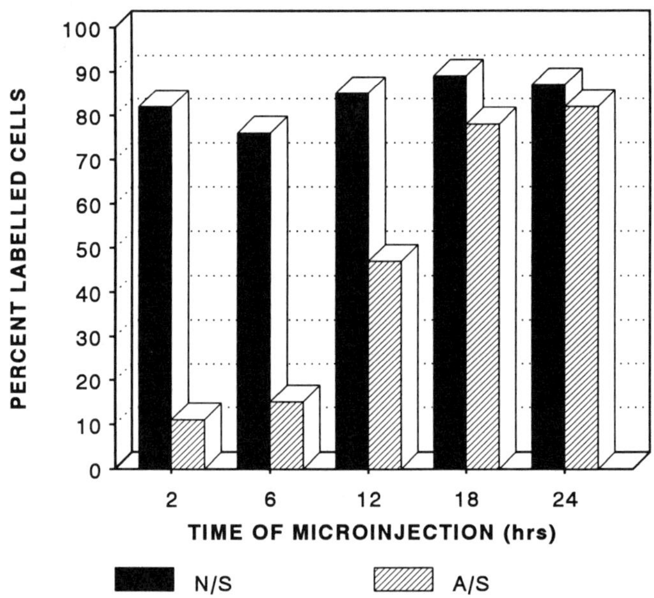

FIGURE 6. Time course of inhibition by antisense oligomers microinjected into cells after serum stimulation. Growth-arrested cells were serum stimulated and subsequently microinjected with antisense (A/S) or nonsense (N/S) oligomers (15 μg/μl) at the times indicated in hours (hrs). [^{3}H]thymidine was added at the time of stimulation, and cells were labeled continuously for 30 hours and then fixed and processed for autoradiography. The percentage of labeled cells was determined by light microscopic counting.

after serum stimulation. At 12 hours the inhibition decreased to 44% and at 18 and 24 hours to 11% and 5%, respectively. Although the higher concentration of antisense oligomers appeared to be slightly more effective at inhibiting the entry of cells into S-phase when microinjected at early times after serum stimulation, some difficulty with clogging of the femtotips was encountered.

SUMMARY AND CONCLUSIONS

In this study the effect of antisense oligomers targeted against the mRNA transcripts of p34^{cdc2} kinase on G_1 progression into S-phase was examined. For this purpose, antisense, sense, or nonsense oligomers were introduced directly into the cytoplasm of T98G cells grown in monolayer cultures by glass-capillary microinjection. The microinjection of antisense oligomers (but not sense or nonsense oligomers) into growth-arrested cells before serum stimulation inhibited G_1 progression into S-phase. This inhibition was correlated with a reduction in the steady-state levels of nuclear $p34^{cdc2}$ protein. Microinjection of antisense oligomers into cells at 2 and 6 hours after serum stimulation also resulted in a marked inhibition in the ability of cells to enter S-phase. The inhibitory effect decreased when cells were microinjected at 12 hours after serum stimulation. When cells were microinjected at 18 and 24 hours after serum stimulation, only a slight inhibition was observed. As the antisense oligomers were introduced directly into the cytoplasm of cells at each of the time points examined, the observed differences in the inhibitory effects of the antisense oligomers at later times after serum stimulation cannot be explained by differences in uptake. An alternative explanation is that after a certain threshold level of nuclear $p34^{cdc2}$ protein is reached in late G_1 phase; no further increase is necessary, because the cells become committed to enter S-phase. In yeast, $p34^{cdc2}$ appears to play an important role in the G_1/S-phase transition at a control point in late G_1 phase called START (reviewed by Lewin[5]). In mammalian cells a control point that could be equivalent to START is the "restriction point" which is defined as the time after which inhibition of protein synthesis fails to block entry into S-phase (reviewed by Pardee[23]). The effects observed with antisense oligomers to $p34^{cdc2}$ kinase are strikingly similar to what is observed when low concentrations of the drug cycloheximide are added to these cells at different times after serum stimulation; entry into S-phase is significantly inhibited when cycloheximide is added up to 12 hours poststimulation.[24] Thus, the results reported in this study are in agreement with the idea that $p34^{cdc2}$ kinase plays a role in the G1/S phase transition in mammalian cells.[8,9]

Finally, introduction of antisense oligomers directly into the cytoplasm of cells grown in monolayer cultures by glass-capillary microinjection appears to be a viable alternative to simply adding the oligomers to the culture medium. The microinjected antisense oligomers are nontoxic and can efficiently inhibit the expression of the gene to which they are targeted. The advantages of this approach are that the concentration of the oligomers and the time of delivery into cells can be controlled by the investigator. In addition, cells into which the oligomers have been properly delivered can be readily identified following injection by inclusion of a fluorescent marker. On the other hand, this approach has the disadvantages of requiring specialized microinjection equipment and a certain degree of technical expertise, and not all cell types are amenable to microinjection.

REFERENCES

1. Calabretta, B. 1991. Cancer Res. **51:** 4505–4510.
2. Jaskulski, D., J. K. DeRiel, W. E. Mercer, B. Calabreta & R. Baserga. 1988. Science **240:** 1544–1546.
3. Travali, S., K. Reiss, A. Ferber, S. Petralia, W. E. Mercer, B. Calabretta & R. Baserga. 1991. Mol. Cell. Biol. **11:** 731–736.
4. Surmacz, E., P. Nugent, Z. Pietrzkowski & R. Baserga. 1992. Exp. Cell Res., in press.
5. Lewin, B. 1990. Cell **61:** 743–752.
6. Lee, M. G. & P. Nurse. 1987. Nature (Lond.) **327:** 31–35.
7. Riabowol, K., G. Draetta, L. Brizuela, D. Vandre & D. Beach. 1989. Cell **57:** 393–401.
8. D'Urso, G., R. L. Marraccino, D. R. Marshak & J. M. Roberts. 1990. Science **250:** 786–791.

9. FURUKAWA, Y., H. PIWNICA-WORMS, T. J. ERNST, Y. KANAKURA & J. D. GRIFFIN. Science **250:** 805–808.
10. STEIN, G. 1979. J. Cell Physiol. **99:** 43–54.
11. MERCER, W. E., M. T. SHIELDS, M. AMIN, G. J. SAUVE, E. APPELLA, J. W. ROMANO & S. J. ULLRICH. 1990. Proc. Natl. Acad. Sci. USA **87:** 6166–6170.
12. KACZMAREK, L., M. R. MILLER, R. A. HAMMOND & W. E. MERCER. 1986. J. Biol Chem. **261:** 10802–10807.
13. MERCER, W. E., M. T. SHIELDS, D. LIN, E. APPELLA & S. J. ULLRICH. 1991. Proc. Natl. Acad. Sci. USA **88:** 1958–1962.
14. CHOMCZYNSKI, P. & N. SACCHI. 1987. Anal. Biochem. **163:** 156–159.
15. MANIATIS, T., E. F. FRITSCH & J. SAMBROOK. 1982. *In* Molecular Cloning: A Laboratory Manual. Cold Spring Harbor Laboratory. Cold Spring Harbor, NY.
16. LEE, M. G., C. J. NORBURY, N. K. SPURR & P. NURSE. 1988. Nature (Lond.) **333:** 676–679.
17. FEINBERG, A. P. & B. VOGELSTEIN. 1983. Anal. Biochem. **132:** 6–13.
18. MCGOWAN, C., P. RUSSELL & S. I. REED. 1990. Mol. Cell. Biol. **10:** 3847–3851.
19. DURKACZ, B., A. CARR & P. NURSE. 1986. EMBO J. **5:** 558–560.
20. LAEMMLI, U. K. 1970. Nature (Lond.) **227:** 680–685.
21. PEPPERKOK, R., M. ZANETTI, R. KING, D. DELIA, W. ANSORGE, L. PHILIPSON & C. SCHNEIDER. 1988. Proc. Natl. Acad. Sci. USA **85:** 6748–6752.
22. DRAETTA, G. & D. BEACH. 1988. Cell **54:** 17–26.
23. PARDEE, A. B. 1989. Science **246:** 603–608.
24. LIN, D., M. T. SHIELDS, S. J. ULLRICH, E. APPELLA & W. E. MERCER. 1992. Proc. Natl. Acad. Sci. USA, in press.

Antisense Approaches to the Function of Glial Cell Proteins[a]

LINDA J. VAN ELDIK,[b,c,d] STEVEN W. BARGER,[c]
AND MICHAEL J. WELSH[e]

Departments of [b]Pharmacology and [c]Cell Biology
Vanderbilt University
Nashville, Tennessee 37232

[e]Department of Anatomy and Cell Biology
University of Michigan Medical School
Ann Arbor, Michigan 48109

S100 refers to a protein fraction originally isolated from bovine brain over 25 years ago[1] and termed S100 to denote its partial solubility in 100% saturated ammonium sulfate. The bovine brain S100 fraction is composed primarily of two small (M_r = 10,000), acidic, calcium binding proteins called S100α and S100β.[2] These proteins share approximately 50% identity in amino acid sequence and can form homologous and heterologous dimers of the form αα ($S100a_o$), αβ (S100a), and ββ (S100b). The S100α and S100β polypeptides are also characterized by the presence of regions of amino acid sequence that could form calcium binding structures: a conventional helix-loop-helix (EF-hand) calcium binding site in their COOH-terminal region and an NH_2-terminal calcium binding site that resembles the site found in calbindin.[3,4] S100α and S100β are structurally conserved in different species. For example, the amino acid sequences of bovine, rat, and human S100β differ by only 3–4 residues (see ref. 5). Less information is available about the structural conservation of the S100α polypeptide, but the bovine and rat proteins show only four amino acid differences (see ref. 6).

In addition to S100α and S100β, several proteins or cDNA sequences have been discovered in recent years that have 30–50% sequence identity to the S100s. This family of S100-like proteins includes calcium binding proteins, proteins that are induced in cells after growth factor or serum stimulation, proteins increased on differentiation or transformation of cells, a protein subunit of a cytoskeletal protein complex, and serum proteins associated with diseases such as cystic fibrosis or rheumatoid arthritis (see ref. 7 for references). Even though these proteins share extensive sequence homology, little is known about their functional roles. In this chapter, we will concentrate on the S100β polypeptide, the principal member of the S100 family in brain.

In both developing and mature vertebrate nervous system, S100β is expressed primarily by glial cells. Although S100β was thought to be brain specific for many years, it is known now that the protein is not restricted to the brain or even to nervous tissue. S100β has been localized to a variety of peripheral tissues (see refs. 5, 7, and 8), and reports about the use of S100β localization in diagnostic pathology abound in

[a]Our recent antisense studies reported here were supported in part by funds from the Muscular Dystrophy Association and National Institutes of Health grant NS29215 (L.V.E.) and National Institutes of Health grant HD17121 (M.J.W.).

[d]Address for correspondence: Dr. Linda J. Van Eldik, 406 MRB, Department of Pharmacology, Vanderbilt University, Nashville, TN 37232–6600.

the literature. In contrast to the extensive literature about the structural properties and localization of S100β, the biological roles of this protein have only recently begun to be elucidated. A number of intracellular functions for S100β in glial cells have been suggested by *in vitro* studies. For example, S100β has been reported to stimulate or inhibit phosphorylation of various proteins, to inhibit microtubule assembly, and to interact with various enzymes and cytoskeletal proteins (see ref. 7 for references).

In addition to these multiple intracellular activities, an increasing body of evidence indicates that S100β also has extracellular roles. S100β is known to be released from glial cells, being detected in brain extracellular fluid[9] and in conditioned media from glial cells.[10,11] Release of S100β from astrocytes can be stimulated by serotonin acting through the astrocyte 5-HT_{1A} receptor.[12] A disulfide-linked, dimeric form of S100β has shown at least two extracellular activities on cultured cells: neurotrophic activity on select neuronal populations and mitogenic activity on astrocytes. S100β stimulates enhanced survival of embryonic cortical neurons in culture[13] and neurite outgrowth from cortical neurons,[13–15] serotonergic neurons of the mesencephalic raphe,[16] cells of the optic tectum,[17] dorsal root ganglia,[18] and spinal cord,[19] and the neuro-2A neuroblastoma cell line.[20] Nanomolar concentrations of dimeric S100β stimulate proliferation and increase in the steady-state levels of c-*myc* and c-*fos* protooncogene mRNAs in glial cells.[21] S100β was recently shown to be a trophic factor *in vivo,* being able to prevent the late naturally occurring motoneuron cell death during embryonic chick development as well as the deafferentation-induced death of motoneurons of the embryonic chick spinal cord.[22,23]

Evidence to date supports the idea that S100β is a multifunctional protein that plays important roles in nervous system development and maintenance. Therefore, the mechanisms involved in regulating the intracellular versus extracellular levels of S100β in the glial cell and the consequences of alterations in S100β levels need to be elucidated. S100β levels in the cell can be altered under certain conditions. For example, intracellular S100β levels increase as cells become confluent and at specific points in the cell cycle, in response to cAMP, and on treatment with various agents (see refs. 7 and 24 for references). In addition, S100β levels were reported to be increased in reactive glial cells of patients with Down syndrome and Alzheimer's disease,[25,26] a finding consistent with the localization of the gene for human S100β to the Down syndrome region of chromosome 21.[27,28] Recent studies with transgenic mice overexpressing S100β have demonstrated increased astroglial proliferation in specific areas of the brain,[29] suggesting that S100β may serve as a glial mitogen *in vivo.*

To directly address how perturbation of S100β levels in the glial cell correlates with changes in cell phenotype, we developed a biological system in which the S100β levels could be decreased in a specific and reproducible manner.[24] Specifically, we used antisense strategies to produce a selective decrease in S100β levels in rat C6 glioma cells. Two separate antisense approaches were used for inhibition of S100β production: (1) generation of clonal isolates of C6 cells stably transfected with an S100β antisense minigene under the control of a dexamethasone-inducible promoter, and (2) analysis of C6 cells treated with S100β antisense oligodeoxynucleotides. Both antisense procedures resulted in a decrease in intracellular S100β levels as measured by radioimmunoassay. This lowering of S100β levels correlated with three alterations in cellular phenotype: (1) a more flattened appearance, with a three- to fourfold increase in cellular area, (2) a more organized microfilament cytoskeletal network, and (3) a decrease in cellular proliferation rate. Similar results obtained with two distinct antisense methods provide strong evidence that S100β has important roles in glial cell morphology, cytoskeletal organization, and cell proliferation.

In this chapter, we briefly review the previous findings and report recent results examining in greater detail the effects of inhibition of glial S100β production on cell morphology.

MATERIALS AND METHODS

Cell Culture. Rat C6 glioma cells were used throughout these studies and were obtained and grown as previously described.[24] These cells contain high levels of S100β and have been used for many years in S100β studies as a biological model for glial cells.

Selection of Stable Clones Containing S100β Antisense Minigene. A plasmid (pS100AS) containing the S100β antisense minigene under the control of a dexamethasone-inducible promoter was constructed by using a synthetic S100β gene[15] as described.[24] C6 cells were transfected with pS100AS, and colonies were selected in G418, subcloned, and expanded. A number of stable clones containing the S100β antisense gene construct were developed, but the majority of experiments to date have used two clones, termed C6-AS1 and C6-AS2. Studies with these two clones yielded similar results and are used interchangeably in this chapter.

To induce expression of the S100β antisense gene, cells were treated for various times with 1 μM dexamethasone (10 μl of 1 mM dexamethasone in ethanol per 10 ml medium). Controls received diluent alone (10 μl of ethanol per 10 ml medium). We found that the endogenous S100β levels in the C6-AS1 and C6-AS2 clones are lower than those in the parental C6 cells. This may merely reflect selection of clones with lower amounts of S100β compared to the C6 cells, or the promoter may be somewhat "leaky," that is, there may be a low level of expression of the antisense gene in the absence of dexamethasone. In this regard, we found that it is important to use lots of fetal calf serum that do not contain high levels of steroid or to charcoal treat the serum before using in order to deplete the serum of steroids.

Treatment of C6 Cells with S100β Antisense Oligonucleotides. Oligodeoxynucleotides were synthesized by β-cyanoethyl phosphoramidite chemistry as described[24] and were used either directly from the synthesizer or after further purification using Sephadex G-25 spin columns. Cells were treated with oligonucleotides as described.[24] Briefly, before treatment, serum-containing medium was aspirated from the cells, and the cells were washed two times in serum-free medium. Cells were then incubated for 2 hours at 37°C in serum-free medium containing 30 μM oligonucleotide. After the 2-hour incubation, fetal calf serum was added to the wells (without changing the medium) to a final serum concentration of 2.5%. For our studies, cells received only a single addition of oligonucleotide at the beginning of the time course, and the medium was not changed during the experiment. In our experiments, we did not find it necessary to use modified oligonucleotides. An important factor, however, is the use of lots of fetal calf serum containing low levels of nuclease activity. Before embarking on an oligonucleotide experiment, we carried out experiments essentially as described[30] to test various lots of fetal calf serum for nuclease activity. We have had excellent results with fetal calf serum obtained from Hyclone Laboratories.

Measurement of S100β Levels. To confirm that S100β production was being inhibited in our antisense experiments, S100β levels were measured in cell extracts by radioimmunoassay, as described.[24]

Analysis of Phenotypic Changes. Morphological examination of cells was performed as described[24] or by phase-contrast microscopy on a Leitz Diavert inverted microscope. The cytoskeletal proteins actin, tubulin, and vimentin were localized by fluorescence methods as described.[24] Cell counts were determined with a hemacytometer.

Time-lapse video microscopy was conducted on cells grown in a T-25 flask. The flask was sealed with a rubber stopper to maintain the 95% air, 5% CO_2 atmosphere. Extended through the stopper was a wire (sealed in the stopper with silicone rubber adhesive) connected to a small thermistor that served as a temperature sensor to regulate an infrared source (Opti-quip Red Beam Incubator) that maintained temperature in the flask at 37.0 ± 0.1°C. To allow injection of dexamethasone via a hypodermic needle and syringe, a small hole was melted in the flask top before culturing cells, using a 2-mm diameter metal rod that had been heated over a bunsen burner. The hole was immediately sealed with silicone rubber adhesive. The adhesive cured as cells were cultured in the flask. The flask was placed on the stage of a Leitz Diavert inverted microscope, and cells were imaged using a long working distance 40× phase contrast objective lens. A Dage-MTI newvicon video camera was used to obtain images, and the images were recorded by a Panasonic AG6750 time-lapse video recorder. For most studies, the recorder was set for a time compression of 240×.

RESULTS

As reported previously,[24] we developed two antisense methods for inhibiting the production of S100β in C6 glioma cells. One method used a eukaryotic expression vector constructed to produce RNA complementary to the coding sequence of an S100β gene. There are several key features in this S100β antisense vector. First, the S100β coding region was derived from a synthetic S100β gene[15] that contains 77% homology with the coding region of rat S100β cDNA,[31] and then inserted into the polylinker region of the pMSVneo eukaryotic expression vector.[32] Second, the pMSVneo vector allows the inducible expression of inserted genes, because of the presence of the dexamethasone-inducible promoter in the mouse mammary tumor virus-long terminal repeat (MMTV LTR). This was advantageous because antisense-containing clones could be selected before S100β inhibition occurred, thus avoiding potential cloning problems due to decreases in cell growth rate. Third, pMSVneo contains an ampicillin resistance gene and pBR322 origin of replication, allowing initial cloning in bacterial cells.

The second antisense method used S100β antisense oligodeoxynucleotides. An S100β antisense oligonucleotide was prepared whose sequence is the inverse complement of 15 bases of the rat S100β cDNA sequence,[31] beginning at the ATG that corresponds to the initiator methionine. The antisense oligonucleotide had the sequence (reading 5′ to 3′): CTCCAGCTCAGACAT. Controls included the sense oligonucleotide that is equivalent to the coding strand of the cDNA in this region (ATGTCTGAGCTGGAG), and a mismatch oligonucleotide that is equivalent to the antisense sequence in which two bases have been changed (CTCCACCTCAGAGAT). In designing the oligonucleotides, we searched sequence databases to confirm that no other known nucleotide sequence matched the sequence of our oligonucleotides. We also kept the base composition of the mismatch oligonucleotide the same as the antisense oligonucleotide, making a G to C change and a C to G change.

We confirmed by radioimmunoassay that both of these antisense methods resulted in an inhibition of S100β production in the glial cell. FIGURE 1 summarizes

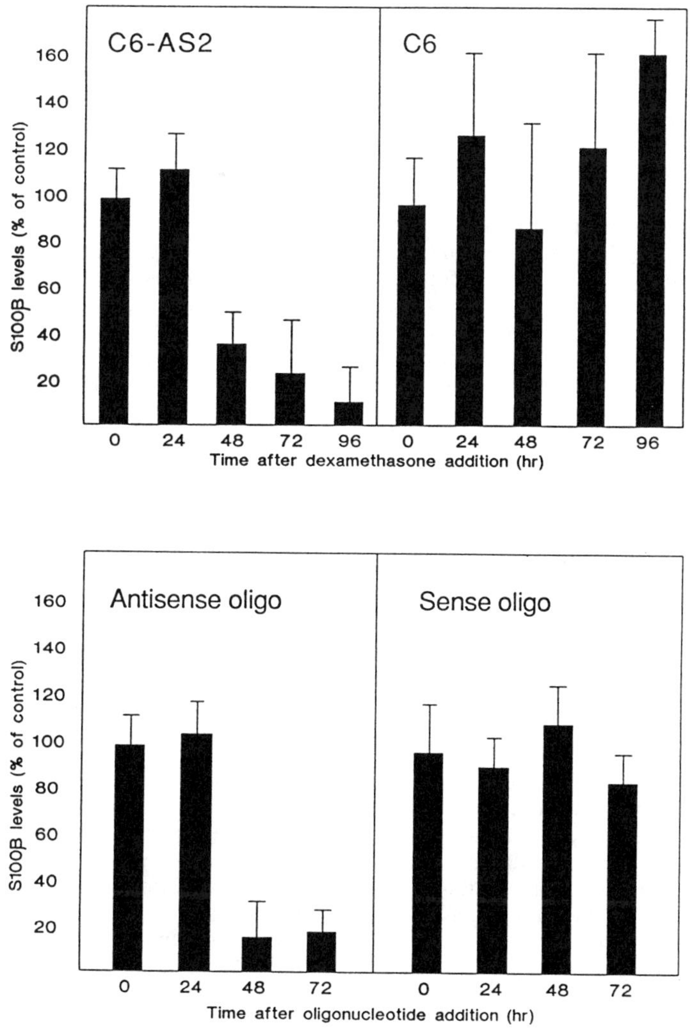

FIGURE 1. Inhibition of S100β production in a C6 clone containing the S100β antisense gene or in C6 cells treated with antisense oligonucleotides. In the top panel, the C6-AS2 clone (*left*) or untransfected C6 cells (*right*) were grown in the presence of dexamethasone for various lengths of time. S100β levels in cell extracts were determined by radioimmunoassay, and calculated as ng S100β/μg total protein. S100β levels are expressed as a percentage of the control levels at the 0-hour time point (i.e., in the absence of dexamethasone). Bars represent the mean ± SEM of four determinations (two separate experiments analyzing two wells per time point). In the bottom panel, C6 cells were treated with 32 μM antisense oligonucleotide (left) or sense oligonucleotide (right) for various lengths of time. S100β levels were determined as above, and expressed as a percentage of the control levels (levels in untreated C6 cells at each time point). Bars represent the mean ± range of duplicate cultures from a representative experiment. (Figure adapted from Selinfreund.[24])

these data, expressed as a percentage of the control S100β levels. It should be noted that by both antisense methods, S100β levels were not significantly lower than the controls until between 24 and 48 hours after antisense gene induction or the addition of antisense oligonucleotide. This is not an unexpected finding, as S100β is a relatively long-lived protein. We also could not completely inhibit S100β production by either method; however, S100β levels were only 15–20% of control values after 72–96 hours of treatment with dexamethasone or antisense oligonucleotide. Even though some S100β could be detected by radioimmunoassay at the 72- and 96-hour time points, we still obtained dramatic phenotypic effects of S100β inhibition, as will be discussed in more detail.

A readily observable effect of selective inhibition of S100β production in C6 cells was a change in cellular morphology. An example of this morphological change is shown in FIGURE 2 for one of the C6 clones treated with diluent alone (**a**) or treated with dexamethasone for 48 hours (**b**). The C6 clone in the absence of dexamethasone (**a**) show a bipolar stellate shape typical of untransfected C6 cells, with a rounded and refractile appearance as the cells become confluent. After treatment of cells with antisense oligonucleotide (data not shown) or induction of the S100β antisense gene with dexamethasone (**b**), the majority of cells exhibit a flattened, enlarged, and less refractile appearance. This change in cellular morphology was also accompanied by alterations in cytoskeletal organization and a decrease in cellular growth rate.[24] These phenotypic changes were not observed in untransfected C6 cells treated with dexamethasone or in C6 cells treated with the sense or mismatch oligonucleotides.[24]

To further analyze the time course of the morphological changes that occur on S100β inhibition, we examined S100β antisense-containing clones by time-lapse video microscopy at various times after the addition of dexamethasone (FIG. 3). Before dexamethasone treatment (panels **a–c**), the cells exhibit rapid undulations or "boiling" of the cell surfaces, along with numerous areas of membrane ruffling at the cell edges. Cells appear small and rounded, and each cell appears to have only one nucleolus. Mitoses are frequent, with an average cell-cycle time of 12–14 hours calculated from the video images.

After 1 day of dexamethasone exposure (panels **d–f**), the cell surface becomes more quiescent with slower undulations and fewer regions of membrane ruffling. The cells still undergo frequent mitoses. After 2–3 days in the presence of dexamethasone (panels **g–i**), the cells exhibit much reduced membrane activity, with very few surface movements or membrane ruffling. The cells are more flattened and spread, with larger nuclei often containing many nucleoli. Mitoses are much less frequent. From examination of the video images, it appears that cells may enter mitosis, but often do not complete cytokinesis. The resultant cells exhibit an enlarged cytoplasm and large nuclei containing multiple nucleoli. Estimates of cell cycle time from the few cells that underwent mitotis gave a value much greater than 30 hours. This lengthening of cell cycle time is consistent with our previous observations[24] of a reduced rate of proliferation after inhibition of S100β production.

DISCUSSION

Our studies demonstrate that selective inhibition of S100β production in glial cells produced by using either an S100β antisense minigene construct or S100β antisense oligonucleotides is correlated with at least three phenotypic changes: (1) a more flattened cellular morphology, with enlarged cytoplasmic area and reduced membrane ruffling; (2) a more organized cytoskeleton at the level of the microfilament network; and (3) a decrease in the cellular proliferation rate. These data

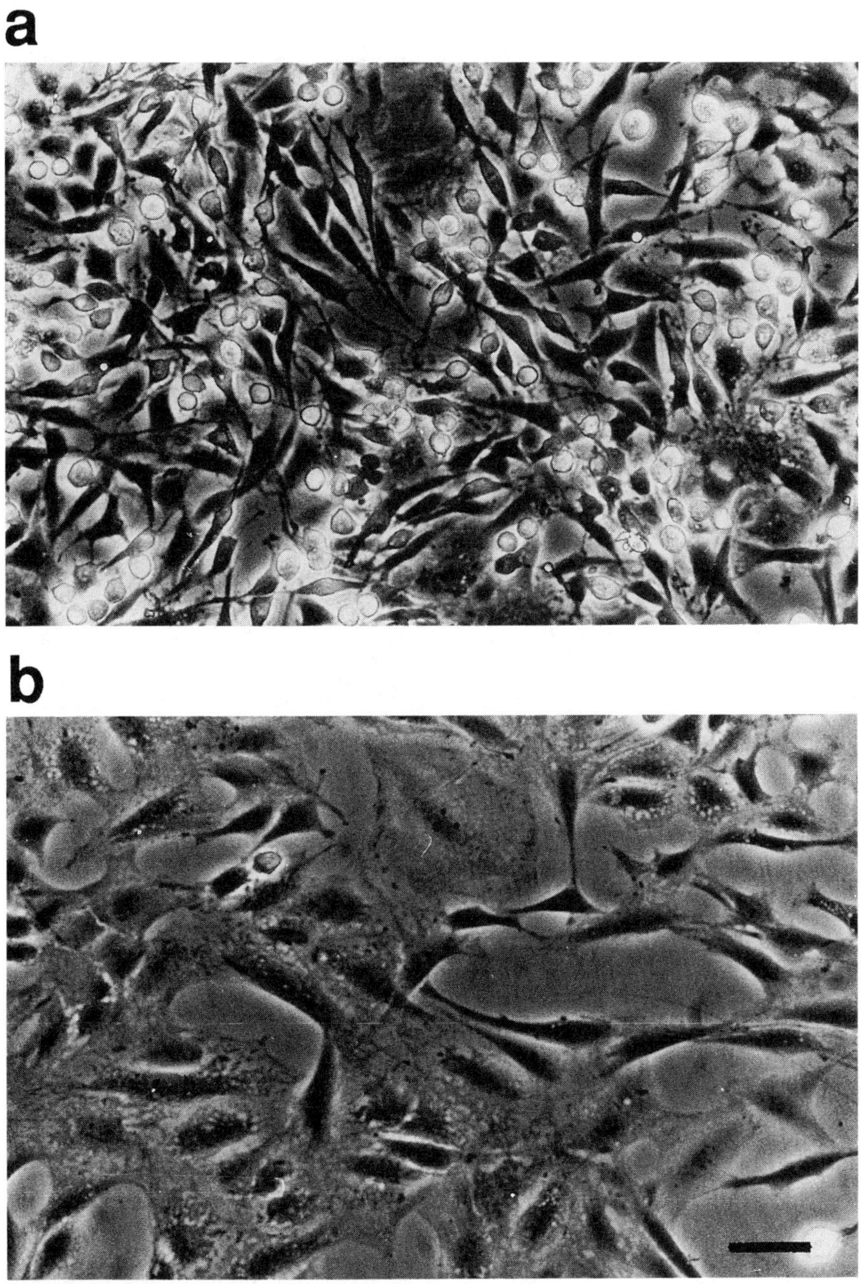

FIGURE 2. Effects of S100β antisense gene induction on cellular morphology. A C6 clone (C6-AS2) containing the S100β antisense gene under the control of a dexamethasone-inducible promoter was grown for 48 hours in the absence (**a**) or presence (**b**) of 1 μM dexamethasone. Cells were examined by phase contrast microscopy.

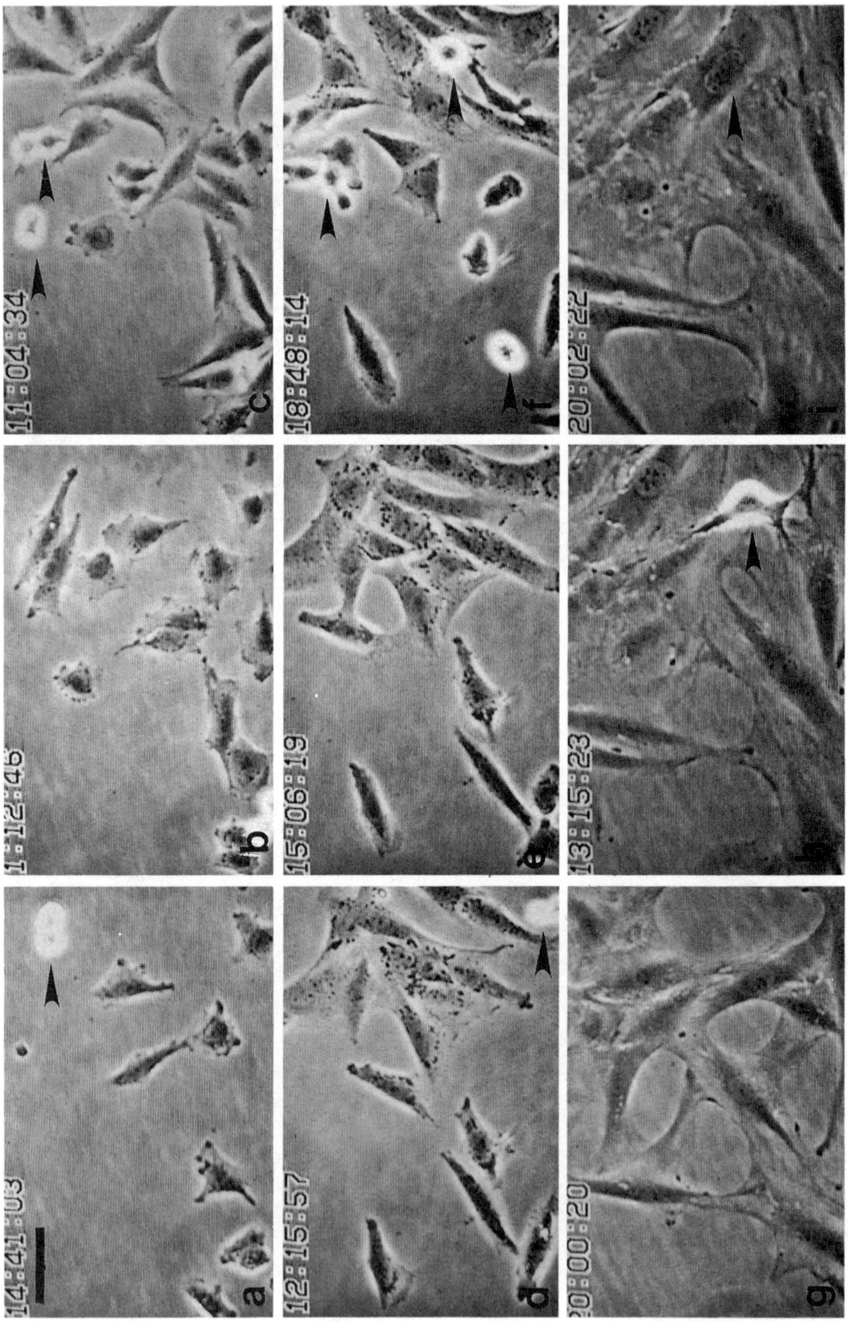
14:41:03
a
1:12:45
b
11:04:34
c
12:15:57
d
15:06:19
e
18:48:14
f
20:00:20
g
13:15:23
h
20:02:22

FIGURE 3.

support the idea that S100β has important *in vivo* roles in regulation of glial cell morphology, cytoskeletal organization, and cell proliferation.

Antisense strategies are becoming increasingly important tools in attempting to understand the function of specific gene products by inhibiting the production of the protein and determining the functional consequences. It should be noted that the S100β gene[15] used in our studies for construction of the antisense minigene has only a 77% homology with the coding region of rat S100β cDNA. Even without complete homology, the antisense gene construct was able to be used successfully to inhibit S100β production in rat C6 cells, as confirmed by direct demonstration of decreased S100β levels by radioimmunoassay. This finding provides a precedent for potential utilization of DNA sequences from one species to inhibit protein production in cells from another species. These data are also consistent with previous studies[33] showing that complete DNA sequence homology is not required for inhibition of gene function by antisense constructs.

The use of antisense approaches has allowed us to begin to address the *in vivo* roles of S100β in glial cells. Our findings suggest that S100β plays a role in regulation of glial cell morphology, cytoskeletal structure, and cell proliferation. The molecular mechanisms by which alterations in S100β levels lead to these phenotypic changes are just beginning to be explored. Our initial findings[24] that a decrease in intracellular S100β levels resulted in a decrease in C6 cell growth rate, along with the detection of S100β in conditioned media of rapidly proliferating C6 cultures,[11] suggested that S100β may be secreted from C6 cells to act as an autocrine growth factor. To test this possibility, we examined the effects of exogenous S100β on C6 cell growth rate. We found[21] that exogenous S100β could stimulate C6 cell proliferation, as measured by both increases in (^{3}H)-thymidine incorporation and cell number. More recent studies[34] have shown that S100β evokes a transient rise in intracellular calcium and stimulates phosphoinositide turnover, which is consistent with the signal transduction of other extracellular mitogenic factors. The hypothesis that S100β antisense procedures slow the growth rate by decreasing production of an autocrine growth factor should be further tested by measuring the amount of S100β released from C6

←

FIGURE 3. Time-lapse video analysis of cellular morphology in a C6 clone containing the S100β antisense gene. Cells were cultured for video recording as described in Methods, and grown in the absence (**a–c**) or presence (**d–i**) of dexamethasone for various lengths of time. In panels **a–c,** the same region of the culture substrate is shown for cells grown in the absence of dexamethasone for ~3, 10, and 20 hours, respectively. In panels **d–f,** the same group of cells are shown after treatment with dexamethasone for ~24, 27, and 31 hours, respectively. In panels **g–i,** cells have been treated with dexamethasone for ~56, 73, and 80 hours, respectively. In the absence of dexamethasone, cells undergo frequent mitoses (*arrows* in panels **a–c**), with a cell cycle time of about 13 hours. Cells treated with dexamethasone for 1 day (panels **d–f**) appear more spread and somewhat larger than untreated cells. Mitosis is still evident in occasional cells (*arrows* in **f**); however, the cell cycle time is significantly longer than that of untreated cells. After 2–3 days of dexamethasone treatment (**g–h**), cells are much larger and more spread than untreated cells; mitosis is very infrequent (*arrow* in **h**). On those rare occasions when a cell does undergo mitosis, it may not go through cytokinesis. The cell indicated by the *arrow* in **h** is the single cell indicated by the *arrow* in **i.** After this cell went through mitosis, it did not undergo cytokinesis but became a daughter cell with two nuclei initially. The two nuclei then fused to form one larger nucleus. The infrequency of mitosis in cells treated with dexamethasone for 2 or more days indicates that the cell cycle time of these cells increases to much longer than 30 hours. Assuming that the cells are still growing, the lack of mitosis may account for the large size of the cells and their nuclei as time in culture in the presence of dexamethasone increases. Bar = 50 μm.

cells before and after inhibition of intracellular S100β production, and by determining if the addition of exogenous S100β can reverse or delay the phenotypic effects of antisense inhibition of S100β production.

The precise role of S100β in cell proliferation is not known, and it is not possible to exclude the involvement of intracellular functions of the protein. One of the most striking morphological changes resulting from a decrease in S100β levels was the establishment of stress fibers, indicating a change in microfilament organization.[24] These effects may be related to the ability of S100β to alter interactions of actin with caldesmon *in vitro*.[35] Effects of S100β on cytoskeletal elements presumably could alter cell division rates. In this regard, we observed in video images of S100β-antisense clones treated with dexamethasone for 2–3 days (i.e., at times when S100β levels are decreased significantly) that some of the cells appeared to enter the mitotic cycle, but then did not complete cytokinesis (FIG. 3). This apparent arrest of cell division resulted in enlarged cells containing large nuclei with multiple nucleoli, and would help explain the reduced proliferation rate observed after S100β inhibition. We have not yet pursued these observations further, but deficiencies in cytokinesis are very likely to result from alterations of cytoskeletal kinetics. A previous report[36] has also implicated S100β in regulation of progression through the cell cycle. Thus, it is possible that S100β plays regulatory roles during specific points in the cell cycle, but eludication of the mechanism of S100β regulation will require further study.

In summary, the available data suggest that homeostasis of S100β levels in the glial cell may be critical to normal brain development and function. The ability to manipulate glial cells by antisense strategies to produce selective decreases in S100β levels provides a useful biological system for directly testing the contribution of S100β to biological processes in the brain. Our studies using antisense approaches have provided new insights into possible *in vivo* roles for S100β in glial cell morphology, cytoskeletal organization, and growth regulation. In addition, these data provide the framework for our continuing studies aimed at defining the molecular mechanisms of S100β action in nervous system development and maintenance.

ACKNOWLEDGMENT

We thank Dr. Richard Selinfreund for his contributions to the original antisense data[24] upon which this report is based.

REFERENCES

1. MOORE, B. W. 1965. A soluble protein characteristic of the nervous system. Biochem. Biophys. Res. Commun. **19:** 739–744.
2. ISOBE, T. & T. OKUYAMA. 1981. The amino acid sequence of the α subunit in bovine brain S100a protein. Eur. J. Biochem. **116:** 79–86.
3. VAN ELDIK, L. J., J. G. ZENDEGUI, D. R. MARSHAK & D. M. WATTERSON. 1982. Calcium-binding proteins and the molecular basis of calcium action. Intl. Rev. Cytol. **77:** 1–61.
4. SZEBENYI, D. M. E., S. K. OBENDORF & K. MOFFAT. 1981. Structure of vitamin D dependent calcium-binding protein from bovine intestine. Nature **294:** 327–332.
5. VAN ELDIK, L. J. & D. B. ZIMMER. 1988. Mechanisms of action of the S100 family of calcium modulated proteins. *In* Calcium and Calcium Binding Proteins. C. Gerday, R. Gilles & L. Bolis, eds. :114–127. Springer-Verlag, Berlin.
6. ZIMMER, D. B., W. SONG & W. E. ZIMMER. 1991. Isolation of a rat S100α cDNA and distribution of its mRNA in rat tissues. Brain Res. Bull. **27:** 157–162.

7. DONATO, R. 1991. Perspectives in S-100 protein biology. Cell Calcium **12:** 713–726.
8. DONATO, R. 1986. S-100 proteins. Cell Calcium **7:** 123–145.
9. SHASHOUA, V. E., G. W. HESSE & B. W. MOORE. 1984. Proteins of the brain extracellular fluid: Evidence for release of S100 protein. J. Neurochem. **42:** 1536–1541.
10. SUZUKI, F., K. KATO, T. KATO & N. OGASAWARA. 1987. S-100 protein in clonal astroglioma cells is released by adrenocorticotropic hormone and corticotropin-like intermediate-lobe peptide. J. Neurochem. **49:** 1557–1563.
11. VAN ELDIK, L. J. & D. B. ZIMMER. 1987. Secretion of S-100 from rat C6 glioma cells. Brain Res. **436:** 362–370.
12. WHITAKER-AZMITIA, P. M., R. MURPHY & E. C. AZMITIA. 1990. Stimulation of astroglial 5-HT1A receptors releases the serotonergic growth factor, protein S-100, and alters astroglial morphology. Brain Res. **528:** 155–158.
13. WINNINGHAM-MAJOR, F., J. L. STAECKER, S. W. BARGER, S. COATS & L. J. VAN ELDIK. 1989. Neurite extension and neuronal survival activities of recombinant S100β proteins that differ in the content and position of cysteine residues. J. Cell Biol. **109:** 3063–3071.
14. KLIGMAN, D. & D. R. MARSHAK. 1985. Purification and characterization of a neurite extension factor from bovine brain. Proc. Natl. Acad. Sci. USA **82:** 7136–7139.
15. VAN ELDIK, L. J., J. L. STAECKER & F. WINNINGHAM-MAJOR. 1988. Synthesis and expression of a gene coding for the calcium-modulated protein S100β and designed for cassette-based, site-directed mutagenesis. J. Biol. Chem. **263:** 7830–7837.
16. AZMITIA, E. C., K. DOLAN & P. M. WHITAKER-AZMITIA. 1990. S-100B but not NGF, EGF, insulin or calmodulin is a CNS serotonergic growth factor. Brain Res. **516:** 354–356.
17. MARSHAK, D. R. 1990. S100β as a neurotrophic factor. Prog. Brain Res. **86:** 169–181.
18. VAN ELDIK, L. J., B. CHRISTIE-POPE, L. M. BOLIN, E. M. SHOOTER & W. O. WHETSELL. 1991. Neurotrophic activity of S100β in cultures of dorsal root ganglia from embryonic chick and fetal rat. Brain Res. **542:** 280–285.
19. WINNINGHAM-MAJOR, F., W. O. WHETSELL & L. J. VAN ELDIK. 1988. Recombinant neurotrophic factor promotes survival and stimulates neurite outgrowth in nervous system cultures. J. Cell Biol. **107:** 729a.
20. KLIGMAN, D. & L. J. HSIEH. 1987. Neurite extension factor induces rapid morphological differentiation of mouse neuroblastoma cells in defined medium. Dev. Brain Res. **33:** 296–300.
21. SELINFREUND, R. H., S. W. BARGER, W. J. PLEDGER & L. J. VAN ELDIK. 1991. The neurotrophic protein S100β stimulates glial cell proliferation. Proc. Natl. Acad. Sci. USA **88:** 3554–3558.
22. BHATTACHARYYA, A., R. W. OPPENHEIM, D. PREVETTE, B. W. MOORE, R. BRACKENBURY & N. RATNER. 1992. S100 is present in developing chicken neurons and Schwann cells and promotes motor neuron survival *in vivo.* J. Neurobiol. **23:** 451–466.
23. QIN-WEI, Y., D. PREVETTE, R. W. OPPENHEIM & L. J. VAN ELDIK. 1991. Brain extract and glial-derived trophic agents prevent deafferentation-induced motoneuron death in the chick embryo. Soc. Neurosci. **17:** 1123.
24. SELINFREUND, R. H., S. W. BARGER, M. J. WELSH & L. J. VAN ELDIK. 1990. Antisense inhibition of glial S100β production results in alterations in cell morphology, cytoskeletal organization, and cell proliferation. J. Cell Biol. **111:** 2021–2028.
25. GRIFFIN, W. S. T., L. C. STANLEY, D. LING, L. K. WHITE, V. MACLEOD, L. T. PERROT, C. L. WHITE & C. ARAOS. 1989. Brain interleukin 1 and S-100 immunoreactivity are elevated in Down syndrome and Alzheimer disease. Proc. Natl. Acad. Sci. USA **86:** 7611–7615.
26. MARSHAK, D. R., S. A. PESCE, L. C. STANLEY & W. S. T. GRIFFIN. 1992. Increased S100β neurotrophic activity in Alzheimer disease temporal lobe. Neurobiology of Aging **13:** 1–7.
27. ALLORE, R., D. O'HANLON, R. PRICE, K. NEILSON, H. F. WILLARD, D. R. COX, A. MARKS & R. J. DUNN. 1988. Gene encoding the β subunit of S100 protein is on chromosome 21: Implications for Down syndrome. Science **239:** 1311–1313.
28. DUNCAN, A. M. V., J. HIGGINS, R. J. DUNN, R. ALLORE & A. MARKS. 1989. Refined sublocalization of the human gene encoding the β subunit of the S100 protein (S100β) and confirmation of a subtle t(9;21) translocation using in situ hybridization. Cytogenet. Cell Genet. **50:** 234–235.

29. YAROWSKY, P. J., B. K. KRUEGER, T. MICHAL, J. D. GEARHART, R. H. REEVES & D. C. HILT. 1991. Astroglial proliferation in vivo in S100β transgenic mouse. J. Cell Biol. **115:** 217a.
30. HOLT, J. T., R. L. REDNER & A. W. NIENHUIS. 1988. An oligomer complementary to c-myc mRNA inhibits proliferation of HL-60 promyelocytic cells and induces differentiation. Mol. Cell. Biol. **8:** 963–973.
31. KUWANO, R., H. USUI, T. MAEDA, T. FUKUI, N. YAMANARI, E. OHTSUKA, M. IKEHARA & Y. TAKAHASHI. 1984. Molecular cloning and the complete nucleotide sequence of cDNA to mRNA for S100 protein of rat brain. Nucl. Acids Res. **12:** 7455–7465.
32. CHUNG, F.-Z., C.-D. WANG, P. C. POTTER, J. C. VENTER & C. M. FRASER. 1988. Site-directed mutagenesis and continuous expression of human β-adrenergic receptors. J. Biol. Chem. **263:** 4052–4055.
33. HOLT, J. T., T. VENKAT GOPAL, A. D. MOULTON & A. W. NIENHUIS. 1986. Inducible production of c-fos antisense RNA inhibits 3T3 cell proliferation. Proc. Natl. Acad. Sci. USA **83:** 4794–4798.
34. BARGER, S. W. & L. J. VAN ELDIK. 1992. S100β stimulates calcium fluxes in glial and neuronal cells. J. Biol. Chem. **267:** 9689–9694.
35. FUJII, T., K. MACHINO, H. ANDOH, T. SATOH & Y. KONDO. 1990. Calcium-dependent control of caldesmon-actin interaction by S100 protein. J. Biochem. **107:** 133–137.
36. MARKS, A., D. PETSCHE, D. O'HANLON, P. C. KWONG, R. STEAD, R. DUNN, R. BAUMAL & S. K. LIAO. 1990. S100 protein expression in human melanoma cells. Comparison of levels of expression among different cell lines and individual cells in different phases of the cell cycle. Exp. Cell Res. **187:** 59–63.

Use of Antisense Oligomers to Study the Role of c-*jun* in G_1 Progression[a]

KENNETH J. SOPRANO,[b,d,e] STEPHEN C. COSENZA,[b,d]
GLADYS YUMET,[b,d] AND DIANNE ROBERT SOPRANO[c,d]

[b]*Department of Microbiology and Immunology,*
[c]*Department of Biochemistry, and the*
[d]*Fels Institute for Molecular Biology and Cancer*
Temple University School of Medicine
3400 North Broad Street
Philadelphia, Pennsylvania 19140

Analysis of the temporal patterns of growth-associated gene expression following the stimulation of growth in quiescent or G_0 cells has long been accepted as a valuable tool for studying the genetic regulation of the cell cycle.[1,2] Our laboratory has also shown that similar temporal patterns of growth-associated gene expression occur during the progression of actively proliferating cells out of mitosis, through G_1 into S.[3,4] These results were provocative in that it was the first time a group of growth-associated genes had been shown to be induced at specific time points in G_1 during active growth. However, the time a gene is expressed does not necessarily determine the time that particular gene's function is needed. Thus, an important question raised by these studies is whether the characteristic pattern of expression observed for each growth-associated gene is important for the progression of the cell out of mitosis, through G_1 into S. A series of follow-up experiments were therefore designed to address these questions. Our approach was to use the antisense oligodeoxynucleotides to down-regulate the expression of various growth-associated genes at specific times during the cell cycle and determine its effect (if any) on entry into DNA synthesis and cell division.

The present study focuses on the gene, c-*jun*. c-*jun* is a protooncogene first characterized as the cellular counterpart of the v-*jun* oncogene carried by the avian sarcoma virus ASV17.[5,6] It is a member of a family of nuclear proteins that are growth factor responsive and function as transcriptional transactivators.[7,8] c-*jun* mRNA is rapidly and transiently induced early in G_1 when quiescent fibroblasts are stimulated to enter the cell cycle by treatment with either growth factors or the tumor promoter TPA.[7] Our studies in actively proliferating cells also showed that c-*jun* mRNA was maximally induced early in G_1, immediately after completion of mitosis. However, we also found that a basal level of accumulated c-*jun* mRNA (approximately 30% of the maximum) was maintained throughout the remainder of G_1.[3,4]

To access the functional implications of these patterns of c-*jun* expression, antisense oligomers specific to c-*jun* were added to either actively proliferating or synchronized Swiss 3T3 cells, and their ability to inhibit DNA synthesis and division was determined. Our results show that if Swiss 3T3 cells are treated with anti–c-*jun*

[a]This work was supported by grants from the W. W. Smith Charitable Trust (C9104) and the National Institutes of Health (NCI) (CA 56309).

[e]Address for correspondence: Dr. Kenneth J. Soprano, Department of Microbiology and Immunology, Temple University School of Medicine, 3400 North Broad Street, Philadelphia, PA 19140.

while actively growing or at any time during G_1 following completion of mitosis, they exhibit a reduction in ability to enter S and subsequently divide. These results demonstrate that the regulation of G_1 progression following mitosis is dependent at least in part on the expression and function of the protooncogene c-*jun*.

MATERIALS AND METHODS

Cell Culture

Stock cultures of Swiss 3T3 cells were maintained in Dulbecco's modified Eagle's medium (DMEM) supplemented with 10% fetal bovine serum (FBS), L-glutamine, and antibiotics (Hazelton, Denver, Pennsylvania) at 37°C in a humidified atmosphere containing 10% CO_2.

Oligodeoxynucleotides

The specified antisense and sense oligodeoxynucleotides were synthesized by an ABI 380B synthesizer on a scale of 1.0 μmol per synthesis. The oligodeoxynucleotides were all 18 mers corresponding to codons 2 to 7. The following chart shows the sequence of each oligodeoxynucleotide and its designation of either antisense, sense, or mismatch. The sequence is written as 5′ to 3′, and all are of mouse origin:

c-*jun* antisense	5′-CGT TTC CAT CTT TGC AGT-3′
c-*jun* sense	5′-GCA AAG GTA GAA ACG TCA-3′
c-*jun* mismatch	5′-CGT GTC AAT ATT GGC AGT-3′

The underline bases are the mismatched bases; the other sequences are identical to the antisense molecule.

The oligodeoxynucleotides were provided in a glass vial containing approximately 1.4 ml of material. The following protocol was used to prepare the oligodeoxynucleotides for experimental use. The original container was placed without its cap in a chemical fume hood overnight to evaporate the NH_4OH. The volume was reduced slightly, and the remaining fluid was moved into a siliconized 1.5-ml eppendorf tube. The oligodeoxynucleotides were dried down in a speed vac under heat. The pellet was resuspended in 300 μl of water and the tube was quickly centrifuged to remove nondissolvable debris. Oligodeoxynucleotides were precipitated by adding 3 volumes of ethanol and 1⁄10 volume of 3M Na acetate (pH 7.0). Precipitation was carried out in a dry-ice ethanol bath. The tubes were centrifuged at 14 K at 4°C, and the pellet was resuspended in 200 μl TE (10 mm Tris, 1 mm EDTA [pH 7.2]). The nucleotide concentration was determined by reading the OD_{260} after diluting 2 μl of the stock oligodeoxynucleotide in 1 ml of H_2O. The concentration of single-stranded DNA was determined using the calculation: 1 OD unit = 25 μg. The normal recovery was 2.5–4.5 mg.

Oligodeoxynucleotides were added directly to the medium at a concentration of 80 μg/ml. This equals a final concentration of approximately 10 μM.

Synchrony

The large scale mitotic shake-off procedure used to obtain highly synchronized Swiss 3T3 cells was published in detail previously.[3] This previous report also provides information concerning the extent of synchrony and the kinetics of G_1 progression and entry into S after plating. For the studies described in this report, mitotic Swiss 3T3 cells were plated onto laminin-treated coverslips contained in 60-mm tissue culture dishes. Extent of synchrony was routinely assayed for each experiment by determining the mitotic index of cells at the time of harvest and at various times after plating.

Determination of Entry into DNA Synthesis

Entry into S was determined by pulse labeling of cells plated on glass coverslips for 1 hour with 7.5 μCi/ml (^{3}H)-thymidine. After the labeling period, coverslips were harvested and fixed in cold absolute methanol and prepared for autoradiography as described previously.[3] The percentage of labeled nuclei from at least 500 cells was determined for each coverslip.

Determination of Cell Growth

Cell growth and viability were determined by trypsinizing the cells, staining them with trypan blue, and performing triplicate cell counts using a hemocytometer.

RESULTS

c-**jun**–*Specific Antisense Oligodeoxynucleotide Treatment Alters the Growth of Actively Proliferating Swiss 3T3 Cells*

The first set of experiments was designed to determine if the addition of antisense oligodeoxynucleotides specific for c-*jun* to actively proliferating Swiss 3T3 cells would result in inhibition of cell proliferation. This was accomplished by determining the effect of antisense treatment on increases in cell number.

Briefly, antisense oligonucleotides (18 mer) that were complementary to the sense strand of codons 2 through 7 of mouse c-*jun* were synthesized. The antisense oligodeoxynucleotides were designed to begin at least one codon down from the ATG, because it is possible that incorporation of this sequence might interfere with other unrelated transcripts. Two different types of control oligodeoxynucleotides (18 mer) were also used for these studies. A sense oligodeoxynucleotide was synthesized by making an oligodeoxynucleotide complementary to the antisense strand of c-*jun*. The second type of control consisted of a c-*jun* antisense oligodeoxynucleotide that contained four mismatches. The sense oligo was used to control for any nonspecific effect on cell growth caused by the addition of large amounts of oligodeoxynucleotides to Swiss 3T3 cells in culture. The mismatch control was used to determine the specificity of the antisense treatment. The exact sequence of each oligodeoxynucleotide can be found in Materials and Methods.

To determine the effects of these oligodeoxynucleotides on cell growth, 2.0×10^5 Swiss 3T3 cells were plated into 60-mm tissue culture dishes in fresh DMEM completed with 10% FBS. The cells were permitted to recover from any effect of the trypsinization procedure. At 24 hours after plating, 10 μM of each oligodeoxynucleotide was added to the plates. A second treatment (10 μM) of oligodeoxynucleotides was added to the remaining plates at 48 hours after plating. At 24-hour intervals after plating, three plates from each experimental group of cells were trypsinized and counted in the presence of trypan blue using a hemocytometer. Regardless of whether the treatment consisted of antisense, sense, or mismatch oligos, the cells demonstrated greater than 98% viability (data not shown).

The results of this experiment are shown in TABLE 1. Clearly, the two treatments of 10 μM anti–c-*jun* had a significant effect on the proliferation of Swiss 3T3 cells. Within 24 hours after the second antisense treatment (3 days after plating), cell number was lower than that of the control treated cells. Moreover, in contrast to the controls, the cell number did not continue to increase significantly during the next 2 days. Therefore, over the 5-day course of the experiment, there was a 44% reduction in cell number when cells were treated with antisense oligodeoxynucleotides specific for c-*jun;* however, when cells were treated with the same amount of sense or mismatch control oligodeoxynucleotides, no significant reduction in cell number was observed. To compare growth inhibition by antisense treatment to a standard method of growth inhibition, we also determined the effect of switching the cells to medium containing a nonmitogenic amount of serum (0.5%). TABLE 1 shows that by 24 hours after treatment with low serum (2 days after plating), growth was inhibited. As a result of this more rapid effect, low serum treatment reduced the final cell number to a greater extent (83%) than did c-*jun* antisense treatment.

c-**jun** *Antisense Oligodeoxynucleotide Treatment Reduces Entry into DNA Synthesis of Actively Growing Swiss 3T3 Cells*

We also determined the effect of antisense treatment specific for c-*jun* on the entrance of actively proliferating Swiss 3T3 cells into DNA synthesis. Actively growing Swiss 3T3 cells were plated onto coverslips at a cell density of 1,000 cells/mm^2 in fresh DMEM containing 10% FBS. The cells were treated with either 0.5% FBS or the appropriate oligodeoxynucleotide. At the indicated times, coverslips were transferred and pulsed for 1 hour with 7.5 μCi (^{3}H)-thymidine/ml and then harvested and prepared for autoradiography, as we described previously.[3]

The results of this experiment are shown in TABLE 2. In all three sets of controls, the percentage of labeled nuclei (or labeling index) at any time during this experiment averaged approximately 53%. TABLE 2 shows that within 3 hours after treatment of actively proliferating Swiss 3T3 cells with 10 μM of antisense specific for c-*jun*, there was a significant reduction (8%) in the number of cells entering S-phase.

TABLE 1. Effect of c-*jun*-Specific Antisense Oligodeoxynucleotides on the Growth of Swiss 3T3 Cells[a]

	Cell Number (× 10^5)						
Treatment	Day 0	Day 1	Day 2	Day 3	Day 4	Day 5	% Untreated on Day 5
No oligo	2.0	2.8	5.0	11.0	13.0	16.0	100
c-*jun* mismatch	2.0	2.9	5.2	10.0	13.0	15.6	98
c-*jun* sense	2.0	2.7	5.0	11.3	14.2	15.2	98
c-*jun* antisense	2.0	2.8	3.2	7.0	8.1	8.9	56
0.5% serum	2.0	2.7	3.0	3.0	2.8	2.7	17

[a] Swiss 3T3 cells were plated at a concentration of 2.0 × 10^5 cells per 60-mm dish. On days 1 and 2 after plating, medium was removed and replaced with fresh medium containing only 10% fetal bovine serum (no oligo), 10 μM of the appropriate oligo (+ 10% FBS), or 0.5% FBS. Cell counts were performed in triplicate daily.

TABLE 2. Effect of c-*jun*-Specific Antisense Oligodeoxynucleotides on the Entry of Swiss 3T3 Cells into S-Phase[a]

	Percentage of Labeled Nuclei							
	Labeling time (hr)							
Treatment	3	6	8	10	12	16	18	24
No oligo	53	56	52	52	56	54	53	53
c-*jun* mismatch	51	54	54	51	56	53	53	54
c-*jun* sense	53	53	54	54	53	53	52	50
c-*jun* antisense	42	42	38	34	29	22	26	24
0.5% serum	45	38	32	28	18	18	8	10

[a] Actively proliferating Swiss 3T3 cells were plated in fresh medium containing 10% fetal bovine serum onto glass coverslips. At 24 hours after plating, the original plating medium was removed and replaced with fresh medium containing only 10% FBS, 10 μM of the appropriate oligo (+ 10% FBS), or 0.5% serum. At various times after treatment, (^{3}H)-thymidine was added and the cells were labeled for 1 hour, then fixed in absolute methanol, and autoradiography was performed. All data points represent the mean percentage of labeled nuclei from three independent coverslips. At least 500 cells per coverslip were counted.

The extent of this reduction increased steadily and reached maximum (42%) at 16 hours after treatment. Once again, treatment with low serum was used as a positive control for comparison purposes. We observed a similar pattern during the first 16 hours following treatment with low serum, however, the extent of inhibition resulting from treatment with low serum was greater (83%) than that observed following treatment with c-*jun*-specific antisense oligos. Thus, it appears that the reduction in cell growth following treatment with anti–c-*jun* described previously occurs as a result of a block in G_1 progression and entry into S-phase. This is consistent with the G_1-specific pattern of c-*jun* mRNA expression that we reported previously.[3,4]

c-jun–*Specific Antisense Treatment Alters the Entry of Mitotically Synchronized Swiss 3T3 Cells into S-Phase*

As DNA synthesis studies suggested that G_1 progression and entry into S was altered in c-*jun* antisense treated cells, and as maximum c-*jun* expression was shown to occur during M/G_1 transition, it was logical to examine the effects on cell-cycle progression of c-*jun* antisense treatment in mitotically synchronized Swiss 3T3 cells. We reasoned that this system would allow us to determine that inhibition of entry into DNA synthesis was in fact a G_1-specific event. It would also permit more precise mapping of the location during G_1 when treatment with c-*jun* antisense inhibits entry into S-phase and thus enable us to better correlate that time with the observed time of expression of c-*jun*.

For these studies, Swiss 3T3 cells were mitotically synchronized, as we previously described.[3] The mitotic cells were then treated with antisense oligodeoxynucleotides using two different experimental designs. Both the expression data and the antisense studies using actively growing populations of cells suggested that to be effective, the c-*jun*-specific antisense oligos would need to be present in the cell very early after division. Thus, it was necessary to modify the procedure used for plating the mitotic cells to ensure that the mitotic cells had sufficient time to take up the antisense oligomers within 1 hour after division, the time of maximal expression for c-*jun*. The modification consisted of plating the mitotic cells in pre-warmed DMEM that contained the specific oligomers but did not contain serum. Under these plating conditions, the mitotic cells attach but do not complete cytokinesis until serum is added. Thus, a 1-hour treatment with antisense oligo in serum-free DMEM would result in an approximate 1-hour delay in the completion of cytokinesis, thereby allowing sufficient time for antisense oligos to enter the mitotic cells before induction of c-*jun* mRNA during early G_1. We have found that the 1-hour treatment in serum-free DMEM results in only a comparable 1-hour delay in the entrance into DNA synthesis but does not alter the rate of entry into DNA synthesis (data not shown). Thus, exposure to serum-free DMEM for 1 hour does not cause an alteration in the kinetics of G_1 progression.

TABLE 3 shows that when mitotic Swiss 3T3 cells were plated in 10 μM of antisense specific for c-*jun,* there was a delay in the start of DNA synthesis by a time equal to the length of antisense treatment. The sense and mismatch treated control cells started to enter into S-phase between 6 and 8 hours. Cells that had been treated with antisense specific for c-*jun* began to enter into S-phase between 8 and 10 hours. In addition, antisense treatment also resulted in a decrease in the total number of cells entering into S-phase. c-*jun* antisense treatment resulted in an approximately 60% reduction in the number of cells entering into S-phase even by 18 hours after the addition of conditioned medium.

Again, the positive control used for comparison in this experiment was treatment of the mitotic cells with 0.5% serum for 2 hours instead of the oligo treatment. TABLE 3 shows that these cells also exhibited a 2-hour lag before the start of DNA synthesis. However, at variance with observations in previous experiments, approximately the same number of cells were inhibited from entering into S-phase as those observed after antisense treatment specific for c-*jun.*

These data show that c-*jun* expression and function during the first 2 hours after mitosis are required for cells to progress through G_1 and enter S.

c-jun Antisense Treatment of Mitotically Synchronized Swiss 3T3 Cells during mid-G_1 Alters Entry into S-Phase

As c-*jun* mRNA was shown to accumulate to maximal levels very early in G_1 following division, it was not surprising that c-*jun*–specific antisense treatment of cells during the M/G_1 transition altered G_1 progression and entry into S-phase.

TABLE 3. Entry into S-Phase of Swiss 3T3 Cells Treated with c-*jun*-Specific Antisense Oligodeoxynucleotides during Mitosis[a]

	Percentage of Labeled Nuclei						
	Time after Plating (hr)						
Treatment	4	6	8	10	12	16	18
No oligo	0	0	26	35	54	63	70
c-*jun* mismatch	0	0	24	32	51	66	68
c-*jun* sense	0	0	24	30	48	59	69
c-*jun* antisense	0	0	3	16	25	27	27
0.5% serum	0	0	2	12	16	21	30

[a] Mitotically synchronized Swiss 3T3 cells were collected by large-scale mitotic shake-off and plated onto laminin-treated coverslips in serum-free medium (no oligo) or serum-free medium containing either 10 μM of the appropriate oligo or 0.5% serum. After 2 hours, the medium was removed and replaced with complete medium containing 10% fetal bovine serum and (^{3}H)-thymidine. At the indicated time points, three coverslips were harvested, fixed, and autoradiographed. Each data point represents the mean of three independent coverslips. At least 500 cells were counted on each coverslip.

TABLE 4. Entry into S-Phase of Swiss 3T3 Cells Treated with c-*jun*-Specific Antisense Oligodeoxynucleotide during Mid G_1[a]

Treatment	% Labeled Nuclei	% Untreated Control
No oligo	66 ± 2	100
c-*jun* mismatch	64 ± 3	97
c-*jun* sense	64 ± 2	97
c-*jun* antisense	45 ± 2	69

[a] Mitotically synchronized Swiss 3T3 cells were collected by large-scale mitotic shake-off and plated onto laminin-treated coverslips in complete medium containing 10% fetal bovine serum. After 3 hours, the medium was removed and replaced with complete medium containing (^{3}H)-thymidine and 10% fetal bovine serum (no oligo), or (^{3}H)-thymidine, 10% FBS, and 10 μM of the appropriate oligo. At 12 hours after plating, three coverslips were harvested, fixed, and autoradiographed. Each data point represents the mean of three independent coverslips. At least 500 cells were counted on each coverslip.

However, as a basal level of c-*jun* mRNA was maintained throughout the remainder of G_1, it was also possible that the addition of antisense after the time of maximum expression could result in a reduction in the number of cells entering into S-phase if continued expression of c-*jun* throughout G_1 is necessary for entrance into DNA synthesis. To test this possibility, we plated mitotically synchronized Swiss 3T3 cells into pre-warmed conditioned medium. Within 30 minutes after plating, 95% of the cells had completed mitosis and presumably entered G_1. After 3 hours, just at mid-G_1, antisense specific for c-*jun* and (^{3}H)-thymidine was added. The cells were harvested for autoradiography 9 hours later, at a time corresponding to the end of S-phase.

TABLE 4 shows that the number of cells that entered S-phase after the addition of 10 μM of antisense oligomers specific for c-*jun* during mid-G_1 was modestly reduced (30%). As this experiment was repeated three independent times with comparable results, we can conclude with confidence that treatment of mid-G_1 cells with c-*jun* antisense did inhibit progression into S-phase. These results are consistent with the possibility, suggested by our previous gene expression studies, that c-*jun* is required not only during M/G_1 transition, but also throughout G_1 in order for Swiss 3T3 cells to progress through G_1 and enter S-phase.

DISCUSSION

Analysis of growth-associated gene expression during the cell cycle has long been accepted in G_0/G_1 model systems as a useful tool for studying cell cycle progression. As we had shown a temporal pattern of growth-associated gene expression during active proliferation, it became important to determine if the patterns of growth-associated gene expression play a role in regulating the ability of an actively proliferating cell to enter into DNA synthesis and divide. To determine the functional significance of the observed patterns of growth-associated gene expression during active proliferation, we used antisense oligodeoxynucleotides specific for one of the growth-associated genes analyzed in the previous studies, c-*jun,* to specifically down-regulate the level of c-*jun* protein and thus determine its effect on cellular proliferation. Our results provide the following new information concerning the role of c-*jun* in cell-cycle progression: (1) Treatment of actively proliferating cells with antisense oligodeoxynucleotides specific for c-*jun* reduced the ability of Swiss 3T3 cells to divide. (2) Reduction of c-*jun* by antisense treatment led to rapid inhibition of DNA synthesis, suggesting that c-*jun* expression is required during G_1, just prior to S-phase. (3) In mitotically synchronized Swiss 3T3 cells, the presence of antisense specific for c-*jun* in mitosis or up to 2 hours following division decreased the number of cells entering S-phase. (4) Treatment of mitotically synchronized Swiss 3T3 cells with antisense specific for c-*jun* in mid G_1, 3 hours after division (1 hour after the time of maximum expression), also altered entry into S-phase. Thus, regulation of G_1 progression following mitosis depends on expression of c-*jun*.

In all of our experiments, the extent of inhibition by antisense treatment was less than that of treatment with reduced serum. This is likely due to a variety of factors including uptake of the oligos, stability of the oligos once they have entered the cell, and efficiency of the oligos in neutralizing the gene product.[9] We have not yet completed experiments to monitor these parameters and establish conditions to optimize them. However, the discrepancy may also reflect the existence of a certain amount of cellular growth-associated gene redundancy. Antisense treatment reduces only a single gene product. It is likely that more than one gene can carry out the function of c-*jun,* but perhaps not as well. Hence, a reduced level of proliferation is observed. In contrast, low serum treatment results in the depletion of many gene products at once, including c-*jun* and the other gene products that can substitute for c-*jun.* We predict that simultaneous treatment with two or more antisense oligos to important growth-associated genes may inhibit growth to an extent more comparable to that of low serum. Obvious candidates for these other important genes include c-*fos,* c-*myc,* or even other members of the *jun* gene family (*jun* B or *jun* D). Nevertheless, treatment with antisense specific for a single growth-associated gene, c-*jun,* consistently resulted in a significant reduction in cell number and entrance into DNA synthesis.

It was not surprising that our analysis of c-*jun* function during active proliferation revealed many similarities to the results with c-*fos* observed by other investigators.[10–12] As it was shown that c-*fos* and c-*jun* are required to form a heterodimer to obtain maximum transcriptional transactivating function,[5] down-regulating one of the proteins should produce phenotypes identical to those expected when the other or both are down-regulated. Our data indicate that the time of expression and the function of these genes act synergistically during early G_1.

It should be noted that the similarities in the requirements for c-*jun* and c-*fos* during active proliferation diverge as the cells enter into mid-G_1. The literature clearly indicates that c-*fos* is not needed during mid- to late G_1.[10–12] Yet, our results with c-*jun*-specific antisense treatment during mid-G_1 showed a significant inhibition of cells entering DNA synthesis. This implies that c-*jun* is also needed during mid- to late G_1. Although most of the literature on c-*jun* focuses on c-*jun* function during G_0/G_1 transition, two studies suggest that c-*jun* functions later in G_1 as well. Carter *et al.*[13] observed a second peak of c-*jun* mRNA induction just before quiescent WI-38 and hamster ts13 cells enter into S-phase after serum stimulation and during active proliferation. Likewise, Kovary and Bravo[14] showed that microinjection of antibodies specific to c-*jun* during mid-G_1, as long as 7 hours after serum stimulation of Swiss 3T3 cells, led to inhibition of entry into S-phase.

As a number of fos-related (fra) proteins have been described and shown to be expressed later in G_1,[15] it is possible that one or more of these may complex with c-*jun* during mid- and late G_1, thereby altering the transcriptional transactivation specificity of c-*jun* from early G_1 genes to late G_1 or S-phase-specific genes. The specific mid- and late G_1 cofactor or cofactors remain to be identified. Likewise, the precise genes that c-*jun* transcriptionally activates during mid- and late G_1 also await further study. Once this information is determined, we will know more about the reason why c-*jun* expression and function are required throughout G_1.

SUMMARY

In actively proliferating Swiss 3T3 fibroblasts, expression of the protooncogene c-*jun* is maximally induced early in G_1, immediately after completion of mitosis. Within 2 hours, c-*jun* mRNA levels drop to a basal amount that is approximately 30% of the maximum. This is maintained throughout the remainder of G_1. To access the functional implications of this pattern of c-*jun* expression, antisense oligomers specific to c-*jun* were added to either actively proliferating or synchronized Swiss 3T3 cells, and their ability to inhibit DNA synthesis and division was determined. Our results show that if Swiss 3T3 cells are treated with anti–c-*jun* while actively growing or at any time during G_1 after completion of mitosis, they exhibit a reduced ability to enter S-phase and subsequently divide. These results demonstrate that the regulation of G_1 progression following mitosis depends on the expression and function of the protooncogene c-*jun*.

REFERENCES

1. Denhardt, D. T., D. Edwards & C. Parfett. 1987. Gene expression during the mammalian cell cycle. Biochim. Biophys. Acta **865:** 83–125.
2. Almendral, J. M., D. Sommer, H. Macdonald-Bravo, J. Buckhardt, J. Pereara & R. Bravo. 1988. Complexity of the early genetic response to growth factors in mouse fibroblasts. Mol. Cell. Biol. **8:** 2140–2148.
3. Cosenza, S. C., R. Carter, A. Pena, A. Donigan, M. Borrelli, D. R. Soprano & K. J. Soprano. 1991. Growth-associated gene expression is not constant in cells traversing G-1 after exiting mitosis. J. Cell Physiol. **147:** 231–241.
4. Soprano, K. J. & S. C. Cosenza. 1992. Cell cycle and cell-growth control. *In* Molecular and Cellular Approaches to the Control of Proliferation and Differentiation. G. Stein & J. Lian, eds. :73–114. Academic Press Inc. San Diego, CA.
5. Nishimura, T. & P. Vogt. 1988. The avian cellular homolog of the oncogene Jun. Oncogene **3:** 659–663.
6. Maki, Y., T. Bos, C. Davis, M. Starbuck & P. Vogt. 1987. Avian sarcoma virus 17 carries the jun oncogene. Proc. Natl. Acad. Sci. USA **84:** 2848–2852.
7. Lamph, W., P. Wamsley, P. Sassone-Corsi & I. Verma. 1988. Induction of proto-oncogene JUN/AP-1 by serum and TPA. Nature **334:** 629–631.
8. Bochman, D., T. Bos, A. Admon, T. Nishimura, P. Vogt & R. Tjian. 1987. Human proto-oncogene c-jun encodes a DNA binding protein with structural and functional properties of transcription factor AP-1. Science **238:** 1386–1392.
9. Marcus-Sekura, C. J. 1988. Techniques for using antisense oligodeoxynucleotides to study gene expression. Analyt. Biochem. **172:** 289–295.
10. Holt, J. T., T. Gopal, A. Moulton & A. Nienhuis. 1986. Inducible production of c-fos antisense RNA inhibits 3T3 cell proliferation. Proc. Natl. Acad. Sci. USA **83:** 4794–4798.
11. Nishikura, K. & J. Murray. 1987. Antisense RNA of proto-oncogene c-fos blocks renewed growth of quiescent 3T3 cells. Mol. Cell. Biol. **7:** 639–649.
12. Riabowol, K. T., R. Vosatka, E. Ziff, N. Lamb & J. Feramisco. 1988. Microinjection of fos-specific antibodies blocks DNA synthesis in fibroblast cells. Mol. Cell. Biol. **8:** 1670–1676.
13. Carter, R., S. C. Cosenza, A. Pena, K. Lipson, D. R. Soprano & K. J. Soprano. 1991. A potential role for c-jun in cell cycle progression through late G-1 and S. Oncogene **6:** 229–235.
14. Kovary, K. & R. Bravo. 1991. The jun and fos protein families are both required for cell cycle progression in fibroblasts. Molec. Cell Biol. **11:** 4466–4472.
15. Cohen, D., P. Ferreira, R. Gentz, B. Franza & T. Curran. 1989. The product of a fos-related gene fra-1, binds cooperatively to the AP-1 site with Jun: Transcription factor AP-1 is comprised of multiple protein complexes. Genes Dev. **3:** 173–184.

Modulation of Double-Stranded RNAs *in Vivo* by RNA Duplex Unwindase[a]

KAZUKO NISHIKURA

The Wistar Institute
3601 Spruce Street
Philadelphia, Pennsylvania 19104

The presence of an activity capable of specifically unwinding RNA duplexes in *Xenopus laevis* eggs and early embryos was first reported in 1987.[1,2] The activity, originally called "double-stranded RNA (dsRNA) unwindase" or "denaturase," was thought to be responsible for failures in the application of antisense RNA in the amphibian embryo.[1,2] As we were engaged in experiments that included antisense transformation of various somatic cell lines, we decided to examine if a similar unwinding activity could also be detected in mammalian somatic cells. It was found not only that the activity is present ubiquitously in a wide range of mammalian somatic cells,[3,4] but also that the intrinsic nature of the unwinding mechanism is unusual and unprecedented; many adenosine residues are converted to inosines during the unwinding reaction.[5,6] Accordingly, the activity is now called the dsRNA unwinding/modifying activity.[5,6] Although the activity is an unlikely cause for failure of some of the antisense RNA techniques, the activity itself seemed to deserve further investigation. Indeed, in the last several years, interesting instances of its involvement in specific eukaryotic and viral gene systems *in vivo* have been reported, and although still speculative at this stage, several hypotheses have emerged proposing possible biological functions of this dsRNA unwinding/modifying activity. In this article, I first briefly review what is presently known about dsRNA unwinding/modifying activity and then summarize the current hypotheses of the biological functions and substrate RNAs, as well as our attempts to isolate the gene coding for this activity. Finally, its relation to the antisense RNA technique will be discussed.

dsRNA UNWINDING/MODIFYING ACTIVITY

dsRNA-Specific Unwinding Activity in Mammalian Cells

We found that an activity capable of specifically dissociating RNA duplexes is also present in mammalian cells[3] similar to that originally reported in *Xenopus* eggs and embryos.[1,2] A protein is involved, as treatment with proteinase K abolishes the activity.[2,3] ATP or any other nucleoside phosphates and divalent cations are not required for the activity.[2,3] The activity is very specific for dsRNA and appears to bind strongly to dsRNA.[2,3] The activity unwinds different dsRNAs made from various *in vitro* synthesized sense and antisense RNAs containing β-globin, c-*myc,* and chloramphenicol acetyl transferase sequences, suggesting that the activity recognized a stretch of RNA double helix structure without strict sequence specificity.

[a] This project was supported by National Cancer Institute grants CA46676 and CA10815 and by National Institutes of Health grant GM40536.

dsRNA Unwindase Deaminates Adenosine Residues to Inosines

Structural alteration of the unwound RNA was suspected, because RNA is incapable of rehybridizing back to dsRNA.[2,5,6] The nature of the alteration in the unwound RNA was elucidated by subjecting the unwound and then strand-separated RNAs to analysis for modified bases. By thin-layer chromatography and high-pressure liquid chromatography analysis, we found that a large fraction of adenosine residues of the unwound RNA is converted to inosine.[5] Both sense and antisense strands of unwound RNA were found to contain adenosine-to-inosine modifications.[5] A similar finding was reported with *Xenopus* dsRNA unwindase.[6]

Our results clearly indicate that there is an activity in mammalian cells that may change the secondary structure of dsRNA by deamination of adenosines to inosines. This is the first, and so far the only, known RNA unwinding activity that results in an accompanying base modification on the substrate RNA. The competition experiments also suggest that the activity binds specifically to dsRNA.[2,3] Thus, the dsRNA unwinding/modifying activity is different from any of the previously known dsRNA unwinding activities or RNA helicases, which have consensus amino acid sequences (DEAD boxes) and which unwind double-stranded RNAs by virtue of binding to single-stranded regions of the RNAs.[7] The dsRNA unwinding/modifying activity is

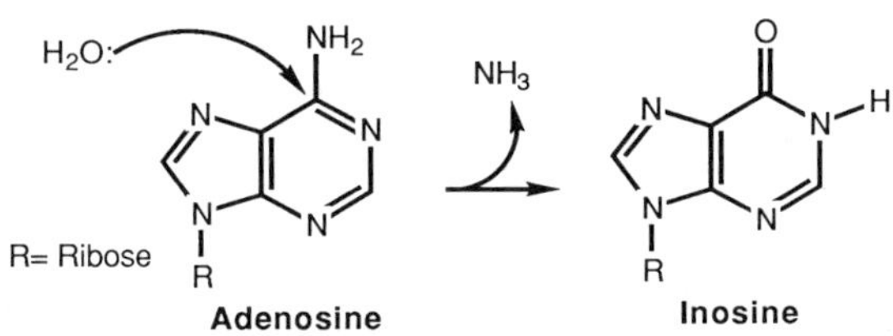

FIGURE 1. Deamination of adenosine to inosine. An amino group at the 6 position of the adenine ring is removed and replaced with a hydroxyl group by deamination processes.

also different from any other known deaminases. Several enzymes involved in purine metabolism are known to catalyze the conversion of adenine to hypoxanthine (the free base of inosine) via deamination at position 6 of the adenine ring (FIG. 1). However, such enzymes, including adenosine deaminase and 5′-adenylic acid deaminase, are known to have strict substrate specificities. These enzymes were found to be incapable of modifying adenosines in dsRNA, and conversely, the dsRNA unwinding/modifying activity was found not to deaminate adenosine or adenylic acid (ref. 6 and Wagner and Nishikura, unpublished results). Thus, the dsRNA unwinding/modifying activity is a novel adenosine deaminase specific for adenosine residues in double-stranded RNAs. In addition to these deaminases, transfer RNA (tRNA)-hypoxanthine ribosyltransferase is also known to introduce inosine in RNA. The enzyme posttranscriptionally substitutes a hypoxanthine for the preexisting base, including adenosine, in the first position of an anticodon of several tRNAs.[8] We previously showed that adenosine residues labeled with ATP carrying a ^{14}C at position 8 of the adenine ring remained labeled after conversion to inosine, suggesting that no replacement of the adenosine base takes place during the modifying reaction.[5] Thus, an enzyme that removes bases or nucleoside residues and replaces them with inosine is unlikely to contribute to the unwinding/modifying mechanism of the dsRNA unwindase.

Substrate Specificity of the Enzyme

Although there is no strict sequence requirement for the substrate dsRNAs, some base preference for the 5′ neighbor of the modified adenosine was reported.[9] When the *in vivo* unwound/modified basic fibroblast growth factor (bFGF) messenger RNA (mRNA) was examined after polymerase chain reaction amplification and cDNA cloning, it was found that adenosines having adenosines or uracils as 5′ neighbors were modified more frequently than were those with guanosines or cytosines as neighbors.[9] Thus, within a given dsRNA with random sequence, adenosine residues in a stretch of consecutive As or an AT-rich region might be preferentially attacked by the dsRNA unwinding/modifying activity.[9]

In addition, investigations on substrate specificity for the dsRNA unwinding/modifying activity have been conducted using an *in vitro* assay system and artificially synthesized RNA substrates. Intermolecular dsRNAs having blunt ends, 5′ overhangs, and 3′ overhangs, as well as partial dsRNAs, were equally well modified.[5,6] The activity also did not distinguish between capped and noncapped RNAs.[5,6] Both intermolecular and intramolecular dsRNAs seem to serve as substrates for the dsRNA unwinding/modifying activity.[10] The dsRNA unwinding/modifying activity appears to require a double-stranded region of at least 15–20 base pairs for substrate recognition.[10] Finally, modification efficiency was found to be critically dependent on the length of the double-stranded region: as the size decreased below 100 base pairs, it dropped precipitously.[10] Efficient modification occurs, at least *in vitro,* only with relatively long (<100 base pairs) dsRNA, perhaps because multiple copies of the enzyme must be bound.[10] Thus, naturally occurring viral and eukaryotic gene transcripts, which usually contain relatively short and often imperfect double-stranded regions, may not be efficient substrates. Alternatively, the naturally occurring hairpins in these RNAs may be stabilized or in some way altered *in vivo* to form more helix-like structures, and thus allow the dsRNA unwinding/modifying activity to recognize them as substrate RNAs.

STRUCTURAL ALTERATION OF dsRNA BY THE UNWINDING/MODIFYING ACTIVITY

Although it is not known precisely what happens to the double helical structure of the RNA substrate during the modification reaction, several characteristic features of the structural alteration of modified RNA products have been observed. The modified RNA products migrate anomalously on a native polyacrylamide gel and also become sensitive to single-strand–specific ribonuclease treatment after the modification reaction. The adenosine-to-inosine transition results in replacement of an A-U base pair with an I-U base pair. The inosines are still capable of forming a so-called wobble base pair with uracils. This wobble I-U base pair is thermodynamically less stable than either the original A-U or an I-C Watson-Crick base pair (FIG. 2). A series of I-U mismatches accumulated in dsRNA appears to be responsible for the anomalous migration of the reaction products on a native polyacrylamide gel. Increased susceptibility of the reaction products to single-strand–specific ribonucleases such as RNaseA is also likely due to these I-U base pairs. The I-U mismatch exposes the U base to nucleophilic attack by RNaseA, an enzyme known to cleave mismatched pyrimidines. After a number of I-U base pairs accumulate, dsRNA may become largely single-stranded because of local unwinding and looping (FIG. 3).

Approximately 50% of the total adenosine residues of both sense and antisense strands of dsRNAs can be modified with the current *in vitro* assay conditions. This means an average of one out of two A-U base pairs, or an average of one out of four base pairs of all types, ends up wobble base-paired. At this point, the substrate retains only very short stretches of perfectly matched base-pairing. This probably accounts for the failure to achieve 100% conversion of adenosine to inosine, as single-stranded RNA is not a substrate for the enzyme.

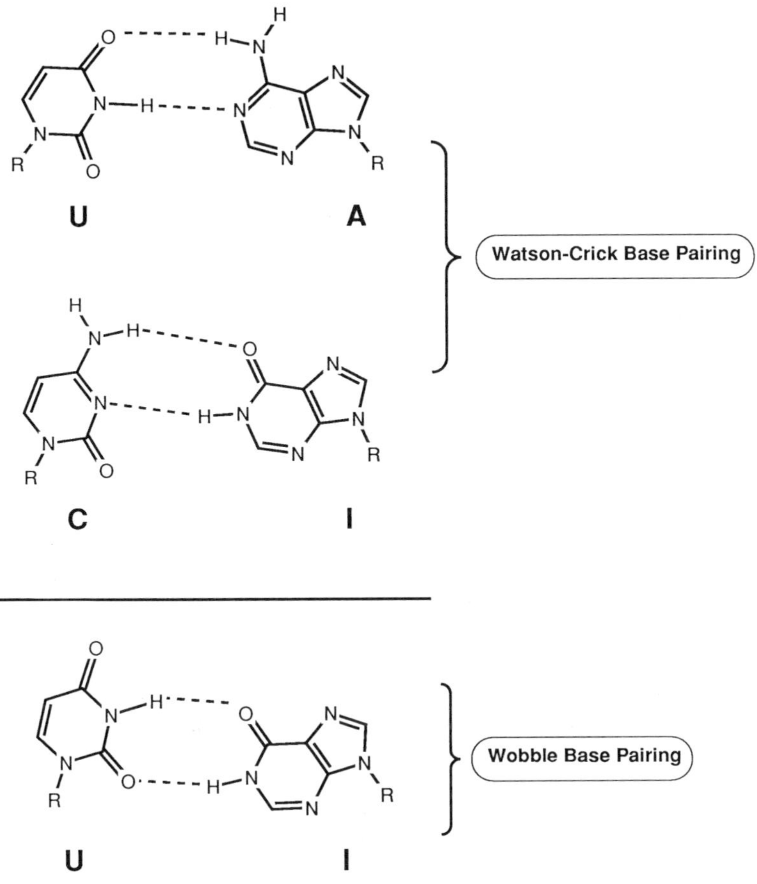

FIGURE 2. Watson-Crick base-pairings of U-A and C-I and wobble base-pairing of I-U. The adenosine-to-inosine conversion results in a mismatched I-U base pair that can adopt a thermodynamically less stable, wobble hydrogen-bonding configuration. Inosine prefers to base-pair with cytosine in a Watson-Crick hydrogen-bonding configuration.

NUCLEAR LOCALIZATION OF THE ACTIVITY

The dsRNA unwinding/modifying activity is localized exclusively in the nucleus of mammalian somatic cells.[4] Interestingly, the activity in *Xenopus laevis,* which has been reported to reside in the nucleus of oocytes, is released into the cytoplasm during meiotic maturation and remains in the cytoplasm during fertilization and early embryogenesis (B. Bass, personal communication). It is currently not known exactly when the dsRNA unwinding/modifying activity returns to the nucleus of somatic cells during *Xenopus* embryogenesis. Neither is it clear at this point if the dsRNA unwinding/modifying activity, though localized in the somatic cell nucleus, is destined to function in this cellular compartment. In *Xenopus laevis* oocytes undergoing the meiotic maturation process, the dsRNA formed between sense and antisense bFGF RNAs (see below) is modified in the cytoplasm by the activity released from the nucleus after the germinal vesicle breakdown that accompanies meiotic maturation.[9] It is plausible that in somatic cells the biological substrate RNAs for the dsRNA unwinding/modifying activity also reside in the cytoplasm, but are attacked only during part of the cell cycle, possibly during or shortly after mitosis when the activity moves from the nucleus into the cytoplasm.

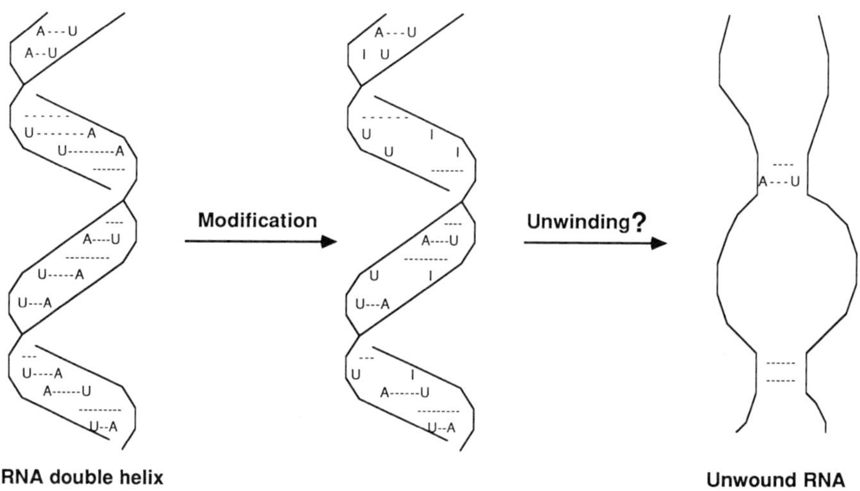

FIGURE 3. Unwinding and modification of dsRNA by the unwindase. Unwinding of dsRNA by the unwindase via A to I conversion and replacement of A-U base pairs with I-U base pairs is schematically shown. It is currently not known whether the modified and partly denatured reaction products are dissociated into two strands *in vivo*.

UBIQUITOUS EXPRESSION OF THE ACTIVITY

The dsRNA unwinding/modifying activity seems to be a typical housekeeping gene, based on its ubiquitous expression in different tissues and among different species.[4] It appears that this activity is of fundamental importance to eukaryotic cells. We have surveyed a variety of primary tissues and transformed cell lines derived from different tissues to see if any difference in the level of activity can be detected.[4] Rat liver, kidney, spleen, testis, and lymph node tissues, as well as human natural killer and T cells isolated from peripheral blood, were all found to contain the activity.[4]

In addition to primary tissues and cells, a variety of transformed cell lines commonly used, including Cos, HeLa, and 3T3, were also analyzed and were all found to contain the activity. Quiescent mouse fibroblast 3T3 cells increased their unwinding/modifying activity threefold after stimulation with serum. In addition, the level of the activity decreased fourfold during *in vitro* induction of the mouse muscle cell line C_2C_{12} from myoblast to myotube, but the activity was still detectable in the differentiated postmitotic myotubes.

The involvement of dsRNA unwinding/modifying activity has been suggested in a highly biased hypermutation of measles virus genomes that occurs in some human central nervous system (CNS) diseases (see below).[11–13] Therefore, we also examined various tissues and cell lines derived from the CNS and from the peripheral nervous system. Extracts of human, calf, rat, or mouse brain contained high levels of the dsRNA unwinding/modifying activity, as did bovine oligodendrocytes and rat astrocytes. In addition to these primary tissues, various transformed cell lines derived from CNS and peripheral nervous system contained high levels of the activity. It was concluded that the dsRNA unwinding/modifying activity is detectable in all cells and tissues, including CNS, regardless of the differentiated, mitotic, or transformed state of the cell.[4]

POSSIBLE BIOLOGICAL FUNCTION

Regulation of RNA Turn-over

The full range of biological functions, as well as natural substrate RNAs, of this newly discovered activity is not yet known. However, several illuminating facts have emerged from recent studies. One of our most plausible hypotheses is that it may be involved in a mechanism that degrades dsRNAs formed *in vivo,* as schematically shown in FIGURE 4. Double-stranded RNAs may be degraded more quickly by ribonucleases after introduction of I-U base mismatches *in vivo* as already shown *in vitro.*[1,5,6] An example of such a case has been reported by Kimelman and Kirschner.[9] In maturing *Xenopus* oocytes, rapid degradation of dsRNA formed between maternal bFGF mRNAs and antisense transcripts derived from the opposite strand of the bFGF gene is proposed to be regulated by the dsRNA unwinding/modifying activity.[9] Increasingly, antisense transcription of a wide range of eukaryotic genes is being reported in somatic cells also.[14,15] Thus, duplexes of these naturally occurring antisense transcripts and sense mRNAs, together with the dsRNA unwinding/modifying activity, may be involved in a widespread RNA degradation regulatory mechanism (FIG. 4A).

This type of RNA degradation control may also act on intramolecular dsRNAs (FIG. 4B). Several examples of cytoplasmic mRNA degradation in which the stem and loop type RNA secondary structures are involved were reported.[16] Two of the best studied cases are histone H3 and transferrin receptor mRNAs. Rapid degradation of both histone H3 mRNA, which is DNA replication-dependent, and transferrin receptor mRNA, which is regulated by iron, has been shown to be promoted through a stem-loop structure located within their 3′ nontranslated regions.[17,18] It would be interesting to determine if the dsRNA unwinding/modifying activity is involved in these two previously known cytoplasmic RNA degradation mechanisms by interacting with their stem structures. It is tempting to propose that the dsRNA unwinding/modifying activity is a part of a eukaryotic gene regulatory mechanism that acts on either intermolecular or intramolecular dsRNAs (FIG. 4). However, finding other mRNAs after modification in somatic cells *in vivo* may be difficult, because such RNAs are destined for immediate degradation. The aforementioned modified bFGF mRNAs were more easily detected,[9] because the degradation mechanism might operate more slowly in maturing *Xenopus* oocytes.

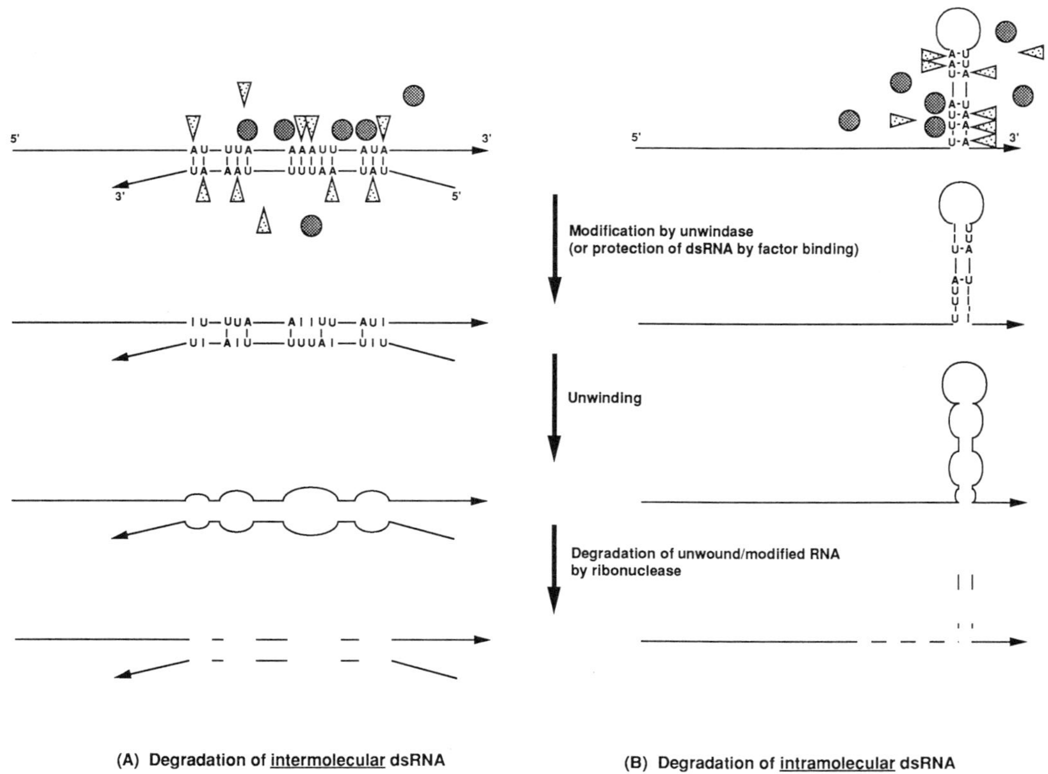

FIGURE 4. Model for the RNA degradation mechanism mediated by the dsRNA unwinding/modifying activity. The proposed model requires either intermolecular (**A**) or intramolecular (**B**) dsRNAs, the dsRNA unwinding/modifying activity, and a ribonuclease, possibly specific for RNA containing I-U mismatched base pairs introduced by the activity. The degradation of targeted RNAs may be regulated by compartmentalization of the unwindase/ribonuclease or by sequence specific cellular factor(s) (*stippled circles*), which can compete with the unwinding/modifying activity (*stippled triangles*) for binding to the dsRNAs.

Involvement in Biased Hypermutation of RNA Virus Genes

Recently, it was suggested that several cases of biased hypermutation of RNA viruses are caused by the dsRNA unwinding/modifying activity.[11–13] In the matrix gene of a defective measles virus, isolated from the brain tissue of a patient with measles inclusion body encephalitis, 50% of the A residues (132 of 266A) of the minus-strand RNA genome were found to be mutated to G by cDNA cloning and sequencing of the viral RNAs.[13] (The A to I change is translated into an A to G change of the cDNA by reverse transcriptase; FIG. 2.) These mutations could be introduced by the dsRNA unwinding/modifying activity. This extensive mutation drastically changed the coding capacity of the matrix gene, resulting in suppression of its expression.[13] The lack of matrix gene expression may help the virus escape immune system recognition and thus allow persistent infection.[13] Two more examples of highly biased mutation (A → G mutations) have been found in the hemagglutinin gene of another strain of measles virus isolated from a patient with subacute sclerosing panencephalitis[12] and in a defective interfering particle of vesicular stomatitis virus.[13]

It may be that biased hypermutation caused by the dsRNA unwinding/modifying activity occurs relatively frequently, but it is propagated and detected only in the genome of persistent viruses or defective-interfering particles because of the lower selective pressures operating in these infections.[12] The frequency and extent of biased hypermutation detected may be related to the dsRNA unwinding/modifying activity level present in different tissues and the life cycle of individual RNA viruses.[12]

RNA Editing Mechanism

The dsRNA unwinding/modifying activity could alter the expression of certain mRNAs posttranscriptionally by converting adenosine to inosine and consequently changing their coding capacity (FIG. 2). At least two reports suggest that the dsRNA unwinding/modifying activity may indeed be involved in such an RNA editing mechanism. Hepatitis delta virus (HDV) carries a single-stranded circular RNA genome that folds on itself by intramolecular base-pairing to form a rod structure with approximately 70% of the bases paired. In addition, the virus synthesizes a complementary circular RNA molecule, the so-called antigenomic RNA. It was recently reported that the dsRNA unwinding/modifying activity may be responsible for specific base transitions (A to G change) of the HDV genome that occur during viral replication.[19] In this case, however, the number and location of A to G mutations are rather restricted, resulting in one instance in replacement of a translation termination codon (UAG) with a tryptophane codon (UGG), thereby generating the longer form of HDV antigen.[19] Thus, in this particular case, the activity could indeed serve as a specific editing mechanism for the virus, which may be exploiting a host cell mechanism (dsRNA unwinding/modifying activity) to modify its genome so as to allow the synthesis of the additional larger form of the delta antigen.

The second example of possible involvement of the dsRNA unwinding/modifying activity in RNA editing was recently reported for glutamate-gated ion channel subunit gene transcripts.[20] Subunit components of two related classes of glutamate receptor channels were compared by cDNA cloning. It was found that the two classes carry either an arginine or a glutamine residue in a defined position of their putative channel-forming segment. The arginine residue in this segment was shown to greatly alter the properties of ion flow. After careful examination of the cDNA and genomic DNA sequences, it was concluded that the glutamine codon, CAG, encoded in the genomic sequences, is posttranscriptionally altered to the arginine codon, CGG, in some of the transcripts by RNA editing.[20] This A to G base modification in the cDNA sequences might be introduced as A to I modification in their mRNAs by the dsRNA unwinding/modifying activity, as in the case of HDV antigen already described.

BIOCHEMICAL PURIFICATION AND CLONING OF THE dsRNA UNWINDING/MODIFYING ACTIVITY

In addition to the limitations imposed by our lack of basic knowledge of the dsRNA unwinding/modifying activity, which was only recently discovered, we suffer from a lack of two crucial tools, specific antibodies and oligonucleotide probes, that are necessary for investigation of this activity at the molecular level. These can be produced only after purification of the activity, at least to the extent of having an identified single peptide on a gel for use as an antigen and for determination of partial amino acid sequence.

To this end, we purified the activity biochemically from bovine liver nuclear extracts. The final fraction, approximately 5,000-fold enriched, contains two polypeptides, of 100 and 95 kD, detected on SDS-PAGE by silver staining. NH_2-terminal sequences of these peptides were determined, and using their amino acid sequence information, we generated short stretches of cDNAs corresponding to the NH_2-terminal amino acid sequences by reverse transcription-polymerase chain reaction. With these cDNAs as probes, we recently isolated longer cDNA clones as well as genomic clones of dsRNA unwinding/modifying activity (Kim and Nishikura, unpublished results). In addition to characterizing the DNA sequences of these clones, we are planning to clone the unwindase gene clones from other species. We are also in the process of preparing specific antibodies against the activity.

With such specific tools, we should be able to try more direct approaches when addressing the question of the biological roles played by this novel activity. For instance, creating a null-mutant phenotype of the unwindase gene by using the antisense RNA technique or making transgenic mice with a nonfunctional unwindase gene through homologous DNA recombination is certainly worth trying. We also plan to investigate the spatial and temporal localization/expression of the activity using the specific antibodies. We anticipate that these types of experiments will be possible in the near future, which, in turn, will bring new perspectives to the biological functions and substrates of the dsRNA unwinding/modifying activity.

IMPLICATIONS FOR ANTISENSE RNA TECHNIQUES

In recent years, the use of specific antisense RNAs to block expression of selected genes has proven to be a powerful tool in elucidating gene functions.[21–23] It has been presumed that duplex formation of sense mRNAs and antisense RNA is essential for the mechanism. Possible sites of inhibition include transcription, splicing, RNA transport, RNA degradation, and translation, but the exact mechanism of antisense RNA inhibition of expression of successfully targeted genes is still not known.[21,22] Moreover, despite some spectacular successes, it is generally agreed that the effectiveness of the techniques varies for unknown reasons. As previously mentioned, the dsRNA unwinding/modifying activity was discovered during the search for the cellular factors that interfere with antisense RNA techniques. However, several considerations now make it unlikely that the activity is a direct cause of interference. First, the dsRNA unwinding/modifying activity is expected to accelerate the degradation of the duplex formed between antisense RNAs (endogenous or artificially introduced) and targeted sense mRNAs, as described in the previous section (FIG. 4). Second, even if unwound/modified dsRNAs were not degraded immediately, the coding of such sense mRNAs will be completely altered, as inosine residues converted from adenosines will be recognized as guanosines by the translation machinery (FIG. 2). Thus, on the contrary, the dsRNA unwinding/modifying activity may in fact have contributed to the success of some of the antisense RNA techniques.

In the last several years, many attempts to develop "gene therapies" using the antisense RNA approach to block viral infection or to treat tumors have been reported. Such clinical application of antisense RNA technology would be more efficient if based on precise knowledge of the fate of dsRNA *in vivo.* However, we currently have only a fragmentary idea of how cells deal with dsRNA. Thus, an important side benefit of our efforts to understand the dsRNA unwinding/modifying activity will be increased understanding of the metabolism of dsRNAs. Future experiments concerning the activity will build the knowledge base required for more effective utilization of the antisense approach.

SUMMARY

The double-stranded RNA (dsRNA) unwinding/modifying activity is a recently discovered cellular activity capable of unwinding or denaturing dsRNAs by modifying multiple adenosine residues to inosines and creating I-U mismatched base-pairings. The biological functions of this activity, which can potentially mutate the coding capacity of messenger RNAs (mRNAs), are presently not known. However, this unwinding/modifying activity is likely to affect the secondary structures, processing, and turn-over of various eukaryotic as well as viral transcripts. Although the activity was originally found and proposed as a cellular factor that interfered with the use of antisense RNA, it now appears more likely that the activity in fact may participate in antisense RNA suppression of target genes, either by altering the coding capacity of the sense mRNAs or by accelerating the degradation of duplex RNAs. Further understanding of this novel enzymatic activity, and thus, in turn, of the metabolism of dsRNAs *in vivo,* should allow us to derive a better strategy for designing antisense RNA.

ACKNOWLEDGMENTS

I thank my colleagues, R. W. Wagner, U. Kim, and C. Yoo, who contributed to this work, and J. M. Murray and F. J. Rauscher for helpful discussion and critical reading of this manuscript. I also thank the Editorial Services Department of The Wistar Institute for preparing the manuscript.

REFERENCES

1. REBAGLIATI, M. R. & D. A. MELTON. 1987. Antisense RNA injections in fertilized frog eggs reveal an RNA duplex unwinding activity. Cell **48:** 599–605.

2. BASS, B. L. & H. WEINTRAUB. 1987. A developmentally regulated activity that unwinds RNA duplexes. Cell **48:** 607–613.
3. WAGNER, R. W. & K. NISHIKURA. 1988. Cell cycle expression of RNA duplex unwindase activity in mammalian cells. Mol. Cell. Biol. **8:** 770–777.
4. WAGNER, R. W., C. YOO, L. WRABETZ, J. KAMHOLZ, J. BUCHHALTER, N. F. HASSAN, K. KHALILI, S. U. KIM, B. PERUSSIA, F. A. MCMORRIS & K. NISHIKURA. 1990. Double-stranded RNA unwinding and modifying activity is detected ubiquitously in primary tissue and cell lines. Mol. Cell. Biol. **10:** 5586–5590.
5. WAGNER, R. W., J. E. SMITH, B. S. COOPERMAN & K. NISHIKURA. 1989. A double-stranded RNA unwinding activity introduces structural alterations by means of adenosine to inosine conversions in mammalian cells and *Xenopus* eggs. Proc. Natl. Acad. Sci. USA **86:** 2647–2651.
6. BASS, B. L. & H. WEINTRAUB. 1988. An unwinding activity that covalently modifies its double-stranded RNA substrate. Cell **55:** 1089–1098.
7. WASSARMAN, D. A. & J. A. STEITZ. 1991. Alive with DEAD proteins. Nature **349:** 463–464.
8. ELLIOTT, M. S. & R. W. TREWYN. 1984. Inosine biosynthesis in transfer RNA by an enzymatic insertion of hypoxanthine. J. Biol. Chem. **259:** 2407–2410.
9. KIMELMAN, D. & M. W. KIRSCHNER. 1989. An antisense mRNA directs the covalent modification of the transcript encoding fibroblast growth factor in *Xenopus* oocytes. Cell **59:** 687–696.
10. NISHIKURA, K., C. YOO, U. KIM, J. M. MURRAY, P. A. ESTES, F. E. CASH & S. A. LIEBHABER. 1991. Substrate specificity of the dsRNA unwinding/modifying activity. EMBO J. **10:** 3523–3532.
11. BASS, B. L., H. WEINTRAUB, R. CATTANEO & M. A. BILLETER. 1989. Biased hypermutation of viral RNA genomes could be due to unwinding/modification of double-stranded RNA. Cell **56:** 331.
12. CATTANEO, R., A. SCHMID, P. SPIELHOFER, K. KAELIN, K. BACZKO, V. MEULEN, J. PARDOWITZ, S. FLANAGAN, B. K. RIMA, S. A. UDEM & M. A. BILLETER. 1989. Mutated and hypermutated genes of persistent measles viruses which caused lethal human brain diseases. Virology **173:** 415–425.
13. CATTANEO, R., A. SCHMID, D. ESCHLE, K. BACZKO, V. MEALLEN & M. A. BILLETER. 1988. Biased hypermutation and other genetic changes in defective measles viruses in human brain infections. Cell **55:** 255–265.
14. NEPVEU, A. & K. B. MARCU. 1986. Intragenic pausing and antisense transcription within the murine c-*myc* locus. EMBO J. **5:** 2859–2865.
15. MIYAJIMA, N., R. HORIUCHI, Y. SHIBUYA, S. FUKUSHIGE, K. MATSUBARA, K. TOYOSHIMA & T. YAMAMOTO. 1989. Two *erb*A homologs encoding proteins with different T_3-binding capacities are transcribed from opposite DNA strands of the same genetic locus. Cell **57:** 31–39.
16. BRAWERMAN, G. 1989. mRNA decay: Finding the right targets. Cell **57:** 9–10.
17. LEVINE, B. J., N. CHODCHOY, W. F. MARZLUFF & A. SKOULTCHI. 1987. Coupling of replication type histone mRNA levels to DNA synthesis requires the stem-loop sequence at the 3′ end of the mRNA. Proc. Natl. Acad. Sci. USA **84:** 6189–6193.
18. MÜLLNER, E. W. & L. C. KÜHN. 1988. A stem-loop in the 3′ untranslated region mediated iron-dependent regulation of transferrin receptor mRNA stability in the cytoplasm. Cell **53:** 815–825.
19. LUO, G., M. CHAO, S.-Y. HSIEH, C. SUREAU, K. NISHIKURA & J. TAYLOR. 1990. A specific base transition occurs on replicating hepatitis delta virus RNA. J. Virol. **64:** 1021–1027.
20. SOMER, B., M. KÖHLER, R. SPRENGEL & P. H. SEEBURG. 1991. RNA editing in brain controls a determinant of ion flow in glutamate-gated channels. Cell **67:** 11–19.
21. IZANT, J. G. & H. WEINTRAUB. 1984. Inhibition of thymidine kinase gene expression by antisense RNA: A molecular approach to genetic analysis. Cell **36:** 1007–1015.
22. IZANT, J. G. & H. WEINTRAUB. 1985. Constitutive and conditional suppression of exogenous and endogenous genes by antisense RNA. Science **229:** 345–352.
23. NISHIKURA, K. & J. M. MURRAY. 1987. Antisense RNA of protooncogenes c-*fos* blocks renewed growth of quiescent 3T3 cells. Mol. Cell. Biol. **7:** 639–649.

Antisense RNA

History and Perspective

SIDNEY PESTKA

Department of Molecular Genetics and Microbiology
UMDNJ-Robert Wood Johnson Medical School
675 Hoes Lane
Piscataway, New Jersey 08854-5635

Since our first report in 1984 of the use of antisense RNA to block the translation of specific messenger RNA (mRNA) molecules,[1] a huge number of reports have appeared. It is worth noting that other reports appeared during the same year initiating this avenue of research,[2–4] a result that indicates that when the time for an idea has come, it may blossom independently from different seeds. The seeds for this effort were formed and began germinating over many years. The first definitive demonstration of a block in translation due to the use of antisense RNA in cell-free extracts was carried out in 1963 by Singer, Jones, and Nirenberg.[5] They demonstrated that the synthesis of polyphenylalanine in response to polyuridylic acid as template in cell-free extracts was completely blocked if polyadenylic acid was present in the translation reaction mixture. Several years later, other reports appeared indicating that RNA complementary to various cellular RNA species could block translation. (See additional references in Daugherty *et al.*[6]). The overall observation that translation of mRNA could be blocked by complementary RNA was cleverly used by Paterson *et al.*[7] in 1977 and Hastie and Held[8] in 1978 for the screening of clones by a technique designated hybrid arrested translation.

Later, clues began to appear that indeed complementary RNA was able to modulate cellular events. Complementary RNA was shown to be an important regulatory mechanism *in vivo.* Plasmid replication and copy number seem to be controlled by complementary RNA transcripts of *Escherichia coli* ColE1 plasmids.[9–15] Similarly, *E. coli* plasmid IncF11[16–18] and *Staphylococcus aureus* plasmid pT181[19] appear to exhibit control by complementary RNA. Translation of the *E. coli* Tn10 transposase mRNA can be inhibited by an RNA complementary to its 5′ end.[20,21] The expression of OmpF in *E. coli* is regulated by an antisense RNA[3,22] complementary to the ribosome binding site[23] and initiation codon. More recently, other examples of antisense RNA regulation of gene expression have been reported.[24,25] In addition, RNA is spliced by small nuclear ribonucleoprotein particles (snRNPs) comprising an RNA complementary to the splice junctions.[26–28] Furthermore, evidence for the presence of multiple processed mRNAs transcribed from opposite strands of the same DNA sequence[29,30] has important implications for the control of gene expression in eukaryotes.

Our initial experiments were carried out to develop a general method to block translation of specific mRNA molecules as well as to develop cloning strategies independent of the knowledge of the protein sequences themselves. Although our results dramatically showed that production of antisense RNA could block translation of specific mRNA molecules,[1,6] it was clear that the procedure would not be highly efficacious unless the effects could be generated by a lower ratio of antisense to sense RNA. Since our initial reports, other approaches have been developed to generate catalytic RNAs (ribozymes) to achieve this goal (for example, see Rossi *et*

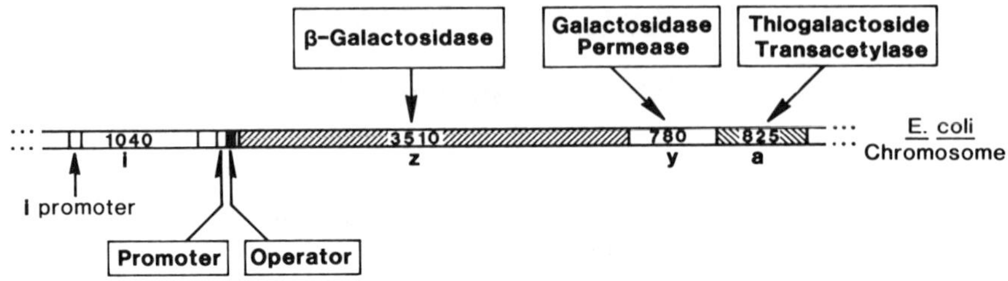

FIGURE 1. The *lac* operon.

al.,[31] Weerasinghe *et al.*,[32] Goodchild and Kohli,[33] Sioud and Drlica,[34] Lorentzen *et al.*,[35] Tabler and Tsagris,[36] and Branch and Robertson[37]). In this symposium many examples of such approaches have been described with noncatalytic (see articles in this volume by Denhardt *et al.*, Knecht *et al.*, Lawrence *et al.*, Holt *et al.*, Sklar *et al.*, Caskey *et al.*, Matthieu *et al.*, Jacobs-Lorena *et al.*, Izant *et al.*, Van Eldik *et al.*, Sczakiel *et al.*, Lawlor *et al.*, Shillitoe and Steele, Freire-Moar *et al.*, and Graeven *et al.*) and catalytic (Zaia, Rossi, *et al.*, Ge and Seitz, *et al.*, Ho and Britton, *et al.*, Chen *et al.*, James and Crisell, Ohtsuka *et al.*, Dinter-Gottlieb *et al.*, and Blumenfeld *et al.*) RNAs. In addition, the use of exogenously supplied oligonucleotides has been effective in blocking viral replication,[38–40] protein synthesis,[41] and cell proliferation.[42] Many more examples of inhibition of gene expression by oligonucleotides are described in this symposium. The rest of this article summarizes our own results with the use of antisense RNA in blocking gene expression in *E. coli.* A number of general reviews concerning antisense RNA have appeared.[13,43–45]

THEORETICAL CONSIDERATIONS

The studies were designed to target the *lac* operon of *E. coli* (FIG. 1). In the absence of the inducer IPTG of the *lac* operon, β-galactosidase and the other enzymes are not synthesized. When IPTG is added to the culture, synthesis of β-galactosidase is initiated. The structure of the vector for synthesis of antisense RNA is shown in FIGURE 2. Synthesis of antisense RNA is controlled by the temperature-sensitive repressor cIts857.[46] Thus, when cells are grown at 30°C, antisense RNA is not synthesized. On transfer of the culture to 42°C, antisense RNA is produced. Accordingly, the system is designed with an on/off switch for synthesis of both β-galactosidase and antisense RNA.

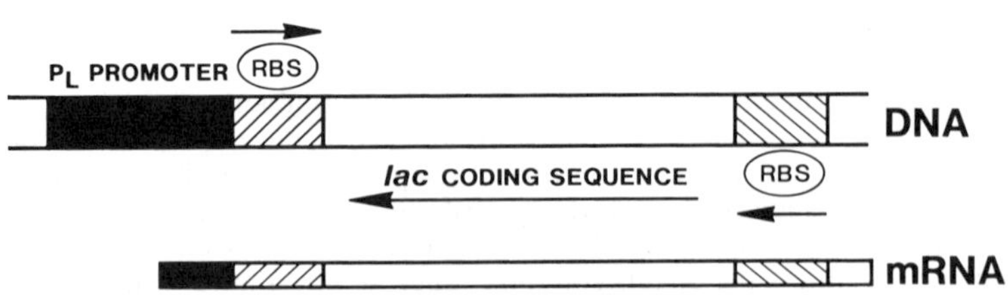

FIGURE 2. Structure of the vector to generate antisense RNA in *E. coli.* The P_L promoter was used to control expression of the antisense RNA under control of the temperature-sensitive repressor cIts857.[46] Constructions of the vectors were described in detail previously.[1,6]

EFFECTIVENESS OF ANTISENSE RNA IN BLOCKING β-GALACTOSIDASE SYNTHESIS

To determine requirements for optimal inhibition by antisense RNA, different regions of sense RNA were targeted (FIG. 3; TABLE 1). Furthermore, various sizes and regions of the coding region were targeted for antisense inhibition.

FIGURE 4 presents data on the effect of antisense RNA produced by several vectors. Inhibition of β-galactosidase synthesis by vectors pBD9 and pBD7 is almost total, whereas inhibition by vectors pBD6, pBD8, and pBD10 is about 50–70%. As shown in FIGURE 5, the switch is reversible. When the culture was placed at 42°C at the time of induction of β-galactosidase by IPTG, β-galactosidase was not synthesized for 60 minutes in the presense of antisense RNA (pBD9, FIG. 5), but control cultures synthesized enzyme (pRC23). When one half the pBD9 culture at 42°C was

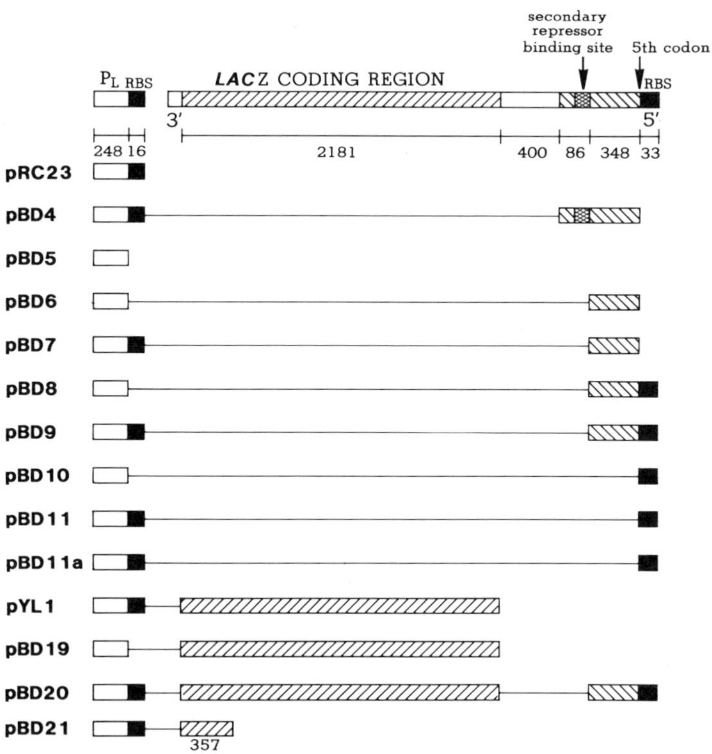

FIGURE 3. Structure of plasmids containing the components noted in TABLE 1. In addition, various sizes and regions of the coding region were targeted by the antisense RNA vectors. The *lacZ* coding region is shown in reverse orientation. Vector pBD9, for example, produces an antisense RNA that contains the functional P_L ribosomal binding site (RBS) and antisense RNA targeted to the *lac* RBS and the 348 base segment of the 5′-coding region of β-galactosidase mRNA. Details of the constructions were previously reported.[1,6] Figure adapted from ref. 6.

TABLE 1. Possible Anti-mRNA Constructions[a]

P_L RBS	*Lac* RBS	*Lac* Coding
−	−	−
−	+	+
−	+	−
−	−	+
+	−	−
+	+	−
+	+	+
+	−	+

[a]Combinations of various regions incorporated into the antisense RNA vectors shown. The P_L RBS represents the functional ribosomal binding site downstream from the P_L promoter. The *lac* RBS represents the ribosomal binding site of the β-galactosidase mRNA. The table illustrates that various combinations of the coding region, *lac* RBS, and functional P_L RBS were included in various vectors as described in more detail in FIGURE 3.

transferred to 30°C, shutting off antisense RNA synthesis, β-galactosidase synthesis proceeded at the same rate as that in control cultures. Little or no β-galactosidase synthesis occurred in cultures remaining at 42°C (FIG. 5). To prove that the switch was effective once enzyme synthesis had already begun, one half the culture at 30°C synthesizing β-galactosidase actively was transferred to 42°C (FIG. 6). As can be seen (FIG. 6), β-galactosidase synthesis was blocked immediately when the culture was switched to the higher temperature.

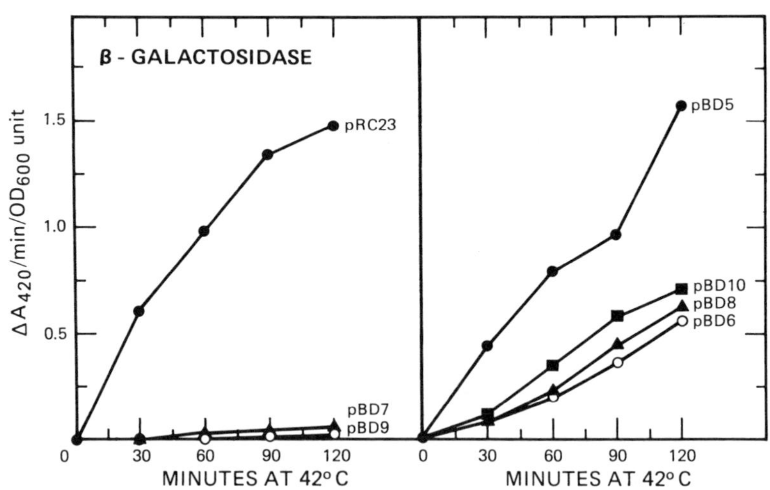

FIGURE 4. Effect of anti-lacZ mRNA on β-galactosidase synthesis. Cells were grown logarithmically at 30°C. To initiate the experiment, at time = 0, the cells were transferred to a 42°C bath, and at the same time IPTG was added to a final concentration of 1 mM. The enzyme level of β-galactosidase was determined as a function of time after cells were transferred to 42°C in the presence of IPTG. Left and right panels show β-galactosidase levels in cells containing derivatives of plasmids pRC23 and pBD5, respectively. Enzyme activity (β-galactosidase activity) was determined as previously reported.[1,6] *Left panel:* ●, pRC23; ○, pBD9; ▲, pBD7. *Right panel:* ●, pBD5; ○, pBD6; ▲, pBD8; ■, pBD10. Plasmids pRC23 and pBD5 represent control plasmids that do not generate antisense RNA. Figure taken from ref. 6.

The effect of the functional ribosomal binding site on the efficacy of antisense RNA is summarized in TABLE 2. Introduction of the functional P_L RBS increases the effectiveness of antisense RNA. The effect of introducing a transcription terminator in the vector just downstream of the antisense insert is summarized in TABLE 3. Introduction of the transcription terminator usually enhances inhibition, so that with some vectors (pBD7 and pBD9) virtually 100% inhibition of enzyme synthesis is obtained by antisense RNA. It is noteworthy that in addition to inhibition of β-galactosidase the two downstream enzymes of the *lac* operon, galactoside permease and thiogalactoside transacetylase, were also inhibited by these same vectors.[6]

Sense and antisense RNA levels were examined in cells with various vectors grown at 42°C. Apparently the more effective vectors produce more antisense RNA

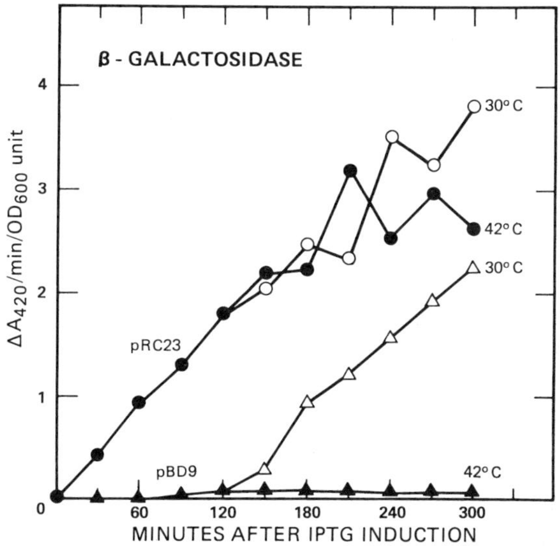

FIGURE 5. Reversibility of the effect of antisense RNA on β-galactosidase synthesis. Enzyme levels were determined as a function of time after cells were transferred to 42°C in the presence of IPTG. At 120 minutes after induction, one half of each culture was transferred to 30°C to shut off synthesis of antisense RNA and enzyme levels followed for an additional 180 minutes. Unfilled symbols, 30°C; filled symbols, 42°C; triangles (△, ▲), pBD9, the antisense plasmid; circles (○, ●), pRC23, the control plasmid.

(FIG. 7). Specifically, introduction of the transcription terminator into vectors pBD8 and pBD9 (pBD8t and pBD9t) leads to the production of more antisense RNA. The level of sense RNA was inversely proportional to the level of antisense RNA present at steady-state levels.

CONCLUSIONS AND COMMENTS

The results presented here indicate that maximum inhibition of protein synthesis *in vivo* can be obtained by optimal configuration and targeting of the anti-mRNA. The inhibition of translation is likely to be the direct result of duplex formation between the anti-mRNA and the normal mRNA. The double-stranded RNA seg-

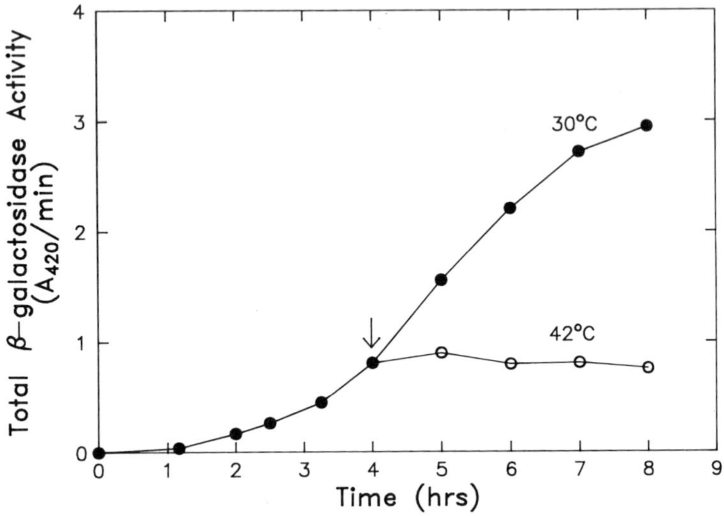

FIGURE 6. Ability of antisense mRNA to shut off β-galactosidase activity in IPTG-induced cells. Cells containing pBD9t[6] were grown logarithmically at 30°C. To initiate the experiment, at time = 0, IPTG was added to a concentration of 1 mM. After growth for 4 hours at 30°C, half the culture was shifted to 42°C (*arrow*). The total β-galactosidase level was determined as a function of time. ●, 30°C; ○, 42°C. Controls in which IPTG was removed by washing followed by incubation at 30°C or 42°C showed curves similar to the 42°C points in the presence of IPTG (*open circles*). Figure taken from ref. 6.

ment blocks protein synthesis. For optimal inhibition the anti-mRNA must contain a functional (sense) ribosome binding site on the 5′ end and be complementary to the 5′ end of the mRNA. In addition, a transcription terminator following the antisense DNA sequence enhances the efficiency of antisense RNA synthesis. The anti-

TABLE 2. Effects of P_L RBS on Anti-mRNA Efficiency (Inhibition of β-Galactosidase Synthesis)[a]

			Inhibition (%)	
			P_L RBS	
Plasmid	*lac*Z	*lac*Z RBS	No	Yes
pBD4	5′	−	ND	98
pBD5/23	—	−	0	0
pBD6/7	5′	−	64	96
pBD8/9	5′	+	60	98
pBD10/11a	None	+	55	94
pBD19/pYL1	3′	−	68	81
pBD20	5′ + 3′	+	ND	80

[a]The plasmid constructions are shown in Figure 3. The column headed *lac*Z refers to the coding region targeted by the antisense RNA. The column headed *lac*Z RBS refers to the presence or absence of the *lac*Z RBS on the antisense RNA to target this area. Inhibition of β-galactosidase in the presence or absence of a functional ribosome binding site (P_L RBS) at the 5′ end of the antisense RNA is shown as a percentage of the control. ND = not done.

TABLE 3. Effect of Transcription Termination on Inhibition of β-Galactosidase Synthesis by Anti-*lac*Z mRNA[a]

		Inhibition (%)	
		Transcription Terminator	
Plasmid	Target	No	Yes
pBD6	5′	64	70
pBD7	5′	96	100
pBD8	5′	60	71
pBD9	5′	98	100
pBD11	5′	(Lethal)	91
pBD11a	5′	89	94
pYL1	3′	81	81
pBD19	3′	68	68

[a] The plasmid constructions are shown in FIGURE 3. The column headed "Target" refers to the coding region targeted by the antisense RNA. Inhibition (%) represents that of β-galactosidase synthesis in the presence or absence of the transcription terminator in the constructs shown.

mRNAs produced in *E. coli* 294/pRKcIts248, which exhibited the greatest inhibition of β-galactosidase synthesis, were transcribed from plasmids containing the λ P_L promoter and a synthetic functional ribosome binding site followed by an initiation codon ATG. The presence of a functional ribosome binding site and an ATG on the

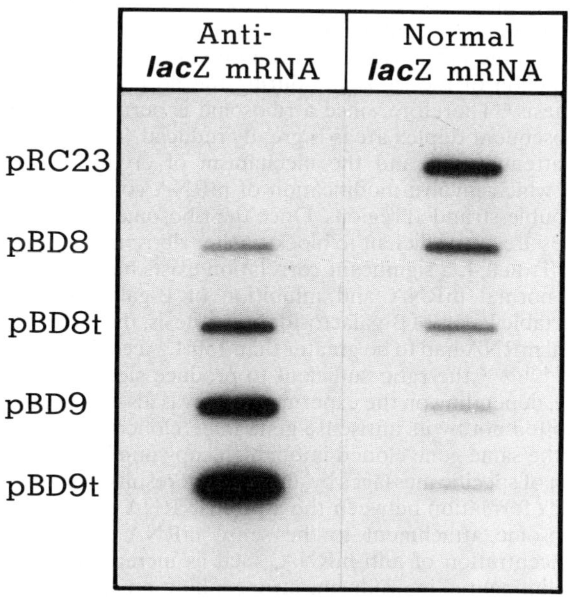

FIGURE 7. Effect of insertion of a transcription terminator on the level of antisense and sense β-galactosidase RNA. The amount of total RNA loaded was 25 ng for detection of antisense RNA and 100 ng for detection of normal (sense) *lac*Z mRNA. The probes used were described.[6] X-ray films were exposed for 2 hours at −70°C.

anti-mRNA favors functional interaction of ribosomes and inteferes with the process of mRNA degradation. This is consistent with previous results[47] which indicated that specific mRNAs can be stabilized *in vivo* by either a ribosome covering the initiation site or the preservation of the polysome configuration on the mRNA. Measurement of levels of *lac*Z mRNA and anti-*lac*Z mRNA in *E. coli* containing these plasmids is consistent with these results.

Only an anti-mRNA that is complementary to the 5′ end of the mRNA can inhibit translation maximally. This region of the gene codes for the NH_2-terminus of the protein as well as the ribosome binding site. Base pairing of this region of the message with a complementary anti-mRNA probably prevents ribosomes from initiating protein synthesis. It is interesting that two plasmids, pBD7 and pBD9, show inhibition of β-galactosidase synthesis of 96% and 98%, respectively, even without a transcription terminator (TABLE 3). Plasmid pBD7 transcribes an anti-mRNA that is complementary to codons 5–119 of β-galactosidase. Plasmid pBD9 transcribes an anti-mRNA that is complementary to codons 1–119 as well as the initiation codon ATG and an additional 12 untranslated nucleotides that include the lacZ ribosome binding site. These two plasmids show a difference in inhibition of β-galactosidase synthesis of only 2%. As it is estimated[48,49] that a ribosome can cover about 40 nucleotides of the mRNA, a duplex including the fifth codon and distal downstream portion of the mRNA may prevent the ribosome from binding efficiently. The necessity that the anti-mRNA be complementary to the 5′ end of the gene has also been deduced by Coleman *et al.*[50] and Melton.[51]

When an anti-mRNA complementary to codons 278–1005 of β-galactosidase is produced (pBD19, pYL1), a lower level of inhibition (80%) is observed. This implies that ribosomes can bind and initiate translation, but they are slowed down after reaching the area of the RNA-RNA duplex. However, as there is a significant level of inhibition, it seems that the duplex is capable of forming, but the ribosome may be capable of stripping the anti-mRNA, as proposed also by Melton.[51] We postulated that this is a function of the ribosome during the normal translocation events of protein biosynthesis.[6] Therefore, once a ribosome is permitted to bind to mRNA, inhibition by subsequent duplex areas is greatly reduced. This is consistent with the mechanism of attenuation[52] and the mechanism of erythromycin-induced resistance,[53] both of which involve modification of mRNA conformation by ribosomes that can open double-stranded regions. Once the ribosomes bind to the mRNA, the adjacent duplexes are not sufficient to block further ribosome movement.

As shown in TABLE 4, a significant correlation exists between the molar ratio of anti-mRNA to normal mRNA and inhibition of β-galactosidase synthesis. To produce undetectable levels of β-galactosidase synthesis, the molar ratio of anti-lacZ mRNA to normal mRNA had to be greater than 150:1. According to other reports, in eukaryotic cells[2,4,51,54–58] the ratio sufficient to produce significant inhibition varied from 1:1 to 600:1, depending on the experiments. It was also shown that expression of OmpF was inhibited not by its antisense gene micF cloned into a low copy number plasmid, but by the same gene cloned into a high copy number plasmid.[59] Inhibition of the translation of specific messages by anti-mRNA results from the competition of the rate of duplex formation between the antisense RNA and the sense RNA with the rate of ribosome attachment to the sense mRNA. Therefore, factors that increase the concentration of anti-mRNA, such as increased stability and rate of synthesis, will be influential in producing strong inhibition of the expression of sense mRNA.

As stated, the presence of a functional ribosome binding site, followed by an ATG on the anti-mRNA, enables ribosomes to bind, probably protecting anti-mRNA from degradative enzymes. Thus, the level of the anti-message may remain sufficiently high for rapid formation of the duplex with normal mRNA. A strong correlation exists between the level of anti-lacZ mRNA and the inhibition of β-galactosidase synthesis. The clones (pBD7 and pBD9) that provided anti-lacZ mRNAs with a functional ribosome binding site at the 5′ end produced very high levels of anti-message and consequently extensive inhibition of β-galactosidase synthesis, compared to the clones (pBD6 and pBD8) that produced anti-message without this functional ribosome binding site[6] (FIG. 7). Furthermore, measurement of the decay rate revealed that the anti-mRNAs with the ribosome binding site at the 5′ end were degraded at one half the rate as an anti-mRNA without the ribosome binding site.[6] This decreased decay rate explains the higher level of anti-message in the clones producing an anti-message with the ribosome binding site at the 5′ end.

An interesting finding is that both plasmids pYL1 and pBD20 (FIG. 3) show virtually identical levels of inhibition of β-galactosidase synthesis (81% and 80%, respectively). This was surprising, as pBD20 contains the 5′ terminal lacZ fragment present in pBD9 as well as the 3′ terminal fragment present in pYL1. Plasmid pYL1 produces an anti-mRNA that is complementary to codons 278–1005 of lacZ. Plasmid pBD20 produces an anti-mRNA that is complementary to codons 1–1005 but that lacks the segment complementary to codons 120–277[1,6] (the region of the secondary repressor binding site). As the transcript generated by pBD20 is about 2.5 times the size of the short transcript of pBD9, its rate of hybridization is somewhat slower than that of the short transcript.[60] In addition, the large size may partly explain a lower level of anti-mRNA. The measurement of the level of anti-mRNA produced by pBD20 was one fifth that produced by pBD9. This could be explained either by a reduced rate of RNA synthesis or an increased rate of decay. Thus, a lower steady-state level of anti-mRNA in pBD20 relative to pBD9 can account for a reduced inhibition of β-galactosidase synthesis. It is noteworthy that the level of normal *lacZ* mRNA seemed to be inversely proportional to the level of anti-*lacZ* mRNA (FIG. 7). It was also shown that *ompF* mRNA[3] and *lpp* mRNA[50] were decreased on induction of *micF* RNA and *mic(lpp)* RNA, respectively. It is likely that the hybrid formed between the normal mRNA and the antisense RNA might destabilize the normal mRNA or cause transcription to terminate prematurely.

TABLE 4. Ratio of Antisense/Sense mRNA and Inhibition of β-Galactosidase Synthesis

Plasmid	Molar Ratio Anti-mRNA:Normal mRNA	Percentage of Inhibition of β-galactosidase Synthesis
pBD6	10-20:1	60
pBD8	10-20:1	60
pBD20	34:1	80
pBD7	100:1	96
pBD9	150:1	98
pBD9t	182:1	100

[a]The constructions are shown in FIGURE 3. Plasmids pBD9 and pBD9t represent the same construction without and with the transcription terminator, respectively. With plasmid pBD9t, β-galactosidase was not detectable in the cultures.[6]

As shown, plasmids containing the phage T4 transcription terminator showed a higher level of inhibition of β-galactosidase synthesis with the exception of pBD19 and pYL1 (TABLE 3). Plasmids pBD7t and pBD9t resulted in undetectable levels of β-galactosidase synthesis in cells containing these plasmids. Hybridization analysis revealed that the anti-mRNA from the plasmids with the terminator present was smaller in size and exhibited a decay rate approximately equal to that of the anti-message from the plasmids without the terminator sequence.[6] Therefore, the increase in inhibition of β-galactosidase synthesis due to these plasmids is most likely a result of the greater rate of hybridization of the short transcripts[60] as well as their quantitative increase.

We have discussed in detail the effect of anti-lacZ mRNA on the translation of β-galactosidase *in vivo.* The manipulation of the *lac* operon in which different segments were inserted in opposite orientation with respect to the lambda P_L promoter provides detailed knowledge about the optimal configuration of anti-sense RNA. It should be interesting to apply this technique to answer specific questions in bacteria and other organisms as well. It is a powerful technique when applied to animal cells where double mutants of a single allele are difficult to obtain routinely. (For examples, see Chang and Stoltzfus,[61] To *et al.,*[62] Izant and Weintraub,[2,54] Kim and Wold,[63] Weintraub *et al.,*[56] Rubenstein *et al.,*[4] Melton,[51] Harland and Weintraub,[64] Wormington,[65] Rosenberg *et al.,*[66] McGarry and Lindquist,[67] Crowley *et al.,*[68] Ch'ng *et al.,*[57] and Stout and Caskey.[58].) Classical mutations alter the structure of the translated product itself. With anti-mRNA, translation (or another process; see below) is blocked by the presence of an untranslatable mRNA template. Moreover, the fact that anti-mRNA can be controlled by an inducible promoter, in effect, turning on and off anti-mRNA at will, provides a powerful tool to examine the function of specific proteins in the physiology and architecture of the cell. Because anti-RNA may be directed to RNA molecules that are not translated (ribosomal RNA, tRNA, intron-exon junctions, RNA precursors, etc.), it is possible that these directed transcripts might serve to inhibit or promote certain other events, such as processing and nuclear-cytoplasmic transport of RNAs. Further studies are delineating these possibilities in many laboratories.

For maximal inhibition of anti-sense RNA in *E. coli,* targeting the 5′ region is best, and covering the ribosome binding site provides further effective enhancement.[50,51,64,69–71] Shorter RNA segments that have a greater rate of hybridization appear to be more effective than longer constructs, but quite large constructs have proved to be effective in animal cells (see this volume). Stabilization of the anti-mRNA by inclusion of a functional ribosome binding site followed by an ATG is necessary to produce a high level of anti-mRNA relative to normal mRNA (100:1) which is an important factor in producing maximal inhibition. By combining all these factors, we have been able to generate anti-sense RNA to reduce β-galactosidase synthesis to virtually undetectable levels (FIGS. 4–6; TABLES 2 and 3). Accordingly, the synthesis of proteins should be able to be effectively blocked by anti-mRNA. Such results can be used to delineate the function of proteins as well as to clone genes when anti-RNA constructs can provide predictable phenotypic changes that could be used for screening or selection. A number of such studies have been reported. The use of anti-RNA in a wide variety of areas is increasing dramatically as scientists envision new targets and new ways to use this technology.

SUMMARY

Antisense RNA was first an *in vitro* curiosity that was found to shut off protein synthesis in cell-free extracts. It was later shown to function in prokaryotic cells as a natural modulator of the synthesis of some proteins. Artificial antisense constructs can inhibit protein synthesis in prokaryotic and eukaryotic cells. To inhibit synthesis of proteins effectively, high ratios of antisense to sense RNAs are required. Thus, the challenge is to develop strategies to locate suitable targets and provide for amplification of the antisense RNA. This report provides a summary of our original work on antisense RNA.

ACKNOWLEDGMENTS

I thank Robert Crowl, Malcolm Casadaban, and Bernard Allet and Walter Fiers for providing, respectively, the following plasmid vectors: pRC21 and pRC23; pMC874; and phage T4-protein 32 transcription terminator. The work from my laboratory described in this summary was carried out by Bruce Daugherty, Kunimoto Hotta, Robert Pestka, Vincent Jung, Chandrika Kumar, Jingdong Zhu, Young-Ik Lee, Paul Gold, and Yong Ho Ahn.

REFERENCES

1. PESTKA, S., B. L. DAUGHERTY, V. JUNG, K. HOTTA & R. K. PESTKA. 1984. Proc. Natl. Acad. Sci. USA **81:** 7525–7528.
2. IZANT, J. G. & H. WEINTRAUB. 1984. Cell **36:** 1007–1015.
3. MIZUNO, T., M.-Y. CHOU & M. INOUYE. 1984. Proc. Natl. Acad. Sci. USA **81:** 1966–1970.
4. RUBENSTEIN, J. L. R., J.-F. NICOLOS & F. JACOB. 1984. C. R. Acad. Sci. (Series 3) **299:** 271–274.
5. SINGER, M. F., O. W. JONES & M. W. NIRENBERG. 1963. Proc. Natl. Acad. Sci. USA **49:** 392–399.
6. DAUGHERTY, B. L., K. HOTTA, C. KUMAR, Y. H. AHN, J. ZHU & S. PESTKA. 1989. Gene Analysis Techniques **6:** 1–16.
7. PATERSON, B. M., B. E. ROBERTS & E. L. KUFF. 1977. Proc. Natl. Acad. Sci. USA **74:** 4370–4374.
8. HASTIE, N. D. & W. A. HELD. 1978. Proc. Natl. Acad. Sci. USA **75:** 1217–1221.
9. TOMIZAWA, J., T. ITOH, G. SELZER & T. SOM. 1981. Proc. Natl. Acad. Sci. USA **78:** 1421–1425.
10. TOMIZAWA, J. I. & T. ITOH. 1982. Cell **31:** 575–583.
11. LACATENA, R. M. & G. CESARENI. 1981. Nature **294:** 623–626.
12. CESARENI, G., M. HELMER-CITTERICH & L. CASTAGNOLI. 1991. Trends Genet. **7:** 230–235.
13. EGUCHI, Y., T. ITOH & J. TOMIZAWA. 1991. Annu. Rev. Biochem. **60:** 631–652.
14. LIN-CHAO, S. & S. N. COHEN. 1991. Cell **65:** 1233–1242.
15. NORDSTROM, K. 1990. Cell **63:** 1121–1124.
16. BRADY, G., J. FREY, H. DANBARA & K. N. TIMMIS. 1983. J. Bacteriol. **154:** 429–436.
17. LIGHT, J. & S. MOLIN. 1983. EMBO J. **2:** 93–98.
18. NORDSTROM, K., S. MOLIN & J. LIGHT. 1984. Plasmid **12:** 71–90.
19. KUMAR, C. C. & R. P. NOVICK. 1985. Proc. Natl. Acad. Sci. USA **82:** 638–642.
20. SIMONS, R. W. & N. KLECKNER. 1983. Cell **34:** 683–691.
21. SIMONS, R. W., B. C. HOOPES, W. R. MCCLURE & N. KLECKNER. 1983. Cell **34:** 673–682.
22. MIZUNO, T., M.-Y. CHOU, & M. INOUYE. 1983. Proc. Jpn. Acad. **59:** 335–338.
23. SHINE, J. & V. DELGARNO. 1974. Proc. Natl. Acad. Sci. USA **71:** 1342–1346.
24. POULSEN, L. K., A. REFN, S. MOLIN & P. ANDERSSON. 1991. Mol. Microbiol. **5:** 1639–1648.
25. MUNROE, S. H. & M. A. LAZAR. 1991. J. Biol. Chem. **266:** 22083–22086.
26. MURRAY, V. & R. HOLLIDAY. 1979. FEBS Lett. **106:** 5–7.
27. LERNER, M. R., J. A. BOYLE, S. M. MOUNT, S. L. WOLIN & J. A. STEITZ. 1980. Nature **283:** 220–224.
28. ROGERS, J. & R. WALL. 1980. Proc. Natl. Acad. Sci. USA **77:** 1877–1879.
29. SPENCER, C. A., R. D. GIETZ & R. B. HODGETTS. 1986. Nature **322:** 279–281.
30. WILLIAMS, T. & M. FRIED. 1986. Nature **322:** 275–279.
31. ROSSI, J. J., E. M. CANTIN, J. A. ZAIA, P. A. LADNE, J. CHEN, D. A. STEPHENS, N. SARVER & P. S. CHANG. 1990. Ann. N. Y. Acad. Sci. **616:** 184–200.

32. WEERASINGHE, M., S. E. LIEM, S. ASAD, S. E. READ & S. JOSHI. 1991. J. Virol. **65:** 5531–5534.
33. GOODCHILD, J. & V. KOHLI. 1991. Arch. Biochem. Biophys. **284:** 386–391.
34. SIOUD, M. & K. DRLICA. 1991. Proc. Natl. Acad. Sci. USA **88:** 7303–7307.
35. LORENTZEN, E. U., U. WIELAND, J. E. KUHN & R. W. BRAUN. 1991. Virus Genes **5:** 17–23.
36. TABLER, M. & M. TSAGRIS. 1991. Gene **2:** 175–183.
37. BRANCH, A. D. & H. D. ROBERTSON. 1991. Proc. Natl. Acad. Sci. USA **88:** 10163–10167.
38. STEPHENSON, M. L. & P. C. ZAMECNIK. 1978. Proc. Natl. Acad. Sci. USA **75:** 285–288.
39. ZAMECNIK, P. C. & M. L. STEPHENSON. 1978. Proc. Natl. Acad. Sci. USA **75:** 280–284.
40. ZAMECNIK, P. C., J. GOODCHILD, Y. TAGUCHI & P. S. SARIN. 1986. Proc. Natl. Acad. Sci. USA **83:** 4143–4146.
41. JAYARAMAN, K., K. MCPARLAND, P. MILLER & P. O. P. TS'O. 1981. Proc. Natl. Acad. Sci. USA **78:** 1537–1541.
42. WICKERSTROM, E. L., T. BACON, A. GONZALEZ, G. H. LYMAN, D. L. FREEMAN & E. WICKSTROM. 1988. Proc. Natl. Acad. Sci. **85:** 1028–1032.
43. GREEN, P. J., O. PINES & M. INOUYE. 1986. Ann. Rev. Biochem. **55:** 569–597.
44. BASERGA, R. 1990. Cancer Res. **50:** 6769–6771.
45. COLMAN, A. 1990. J. Cell. Sci. **97:** 399–409.
46. CROWL, R. 1986. *In* Methods in Enzymology. S. Pestka, ed. Vol. **119:** 376–383. Academic Press. New York, NY.
47. WALKER, A. C., M. L. WALSH, D. PENNICA, P. S. COHEN & H. L. ENNIS. 1975. Proc. Natl. Acad. Sci. USA **73:** 1126–1130.
48. STEITZ, J. A. 1969. Nature **224:** 957–964.
49. PLATT, T. & C. YANOFSKY. 1975. Proc. Natl. Acad. Sci. USA **72:** 2399–2403.
50. COLEMAN, J., P. J. GREEN & M. INOUYE. 1984. Cell **37:** 429–436.
51. MELTON, D. A. 1985. Proc. Natl. Acad. Sci. USA **82:** 144–148.
52. YANOFSKY, C. 1981. Nature **289:** 751–758.
53. HORINOUCHI, S. & B. WEISBLUM. 1980. Proc. Natl. Acad. Sci. USA **77:** 7079–7083.
54. IZANT, J. G. & H. WEINTRAUB. 1985. Science **229:** 345–352.
55. HOLT, J. T., T. V. GOPAL, A. D. MOULTON & A. W. NIENHUIS. 1986. Proc. Natl. Acad. Sci. USA **83:** 4794–4798.
56. WEINTRAUB, H., J. G. IZANT & R. M. HARLAND. 1985. Trends Genet. **1:** 22–25.
57. CH'NG, J. L., R. C. MULLIGAN, P. SCHIMMEL & E. W. HOLMES. 1989. Proc. Natl. Acad. Sci. USA **86:** 10006–10010.
58. STOUT, J. T. & C. T. CASKEY. 1990. Somat. Cell. Mol. Genet. **16:** 369–382.
59. MATSUYAMA, S. & S. MIZUSHIMA. 1985. J. Bacteriol. **162:** 1196–1202.
60. WETMUR, J. G. & N. DAVIDSON. 1968. J. Mol. Biol. **31:** 349–370.
61. CHANG, L.-J. & C. M. STOLTZFUS. 1985. Mol. Cell. Biol. **5:** 2341–2348.
62. TO, R. Y.-L., S. C. BOOTH & P. E. NEIMAN. 1986. Mol. Cell. Biol. **6:** 4758–4762.
63. KIM, S. K. & B. J. WOLD. 1985. Cell **42:** 129–138.
64. HARLAND, R. & H. WEINTRAUB. 1985. J. Cell Biol. **101:** 1094–1099.
65. WORMINGTON, W. M. 1986. Proc. Natl. Acad. Sci. USA **83:** 8639–8643.
66. ROSENBERG, U. B., A. PREISS, E. SEIFERT, H. JACKLE & D. C. KNIPPLE. 1985. Nature **313:** 703–706.
67. MCGARRY, T. J. & S. LINDQUIST. 1986. Proc. Natl. Acad. Sci. USA **83:** 399–403.
68. CROWLEY, T. E., W. NELLEN, R. H. GOMER & R. A. FIRTEL. 1985. Cell **43:** 633–641.
69. GREEN, P. J., C. MITCHELL & M. INOUYE. 1985. Biochimie **67:** 763–767.
70. GREEN, P. J., O. PINES & M. INOUYE. 1986. Ann. Rev. Biochem. **55:** 569–597.
71. HIRASHIMA, A., S. SAWAKI, Y. INOKUCHI & M. INOUYE. 1986. Proc. Natl. Acad. Sci. USA **83:** 7726–7730.

Cleavage of Inhibin α Subunit mRNA by Engineered Ribozyme

L. GE AND A. W. SEITZ

Department of Veterinary and Animal Sciences
University of Massachusetts
Amherst, Massachusetts 01003

Ribozymes are small RNA molecules that cleave specific RNA molecules enzymatically. Haseloff and Gerlach[1] constructed ribozymes based on the hammerhead model of self-cleaving RNAs, in which most of the conserved nucleotides were arranged on the ribozyme, leaving only three required nucleotides, GUC, GUA, or GUU, on the RNA substrate. Such engineered hammerhead ribozymes have been shown to cleave chloramphenicol acetyltransferase (CAT) mRNA and HIV-1 gag mRNA in both cell-free systems and cell cultures.[1-3]

We designed and constructed a ribozyme to regulate the expression of inhibin. Inhibin is a peptide hormone that suppresses synthesis and release of follicle-stimulating hormone (FSH). When inhibin levels in ewes and rats were decreased using immunization methods, FSH levels were elevated and ovulation rates increased. We predict that a decrease in inhibin synthesis will increase FSH levels and oocyte production. Inhibin consists of two subunits: α and β. The α subunit is unique to inhibin. Two β subunits form another peptide hormone, activin, which stimulates FSH release. Our ribozyme was designed to cleave inhibin α subunit mRNA.

The three required nucleotides, GUA, GUC, and GUU, are present in several sites in inhibin α subunit mRNA, allowing a choice of several cleavage sites. Comparison of known sequences in rat, bovine, porcine, and human inhibin α subunit mRNA showed a short identical or highly homologous sequence of about 20 nucleotides flanking a GUC sequence. Our ribozyme was chosen to recognize this site in order to cleave the inhibin α subunit mRNA in other species.

The ribozyme was synthesized as double-stranded oligodeoxynucleotides and cloned into the plasmid pGEM-4Z. *In vitro* transcripts of this cloned ribozyme and the cloned rat inhibin α subunit cDNA were incubated together at 50°C. Northern blot analysis of the incubations revealed two extra bands representing cleavage products of the expected size (FIG. 1). Cleavage was specific, as no cleavage of antisense inhibin α subunit transcript was observed (FIG. 1). The results demonstrate that the flanking sequences of the cleavage site are accessible to the ribozyme.

This ribozyme is expected to be active *in vivo,* as the same cleavage was observed also at 37°C (data not shown).

The ribozyme was designed before mouse inhibin α subunit cDNA was sequenced. The results also showed that this ribozyme cleaves mouse inhibin α subunit mRNA in ovarian mRNA *in vitro* (data not shown). It suggests that it is possible to design a ribozyme to target conserved sequences of a given gene in order to cleave the transcripts of the same gene in other species in which the DNA sequence is unknown.

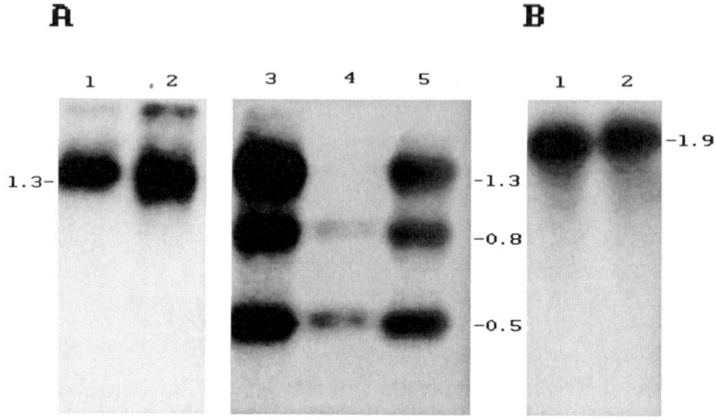

FIGURE 1. Northern blot of *in vitro* cleavage. The *in vitro* transcript of Mlu I linearized plasmid rINA-13,[4] a cDNA clone containing rat inhibin α cDNA (a kind gift from Dr. Kelly Mayo), is about 1.3 kb. The expected cleavage products are approximately 0.8 and 0.5 kb. The ribozyme and inhibin α subunit *in vitro* transcripts were incubated in 50 mM Tris.Cl (pH 7.5) and 20 mM $MgCl_2$ at both 50°C and 37°C for 1 hour. The cleavage products were separated on 3.5% Nusieve GTG agarose/formaldehyde gel, transferred to nylon membranes, and hybridized with ^{32}P-labeled rat inhibin α subunit cDNA. (**A**) *Lanes 1 and 2:* substrate incubated without the ribozyme; *lanes 3, 4, and 5:* substrate incubated with the ribozyme. The molar ratio of ribozyme to the substrate was 30:1. (**B**) *Lane 1:* antisense inhibin α subunit mRNA incubated without the ribozyme; *lane 2:* antisense inhibin α mRNA incubated with the ribozyme. All samples were incubated at 50°C for 1 hour.

REFERENCES

1. Haseloff, J. & W. L. Gerlach. 1988. Nature **334:** 585–591.
2. Cameron, F. H. & P. A. Jennings. 1989. Proc. Natl. Acad. Sci. USA **86:** 9139–9143.
3. Sarver, N., E. M. Cantin, P. S. Chang, J. A. Zaia, P. A. Ladne, D. A. Stephens & J. J. Rossi. 1990. Science **247:** 1222–1225.
4. Woodruff, T. K., G. Meunier, P. B. Jones, A. J. W. Hsueh & K. E. Mayo. 1987. Mol. Endocrinol. **1:** 561–568.

Catalytic RNA Molecules and Their Cleavage of Viral mRNA

SIEW PENG HO AND DUSTIN H. O. BRITTON

Du Pont Merck Pharmaceuticals
P.O. Box 80328
Wilmington, Delaware 19880–0328

Hammerhead ribozymes are catalytic RNA molecules originally discovered in plant pathogens. Although these molecules are self-cleaving in nature, fundamental studies by others have led to the development of small *trans*-acting ribozymes with full activity.[1,2]

Hammerhead ribozymes can easily be designed to target and cleave other RNA molecules, and our goal is to develop such ribozymes into tools and therapeutic agents. Our initial targets include RNA from human immunodeficiency virus (HIV) and cytomegalovirus (CMV). In our HIV work, we first looked at cleavage by two different ribozyme designs (FIG. 1) on RNA from a region in the gag gene of HIV. Our initial targets were small, chemically synthesized RNA oligonucleotides. We found the Uhlenbeck design[1] to be totally inactive, whereas the Gerlach design[2] showed good activity. A subsequent report showed that a crucial base pair (in bold) had at least a 100-fold greater activity if it was a C-G compared to an A-U base pair.[3] Therefore, the presence of the A-U base pair at this site accounts for the inactivity of the Uhlenbeck design ribozyme.

We next looked at cleavage by this active 32-nt Gerlach design ribozyme (R1) on several HIV gag transcripts. With T7 RNA polymerase and a plasmid linearized with either *Pst*1 or *Hind*III, internally labeled transcripts (800 and 465 nts) containing the cleavage site of interest were made. These transcripts were allowed to react with R1 at 37°C for 3 hours. Specific cleavage was observed, as demonstrated by the two product fragments obtained from each transcript (FIG. 2). RNA molecular weight markers enabled size estimates of the cleavage fragments to be made, and this confirmed cleavage at the expected sites. Suboptimal concentrations of transcript led to slower cleavage kinetics (compared to smaller substrates).

We designed another ribozyme (R2) targeted to a site that is 15 nts downstream of the start codon of the gag gene. This ribozyme was also active, again producing only two cleavage fragments of the expected size (gel not shown). When R1 and R2 were mixed together, a fifth band in addition to the four expected fragments was observed. This extra band most likely resulted from cleavage of the 331-nt fragment

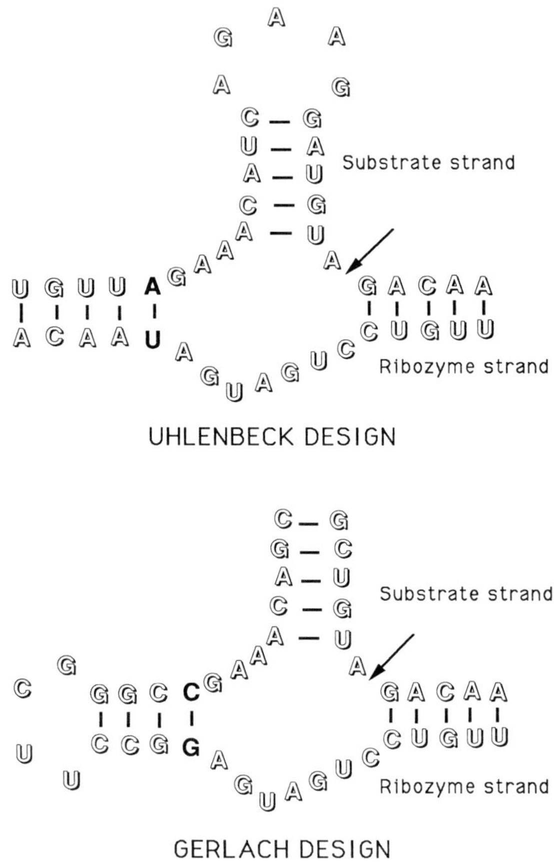

FIGURE 1. Uhlenbeck and Gerlach ribozyme designs. The substrate strand sequence is from a portion of the gag gene of HIV. The position of cleavage is indicated by the *arrow.* The crucial base pair is shown in *bold.*

by R2, as this fragment contains the R2 cleavage site. This hypothesis is confirmed by the relative intensity of the 331-nt fragment which is unexpectedly less intense than the other fragments.

We also demonstrated multiple substrate turnover (shown with various CMV sequences). At conditions that were far from saturating ribozyme concentrations, we were able to obtain 21 turnovers of substrate per ribozyme per hour. We are currently developing HIV ribozymes for cell culture experiments.

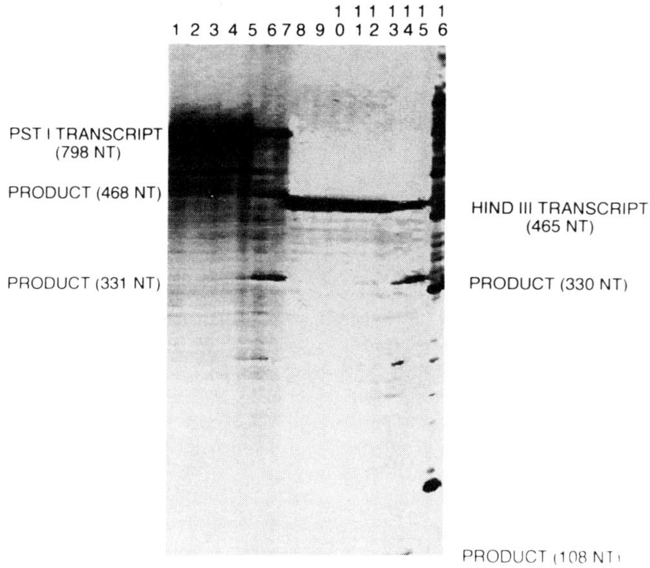

FIGURE 2. Cleavage of 800- and 465-nt transcripts by R1. *Lanes 1,8:* transcript + R1 without Mg^{2+}. *Lanes 2–7, 9–15:* transcript + R1 + varying amounts of Mg^{2+}. Cleavage is seen only in *lanes 6–7 and 14–15* where the Mg^{2+} concentrations are at or above 10 mM. *Lane 16:* RNA ladder. Reactions are done at 37°C, pH 7.5. Ribozyme concentration is 0.1 μM, whereas transcript concentrations are considerably lower.

REFERENCES

1. Uhlenbeck, O. C. 1987. Nature **328:** 596–600.
2. Haseloff, J. & W. L. Gerlach. 1988. Nature **334:** 585–591.
3. Ruffner, D. E., G. D. Stormo & O. C. Uhlenbeck. 1990. Biochemistry **29:** 10695–10702.

Antisense RNA-Mediated Inhibition of the Replication of the Human Immunodeficiency Virus Type 1

Discussion of Critical Parameters

GEORG SCZAKIEL, MICHAEL PAWLITA,
KAROLA RITTNER, AND MATTHIAS HOMANN

Deutsches Krebsforschungszentrum
Angewandte Tumorvirologie
Abteilung Genomveränderungen und Cancerogenese
Im Neuenheimer Feld 280
D-6900 Heidelberg, Germany

The acquired immunodeficiency syndrome (AIDS) is associated with infection by human immunodeficiency viruses (HIV)[1] that show tropism predominantly for hematopoietic cell lineages (e.g., lymphocytes and macrophages) which differentiate from bone marrow stem cells. One way of making hematopoietic cells resistant to replication of HIV-1 could be the intracellular expression of an antivirally active RNA or protein, a therapeutic principle that has been termed "intracellular immunization."[2] Tools for this therapeutic principle include antisense nucleic acids (for review see ref. 3), particularly the intracellular expression of antisense RNA.[2,4] In HIV-1, antiviral effects of intracellular antisense RNA expression were demonstrated,[5–7] even though inhibition under most of the experimental conditions was not complete.

After microinjection into human cells of infectious proviral HIV-1 DNA together with antisense RNA expression plasmids or *in vitro* synthesized antisense RNA, the extent of virus production (measured by HIV-1 antigen ELISA) indicates inhibitory effects. Different target regions for antisense RNA ranging in size from about 80 nts to more than 3,000 nts were tested. Sequences from the 5′-leader/gag region and from the first coding tat and rev exons exhibited strongest inhibition in this transient assay as well, when stably expressed in antisense orientation in human T-lymphoid Jurkat cell clones (FIG. 1). In the latter model system, inhibition of HIV-1 replication can be expressed as a shift of virus spread in the culture along the time axis in antisense RNA producing Jurkat clones versus control cells; that is, infection finally broke through under the applied high infective HIV-1 doses.[4] However, when the dose of infectious particles was smaller than 0.1 per cell, we did not observe virus production for up to 5 weeks in one of the pAR6-derived clones.

It is reasonable to assume that two characteristics of the subgenomic or subgenic target sequences contribute to the extent of inhibition in general and specifically in the case of HIV-1: functional elements (e.g., translational start sites, splice sites) and secondary structure motifs that were shown in prokaryotic systems to influence the hybridization kinetics between the antisense and the target strand *in vitro*. This was found to correlate with antisense RNA-mediated biological effects.[8] In analogy, we measured hybridization kinetics *in vitro* between pAR6-derived antisense RNA and the complementary strand (pSR6) as a model target RNA. Comparison of the measured k_2 values in this study with published k_2 values for other antisense RNA-regulated systems (TABLE 1) suggested that for the pSR6/antisense RNA system (pAR6, αY150), k_2 values could be further increased by modifying pAR6 constructs. An increase in k_2 might be linked with improved antiviral effects.

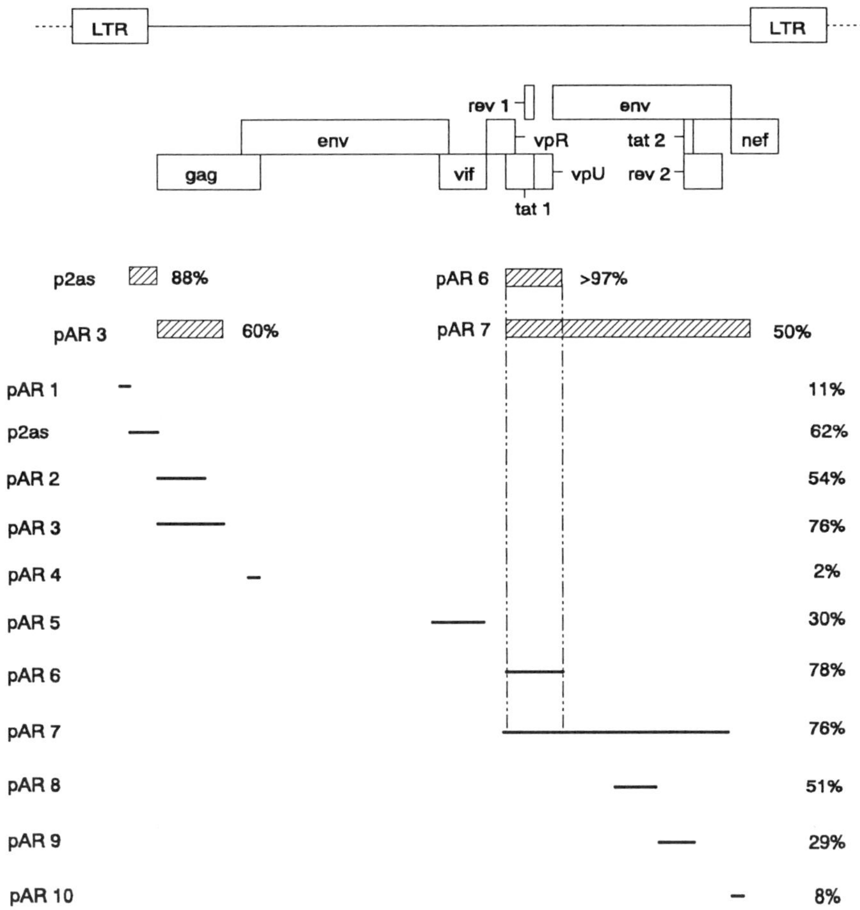

FIGURE 1. Dependence of the extent of antisense RNA-mediated inhibition of HIV-1 replication on the subgenomic viral target region. Antisense RNA constructs were tested in human T-lymphoid Jurkat clones selected for stable integration of the antisense RNA expression cassette (*hatched bars*) or in human epithelioid SW480 cells co-microinjected together with infectious proviral DNA (*solid bars*). Inhibition data were derived from refs. 4 and 11.

For pAR6 RNA and pAR6-derived shortened RNA (αY150) which is the same range of length as that of most naturally occurring antisense RNAs in prokaryotic systems, we assume that those structural elements of antisense and target RNAs that are important for RNA double-strand formation are not optimally "fitted." For this

TABLE 1. Antisense RNA-Mediated Biological Effects and *in Vitro* Determined Time Constants (k_2) for Hybridization Kinetics of Selected Antisense RNA-Regulated Systems

Antisense RNA-Regulated Biological System	Target RNA	Antisense RNA (length [nts])	$k_2[M^{-1} \times s^{-1}]^a$	Reference
Col E1 plasmid replication in *Escherichia coli*	RNA II primer for plasmid replication	RNA I (108)	7×10^5	Tomizawa *et al.*[9]
Transposition of TN10 in *E. coli*	mRNA IN coding for transposase	RNA OUT (70)	3×10^5	Kleckner *et al.*[10]
Inc F II plasmid replication in *E. coli*	mRNA CX (copT) coding for repA, replication initiator	copA RNA (90)	1×10^6	Persson *et al.*[8]
HIV-1 replication in human cells	SR-6 viral subregion (see Fig. 1)	αY150[b] (150)	1.3×10^4	This work
HIV-1 replication in human cells	SR-6 viral subregion (see Fig. 1)	AR-6 (562)	$2\text{-}2.5 \times 10^{4c}$	This work

[a] k_2 determined by hybridization at 37°C *in vitro* under physiological ion strength.
[b] αY150 is a 150-nt subregion of AR-6.
[c] Determined by a modified hybridization assay, which cannot distinguish between intermediates and final products of hybridization. Therefore, the true value for k_2 might be overestimated in this work.

reason we are attempting to analyze and modify structural elements of pAR6-derived antisense RNAs.

REFERENCES

1. BARRE-SINOUSSI, F., J. C. CHERMAN, R. REY, M. T. NUGERYRE, S. CHAMARET, J. GRUEST, C. DAUGET, C. AXLER-BLIN, F. VERZINET-BRUN, C. ROUZIOUX, W. ROSENBAUM & L. MONTAGNIER. 1983. Science **220:** 868–871.
2. BALTIMORE, D. 1988. Nature (Lond.) **335:** 395–396.
3. JAMES, W. 1991. Antiviral Chem. Chemother. **2:** 191–214.
4. SCZAKIEL, G., K. RITTNER & M. PAWLITA. 1991. *In* Prospects for Antisense Nucleic Acids Therapy of Cancer and AIDS. E. Wickstrom, ed.: 179–193. Wiley-Liss. New York, NY.
5. SCZAKIEL, G., A. KLEINHEINZ & M. PAWLITA. 1989. Biochem. Biophys. Res. Comm. **169:** 643–651.
6. RHODES, A. & W. JAMES. 1990. J. Gen. Virol. **71:** 1965–1971.
7. SCZAKIEL, G. & M. PAWLITA. 1991. J. Virol. **65:** 468–472.
8. PERSSON, C., E. G. H. WAGNER & K. NORDSTRÖM. 1988. EMBO J. **7:** 3279–3288.
9. TOMIZAWA, J. 1984. Cell **38:** 861–870.
10. KLECKNER, N., J. D. KITTLE & R. W. SIMONS. 1987. *In* Molecular Biology of RNA: New Perspectives. M. Inouye & B. Dudock, eds.: 413–421. Academic Press. Orlando.
11. RITTNER, K. & G. SCZAKIEL. 1991. Nucl. Acids Res. **19:** 1421–1426.

Inhibition of HIV-1 Replication by Novel Multitarget Ribozymes

CHANG-JIE CHEN, AKHIL C. BANERJEA,
KARL HAGLUND, GEORGE G. HARMISON,
AND MANFRED SCHUBERT

Laboratory of Viral and Molecular Pathogenesis
National Institute of Neurological Disorders and Stroke
National Institutes of Health
Bethesda, Maryland 20892

Ribozymes are small catalytic RNA molecules that can be targeted to cleave a substrate RNA at a specific nucleotide sequence.[1] Expression of a single ribozyme can inhibit HIV-1 replication[2] or revert the transformed phenotype of a cell by inhibiting expression of the fos oncogene.[3] It is anticipated, however, that the effective use of ribozymes, particularly against RNA viruses, will be limited to only a few rounds of viral replication because of the high mutation rate of these viruses, allowing a rapid selection of resistant virus variants, probably even in the presence of the ribozyme. In addition, any therapeutic approach that employs ribozymes also has to be effective against preferably all virus isolates.

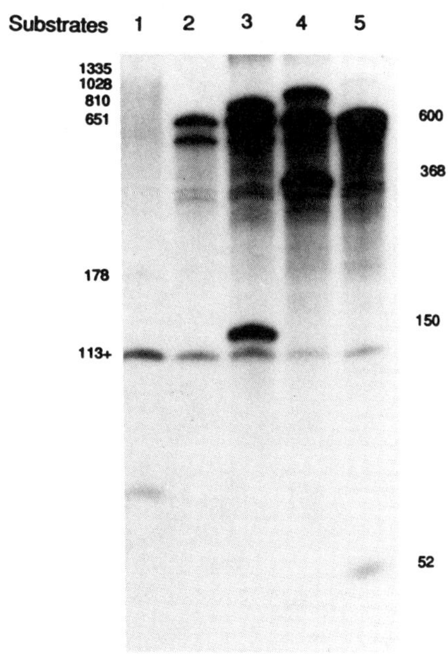

FIGURE 1. A nested set of five different sized T7 RNA polymerase run-off transcripts of the HIV env gene, ranging in size from 178 nt (*lane 1*), 651 nt (*lane 2*), 810 nt (*lane 3*), 1028 nt (*lane 4*) to 1335 nt (*lane 5*), were synthesized. These RNAs were incubated as substrates for the pentaribozyme Rz63, which was targeted to cleave the largest 1335-nt substrate at five different sites. Within the context of our antiviral strategy, all five target sites had been selected from conserved regions within the gp120 region of the HIV env gene. Substrates and ribozyme were combined at a 1:1 ratio and incubated for 60 minutes at 37°C and 15 minutes at 55°C. In *lanes 2 through 4,* the slowest migrating RNA bands correspond to the remaining undigested substrate. Note the complete absence of the largest 1335-nt substrate RNA in *lane 5* and the presence of specific, complete, and partial digestion products. The 1335-nt substrate contains all five target sites and was completely cleaved multiple times.

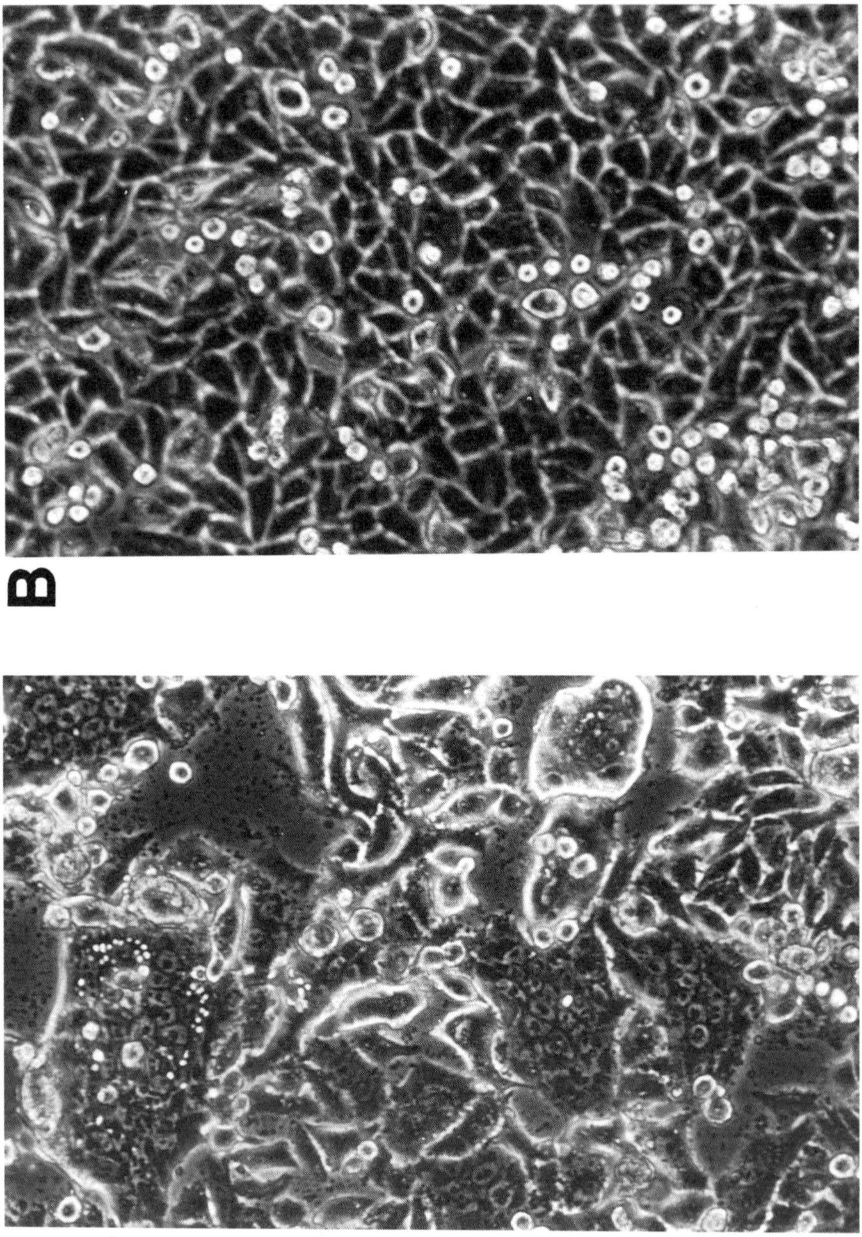

FIGURE 2.

FIGURE 2. Inhibition of HIV-1 induced syncytia formation and p24 antigen release (not shown) by expression of a novel multitarget ribozyme. A nonaribozyme, targeted to cleave within the env region of HIV-1 RNA, was constructed and expressed under control of the HIV-1 LTR. HeLa T4 cells, which express the HIV receptor CD4, were transfected for 4 days with the infectious DNA clone of HIV, pNL 4-3, with (**B**) or without (**A**) the nonaribozyme encoding DNA construct. HIV-1 replication was dramatically inhibited, as evidenced by the dramatic reduction in syncytia formation and the strong reduction in the amount of viral p24 antigen released from these cells (data not shown).

To find answers to these most critical challenges, we constructed mono-, di-, penta-, and nonaribozymes that are targeted to cleave HIV-1 at up to nine different, relatively conserved sites within the env region. These multitarget-hammerhead ribozymes were functional *in vitro,* and they gave rise to multiple anticipated cleavage products, demonstrating specific cleavage. FIGURE 1 shows the cleavage of five different-sized run-off HIV env RNA transcripts (lanes 1–5) by a pentaribozyme that is targeted to cleave this RNA at five different sites. The cleavage patterns demonstrate for the first time not only the functionality and specificity of multitarget ribozymes, but also their increased efficiency against larger-sized substrates. At equal molar ratios, multitarget ribozymes were significantly more efficient than were ribozymes designed to cleave at fewer sites within the same target RNA. The total disappearance of the 1335-nt long substrate RNA in lane 5 indicates that the substrate had been cleaved at least once, which is sufficient to inactivate it. A 400-nt long nonaribozyme, targeted to cleave the HIV-1 env mRNA at nine different sites, also showed specificity and efficiency of cleavage, even when it was part of a 3.3-kb RNA transcript (data not shown). The infectious HIV-1 DNA clone pNL4-3, when transfected into HeLa T4 cells, caused a massive HIV-1 induced cell fusion (FIG. 2A). The nonaribozyme, when coexpressed with the infectious HIV-1 clone pNL4-3 in HeLa T4 cells under control of the HIV-1 LTR and as part of a 3.5-kb transcript, drastically reduced syncytia formation, as shown in FIGURE 2B, which correlated with a decrease in the amount of viral antigen (p24) released from these cells.

We conclude that multitarget ribozymes combine several advantages over monoribozymes: (1) Although they retain the specificity of monoribozymes, (2) they are more efficient per RNA copy, (3) they remain highly active when they are part of a large transcript, and (4) their multitargeting may significantly help prevent the selection of viral escape mutants. We anticipate that these combined features increase the chance for prolonged effectiveness of ribozymes particularly against an RNA virus such as HIV-1.

REFERENCES

1. CECH, T. R., A. J. ZAUG & P. J. GRABOWSKI. 1981. Cell **27:** 487–496.
2. SARVER, N., E. M. CANTIN, P. S. CHANG, J. A. ZAIA, P. A. LADNE, D. A. STEPHENS & J. J. ROSSI. 1990. Science **247:** 1222–1225.
3. SCANLON, K. J., L. JIAO, T. FUNATO, W. WANG, T. TONE, J. J. ROSSI & M. KASHANI-SABET. 1991. Proc. Natl. Acad. Sci. USA **88:** 10591–10595.

Gene Inhibition of HIV-1 Replication

A Comparative and Mechanistic Study

WILLIAM JAMES, PAUL CRISELL,
AND ANDREW RHODES

Sir William Dunn School of Pathology
University of Oxford
Oxford, UK, OX1 3RE

The first requirement of a candidate antiviral drug is that it should bind specifically to some essential component of the pathogen. The second is that the result of this interaction should be to inactivate the pathogen. The antisense principle provides a mechanism for the specific binding of an artificial molecule of RNA to a target RNA of known sequence. To employ this antisense principle effectively in the gene inhibition therapy of viral infection, we need to answer three questions. What is the optimum structure for efficient association between the complementary molecules in the cell? How can this interaction best be harnessed to virus inhibition? How can inhibitory nucleic acids be introduced most effectively into the susceptible cell *in vivo*?

We have addressed these questions using HIV-1 as a model system. We used retroviral vectors to transduce HIV-1–susceptible cells with antisense RNA, ribozyme, and decoy RNA-encoding genes and tested their ability to block the replication HIV. Antisense RNAs inhibited HIV replication by 0–80%.[1] Inhibitory potential did not correlate with possession of obvious features of the primary sequence. However, analysis of the predicted secondary structures of each RNA led us to propose that the factor that determined the inhibitory potential of an AR was its tendency to form intramolecular base pairs; that is, inhibitory molecules were sufficiently single-stranded to be able to hybridize to their target RNA. Perfect complementarity over approximately 50 or more bases between the target and antisense molecules seems to be necessary for efficient inhibition.[2]

We believe that inhibitory AR forms duplexes with the target RNA with sufficient probability (determined by local folding potential; see also refs. 3 and 4) and sufficient stability (determined by the ΔG of duplex formation) to prevent the export of target RNA from the nucleus and entrap it in the vigorous RNA turnover system. A theoretical objection to antisense RNA is that long, single-stranded complementary molecules hybridize at negligible rates *in vitro* because of considerable entropic barriers. However, helix-destabilizing effects of ssRNA-binding proteins of the nucleus increase the rates of association of such molecules by at least three orders of magnitude.[5] We sought to endow the interaction of antisense and target mRNA with a greater inhibitory "punch" by incorporating hammerhead ribozyme sequences.[6] For example, ribozyme 496 gives rise to over 98% inhibition of HIV replication, whereas its parent antisense RNA (AR5) produces insignificant levels of inhibition. As this ribozyme has flanking complementary regions of 43 and 70 bases, it is far from the optimal structure determined by cell-free assays of ribozyme activity. Indeed, it cleaves with a stoichiometry of around 2, implying little turnover. However, as few short ribozymes have proved effective *in vivo,* it is possible that the helix destabilizing environment of the nucleus shifts the optimum length of the complementary regions from around 8 bases to 20 or more. To test this, we made a nested set of deletion derivatives of ribozyme 496 in which the flanking length varies in three base steps from 33 to 6 bases and have begun to test these *in vitro* and *in vivo.*

REFERENCES

1. RHODES, A. & W. JAMES. 1990. J. Gen. Virol. **71:** 1965–1974.
2. RHODES, A. & W. JAMES. 1991. AIDS **5:** 145–151.
3. SCZAKIEL, G. & M. PAWLITA. 1991. J. Virol. **65:** 468–472.
4. RITTNER, K. & G. SCZAKIEL. 1991. Nucl. Acids Res. **19:** 1421–1426.
5. PONTIUS, B. W. & P. BERG. 1990. Proc. Natl. Acad. Sci. USA **87:** 8403–8407.
6. CRISELL, P. & W. JAMES. 1991. VII Int. Conf. AIDS (abstract M. A. 1208).

Design of RNAs That Inhibit the Activated c-Ha-*ras* Gene in Mammalian Cells

MAKOTO KOIZUMI AND EIKO OHTSUKA

Faculty of Pharmaceutical Sciences
Hokkaido University
Sapporo, 060, Japan

To determine the structural bases in self-cleaving RNA domains, we synthesized oligoribonucleotides and their analogs to create a series of "hammerhead ribozymes." We found several essential bases for the enzymatic activity of the catalytic domain and possible target sequences in the substrates.[1] On the basis of the foregoing information, ribozymes that cleave mRNA from mutated c-Ha-*ras* genes, which contain a single mutation at codon 12 (Y = G, normal c-Ha-*ras* mRNA; Y = U, activated c-Ha-*ras* mRNA), were designed (FIG. 1). Experiments *in vitro* show that the ribozymes specifically cleaved the mutated mRNA.[2]

We used NIH-3T3 cells to investigate the ability of ribozymes to cleave c-Ha-*ras* mRNA. Plasmids containing various ribozyme genes were constructed and expressed under the control of the long terminal repeats of Rous sarcoma virus. The ribozymes inhibited the formation of foci in transfected NIH-3T3 cells by approximately 50%.

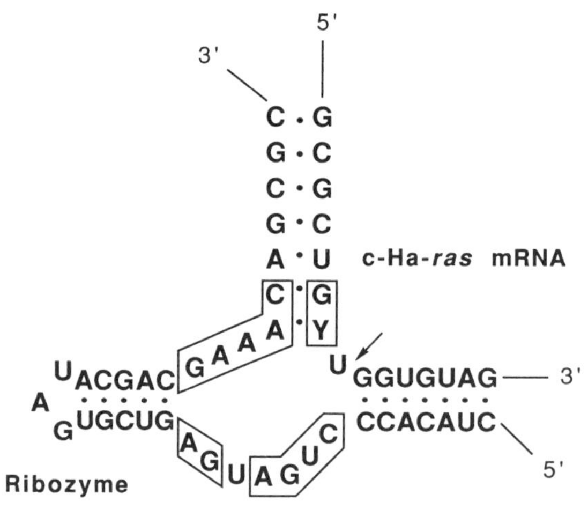

FIGURE 1. Secondary structure of the complex between a designed ribozyme and c-Ha-*ras* mRNA. The consensus sequences are *boxed* and the cleavage site is indicated by an *arrow*.

REFERENCES

1. KOIZUMI, M., S. IWAI & E. OHTSUKA. 1988. FEBS Lett. **228:** 228–230.
2. KOIZUMI, M., Y. HAYASE, S. IWAI, H. KAMIYA, H. INOUE & E. OHTSUKA. 1989. Nucl. Acids Res. **17:** 7059–7071.

Internally Deleted Antigenomic Hepatitis Delta RNA Self-Cleaves Even in 18 M Formamide

Y. PRASAD, M. G. BELINSKY, J. B. SMITH,
AND G. DINTER-GOTTLIEB

Department of Bioscience and Biotechnology
Drexel University
Philadelphia, Pennsylvania 19104

Self-splicing and self-cleaving RNAs have been engineered to function as antisense RNAs by defining separate enzyme and substrate moieties.[1–3] Although these RNAs are efficient in cleaving the substrate RNAs *in vitro,* none has yet been shown to work enzymatically *in vivo.* The hepatitis delta RNAs offer higher efficiency of cleavage and stability than do any other ribozymes.

Hepatitis delta virus is a unique mammalian virus, a single-stranded, circular RNA of only 1679 nucleotides (nts). We detected self-cleavage activity of longer-than-unit-length genomic and antigenomic RNAs.[4,5] Subsequently, both RNAs were truncated to 86 genomic nts[6] and 89 antigenomic nts, without loss of activity. In fact, the antigenomic RNA cleaved more efficiently at 37°C than has any other self-cleaving RNA yet reported, cleaving to completion in less than 15 seconds.[7]

Although the computer-derived structure of the genomic RNA generated a central core in RNAs ranging from 154 to 86 nts (FIG. 1A),[8] the structure of the antigenomic RNA, even in the same size class, differed greatly. As the RNAs are 70% similar in sequence, however, we generated an analogous antigenomic structure by analysis of possible conserved base pairs (FIG. 1B). A striking feature in both molecules was the extended stem 4 structure. To determine if this stem region was required for catalytic activity, deletions were introduced into the cloned cDNA. The resulting RNAs transcribed *in vitro*[9] were assayed for self-cleavage at 37°C in 50 mM Tris buffer, pH 7.5, and 10 mM $MgCl_2$ for 5 minutes.

When the self-cleavage of the internally truncated genomic RNA G73 was assayed (TABLE 1) and compared with the extended stem 4 parent RNA G86 in the absence of formamide, cleavage was reduced sixfold, from 59 to 9%. To test the effect of formamide on the cleavage reaction, first the RNAs were incubated in the presence of formamide at 37°C for 5 minutes, and then 10 mM magnesium was added, followed by an additional 5 minutes of incubation. The formamide enhanced the cleavage of G86, and complete cleavage was seen in 5 minutes. This presumably was due to the denaturing of inactive RNA conformers, allowing formation of the active RNA and self-cleavage. The truncated G73 completely lost activity in the presence of formamide, and no cleavage was seen at all. Rather than allow structure to form, the formamide appeared to denature the active site of the molecule, preventing cleavage. Thus, an extended stem 4 is essential to the active core of the genomic RNA in the presence of formamide.

The antigenomic RNA had previously been noted to be unusually resistant to formamide denaturation even in comparison with the genomic RNA.[7,10,11] We identified a sequence element, GGAGA, at the 3′ terminus of the 89 nucleotide

RNA that was necessary for self-cleavage in any concentration of formamide.[11] Truncation of stem 4 produced an RNA, Ag77, that retained its cleavage in the presence of high formamide (TABLE 1), cleaving to 40% without formamide and increasing to 60% in 18 M formamide, a completely different pattern from that seen with genomic RNA (TABLE 1).

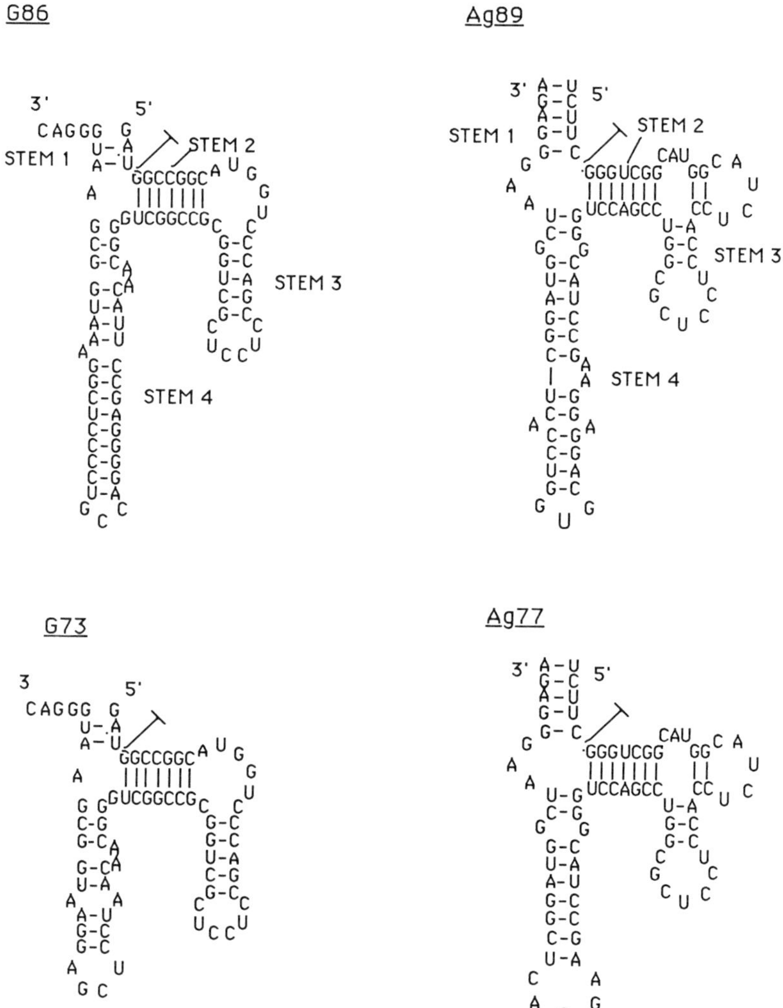

FIGURE 1. Secondary structures for HDV **(left)** genomic and **(right)** antigenomic self-cleaving RNAs. **(Left)** The folding of genomic self-cleaving RNA, G86, was obtained by a computer program, RNAFOLD (Zuker, 1986), and G73 is its derivative having a 13-nt deletion in stem IV. **(Right)** A forced folding of HDV antigenomic self-cleaving RNA, Ag96, based on postulated conserved base pairs. Ag77 has a 12-nt deletion in stem IV.

TABLE 1. Effect of Stem 4 Length on Cleavage in Formamide[a]

	0 M	5 M	10 M	18 M
G86	59	100	100	0
G73	9	0	0	0
Ag89	100	100	100	100
Ag77	40	50	70	60

[a]The percentage of RNA cleaved is shown, with increasing concentrations of formamide.

To define a *trans*-cleaving enzyme-substrate pair of ribonucleotides, it is desirable to trim the RNA to a size at which it is still active, but limited in the formation of inactive conformers. We demonstrate here that the stem 4 region can be truncated in both genomic and antigenomic delta RNAs, but that the internal portion of stem 4 of the antigenomic RNA may be inherently more stable in support of the self-cleavage reaction, even in the presence of formamide. Further truncations will determine the limits of this stability.

REFERENCES

1. ZAUG, A. J. & T. R. CECH. 1986. Science **231:** 470–475.
2. UHLENBECK, O. 1987. Nature **328:** 596–600.
3. HAMPEL, A. & R. TRITZ. 1989. Biochemistry **28:** 4929–4933.
4. SHARMEEN, L. *et al.* 1988. J. Virol. **62:** 2674–2679.
5. KUO, M. *et al.* 1988. J. Virol. **62:** 4439–4444.
6. PERROTTA, A. T. & M. D. BEEN. 1990. Nucl. Acids Res. **18:** 6821–6827.
7. SMITH, J. B. & G. DINTER-GOTTLIEB. 1991. Nucl. Acids Res. **19:** 1285–1289.
8. ZUKER, M. 1989. Science **244:** 48–52.
9. BELINSKY, M. & G. DINTER-GOTTLIEB. 1991. *In* The Hepatitis Delta Virus. J. Gerin, ed.:265–274. Alan R. Liss, Inc. New York, NY.
10. BELINSKY, M. & G. DINTER-GOTTLIEB. 1991. Nucl. Acids Res. **19:** 1285–1289.
11. SMITH, J. B., P. A. GOTTLIEB & G. DINTER-GOTTLIEB. Submitted.

Inhibition of Processing of the Primary Transcript of the Gene Encoding Tissue Inhibitor of Metalloproteinases (TIMP) by Antisense TIMP RNA in Mouse 3T3 Cells[a]

BO FENG AND DAVID T. DENHARDT

Department of Biological Sciences
Rutgers University
Piscataway, New Jersey 08855

The purpose of this study was to determine the mechanism of down-modulation of TIMP mRNA levels in derivatives of Swiss 3T3 cells that contain an expression cassette driving synthesis of an RNA molecule complementary to the TIMP mRNA.[1] These cell lines are of interest because as a consequence of reduced TIMP expression, they have acquired both an invasive and a tumorigenic phenotype.[2] We address here the detailed mechanism of how TIMP expression is down-modulated in light of the finding that two cell lines differing substantially in the number of copies of the expression cassette and the level of cytoplasmic antisense RNA show comparable extents of TIMP expression. We found that although TIMP mRNA levels (cytoplasmic and nuclear) were down-regulated by antisense TIMP RNA, total nuclear TIMP RNA levels were not. Compared with control cells, cells producing antisense TIMP RNA contained a higher level of unprocessed TIMP RNA. The increased amount of unprocessed primary transcripts in the nucleus accompanied by a reduction in the concentration of TIMP mRNA molecules is consistent with the hypothesis that antisense TIMP RNA inhibits processing of the transcript.

RESULTS AND DISCUSSION

LA1 and LA7 contain, respectively, 1 and about 100 copies of an antisense expression cassette stably incorporated into the cell's DNA. FIGURE 1 shows that when RNA preparations from the nucleus and the cytoplasm were probed for the sense transcript, down-modulation of TIMP mRNA could be detected in the nucleus as well as in the cytoplasm; MC2 is a G418-resistant control clone. To detect either the primary TIMP transcript itself or intermediates, we applied the reverse transcription-coupled polymerase chain reaction (RT-PCR)[3] to determine the steady-state level of nuclear TIMP transcripts. We take the abundance of TIMP intron 1 to reflect the abundance of the unprocessed TIMP hnRNA. First, we verified that a linear relation existed between input RNA molecules and the signal detected after PCR amplification. By careful selection of cycle number and quantity of input RNA, we could demonstrate the expected exponential relation between the amount of tem-

[a] This research was supported by the Bureau of Biological Research, Rutgers University, and a grant (CA50183) from the National Institutes of Health.

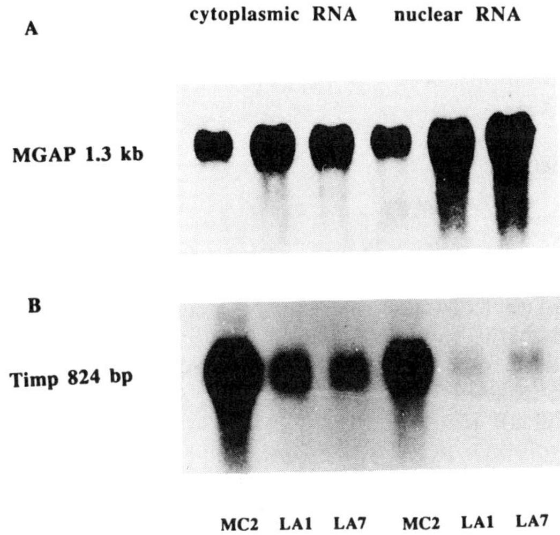

FIGURE 1. Northern blot analysis of sense and antisense TIMP RNA in the cytoplasm and nucleus of various cell lines. Five micrograms of purified nuclear RNA were fractionated by denaturing agarose/formaldehyde gel electrophoresis and transferred to a nitrocellulose filter. The blot was hybridized successively with (**A**) a probe specific for mouse glyceraldehyde-3-phosphate dehydrogenase (MGAP) mRNA and (**B**) a probe specific for TIMP mRNA.

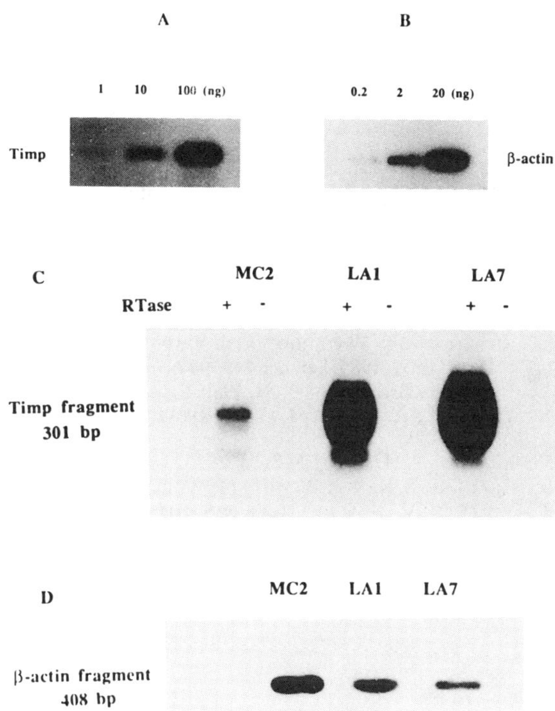

FIGURE 2. RT-PCR of TIMP hnRNA and β-actin mRNA in transfected cell lines. (**A** and **B**) The indicated amount of nuclear RNA from MC2 cells was subjected to the RT-PCR with TIMP intron primers and β-actin mRNA primers. Amplified TIMP PCR fragments (301 bp) and β-actin mRNA fragments (408 bp) were labeled with [α^{32}P]dCTP during the PCR. (**C** and **D**) Autoradiographs of equal portions of PCR product from different cell lines electrophoresed on an agarose gel. Nuclear RNA (4 and 20 ng) was used for amplification of β-actin mRNA and TIMP primary transcripts. (+) and (−) indicate the presence and absence of RTase.

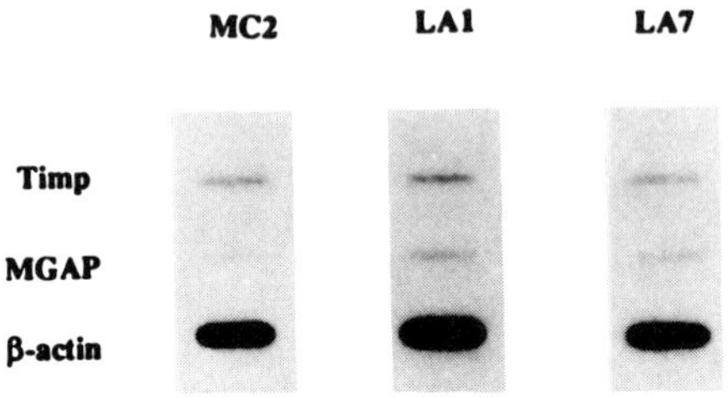

FIGURE 3. Transcriptional activity of the TIMP gene in control and antisense-producing cells assessed in a nuclear run-on assay. The TIMP probe was a DNA fragment within the first intron of the genomic DNA. MGAP and β-actin cDNA probes were used for control.

plate RNA and the signal resulting from the amplified DNA (FIG. 2A and B). Relative to the control MC2 cell line, the RT-PCR methodology revealed a substantially increased level of unprocessed TIMP RNA in both LA1 and LA7 (FIG. 2C). No change in the level of β-actin RNA was observed (FIG. 2D). The ability of antisense TIMP RNA to affect TIMP transcription was ruled out by a number of nuclear run-on experiments (FIG. 3) that showed that the rate of TIMP transcription was the same in control cells and in cells expressing antisense TIMP RNA.

Several studies are consistent with the possibility that antisense RNA in the nucleus can inhibit processing of the primary transcript.[4–7] Our RT-PCR data show an increase in the steady-state nuclear concentration of the first intron, which cannot itself be in a duplex structure as the antisense RNA is composed only of exons. This increased abundance can be explained by an inhibition of either TIMP pre-mRNA splicing or a normal process of degradation of the primary transcript. As the latter seems unlikely and does not explain the decreased level of processed mRNA, we favor the argument that the effect of antisense TIMP RNA in these cells is to interfere with processing of the primary transcript.

ACKNOWLEDGMENT

We thank Dylan Edwards for sharing information regarding the structure of the TIMP gene and generously making available portions of the mouse gene.

REFERENCES

1. KHOKHA, R. & D. T. DENHARDT. 1987. *Anticancer Res.* **7:** 653–660.
2. KHOKHA, R., P. WATERHOUSE, S. YAGEL, P. K. LALA, C. M. OVERALL, G. NORTON & D. T. DENHARDT. 1989. Science **243:** 947–950.
3. LIPSON, K. E. & R. BASERGA. 1989. Proc. Natl. Acad. Sci. USA **86:** 9774–9777.
4. TOSIC, M., A. ROACH, J.-C. D. RIVAZ, M. DOLIVO & J.-M. MATTHIEU. 1990. *EMBO J.* **9:** 401–406.
5. MUNROE, S. H. 1988. EMBO J. **7:** 2523–2532.
6. CRAIG, E., S. KASS & B. SOLLNER-WEBB. 1991. Molec. Cell. Biol. **11:** 458–467.
7. VOLLOCH, V., B. SCHWEITZER & S. RITS. 1991. Biochem. Biophys. Res. Comm. **179:** 1593–1599.

Antisense RNA to the Putative Tumor Suppressor Gene "Deleted in Colorectal Cancer" Transforms Fibroblasts

KENNETH G. LAWLOR,[a] NITIN T. TELANG,[b]
MICHAEL P. OSBORNE,[b] ROEL Q. SCHAAPVELD,[a]
KATHLEEN R. CHO,[c] BERT VOGELSTEIN,[c]
AND RAMASWAMY NARAYANAN[a,d]

[a]*Department of Molecular Genetics*
Hoffmann-La Roche, Inc.
340 Kingsland St.
Nutley, New Jersey 07110

[b]*Breast Cancer Research Laboratory*
Department of Surgery
Memorial Sloan-Kettering Cancer Center
New York, New York 10021

[c]*The Oncology Center*
Johns Hopkins University School of Medicine
Baltimore, Maryland 21231

Allelic deletions involving chromosome 18q occur in over 70% of colorectal cancers.[1] A candidate suppressor gene, termed "deleted in colorectal cancer" (DCC), has been identified in this region.[2] DCC expression was detected in many normal tissues (e.g., brain and colonic mucosa), but it was absent or diminished in most colorectal carcinomas.[2] The predicted amino acid sequence of the DCC cDNA specified a protein with sequence homology to neural cell adhesion molecules and various other related cell surface glycoproteins.[2]

We attempted to establish the tumor suppressor function of DCC by inhibiting its expression in normal cells using antisense RNA technology. We chose a dexamethasone-inducible vector, pMAM Neo CAT (MNC), to produce an inducible chimeric antisense RNA to DCC.[3] When Rat-1 cells were transfected with the activated *ras,* MNC (vector), DCC,^{S} and DCC^{AS} constructs, stable G-418^{R} clones were readily established for the *ras,* MNC, or DCC^{S} transfectants but not for the DCC^{AS} transfectant. We reasoned that expression of antisense RNA to DCC might interfere with the cells' ability to attach to the substratum. Subsequently, we established several independent G-418^{R} clones from DCC^{AS}-transfected cells grown in dishes coated with an extracellular matrix component, laminin. The DCC^{AS} transfectants showed 2.5- to 3-fold increased saturation density, faster growth rate, and anchorage independence in a dexamethasone-dependent manner as well as *in vivo* tumorigenicity in nude mice. These DCC^{AS} transfectants showed a pronounced reduction in DCC gene expression in a dexamethasone-dependent manner as monitored by quantitative reverse transcriptase PCR.[3]

[d]Address all correspondence to: Ramaswamy Narayanan, Ph.D., Department of Molecular Genetics, Hoffmann-La Roche, Inc., 340 Kingsland St., Nutley, NJ 07110.

It was possible to develop stable Rat-1 cell lines expressing DCC^{AS} transcripts only after transfectants were selected for G-418 resistance on an extracellular matrix-coated substratum. This raised the possibility that expression of DCC^{AS} RNA in these cells interfered with their ability to adhere to the substratum. We sought to

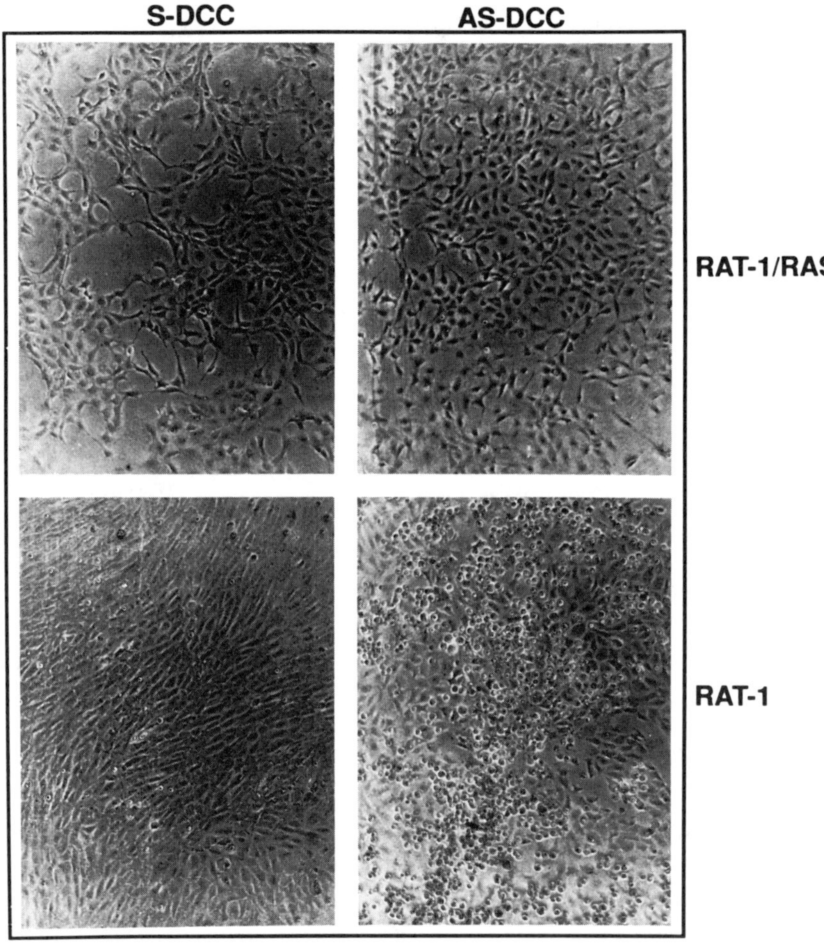

FIGURE 1. Effect of antisense thio oligo to DCC on cellular adhesion. Rat-1 cells or *ras*-transformed Rat-1 cells (5×10^4) were plated onto 48-well Costar dishes and 24 hours later were treated with a daily addition of 30 μM of sense oligo (5′ AAT ATG GAG AAT AGT CTT AGA 3′) and antisense oligo (5′ TCT AAG ACT ATT CTC CAT ATT 3′) for 48 hours. (Magnification ×50.)

confirm this by the use of phosphoro-thio antisense oligos to DCC. When the Rat-1 cells were treated with 30 μM of antisense DCC oligo, more than 50% of the cells were detached from the dishes within 24–48 hours (FIG. 1). This effect was dose dependent, that is, more pronounced with the daily addition of the antisense oligo, seen at subconfluent and confluent culture conditions, and not observed with the sense (control) oligo. The loosely adherent cells generated by the addition of antisense oligo were viable as evidenced by the trypan blue exclusion test; however, they could not be replated onto either uncoated or coated dishes. The detachment of Rat-1 cells by the antisense DCC oligo could be abrogated if these cells were first plated onto an extracellular matrix before treatment with the oligo.[3]

FIGURE 1 further demonstrates that the effect of the antisense DCC oligo is sequence specific: when the Rat-1 cells were transformed with *ras,* these cells were no longer sensitive to the antisense oligo to DCC. The *ras*–Rat-1 cells showed a pronounced reduction in DCC gene expression when compared with that of the parental Rat-1 cells. Similarly, the antisense oligo to DCC had no effect on a bone marrow-derived stromal cell line that does not express DCC.[3]

In summary, we demonstrate that the DCC gene is involved in cellular adhesion to the substratum and has a potential tumor suppressor function in an immortalized Rat-1 cell line, as expression of antisense RNA to DCC results in neoplastic transformation. Furthermore, antisense oligos to DCC cause pronounced cell detachment which can be abrogated by providing an extracellular matrix. Our study provides the first direct biological evidence of the role of DCC in cell transformation.

REFERENCES

1. VOGELSTEIN, B., E. R. FEARON, S. R. HAMILTON, S. E. KERN, A. C. PREISINGER, M. LEPPERT, Y. NAKAMURA, R. WHITE, A. M. SMITS & J. L. BOS. 1988. N. Engl. J. Med. **319:** 525–532.
2. FEARON, E. R., K. R. CHO, J. M. NIGRO, S. E. KERN, J. W. SIMONS, J. M. RUPPERT, S. R. HAMILTON, A. C. PREISINGER, G. THOMAS, K. W. KINZLER & B. VOGELSTEIN. 1990. Science **247:** 49–56.
3. NARAYANAN, R., R. Q. J. SCHAAPVELD, K. R. CHO, B. VOGELSTEIN, P. B.-V. TRAN, M. P. OSBORNE & N. T. TELANG. 1992. Oncogene **7:** 553–561.

Inhibition of the Transformed Phenotype of Carcinoma Cells That Contain Human Papillomavirus

E. J. SHILLITOE AND C. STEELE

Department of Microbiology
University of Texas Dental Branch
Houston, Texas 77225

Human papillomavirus type-18 (HPV-18) is present in many human oral and cervical carcinomas.[1] This virus can transform cells through the action of the E6 and E7 genes, and the E7 gene must be continually expressed for maintenance of the transformed phenotype.[2] von Knebel Doeberitz *et al.*[3] showed that a vector that expressed antisense RNA from the E6/E7 region of the genome could affect the behavior of C4-1 cells. We have now examined the effect of HPV-18 antisense expression vectors and oligonucleotides on a variety of cell lines that were derived from human squamous cell carcinomas and that carried DNA of HPV-18. Expression vectors consisted of the E7 or E6/E7 regions of HPV-18 cloned adjacent to the MMTV promoter of the shuttle vector pMAM-neo-BLUE, and oligonucleotides consisted of 20 mers that spanned the start codons of the E6 and E7 genes.

The transfection efficiency of 1483 cells with the antisense-expression vectors was extremely low, and no cell lines could be established. However, all other cells could be transfected and cell lines were obtained. When expression of antisense RNA was induced with steroids, the carcinoma cells lost many of the parameters of cell transformation (TABLE 1).[4]

The 1483 cells were susceptible to antisense oligonucleotides that targeted the E6 or E7 genes, and especially combinations of the two. The growth rate and plating efficiency of the cells were reduced, but colony formation in soft agar was not affected. Random-sequence oligonucleotides had no effect on any cells tested. Vero

TABLE 1. Inhibition of Parameters of Cell Transformation by HPV-18 Antisense Molecules

Cell	HPV-18 (copies/cell)	Phenotype Inhibited	Antisense Oligos/Vectors
Vero	0	None	O, V
183	0	None	O, V
1483	1	Transfection efficiency	V
"		Growth rate	O
"		Plating efficiency	O
C4-1	1	Transfection efficiency	V
HeLa	50	Growth rate	V
"		Anchorage independence	V
"		Growth in low serum concentration	V
"		Plating efficiency	V

cells and 183 cells, which do not have HPV DNA, were not affected by antisense oligonucleotides or by expression vectors.

These data imply that human carcinomas that carry DNA of HPV-18 are dependent on continuous expression of the E6 and/or E7 genes for their malignant behavior. The inability to transfect 1483 cells with antisense expression vectors and the reduced transfection efficiency of the C4-1 cells implies that the low copy number of HPV-18 DNA in these cells is readily overcome by low levels of antisense molecules. This explanation is consistent with findings in other antisense systems.[5] The present results suggest that gene transfer with antisense constructs could be exploited to develop a form of therapy for malignant tumors that carry DNA of HPV-18, and that tumors with low viral DNA copy numbers would be particularly susceptible.

REFERENCES

1. Steele, C. & E. J. Shillitoe. 1991. Viruses and oral cancer. Crit. Rev. Oral Biol. Med. **2:** 153–175.
2. Crook, T., J. P. Morgenstern, L. Crawford & L. Banks. 1989. Continued expression of HPV-16 E7 protein is required for maintenance of the transformed phenotype of cells co-transformed by HPV-16 plus EJ-*ras*. EMBO J. **8:** 513–519.
3. von Knebel Doeberitz, M., T. Oltersdorf, E. Schwarz & L. Gissman. 1988. Correlation of modified human papilloma virus early gene expression with altered growth properties in C4-1 cervical carcinoma cells. Cancer Res. **48:** 3780–3786.
4. Steele, C., P. G. Sacks, K. Adler-Storthz & E. J. Shillitoe. 1992. Effect on cancer cells of plasmids that express antisense RNA of human papillomavirus type-18. Cancer Res. In press.
5. Melani, C., L. Rivoltini, G. Parmiani, B. Calabretta & M. Colombo. 1991. Inhibition of proliferation by c-*myb* antisense oligodeoxynucleotides in colon adenocarcinoma cell lines that express c-*myb*. Cancer Res. **51:** 2897–2901.

Use of an Antisense RNA Expression System to Explore Functions of PKC Isoenzymes in T-Cell Activation

J. FREIRE-MOAR,[a] A. TSUTSUMI, AND P. BARJA

Department of Molecular Immunology
Syntex Research
Palo Alto, California 94303

A substantial amount of evidence indicates that protein kinase C (PKC) is involved in the induction of interleukin-2 p55 receptor expression and lymphokine production. Studies of PKC isoenzyme expression in lymphocytes have shown the presence of PKCα, PKCβ, PKCδ, PKCϵ, and PKCζ, but not that of PKCγ; it has been suggested that the different PKCs could play unique roles in the activation process of T cells.[1,2] We decided to use antisense technology as an approach to study the function of different PKC isoenzymes in T cells.

PKCα and PKCβ cDNA fragments, corresponding to their amino terminal sequences, were generated by PCR using Jurkat cDNA as target, and cloned in the antisense orientation in the PREP3 episomal vector.[3] Jurkat cells, transfected with the different PREP3 plasmids by electroporation, were selected by culturing in hygromycin-containing media. Western blot analysis of stable transfected cells using specific antibodies against PKCα showed only a slight decrease in the levels of expression of this isoenzyme in cells transfected with antisense PKCα/PREP3 vector than in cells transfected with other PREP3 plasmids.

To study the effect of different plasmids in the activation of Jurkat cells, we analyzed the response of stable transfected Jurkat cells to PMA and anti-CD3 antibodies in terms of IL-2R p55 and lymphokine expression. Induction of IL-2R expression by PMA and anti-CD3 in transfected Jurkat cells was assayed by FACS analysis of cells stained with anti-IL-2R (CD25) antibodies conjugated with phycoerythrin (FIG. 1). Jurkat cells transfected with the antisense PKCα/PREP3 plasmid showed a lower degree of IL-2R p55 expression in response to PMA and anti-CD3 stimulation than did cells transfected with the normal PREP3 plasmid, whereas Jurkat cells transfected with an antisense PKCβ/PREP3 plasmid expressed higher levels of IL-2R p55 than did cells transfected with the normal PREP3 plasmid. Induction of IL-2 and TNFβ mRNA expression by PMA and anti-CD3 antibodies was determined by MAPPing.[4] As in IL-2R expression, cells transfected with antisense PKCα/PREP3 plasmid express lower levels of IL-2 and TNFβ mRNA, in response to PMA and anti-CD3 stimulation, than do cells transfected with PREP3 plasmid alone. Cells transfected with antisense PKCβ/PREP3 plasmid express

[a]Address for correspondence: J. Freire-Moar, Department of Molecular Immunology, Syntex Research S3-6, 3401 Hillview Avenue, Palo Alto, CA 94304.

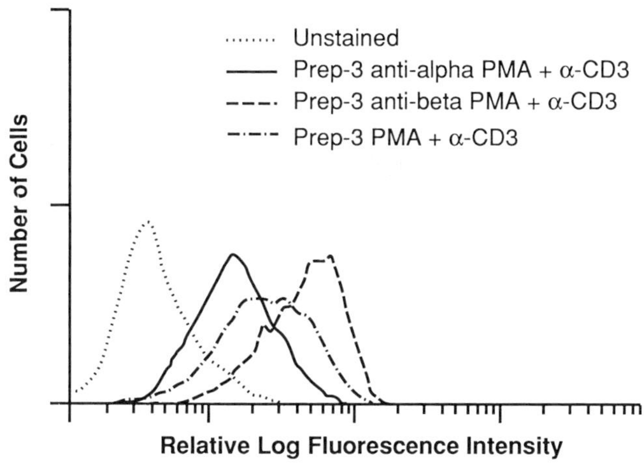

FIGURE 1. Induction of IL-2R p55 expression in transfected Jurkat cells. Stable transfected cells treated for 16 hours with PMA (5 ng/ml) and IOT3 (anti-CD3 monoclonal antibody) (2 μg/ml) were stained by direct immunofluorescence with anti-IL-2R (CD25) phycoerythrin-conjugated antibodies.

higher levels of mRNA for these two lymphokines. Similar results were obtained with cells incubated in the presence of PMA or anti-CD3 antibodies alone.

We also analyzed the levels of expression of CD2 and CD3 antigens constitutively expressed in Jurkat cells and resting T cells. Expression of CD2 and CD3 antigens in transfected Jurkat cells, incubated in the presence or absence of PMA and anti-CD3 antibodies, was determined by FACS analysis of cells stained with anti-Leu-5b (CD2) or with anti-Leu-4 (CD3) fluorescein-conjugated antibodies (TABLE 1). No significant difference in the levels of expression of CD2 and CD3 antigens was detected between Jurkat cells transfected with different PREP3 plasmids.

Although we have not yet been able to show a drastic reduction in the levels of PKCα and PKCβ protein in cells transfected with the respective antisense plasmid, it is tempting to speculate that PKCα isoenzyme is involved in the activation of Jurkat cells, whereas PKCβ isoenzyme, as previously suggested,[2] is not required or alternatively plays a negative regulatory role in this process.

TABLE 1. Expression of CD2 and CD3 Antigens on Transfected Jurkat Cells and Their Modulation by PMA and anti-CD3 Antibodies

	CD2 Expression[a]		CD3 Expression[a]	
Plasmid	Untreated	PMA + IOT3	Untreated	PMA
PREP3	55.1	69.8	61.4	17.3
Anti-PKCα/PREP3	68.4	86.5	61.4	19.1
Anti-PKCβ/PREP3	59.2	70.9	63.6	13.2

[a]Values represent the peak channel of the distribution of the cell population stained with anti-Leu-5b (CD2) or anti-Leu-4 (CD3) fluorescein-conjugated antibodies.

Results are from one representative experiment.

REFERENCES

1. FREIRE-MOAR, J., H. CHERWINSKI, F. HWANG, J. RANSOM & D. WEBB. 1991. Expression of protein kinase C isoenzymes in thymocyte subpopulations and their differential regulation. J. Immunol. **147:** 405–409.
2. KORETZKY, G. A., M. WAHI, M. E. NEWTON & A. WEISS. 1989. Heterogeneity of protein kinase C isoenzyme gene expression in human T cell lines. J. Immunol. **143:** 1692–1695.
3. GROGER, R. K., D. M. MORROW & M. L. TYKOCINSKI. 1989. Directional antisense and sense cDNA cloning using Epstein-Barr virus episomal expression vectors. Gene **81:** 285–294.
4. BRENNER, C. A., A. W. TAM, P. A. NELSON, E. G. ENGLEMAN, N. SUZUKI, K. E. FRY & J. W. LARRICK. 1989. Message Amplification Phenotyping (MAPPing): A technique to simultaneously measure multiple mRNAs from small numbers of cells. Biotechniques **7:** 1096–1103.

A Novel Method to Modulate Desensitization and Truncation of LH Receptors Using Antisense Oligodeoxynucleotides[a]

ANTHONY P. WEST AND BRIAN A. COOKE

Department of Biochemistry
Royal Free Hospital School of Medicine
Rowland Hill Street
London, NW3 2PF, UK

Luteinizing hormone (LH) stimulates testosterone synthesis in the testis Leydig cell via cAMP and other second messengers.[1] Continued exposure of the Leydig cells to LH leads to a desensitization of these same responses.[2] The structure of the LH receptor indicates that it is a member of the seven transmembrane helix-G protein coupled receptor family,[3,4] containing potential phosphorylation sites in the COOH-terminus (FIG. 1). To determine the role of the latter in desensitization, we used an antisense strategy[5,6] to generate receptors lacking COOH-terminal residues. The effects of these procedures on LH-stimulated cAMP and steroidogenesis under stimulatory and desensitizing conditions were investigated.

METHODS

Mouse Leydig tumor (MA10) cells, maintained according to the method of Ascoli,[7] were treated with 0.05% trypsin/0.53 mM EDTA for 2 minutes at 37°C and then cultured in maintenance medium. Antisense oligodeoxynucleotides (AS oligos) (FIG. 1), at a concentration of 2.5 μM, were added during plating down and 24 hours afterwards.[5] Assays[5,8] were carried out after 48-hours of incubation with AS oligos. We previously showed that LH receptors are replenished during this time period.[9]

RESULTS AND CONCLUSIONS

Incubation with oligos 2 and 3 did not alter LH-stimulated cyclic AMP and pregnenolone production, but oligo 4 resulted in a 50% decrease in cAMP production. LH (3.3 nM) and phorbol 12-myristate-13-acetate (PMA)-induced cAMP desensitization was prevented with oligos 2, 3, and 4, but desensitization still occurred with cells treated with oligo 5. PMA (10^{-7} M) and dibutyryl-cyclic AMP (db-cAMP) (1 mM) (but not LH)-induced desensitization of pregnenolone was prevented in cells incubated with oligos 2, 3, and 4. All the AS oligos prevented LH and db-cAMP induced loss of surface binding over 2 hours. The results clearly show

[a]This work was supported by the Science and Engineering Research Council and the Peter Samuel Royal Free Fund.

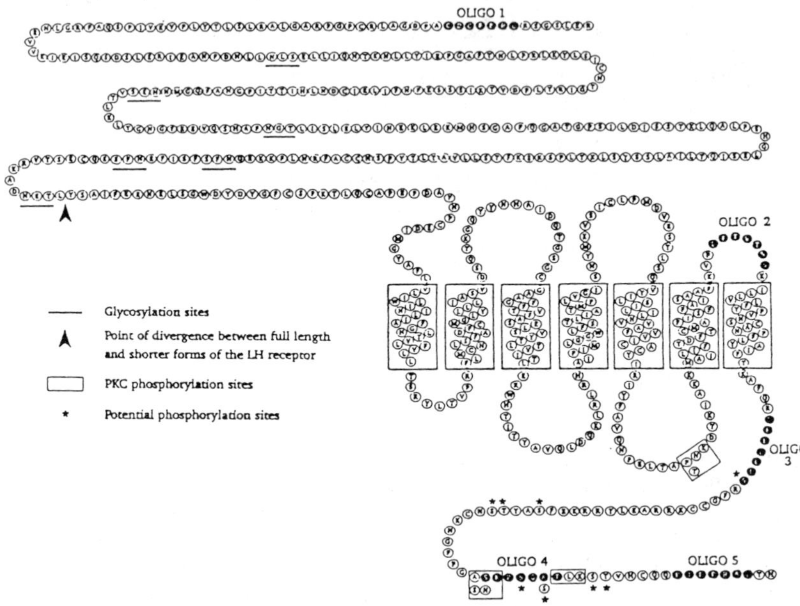

FIGURE 1. Luteinizing hormone receptor structure.[3] Regions encoded for by the antisense oligodeoxynucleotides added to the MA10 cells are shown (oligos 1, 2, 3, 4, and 5).

that desensitization and loss of surface LH receptors can be prevented in MA10 cells by the presence of COOH-terminal AS oligos. Furthermore, the latter either do not or only partially inhibit the stimulatory effects of LH on cyclic AMP production and steroidogenesis.

ACKNOWLEDGMENTS

We are grateful to Dr. M. Ascoli (University of Iowa College of Medicine, Iowa) for the gift of the MA10 cells.

REFERENCES

1. Cooke, B. A. 1990. Mol. Cell. Endocrinol. **69:** C11–C15.
2. Cooke, B. A. & F. F. G. Rommerts. 1988. *In* New Comprehensive Biochemistry: Hormones and Their Actions. B. A. Cooke, R. J. B. King & H. J. van der Molen, eds. Vol. II: 155–162. Elsevier. Amsterdam.
3. McFarland, K. C., R. Sprengel, H. S. Phillips, M. Kohler, N. Rosemblit, K. Nikolics, D. L. Segaloff & P. H. Seeburg. 1989. Science **245:** 494–499.
4. Loosfelt, H., M. Mirashi, M. Atger, R. Salesse, M. T. V. H.-L. Thi, A. Joliuet, A. Guiochon-Mantel, S. Sar, B. Jallal, J. Garnier & E. Milgrom. 1989. Science **245:** 525–528.
5. West, A. P. & B. A. Cooke. 1991. Mol. Cell. Endocrinol. **79:** R9–R14.
6. Weintraub, H. M. 1990. Sci. Am. **262:** 34–40.
7. Ascoli, M. 1981. Endocrinology **108:** 88–95.
8. West, A. P. & B. A. Cooke. 1991. Endocrinology **128:** 363–370.
9. West, A. P., M. P. Rose & B. A. Cooke. 1990. Biochem. J. **270:** 499–503.

Stable Transfection of Human Malignant Melanoma Cells with Basic Fibroblast Growth Factor Antisense cDNA[a]

ULLRICH GRAEVEN,[b] ULRICH RODECK,
ROBERTO WEINMANN, AND MEENHARD HERLYN

The Wistar Institute of Anatomy and Biology
3601 Spruce Street
Philadelphia, Pennsylvania 19104

Among the growth factors constitutively produced by melanoma cells, basic fibroblast growth factor (bFGF) has emerged as an important growth regulator. For *in vitro* growth, normal melanocytes depend on supplementation of culture medium with bFGF, whereas primary and metastatic melanoma cells grow independently of exogenous bFGF. In short-term assays, both neutralizing antibodies to bFGF as well as antisense oligonucleotides targeted against bFGF mRNA have been shown to inhibit melanoma cell growth *in vitro.*[1,2] Besides its role in autocrine growth stimulation, bFGF may have other important functions during melanoma development and progression such as regulation of extracellular matrix formation and stimulation of tumor neoangiogenesis.

To study potential paracrine functions of bFGF, we attempted to suppress or increase bFGF synthesis in melanoma cells by stable transfection of bFGF antisense or sense cDNA. Expression vectors were constructed using a 480-bp bFGF cDNA fragment covering the entire coding region for human bFGF.[3] This fragment was placed in either sense or antisense orientation downstream of the cytomegalovirus immediate early gene enhancer and promoter. For selection of stable transfectants these vectors also contained the neomycin resistance gene under expression control of the SV40 promoter.

Sense and antisense vectors were transfected into the bFGF expressing metastatic melanoma cell line WM852 using liposome-mediated transfection. After transfection, cells were selected for geneticin resistance and grown in the presence of exogenous bFGF (40 ng/ml) to compete for a negative effect on cell proliferation after expression of antisense bFGF mRNA. Individual clones, resistant to 800 μg of G418 per milliliter of medium, were isolated and further analyzed. To confirm the integration of vector DNA, cellular DNA of individual clones was digested with *Hind*III and *Bam*HI and hybridized with a neo probe. All isolated clones showed integration of vector DNA, and the estimated copy number ranged between 5 and 50 copies per genome. Total RNA was analyzed for transcription of the integrated DNA. Hybridization with a bFGF cDNA fragment showed strong overexpression of bFGF mRNA in bFGF-transfected cells. The morphology of sense-transfected cells did not differ from that of untransfected parental cells, whereas antisense-transfected cells showed an increase in cell size with a wide cytoplasm. In monolayer culture, antisense-transfected cells in contrast to untransfected cells did show signs of contact inhibition and did not overgrow. To further study the effects of bFGF

[a]This work was supported by Deutsche Forschungsgemeinschaft Gr 984/2-1.

[b]PRESENT ADDRESS: Dr. med. Ullrich Graeven, Department of Oncology and Hematology, University Hospital Eppendorf, Martinistr. 52, 2000 Hamburg 20, Germany.

sense and antisense expression on cell growth, we measured (^{3}H)thymidine incorporation upon the addition of bFGF to cells that were grown in protein-free medium. As indicated in FIGURE 1, untransfected WM852 cells and sense-transfected cells showed no increase in (^{3}H)thymidine uptake, whereas antisense-transfected cells had 1.4–1.9-fold higher (^{3}H)thymidine incorporation upon the addition of bFGF. After injection into nude mice, bFGF antisense-transfected cells were able to form tumors but showed a longer latency period and slower tumor growth compared to those of parental cells. These bFGF antisense-transfected cell lines will be a valuable tool to further delineate the multiple functions of bFGF for melanoma development and progression.

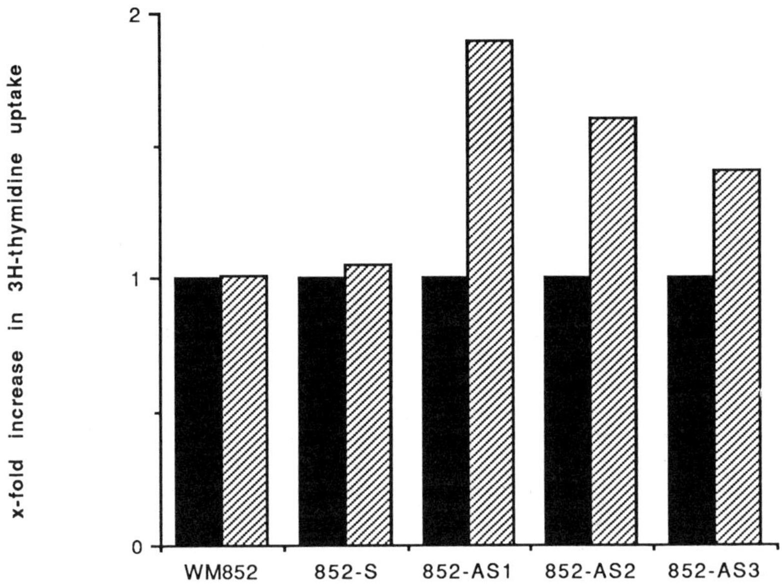

FIGURE 1. Growth response to exogenous bFGF of WM852 clones transfected with bFGF sense or antisense cDNA. Parental melanoma cells (WM852), bFGF sense-transfected melanoma cells (852-S), or bFGF antisense-transfected melanoma cells (852-AS1-3) were first grown in protein-free medium. After 24 hours the medium was changed to either fresh protein-free medium (*solid bars*) or protein-free medium containing 40 ng/ml bFGF (*hatched bars*), and the cells were pulse labeled with (^{3}H)thymidine 16 hours after factor addition. Shown is the x-fold increase in (^{3}H)thymidine uptake over base levels for protein-free growth conditions.

REFERENCES

1. HALABAN, R., B. S. KWON, S. GOSH, P. DELLI BOVI & A. BAIRD. 1988. bFGF as an autocrine growth factor for human melanomas. Oncogene Res. **3:** 177–186.
2. BECKER, D., C. B. MEIER & M. HERLYN. 1989. Proliferation of human malignant melanomas is inhibited by antisense oligodeoxynucleotides targeted against basic fibroblast growth factor. EMBO J. **8:** 3685–3691.
3. SOMMER, A., M. T. BREWER, R. C. THOMPSON, D. MOSCATELLI, M. PRESTA & D. B. RIFKIN. 1987. A form of basic fibroblast growth factor with an extended amino terminus. Biochem. Biophys. Res. Commun. **144:** 543–550.

Early Cell Growth Stimulation Is Inhibited by Casein Kinase II Antisense Oligodeoxynucleotides

WALTER PYERIN,[a] RAINER PEPPERKOK,[b]
WILHELM ANSORGE,[b] AND PETER LORENZ[a]

[a]*Biochemical Cell Physiology*
German Cancer Research Center and
[b]*Biochemical Instrumentation*
European Molecular Biology Laboratory
D-6900 Heidelberg, Germany

Casein kinase II (CKII) is a ubiquitous protein kinase that independently of second messengers phosphorylates proteins at serine and threonine residues neighbored COOH terminally by clusters of acidic residues. CKII has a tetrameric structure containing two catalytically active alpha subunits (CKIIα and/or CKIIα′) and two

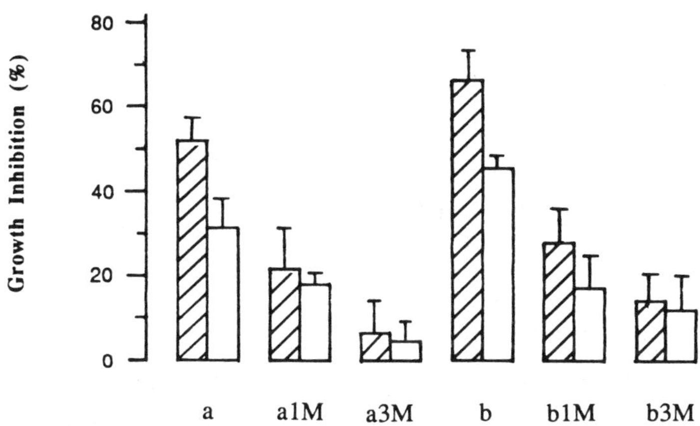

FIGURE 1. Effect of oligodeoxynucleotides complementary to CKIIα and CKIIβ subunits on mitogenic stimulation of cells. Cultured IMR-90 cells (human fibroblasts) rendered quiescent (0.5% serum for 3 days) were supplied with antisense oligodeoxynucleotides (final concentration, 100 μg/ml) 2 hours before growth stimulation by EGF (final concentration, 200 μg/ml; *hatched columns*) or by fetal calf serum (final concentration 20%; *open columns*). Controls received respective sense oligodeoxynucleotides. Growth stimulation was determined via DNA synthesis monitored by indirect immunofluorescence of incorporated 5-bromo-2′-deoxyuridine (BrdU; final concentration, 100 μg/ml; labeling at 18–24 hours after growth stimulation). At least 500 cells were analyzed for DNA synthesis in each experiment, and the oligodeoxynucleotide effect was calculated from the percentages of BrdU-positive cells in experiments with antisense oligodeoxynucleotides and that with respective sense oligodeoxynucleotides.[8] Shown are the means ± standard deviations. **a**, antisense α; **a1M**, antisense α with one mutation at middle position; **a3M**, antisense α with three mutations within middle range; **b**, antisense β; **b1M**, antisense β with one mutation at middle position; **b3M**, antisense β with three mutations within middle range.

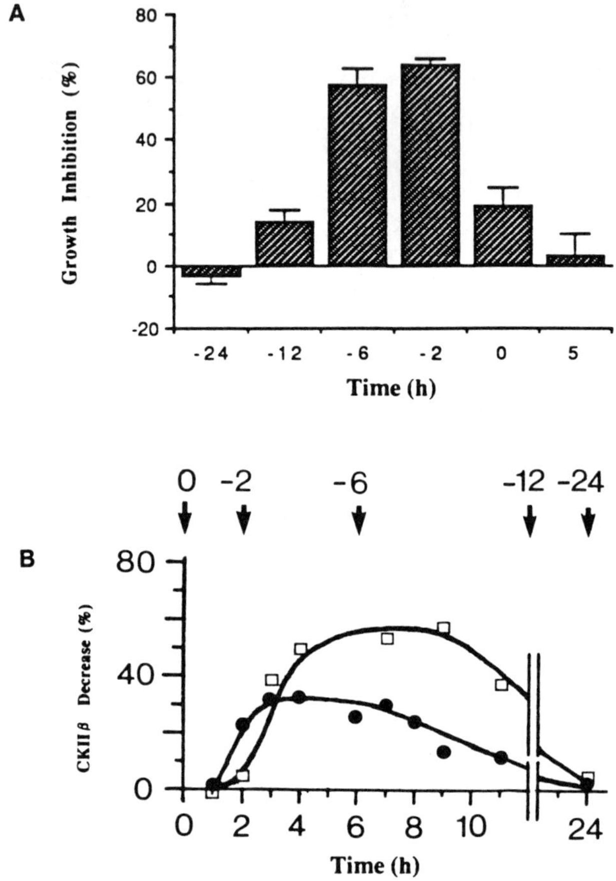

FIGURE 2. Effect of CKII antisense oligodeoxynucleotides on cell growth stimulation and intracellular level of CKIIβ protein. (**A**) Timing of antisense β and cell growth stimulation. Quiescent IMR-90 cells (see legend to FIG. 1) were treated with antisense β (final concentration, 100 μg/ml) before (−24, −12, −6, −2 hours), at same time as (0 hour) or after (5 hours) EGF (final concentration, 200 ng/ml) and DNA synthesis determined by BrdU incorporation (see legend to FIG. 1). Controls received sense β. (**B**) Effect of antisense β on cellular CKIIβ protein. Treatment of cells as in **A** followed by determination of CKIIβ protein at different times after antisense CKII application by quantitative immunofluorescence image analysis of nuclear (□) and cytoplasmic (●) areas. Effect calculated from the protein levels in cells treated with sense β and antisense β as percentage decrease.[8] Shown are the means of at least three independent experiments. *Arrows* indicate intracellular CKIIβ level at time points of EGF addition after antisense β.

regulatory beta subunits (CKIIβ).[1–4] CKII has numerous cellular functions including a possible role in mitogenic signaling; in cultured cells, growth hormones such as epidermal growth factor (EGF) transiently stimulate the cytosolic CKII activity via hyperphosphorylation of subunit CKIIβ which, unlike regulatory subunits of other protein kinases, acts stimulatory.[1,2] The mitogenic signal appears to be transduced further in that CKII acts directly or indirectly on cell growth regulatory proteins present in the cytosol or the nucleus of cells. A number of such proteins is known including the oncogene products Myc, Myb, Fos, SV40 large T antigen, adenovirus E1a, papilloma virus E7, the p53 tumor suppressor protein, the serum response factor, and nucleolin (cited in ref. 4). However, direct evidence is lacking, and involvement of CKII in mitogenic stimulation remains a matter of speculation. Molecular cloning and sequencing of cDNAs of the human CKIIα and CKIIβ[5–7] allowed us to address this question by antisense oligodeoxynucleotide methodology.

Growth-arrested primary human fibroblasts (IMR-90) before growth stimulation with EGF or serum were exposed to oligodeoxynucleotides (20 mers) complementary to the translation start region of mRNAs coding for CKIIα and CKIIβ. Significant inhibition of growth stimulation was observed with both antisense α and antisense β; on the average, 51 and 64% of cells, respectively, did not enter S phase when compared with controls with sense α and sense β at equivalent concentrations (FIG. 1). Introduction of mutations into oligodeoxynucleotides reduced growth inhibition; one mutation reduced the inhibition by half, and three mutations caused total loss of inhibition. Similar results were obtained, although less pronounced, when fetal calf serum was used as growth stimulator (FIG. 1). The inhibition was reversible, excluding toxic effects of antisense oligodeoxynucleotides, and became neutralized by the simultaneous presence of the respective sense oligodeoxynucleotides.[8] Thus, the effect is highly structure-specific.

Decisive for the antiproliferative effect of CKII antisense oligodeoxynucleotides has been the timing of their application with respect to that of the growth stimulator. Maximum inhibition of growth stimulation by EGF was obtained when applied between 2 and 6 hours before EGF (FIG. 2A). As investigated in detail for CKIIβ, this time frame correlated with the down-regulation of CKIIβ protein. The intracellular level of CKIIβ protein was visualized by immunofluorescence and quantitative image analysis. The antibody employed had been raised against a CKIIβ-fusion protein, and its specificity has been proven by Western blotting of cell extracts.[8] CKIIβ protein was localized in the cytoplasm and in the nucleus with at least a five-times higher nuclear level found with both image analysis of whole cells and Western blots of nuclear versus cytoplasmic fractions.[8] Application of antisense β significantly decreased both the cytoplasmic and the nuclear CKIIβ level. The decrease became measurable within 1–2 and 2–3 hours postapplication in the cytoplasm and in the nucleus, respectively; it lasted more than 7 hours and subsequently returned to normal levels (FIG. 2B). Thus, EGF applied to cells 2 or 6 hours after antisense β meets significantly decreased CKIIβ levels. These results demonstrate that CKII is an important early phase mediator of cell growth stimulation (G_0/S-phase transition).

REFERENCES

1. TUAZON, P. T. & J. A. TRAUGH. 1991. Adv. Sec. Messenger Phosphoprot. Res. **23:** 123–164.
2. PINNA, L. A. 1990. Biochim. Biophys. Acta **1054:** 267–284.
3. PYERIN, W., E. BUROW, K. MICHAELY, D. KÜBLER & V. KINZEL. 1987. Biol. Chem. Hoppe-Seyler **368:** 215–227.
4. VOSS, H., U. WIRKNER, R. JAKOBI, N. A. HEWITT, C. SCHWAGER, J. ZIMMERMAN, W. ANSORGE & W. PYERIN. 1991. J. Biol. Chem. **266:** 13706–13711.
5. MEISNER, H., R. HELLER-HARRISON, J. BUXTON & M. P. CZECH. 1989. Biochemistry **28:** 4072–4076.
6. LOZEMAN, F. J., D. W. LITCHFIELD, C. PIENING, K. TAKIO, K. A. WALSH & E. G. KREBS. 1990. Biochemistry **29:** 8436–8447.
7. JAKOBI, R., H. VOSS & W. PYERIN. 1989. Eur. J. Biochem. **183:** 227–233.
8. PEPPERKOK, R., P. LORENZ, R. JAKOBI, W. ANSORGE & W. PYERIN. 1991. Exp. Cell Res. **197:** 245–253.

Behavioral Change after Local Administration of Antisense Sequence for Progesterone Receptor mRNA in Female Rat Hypothalamus

SONOKO OGAWA, URSULA E. OLAZÁBAL,
AND DONALD W. PFAFF

The Rockefeller University
1230 York Ave.
New York, New York 10021

Reproductive behavior of female rats is regulated by synergistic actions of estrogen (E) and progesterone (P) in the brain. Potentiation of reproductive behavior by P is highly correlated with the E-induced increase of progestin receptors in hypothalamic neurons.[1] It has been shown that E increases progesterone receptor (PR) mRNA in the ventromedial nucleus (VMN) of the hypothalamus of genetic females[2] but not males.[3] Therefore, we hypothesized that if PR mRNA increases are correlated with facilitation of reproductive behavior, then effective PR antisense sequences might reduce this behavior.

Antisense oligonucleotides (15 bases) were chosen to span the PR mRNA translation start site. Control oligomers were composed of the same nucleotide bases in scrambled order. Avoidance of possible cross-hybridizations was confirmed. Ovariectomized female rats ($n = 12$) were implanted with bilateral cannulas directed at the VMN. Following 1-week postsurgery recovery, they were tested four times for reproductive behavior, each rat serving as its own control. In each experiment, estradiol benzoate (EB; 1 μg) was injected subcutaneously at hour 0; then antisense or scrambled control oligomers were microinjected in the hypothalamus at hour 12 or 24; finally, P (500 μg) was given subcutaneously at hour 44, four hours before the behavioral test. Starting 48 hours after EB injections, females were tested for lordosis behavior in two behavioral paradigms, a mating behavior test with a male and a lordosis reflex test with manual stimulation. In the mating behavior test, occurrence of proceptive behavior (hopping or darting), rejection behavior (biting or kicking), and vocalization was also recorded.

It was found that lordosis behavior was reduced by antisense for PR mRNA (*vs* scrambled control), whether measured by the mating behavior test (73% ↓, $p < 0.01$) or the lordosis reflex test (44% ↓, $p < 0.05$). Occurrence of proceptive behavior, known to be strongly P dependent, was also greatly reduced (*vs* scrambled control; 80% ↓, $p < 0.01$) by the administration of antisense oligomer. In contrast, rejection behaviors and vocalization were not affected. Effects of antisense were specific with respect to time of administration. Antisense reduced both lordosis behavior and proceptive behavior when given 12 but not 24 hours after EB. This finding is consistent with a previous *in situ* hybridization study which showed that PR mRNA induction is already at peak by 24 hours after estrogen,[2] which would presumably allow time for PR synthesis before the later antisense administration timepoint.

Histochemical experiments to examine possible antisense effects on PR immunoreactivity and [^{3}H]-progestin binding in hypothalamic neurons are under way.

The present study indicates that interrupting mRNA function for PR, a transcription factor, by direct hypothalamic administration of antisense oligomer can have behavioral consequences.

REFERENCES

1. PARSONS, B. & D. W. PFAFF. 1985. Progesterone receptors in CNS correlated with reproductive behavior. *In* Current Topics in Neuroendocrinology. D. Ganten & D. W. Pfaff, eds. Vol. **5:** 103–140. Springer-Verlag. Berlin.
2. ROMANO, G. J., A. KRUST & D. W. PFAFF. 1989. Expression and estrogen regulation of progesterone receptor mRNA in neurons of the mediobasal hypothalamus: An *in situ* hybridization study. Mol. Endocrinol. **3:** 1295–1300.
3. LAUBER, A. H., G. J. ROMANO & D. W. PFAFF. 1991. Sex difference in estradiol regulation of progestin receptor mRNA in rat mediobasal hypothalamus as demonstrated by *in situ* hybridization. Neuroendocrinology **53:** 608–613.

Liposomal Delivery of Antisense Oligodeoxynucleotides

Application to the Inhibition of the Multidrug Resistance in Cancer Cells

ALAIN R. THIERRY[a] AND ANATOLY DRITSCHILO

Department of Radiation Medicine
Vincent T. Lombardi Cancer Research Center
Georgetown University School of Medicine
3800 Reservoir Road
Washington, DC 20007

The potential of antisense oligodeoxynucleotides (oligo(dn)) as a new molecular targeting tool to treat disease is hindered by their instability in a biological environment.[1] Although the use of oligodeoxynucleotide analogs results in improvement of the *in vitro* antisense activity, these oligomers exhibit rapid *in vivo* elimination. Clinical evaluation of this new class of compounds seems to rely on the employment of an effective delivery system. Therefore, we developed a new preparation of liposomally encapsulated oligodeoxynucleotides (Lip-oligo(dn)).[2,3] We examined the potential of liposomal antisense oligo(dn) in modulating the expression of multidrug resistance (MDR). Emergence of MDR in cancer cells involves the overexpression of the P-glycoprotein, a plasma membrane glycoprotein encoded by the *mdr-1* gene, which functions as a drug efflux pump.[4] Inhibition of the P-glycoprotein by modulating *mdr-1* gene expression may be a novel way to sensitize MDR tumor cells. The 15-mer cap oligo(dn) (phosphorothioate modification at the first two bases of both ends) in the antisense orientation to the 5′ end of the coding region of the human *mdr-1* gene was encapsulated in liposomes and delivered to MDR human ovarian carcinoma SKLVB cells and human breast carcinoma MCF-7/ADR cells.

RESULTS AND DISCUSSION

Small unilamellar liposomes formed from cardiolipin, phosphatidylcholine, and cholesterol using the minimal volume entrapment (MVE) technique encapsulate oligo(dn) with high efficiency (80% from a concentration of 10 mg/ml). Furthermore, unmodified oligo(d15) are completely protected from nuclease attack up to 1 week when they are presented into MVE liposomes. In contrast, unmodified oligo(d15) in the 10% serum-containing medium are mostly degraded within 30 minutes.[2,3] In addition to the effective protection from environmental degradation, lip-oligo(dn) have been shown to penetrate much more efficiently than do free modified oligo(dn).[2] Lip-oligo(dn) internalization depends on the cell line used and may be as high as 18 times greater than that of free oligo(dn) as observed in the lymphoma MOLT-3 cells.[3] In our study, a fourfold increase in cellular accumulation was observed in SKVLB cells using Lip-oligo(d15) as compared to free oligo(d15).

[a] To whom correspondence should be sent.

Following a 4-days incubation with 5 μM free or liposomal oligo(dn), treated cells were tested for P-glycoprotein content and doxorubicin cytotoxicity (FIG. 1). Results showed that cap oligo(dn) complementary to the 5′ end coding region of the *mdr-1* gene encapsulated in liposomes effectively circumvented the MDR phenotype in both cell lines. Approximately a fourfold and a twofold reduction of P-glycoprotein content was observed in MCF-7/ADR and SKVLB cells, respectively. A fivefold and a fourfold inhibition of doxorubicin resistance was determined in MCF-7/ADR

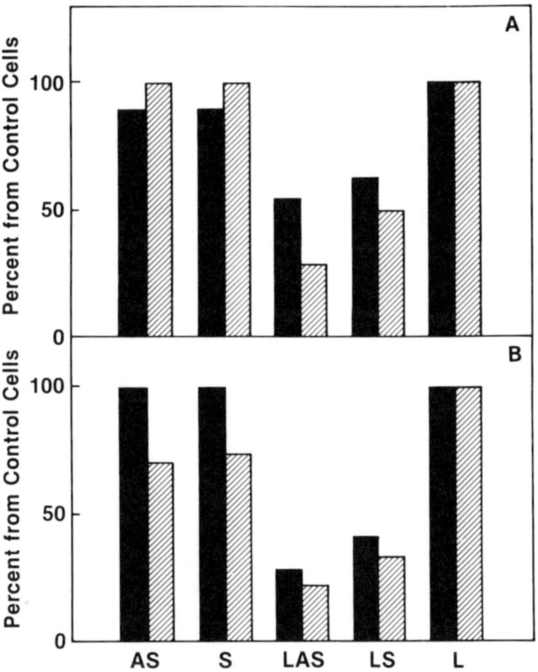

FIGURE 1. Circumvention of the multidrug resistance phenotype in cells treated with free and liposomal oligodeoxynucleotides. SKVLB (**A**) and MCF-7/ADR (**B**) cells were incubated 4 days with 5 μM free antisense (AS) or sense (S) and liposomal antisense (LAS) or sense (LS) oligo(dn). Cellular P-gp content (■) was determined using immuno-flow cytometry and expressed as a percentage of control cells. Doxorubicin resistance (▨) was evaluated after cell treatment for 1 hour with the drug at a concentration corresponding to their IC50 using the clonogenic assay. Results are expressed as a percentage of control cells treated only with doxorubicin.

and SKVLB cells, respectively. These data confirm previous observations of the role of the P-glycoprotein in the emergence of the MDR phenomenon. Empty liposomes and free oligo(dn) showed little or no effect. The effect described here is in part sequence nonspecific, as sense sequence encapsulated in liposomes also exhibited significant but lower modulation of MDR phenotype. Moreover, no effect was observed after treatment in the same conditions with either antisense or sense sequences, in the parental non-MDR SKOV3 cell line from which SKVLB cells are derived (data not shown). This result illustrates the potential of the sense approach and the complexity of the mechanisms of action of the oligo(dn).

In conclusion, our studies demonstrate that the liposomal transport of oligo(dn) may improve the antisense activity and might provide critical benefits for the clinical application of this new class of compounds.

REFERENCES

1. HELENE, C. & J. J. TOULME. 1990. Specific regulation of gene expression by antisense, sense and antigene nucleic acids. Biochem. Biophys. Acta **1049:** 99–125.
2. THIERRY, A. R., A. RAHMAN & A. DRITSCHILO. 1990. Liposomal carrier system as a novel approach to transport oligonucleotides into cells. Proc. Am. Assoc. Cancer Res. **31:** 445.
3. THIERRY, A. R., A. RAHMAN & A. DRITSCHILO. 1992. Liposomal delivery as a new approach to transport antisense oligonucleotides. *In* Gene Regulation: Biology of Antisense RNA and DNA. R. P. Erickson & J. G. Izant, eds. Vol. **1:** 147–161. Raven Press. New York, NY.
4. ENDICOTT, J. A. & V. LING. 1989. The biochemistry of P-glycoprotein-mediated multidrug resistance. Ann. Rev. Biochem. **58:** 137–171.

Design, Synthesis and Assay of Tetrapeptide-Acridine Mimics of Ribonuclease

CHING-HSUAN TUNG AND STANLEY STEIN

Center for Advanced Biotechnology and Medicine / UMDNJ
679 Hoes Lane
Piscataway, New Jersey 08854
and
Chemistry Department
Rutgers University
Piscataway, New Jersey 08855

RNA has become an interesting target for artificial hydrolytic enzymes during the last few years, partly because of possible therapeutic applications.[1] We now report a new type of compound that is capable of cleaving RNA under physiological conditions. On the basis of the active site and cleaving mechanism of ribonuclease A[2], a series of ribonuclease mimics was designed and synthesized. Each of these mimics contains a tetrapeptide and an intercalator (FIG. 1). Every amino acid on the tetrapeptide has its own specific function. The first three amino acid residues from the NH_2-terminus of the peptide are designed to be the active site. The first amino acid is preferred to have a positively charged group. The second position is a spacer which allows the reactive groups on both of its sides to have a better orientation for the reaction. A basic group is suitable at the third position to act as a proton-withdrawing group. The fourth amino acid functions as a tether to connect the active region and the binding region. The intercalator is designed to be the binding domain, and with its help the active site has a better spatial configuration. Various mimics were shown by computer modeling to be capable of assuming the required orientation on a double-stranded RNA substrate. Hydrolysis of the RNA backbone was catalyzed by these tetrapeptide-acridine conjugates.

Random cleavage of RNAs was detected by two different techniques, denaturing electrophoresis and gel-permeation chromatography. Gel electrophoresis could definitively demonstrate the extent of cleavage in a qualitative manner. In gel-permeation chromatography, the samples could be separated according to size, with larger molecules eluting earlier. Although a precise relation between peak shift and number of phosphodiester bond cleavages could not be determined, these measurements allowed a semiquantitative means for comparing different compounds.

Systematic substitution of the amino acid residues in the enzyme mimics was done to determine the possible reaction mechanism and to find the most potent enzyme mimic. The best derivatives are His-Pro-His-Lys(Acridinyl)-NH_2, His-Gly-His-Lys(Acridinyl)-NH_2, and Gly-Gly-His-Lys(Acridinyl)-NH_2. A low concentration of any of these enzyme mimics (200 μM) is sufficient to catalyze rRNA hydrolysis within 24 hours. The reactions were carried out under physiological conditions, with phosphate-buffered saline solution at 37°C.

Different types of RNA were used to test the catalytic capability of the enzyme mimics (FIG. 2). Ribosomal RNA and single-stranded messenger RNA are good substrates for the enzyme mimics, but cleavage could not be detected for transfer

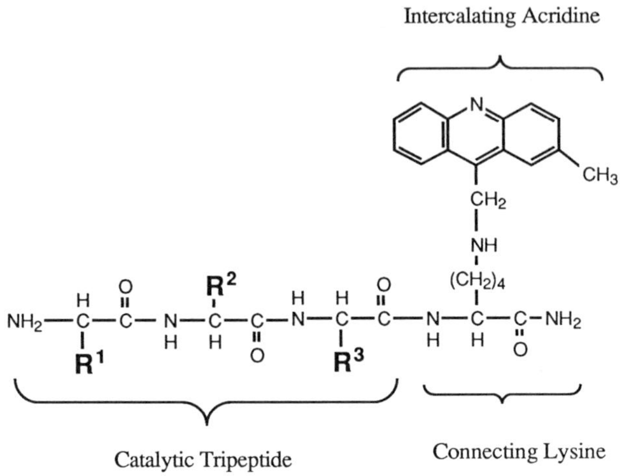

FIGURE 1. Arrangement of the peptide-acridine complex. The tripeptide, which comprises the catalytic structure, was varied in this study.

RNA or double-stranded viral RNA. The substrate specificity is interesting in that these mimics are behaving as endoribonucleases. Apparently, fragments smaller than about 80 nucleotides are resistant to further cleavage. Single-stranded RNA has considerable secondary structures, and we may speculate that the acridine moiety intercalates into double-stranded regions, but cleavage occurs at nearby single-stranded regions. Eventually, these preferred stem-loop, hairpin, or bulge structures would no longer exist for further hydrolysis.

The goal of producing a ribonuclease mimic, using a small peptide for the active site and an intercalator for substrate binding, has been achieved, but the actual

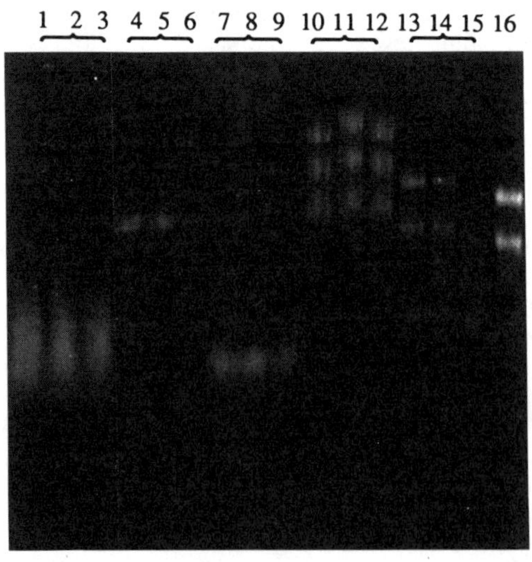

FIGURE 2. RNA cleavage assay (composite from two different gels). *Lanes 1, 4, 7, 10, 13, and 16* are 0-hour negative control (RNA alone), *lanes 2, 5, 8, 11, and 14* are 24-hour negative control (RNA alone), and *lanes 3, 6, 9, 12, and 15* are RNA and Gly-Gly-His-Lys(Acridinyl)-NH_2 with 24 hours of incubation. *Lanes 1–3* are poly(AU), *lanes 4–6* are m transcript ssRNA of the dsRNA segment of killer virus of yeast, *lanes 7–9* are tRNA^{phe}, *lanes 10–12* are reovirus type 3 dsRNA, *lanes 13–15* are 18S/28S rRNA, and *lane 16* is 16S/23S rRNA.

reaction mechanism is still under investigation. Finally, ribonuclease mimics, such as these peptide-acridine conjugates, might be useful in antisense therapeutics. Ribonuclease appended to an oligonucleotide has been shown to cleave its complementary strand,[3] but for *in vivo* applications, it might be preferable to use a lower molecular weight mimic for improved uptake and stability. Much work remains before the present compounds can reach this stage.

REFERENCES

1. HELENE, C. & J.-J. TOULME. 1990. Biochim. Biophys. Acta **1049:** 99–125.
2. BRESLOW, R., D.-L. HUANG & E. ANSLYN. 1989. Proc. Natl. Acad. Sci. USA **86:** 1746–1750.
3. ZUCKERMANN, R. N. & P. G. SCHULTZ. 1989. Proc. Natl. Acad. Sci. USA **86:** 1766–1770.

Chemical Modifications to Improve Uptake and Bioavailability of Antisense Oligonucleotides

M. MANOHARAN,[a] L. K. JOHNSON, D. P. C. McGEE, C. J. GUINOSSO, K. RAMASAMY, R. H. SPRINGER, C. F. BENNETT, D. J. ECKER, T. VICKERS, L. COWSERT, AND P. D. COOK

Isis Pharmaceuticals
2280 Faraday Ave.
Carlsbad, California 92008

The fate and function of antisense oligonucleotides are primarily controlled by their uptake and distribution in the cell.[1] However, the efficiency of uptake is hampered by the negative charge on the backbone and also by the hydrophilic properties of oligonucleotides. Solutions to the uptake problem would be the modification of the antisense oligonucleotide to include: (1) hydrophobic moieties, (2) cationic modifications to overcome charge effects, (3) cell receptor binding molecules, and (4) amphipathic modifications having one or more of the foregoing properties. We have initiated chemical modifications[2,3] aimed at improving uptake of antisense oligonucleotides using these guidelines, and our preliminary results are summarized here.

To confer hydrophobicity to oligonucleotides, cholic acid was activated and conjugated to oligonucleotide DNA phosphodiesters, phosphodiester RNA mimics (2′-OMe analogs) and phosphorothioates at either the 5′ or the 3′ end using the appropriate aminolinker (FIG. 1). In evaluating hybridization properties of cholic acid-conjugated oligonucleotides we observed that these conjugates did not affect the melting temperature of the parent oligomers against both DNA and RNA. Moreover, in the case of diesters, the 3′ conjugation offered significant nuclease stability in fetal calf serum (half-life >24 hours). Thus, the conjugated diesters had a lifetime similar to that of unmodified thioates.

Uptake was monitored either by fluorescent microscopy of the oligonucleotides in cells assessing subcellular distribution or by cellular activity in measuring protein synthesis. Fluorescent microscopy shows cellular localization of oligonucleotides, and protein synthesis assays with nonfluorescent conjugates showed the ultimate performance of these antisense oligonucleotides. Fluorescein was attached to the 5′ end of the oligonucleotide, whereas cholic acid was attached to the 3′ end. Oligonucleotides targeted against human intercellular adhesion molecule-1 (ICAM-1), human immunodeficiency virus (HIV-1), and bovine papillomavirus (BPV-1) were used to study the effects of cholic acid conjugation on antisense activity. The ICAM-1 and BPV-1 oligos were 2′-deoxy phosphorothioates, whereas the antisense HIV-1 oligos were 2′-O-methyl phosphodiesters.

With ICAM-1, we observed localization of cholic acid-conjugated oligonucleotides in the cytoplasm by the fluorescent tag. In the protein synthesis assay, the conjugate did not change the potency of the parent oligonucleotide. However, in the

[a] To whom correspondence should be addressed.

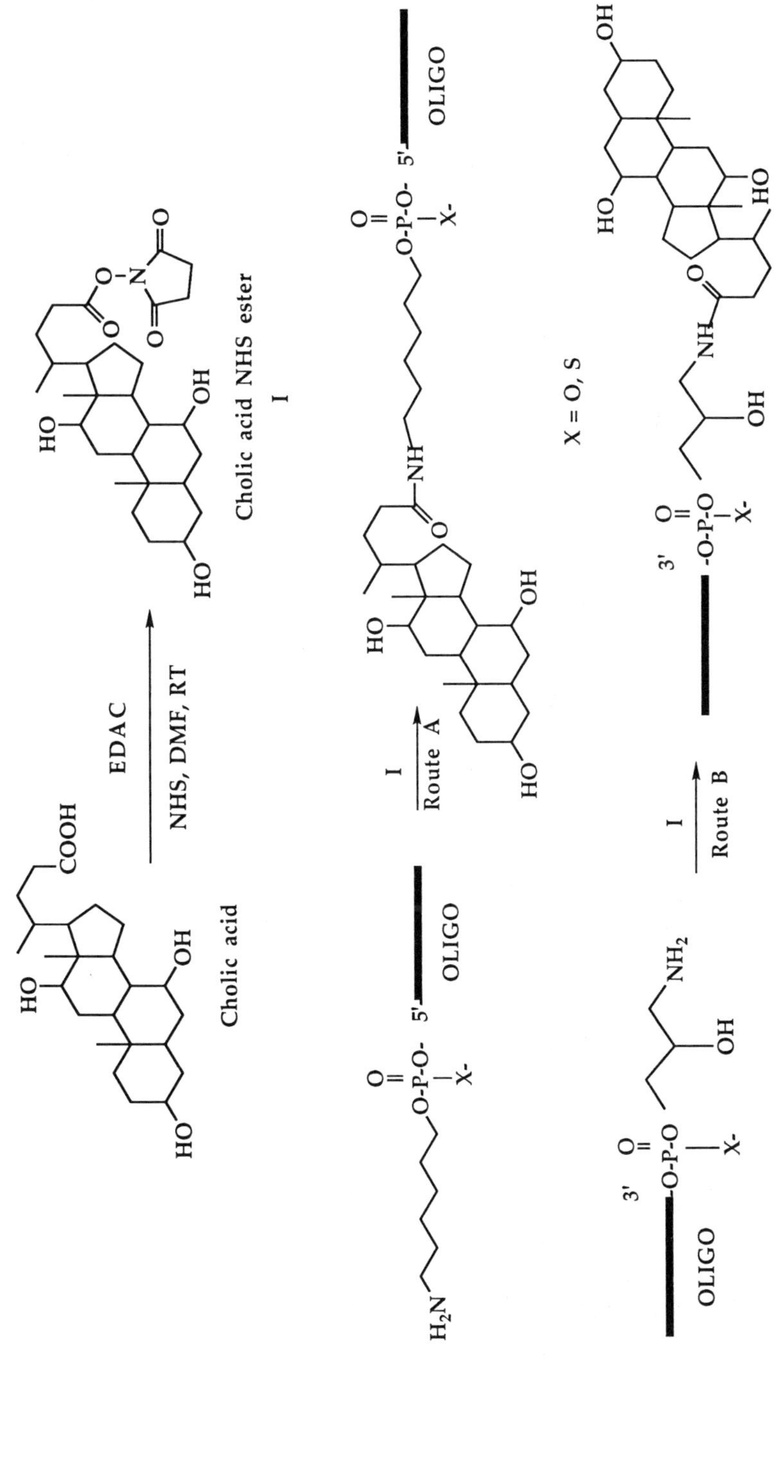

FIGURE 1. Cholic acid conjugation to oligonucleotides at either the 5′ (**route A**) or the 3′ end (**route B**).

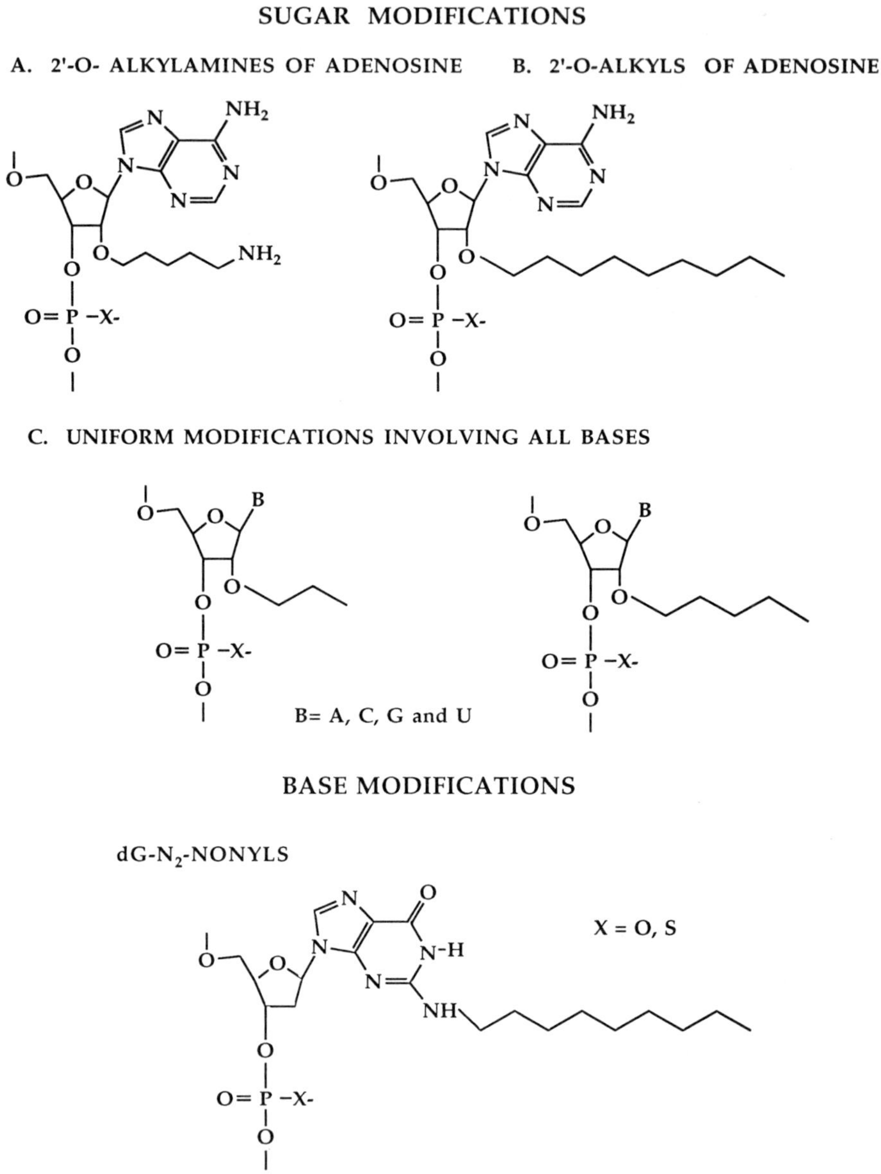

FIGURE 2. Sugar and base modifications incorporated into oligonucleotides.

presence of excess amounts of a cationic lipid, *N*-[1-(2,3-dioleyloxy)propyl-*N,N,N*-trimethylammonium chloride (DOTMA), the conjugate was localized in the nucleus, and the cellular fluorescent intensity was enhanced. Also, protein synthesis was significantly reduced. Our control experiments prove that the antisense effects observed are sequence specific.

The HIV-1 and BPV-1 oligonucleotides were evaluated in transactivation assays. Preliminary results indicate some uptake enhancement for the cholic acid conjugates. In BPV-1 assays, the 3′ conjugate performed better than did the 5′ conjugates. In HIV-1 assays, there is approximately a threefold enhancement of activity for a 5′ conjugated oligomer and a twofold enhancement for a 3′ conjugated oligomer. These experiments are being repeated with increased pre-incubation time during the assays.

In an alternate approach to chemically modify oligonucleotides, nucleosides were synthesized by incorporating alkyl groups and alkylamino groups either in the 2′ position of sugars or at the N_2 position of purines. ICAM-1 oligonucleotides were synthesized, incorporating these modifications either in a site-specific manner in the sugar ring of adenosines (2′-*O*-nonyls and 2′-*O*-pentylamines)[2,3] and N_2 position of guanosine (nonyls) or in a uniform fashion (2′-*O*-propyls and 2′-*O* -pentyls). (FIG. 2). Evaluation of these modified oligonucleotides for uptake enhancement is in progress:

TG*G GAG CCA TAG CGA GG*C (thioate, purine modification)
TGG GA*G CCA* TA*G CGA* GGC (thioate, sugar modification in adenosine)
UGG GAG CCA UAG CGA GG dC (thioate, uniform 2′-sugar modification).

REFERENCES

1. COOK, P. D. 1991. Anti-Cancer Drug Design **6:** 585 (see also references cited therein).
2. GUINOSSO, C. J., G. D. HOKE, S. FRIER, J. F. MARTIN, D. J. ECKER, C. K. MIRABELLI, S. T. CROOKE & P. D. COOK. 1991. Nucleosides Nucleotides **10:** 259.
3. MANOHARAN, M., C. J. GUINOSSO & P. D. COOK. 1991. Tetrahedron Lett. **32:** 7171.

Separation of Synthetic Phosphorothioate Oligonucleotides from Phosphodiester-Defect Species by Strong-Anion Exchange HPLC

B. JOHN BERGOT AND GERALD ZON

Applied Biosystems, Inc., Therapeutics Group
400 Lincoln Centre Drive
Foster City, California 94404

Among the various classes of modified oligonucleotides under investigation as potential antisense drugs,[1] phosphorothioates have received considerable attention. These nuclease-resistant analogs have sulfur in place of one of the nonbridging oxygens at the internucleoside phosphate linkage and are commonly synthesized by automated solid-phase methods employing stepwise sulfurization with, for example, tetraethylthiuram disulfide (TETD).

Despite the high efficiency of TETD and alternative sulfurizing agents, the presence of low levels (<1 mole% per linkage) of randomly distributed phosphodiester "defect" species can result from either incomplete sulfurization, alkaline hydrolysis, or other conceivable mechanisms. For an antisense phosphorothioate, this leads to full-length material having linkages comprised entirely of phosphorothioate (PS) moieties as well as subpopulations containing, randomly, either one phosphodiester (PO), two PO, or more. This mixture of materials follows a binomial distribution derived mathematically in the following manner for a 21 mer having 20 internucleoside linkages:

$$(x + y)^{20} = x^{20} + 20x^{19}y + 190x^{18}y^2 + 1140x^{17}y^3 + \dots$$

$$x = \text{mole-fraction PS}$$

$$y = \text{mole-fraction PO}$$

At 99 mole% overall sulfurization ($x = 0.99$) accompanied by 1 mole% oxygenation ($y = 0.01$), the numerical values of the first four terms in the expansion are 0.818, 0.165, 0.016, and 0.00096, which indicate 81.8, 16.5, 1.6, and 0.096 mole% for the zero PO (all PS), 1 PO, 2 PO, and 3 PO populations, respectively. This indicates the criticality of achieving ultraefficient sulfurization, that is, ≥ 99.5 mole%.

Accurate measurement of sulfur content in a phosphorothioate oligonucleotide is important for determination of product composition in the context of both research and process development as well as drug quality control. Up to this time, such quantitation has only been possible using integrated intensities of PS and PO ^{31}P NMR signals. Interestingly, conditions reported below for strong-ion exchange HPLC have been discovered to afford separation of the aforementioned zero, 1, 2, etc. PO species, regardless of the position of the PO linkage within the oligomer chain.[2] *This direct method of measurement is superior to NMR in that it requires much less sample and very little analysis time, and it is easily automated and uses less expensive equipment commonly found in most laboratories.*

Purified detritylated samples (0.2–0.5 OD) were analyzed on a series 40 B10 LC equipped with an ISS-200 autosampler and 1020 Data System (all Perkin-Elmer). The column was a PL-SAX (Polymer Laboratories), 8–10 μm particle size, 100 Å porosity, 15 cm × 7.5 mm ID. A ternary gradient system was employed, where solvent A was 50 mm phosphate buffer, pH 8.2, containing 5% (v/v) CH_3CN, solvent B was

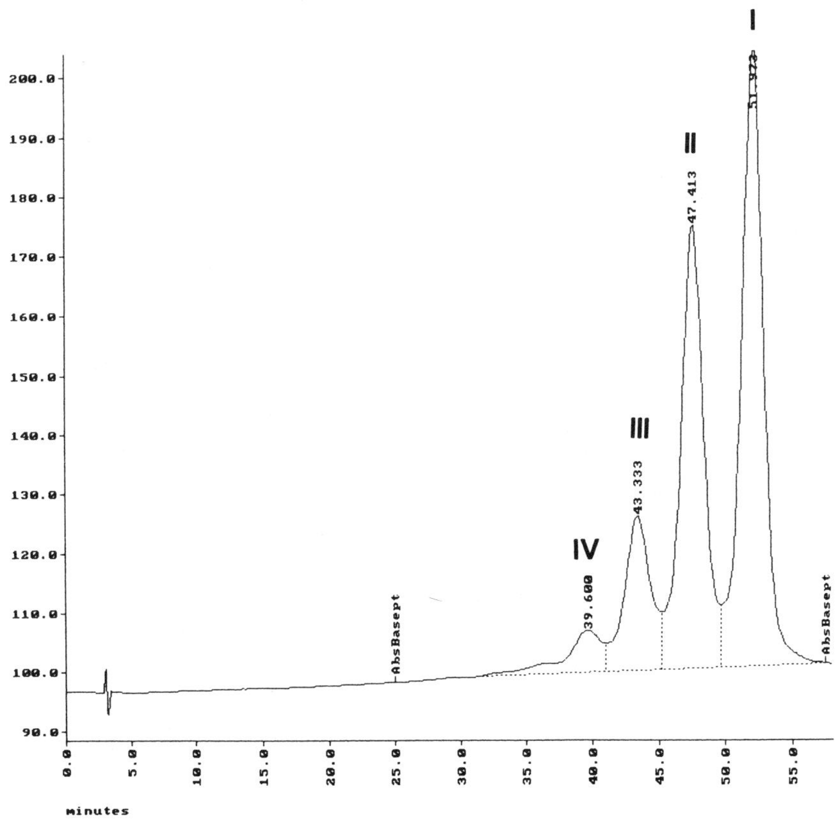

FIGURE 1. HPLC trace obtained for a 21-mer phosphorothioate DNA analog resolution standard having mole% PO/PS = 3.5/96.5, as measured independently by ^{31}P NMR. Peak I = zero PO, peak II = 1 PO, peak III = 2 PO, and peak IV = 3 PO.

1.5 M KBr in 50 mM phosphate buffer, pH 6.7, containing 20% (v/v) CH_3CN, and solvent C was CH_3CN. Flow rate was 1.5 ml/min, and peak monitoring was by UV at 260 nm using a 759A UV Detector (Applied Biosystems). Peak elution times (FIG. 1) are in the order of zero >1 >2 etc. PO, as verified by collection and ^{31}P NMR analysis. Various mixtures were analyzed by both HPLC and ^{31}P NMR in order to validate the accuracy of the method.

Among columns evaluated (Nucleogen-DEAE [Macherey-Nagel], Gen-Pak FAX [Waters], Zorbax Oligo [duPont], Vydac 304 OL [Separations Group], TSK DEAE-NPR [Supelco], DEAE-WAX [Whatman], and PL-SAX [Polymer Labs]), PL-SAX consistently displayed better separation characteristics.

REFERENCES

1. Zon, G. & W. J. Stec. Phosphorothioate oligonucleotides. *In* Oligonucleotides and Their Analogues. F. Eckstein, ed.: 87–108. IRL Press. London.
2. Bergot, B. J. & W. Egan. J. Chromatogr., in press.

Mechanistic Aspects of the Interaction of Polyanionic Oligodeoxynucleotides with HL60 Cells

C. A. STEIN,[a,b] J. TONKINSON,[a] J. GERVASONI,[a]
L. M. ZHANG,[a] S. KRISHNA,[a] R. TAUB,[a] S. FIELDS,[a]
T. DELOHERY,[a] AND S. ROTENBERG[c]

[a]*Division of Medical Oncology*
Columbia University
College of Physicians and Surgeons
630 W. 168 St.
New York, New York 10032

[c]*Department of Chemistry and Biochemistry*
Queens College CUNY
Kissena Blvd.
Queens, New York 11367

Although polyanionic (PA) antisense oligodeoxynucleotides may potentially produce novel therapy[1,2] because of their ability to inhibit gene expression in a sequence-specific manner, little is known about the fundamental interactions of these materials with cells. Loke *et al.*[3] and Yakubov *et al.*[4] recognized that an 80-kD protein exists on the surface of multiple cell types that appears to bind all PA oligomers irrespective of base sequence. However, it is unclear what role this protein plays in the overall internalization of oligomer, or indeed what regulates the rate of oligomer internalization. To study this problem, we synthesized 5′ fluorescently labeled oligomers and used fluorescence-activated cell sorting (FACS) analysis to determine parameters of oligomer cell association (uptake) and internalization.

The reporter oligomer in most experiments was 5′ fluorescein isothiocyanate-OdT 15 (FOdT 15). Because of difficulties in separating this material from free fluorescein, the final product was purified by dialysis, reverse phase HPLC (PRP-1 support), and 7M urea 20% PAGE before gel filtration (Sephadex G25) and lithium perchlorate precipitation. Titration of HL60 cells with FOdT 15 revealed saturation binding at equilibrium, with the value of the dissociation constant (K_d1) being 22 nM. Values of K_d for phosphorothioate (PS) oligomer competitors of FOdT 15 binding were determined. A dramatic oligomer chain length dependence of the values of K_d (300 nM, SdC3, 5: 53 nM, SdC8, 10, 12, 15: 3 nM, SdC18, 21, 28) was observed. The results may imply that PS oligomers bind to the putative cell surface binding protein at three distinct loci, and that the longer oligomers can span all the loci. FOdT 15 can be removed from the cell surface by treatment with SdC28, because of the difference in K_d. This was confirmed by laser-assisted confocal microscopy. Cellular internalization, assessed after removal of cell surface oligomer, revealed a concentration x time-dependent process consistent with pinocytosis. However, the pinocytotic process is inhibited by inhibitors of protein kinase C (PKC). These include H7, 1% DMSO, and the PKC oligopeptide pseudosubstrate. Alternatively, stimulators of

[b]Address all correspondence to this author.

PKC activity, including thrombin and phorbol 12, 13 myristate acetate, increase FOdT15 internalization. Furthermore, SdC28 is a potent inhibitor of the pinocytosis of fluoresceinated albumin and dextran sulfate; however, SdC15 and OdC28 are weak inhibitors, and OdC15 and SdC5 do not inhibit. Also, SdC28 is a direct, potent inhibitor of phosphatidyl serine-stimulated PKC phosphorylation (IC 50 = 1 μM), whereas the IC 50 for OdC28 is approximately 2 logs higher. Thus, it appears that the rate of oligomer internalization is dependent on the activity of PKC and that SdC28, in HL60 cells, blocks its own uptake.

REFERENCES

1. Matsukura, M., G. Zon, K. Shinozuka, M. Robert-Guroff, T. Shimada, C. A. Stein, H. Mitsuya, F. Wong-Staal, J. Cohen & S. Broder. 1989. Proc. Natl. Acad. Sci. USA **86:** 4244–4248.
2. Reed, J., C. A. Stein, C. Subasinghe, S. Haldar, C. Croce, S. Yum & J. Cohen. 1990. Cancer Res. **50:** 6565–6570.
3. Loke, S., C. A. Stein, X. Zhang, K. Mori, M. Nakanishi, C. Subasinghe, J. Cohen & L. Neckers. 1989. Proc. Natl. Acad. Sci. **86:** 3474–3478.
4. Yakubov, L., E. Deeva, V. Zarytova, E. Ivanova, A. Ryte, L. Yurchenko & V. Vlassov. 1989. Proc. Natl. Acad. Sci. USA **86:** 6454–6458.

Cell Physiology of Antisense TGF-β Oligomers in Hepatoma Cells

MITSUHIRO INAGAKI,[a] KAZUMI TOGAWA,[a]
BRIAN I. CARR,[a] KRISHNA GHOSH,[b] AND
JACK S. COHEN[b]

[a]*Department of Surgery*
University of Pittsburgh
Pittsburgh, Pennsylvania 15261

[b]*Department of Pharmacology*
Georgetown University
Washington, DC

To study the role of endogenous transforming growth factor β (TGF-β) in human hepatoma cell lines, we examined the effects of antisense TGF-β oligonucleotides (oligomers) on cell growth and gene expression as well as the nature of the cell binding and uptake of the oligomers.

Unmodified (PO) and phosphorothioate (PS) oligomers were synthesized on an automated synthesizer. The sequence of an antisense TGF-β 25 mer and that of a sense TGF-β 25 mer were as follows: (a) antisense: 5′-CAG CCC GGA GGG CGG CAT GGG GGA G-3′; (b) sense: 5′-CTC CCC CAT GCC GCC CTC CGG GCT G-3′.

Hep 3B cell growth was dramatically suppressed following the addition of an antisense TGF-β PS oligomer (0.5–10 μM) in modified Eagle's medium with 10% fetal bovine serum (FBS) (FIG. 1). This effect appeared to be sequence specific, as a sense TGF-β PS oligomer had little effect on cell growth. Gel electrophoresis of a ^{32}P end-labeled PS oligomer removed from the conditioned medium containing 10% FBS at 37°C indicated that the oligomer was undegraded up to 48 hours.

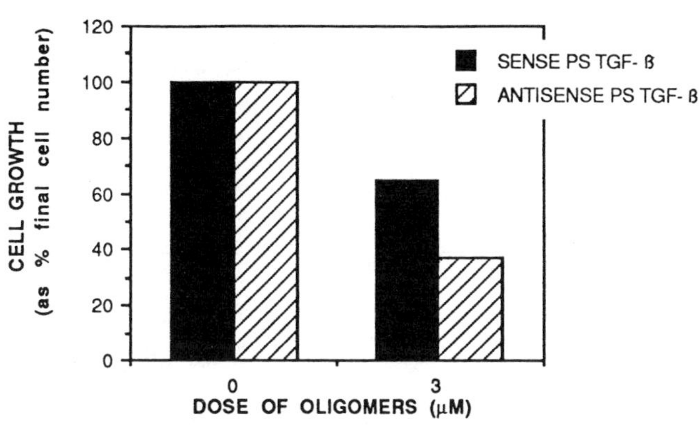

FIGURE 1. Effect of treatment with antisense or sense TGF-β PS oligomers on Hep 3B cell growth (culture day 7). Experiments were performed in duplicate.

From a typical hybridization experiment, we detected the annealing of antisense TGF-β PS oligomer to cell mRNA purified from oligomer-treated cells after 37°C incubation. The antisense and 1 base mismatch TGF-β PS oligomer could be detected as hybridizing with cell mRNA. With subcellular separation techniques, cells incubated with ^{32}P end-labeled oligomer were separated into nuclear and

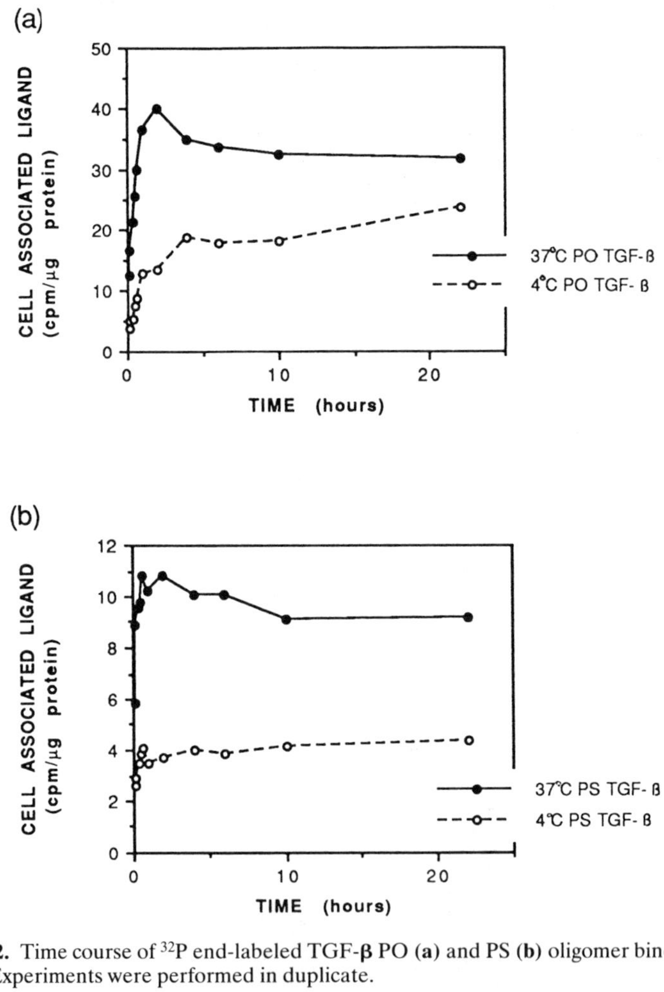

FIGURE 2. Time course of ^{32}P end-labeled TGF-**β** PO (**a**) and PS (**b**) oligomer binding to Hep 3B cells. Experiments were performed in duplicate.

cytosolic fractions. Radiolabeled oligomer was detected predominantly in the cytosolic fraction.

To characterize the oligomer binding to Hep 3B cells, cells were incubated with ^{32}P end-labeled PO oligomer, and cell-associated ligand was counted. Cell-associated radioactivity increased with time at both 4°C and 37°C (FIG. 2). In a saturation binding assay, oligomer binding to Hep 3B cells was saturated over a 100 μM concentration. Scatchard analysis of the data yielded a linear binding plot. The calculated dissociation constant (K_d) was 37 nM and the maximal binding capacity (B_{max}) was 57.8 pmol/mg protein. The F statistics test of the data yielded a best fit consistent with a one-site model.

Surprisingly, expression of TGF-β mRNA treated with antisense TGF-β PS oligomer was increased twofold compared with that of control cells, although no differences in the protein levels could be detected.

REFERENCES

1. JAROSZEWSKI, J. W. & J. S. COHEN. 1991. Cellular uptake of antisense oligonucleotides. Adv. Drug Delivery Rev. **6:** 235–250.
2. LOKE, S. L., C. A. STEIN, X. H. ZHANG, K. MORI, M. NAKANISHI, C. SUBASINGHE, J. S. COHEN & L. M. NECKERS. 1989. Characterization of oligonucleotide transport into living cells. Proc. Natl. Acad. Sci. USA **86:** 3474–3478.

Capped Oligodeoxynucleotide Phosphorothioates

Pharmacokinetics and Stability in Mice

J. TEMSAMANI,[a] J.-Y. TANG,[a] AND SUDHIR AGRAWAL[b]

[a]*Hybridon, Inc.*
One Innovation Drive
Worcester, Massachusetts 01605

[b]*Worcester Foundation for Experimental Biology*
222 Maple Avenue
Shrewsbury, Massachusetts 01545

Antisense oligonucleotide phosphorothioates offer a chemotherapeutic approach to acquired immunodeficiency syndrome (AIDS) by selectively blocking the expression of genes from HIV, the causative agent of AIDS.[1] The oligonucleotide phosphorothioates are effective in inhibiting HIV replication in tissue culture. A pharmacokinetic study of oligonucleotide phosphorothioates[2] showed that oligonucleotide was distributed throughout most tissues and was quite stable in most organs except the liver and kidney. In the liver and kidney, extensive degradation of oligonucleotide phosphorothioate was observed.

In the present study, various "end capped" oligonucleotide phosphorothioates were synthesized (TABLE 1) and studied for their comparative stability in monkey plasma (FIG. 1) and their pharmacokinetics and *in vivo* stability in mice.

Synthesis of the oligonucleotide phosphorothioate was carried out with H-phosphonate chemistry using an automated synthesizer. 3′-, 3′,5′-, and 5′-modifications on oligonucleotide phosphorothioates were carried out using *N-Fmoc-o-DMTr-3-amino-1,2-propanediol reagent* either as H-phosphonate for 5′-end modification or attached to CPG for 3′-end modification. After assembly of the required sequence, CPG-bound oligonucleoside H-phosphonate was oxidized with elemental sulfur to generate phosphorothioate internucleotide linkages. Oligonucleotide phosphorothioates were deprotected and purified as reported earlier.[3] ^{35}S-labeled end-capped oligonucleotide phosphorothioates were synthesized by oxidizing the CPG-bound oligonucleoside H-phosphonate with $^{35}S_8$[2]. Study of the stability of "end-capped" oligonucleotide phosphorothioates in monkey plasma (FIG. 1) showed that 3′-end capping of oligonucleotide phosphorothioate protected the oligonucleotides from degradation, whereas 5′-end capping failed to do so. This indicates that degradation of oligonucleotide phosphorothioate is attributable mainly to 3′-exonucleases.

End capped oligonucleotide phosphorothioates were studied for their pharmacokinetics and stability in mice. A single dose of 30 mg/kg was administered intravenously. The rate of excretion was about 30% at 24 hours postdosing, independent of end-capped modification. Analysis of the stability of oligonucleotide phosphorothioate excreted in the urine by polyacrylamide gel electrophoresis showed that uncapped and 5′-capped oligonucleotide phosphorothioates were degraded extensively, whereas 3′-capped and 3′,5′-capped oligonucleotide phosphorothioate remained almost intact as 20 mers after 24 hours postdosing.

TABLE 1. Structure of the S-Oligonucleotides

Uncapped	ACA CCC AAT TCT GAA AAT GG
5′-capped	**X**-ACA CCC AAT TCT GAA AAT GG
3′-capped	ACA CCC AAT TCT GAA AAT GG-**X**
5′-3′-capped	**X**-ACA CCC AAT TCT GAA AAT GG-**X**

X =
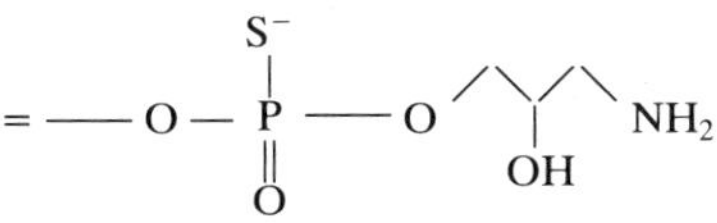

Biodistribution of end capped oligonucleotide phosphorothioate in various organs was not significantly different on the basis of radioactivity measurement. However, of the total bioavailable concentration in uncapped and 5′-capped oligonucleotide, only 50% was intact 20 mers after 24 hours postdosing. The rest was degraded oligonucleotide. In the case of 3′-capped and 3′,5′-capped oligonucleotides, almost the total bioavailable concentration was found to be intact 20 mers.

This preliminary study shows that in mice degradation of oligonucleotide phosphorothioates occurs mainly by 3′-exonuclease activity, which can be slowed down by appropriate chemical modification at the 3′-end of the oligonucleotide.

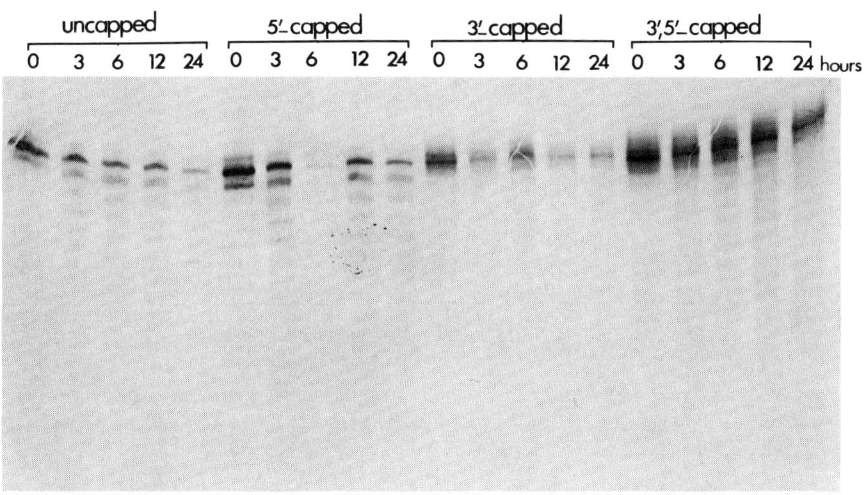

FIGURE 1. Comparative stability of uncapped and capped S-oligonucleotides in monkey plasma. 0.2 μg of ^{35}S-labeled oligonucleotides (specific activity 1.3 μCi/μg) were incubated with 50 μl of heparinized monkey plasma (male cyno monkey) at 37°C. After each time point, the mixture was treated with proteinase K (2 mg/ml) for 1 hour at 37°C, followed by phenol/chloroform extraction and ethanol precipitation. The S-oligonucleotides were analyzed by electrophoresis on a 20% polyacrylamide gel followed by autoradiography. (In lane 6 hours with the 5′-capped oligonucleotide, most of the incubation mixture was accidently lost during handling.)

REFERENCES

1. AGRAWAL, S. 1991. *In* Prospects for Antisense Nucleic Acid Therapy of Cancer and AIDS. :143–158.
2. AGRAWAL, S., J. TEMSAMANI & J.-Y. TANG. 1991. Proc. Natl. Acad. Sci. USA **88:** 7595–7599.
3. AGRAWAL, S., T. IKEUCHI, D. SUN, P. S. SARIN, A. KONOPKA, J. MAIZEL & P. C. ZAMECNIK. 1989. Proc. Natl. Acad. Sci. USA **86:** 7790.

Ion Exchange HPLC Analysis of Oligoribonucleotides and Chimeric Oligoribo-oligodeoxyribonucleotides[a]

VALERI METELEV, KONRAD MISIURA, AND SUDHIR AGRAWAL

Worcester Foundation for Experimental Biology
222 Maple Avenue
Shrewsbury, Massachusetts 01545

There is increasingly widespread use of oligoribonucleotides and their analogs in diverse areas of both basic and applied research. Such uses include potential chemotherapeutic agents as either antisense oligonucleotides or ribozymes.[1]

Oligoribonucleotides can be synthesized using phosphoramidite chemistry[2] or H-phosphonate chemistry[3] using the *tert*-butyldimethylsilyl group for 2′-OH protection. In general, purification of oligoribonucleotides has currently been limited to polyacrylamide gel electrophoresis, which can purify only small amounts of oligoribonucleotides.

We have explored the use of ion exchange HPLC for purification and analysis of oligoribonucleotides. HPLC was carried out using strong (SAX) as well as weak anion exchange (WAX) columns. Oligoribonucleotides and chimeric oligonucleotides containing oligoribo- as well as oligodeoxyribonucleotides were synthesized using H-phosphonate chemistry. Deprotection of the synthesized oligonucleotides was carried out using standard deprotection conditions.

Several oligonucleotides containing a phosphodiester backbone were analyzed using SAX columns (FIG. 1). Oligonucleotide containing all deoxyribonucleotides had retention time of 38.2 minutes. When numbers of ribonucleotide moieties were increased in the same oligonucleotide, retention time increased (41 minutes) as compared to that of oligonucleotide containing all deoxyribonucleotides. The retention time of oligonucleotides depends on the number of ribonucleotide moieties in the oligonucleotide. Oligonucleotides containing increasing numbers of ribonucleotide moieties were retained more than were the oligonucleotides containing deoxyribonucleotide moieties. The retention time of oligoribonucleotides was length dependent. With the aforedescribed chromatographic conditions, oligoribonucleotides up to 25 mers can be purified and analyzed successfully.

Recently, we reported the use of a WAX column for purification and analysis of oligodeoxynucleotides and also for their phosphorothioate analogs.[4] Weak anion exchange chromatographic medium exhibits an interaction with the anion weaker than with strong anion exchange chromatographic medium and also has increased reversed phase characteristics due to its somewhat higher carbon content. Analysis of oligoribonucleotides and chimeric oligonucleotides containing oligoribo- as well

[a]This research was supported by a grant from the G. Harold and Leila Y. Mathers Foundation and NIAID Cooperative Grant U01 A124846.

as oligodeoxyribonucleotides and oligodeoxyribonucleotides on weak anion exchange columns showed that retention times of oligoribonucleotides are lower than those for oligodeoxyribonucleotides (FIG. 2a). The chromatographic profiles contrasted with the strong anion exchange HPLC results, in which oligoribonucleotides had higher retention times than did oligodeoxynucleotides. The separation of oligoribonucleotides using WAX columns was length dependent (FIG. 2b).

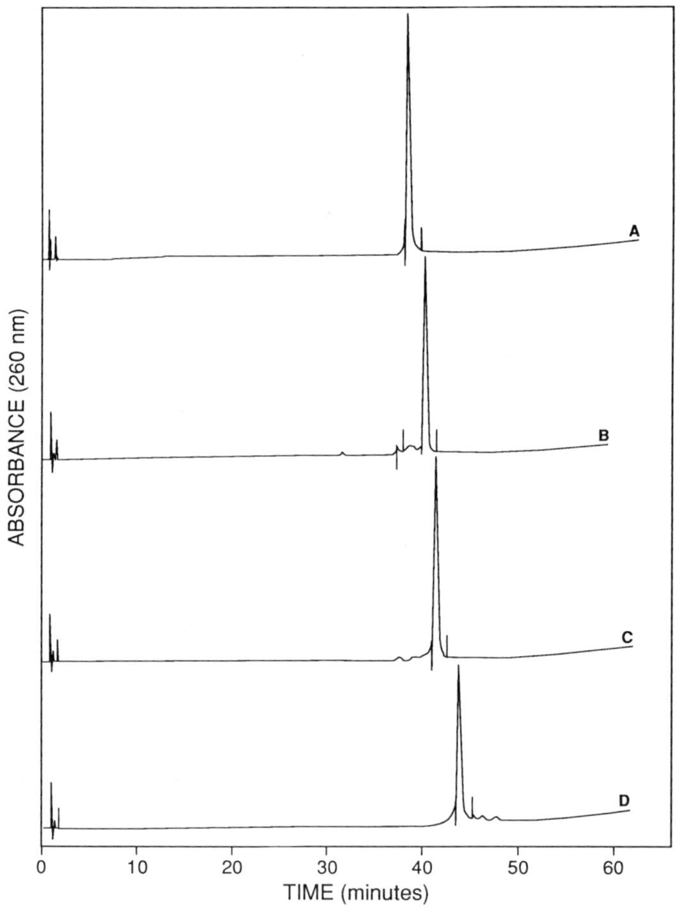

FIGURE 1. HPLC was carried out using a Waters 600E System, 484 variable wavelength detector, 746 data module, and 7125 rheodyne injector. HPLC traces are from oligonucleotides analyzed on a Partisphere SAX column (Whatman; 4.6 mm × 12.5 cm) using a linear gradient of buffer A (1 mM KH_2PO_4 [pH 6.3] in 25% acetonitrile) and buffer B (500 mM KH_2PO_4 [pH 6.3] in 25% acetonitrile). Gradient 0% B for 2′, 0–60% B in 60′, flow rate 1.5 ml/min, 260 nm. **(A)** ACACCCAATTCTGAAAATGG; **(B)** ACACCCAATTCUGAAAAUGG; **(C)** ACACCCAATTCTGAAAATGG; and **(D)** ACACCCAAUUCUGAAAAUGG. The underlined sequences are ribonucleotides.

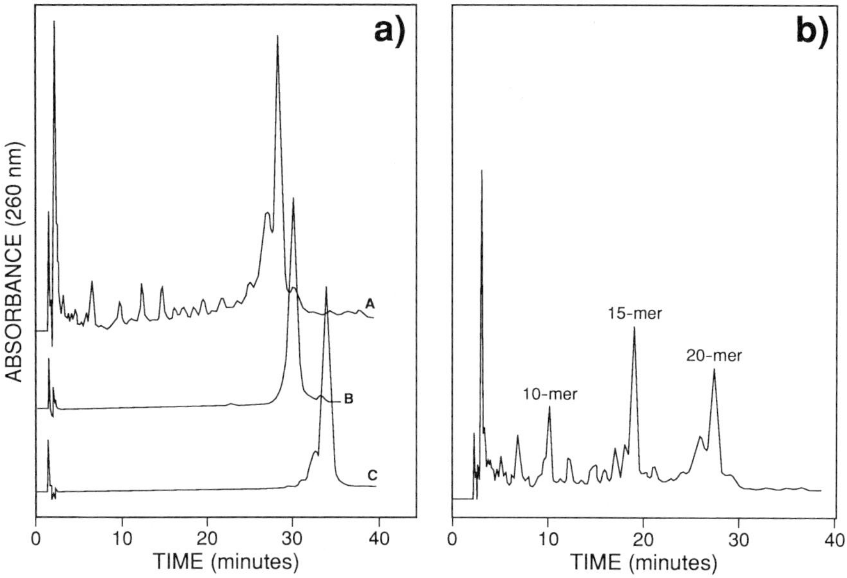

FIGURE 2. HPLC traces of oligonucleotides analyzed on a Partisphere WAX column (Whatman; 4.6 mm × 25 cm) using a linear gradient of buffer A (50 mM KH_2PO_4 in 25% acetonitrile) and buffer B (500 mM KH_2PO_4 in 25% acetonitrile). Gradient 0% B for 2′, 0–20% B in 60′, flow rate 1.5 ml/min, 260 nm. Partisphere WAX column needs to be well equilibrated with gradient of buffer A and B a few times before use. **(a): (A)** CCAUUUUCAGAAUUGGGUGU (crude); **(B)** ACACCCAAUUCUGAAATUGG; **(C)** ACACCCAATTCTGAAAATGG; **(b):** mixture of crude GGGAAGCUUU (10 mer); UUCAGAAUUGGGUGU (15 mer); and CCAUUUUCAGAAUUGGGUGU (20 mer).

ACKNOWLEDGMENT

Authors are thankful to Dr. Paul C. Zamecnik for his support and encouragement.

REFERENCES

1. Erickson, R. P. & J. G. Izant, (Ed.). 1992. Gene Regulation: Biology of Antisense RNA and DNA. Raven Press. New York.
2. Ogilvie, K. K., N. Usman, K. Nichoghosian & R. J. Cedergren. 1988. Proc. Natl. Acad. Sci. USA **85:** 5764–5768.
3. Agrawal, S. & J.-Y. Tang. 1990. *Tet* Lett. **31:** 7541–7544.
4. Metelev, V. & S. Agrawal. 1992. Anal. Biochem. **200:** 342–346.

Enantio- and *Meso*-DNA

KOICHI SHUDO

Faculty of Pharmaceutical Sciences
University of Tokyo
Bunkyo-ku, Tokyo, Japan 113

Gene expression can be controlled by antisense oligonucleotides and their analogs, which specifically bind complementary mRNA sequences. Oligomers with modified sugar backbones, oligomers with modified phosphate backbones, and oligomers derivatized with specific functional groups were studied. One approach to the molecular design of a superior, modified, antisense oligonucleotide analog, which may be classified with the sugar-modified oligomers, involves the use of unnatural *enantio*-deoxynucleotides, that is, DNA with 2-deoxy-L-erythropentose instead of natural D-2-deoxyribose as a sugar backbone. As enzymes generally recognize only naturally occurring enantiomeric isomers, this unnatural DNA is thought to resist nucleases. This paper presents a characteristic of this unnatural *enantio*-DNA.

The enantiomer of 2′-deoxynucleotides was synthesized by direct glycosylation from 2-deoxy-L-erythropentose. Oligonucleotides were prepared by the usual triester method or by DNA synthesizers using phosphoramidites derived from the *enantio*nucleosides.[1]

To investigate the interaction of synthesized L-dA_6 with polynucleotides, we measured ultraviolet mixing curves in 10 mM Tris pH 7.4/10 mM $MgCl_2$ by the method of continuous variations. The UV mixing curve for interaction of L-dA_6 with poly(U) shows maximum hypochromicity ($\sim$22%) at an approximate 1:2 molar ratio (35% L-dA_6) of L-dA_6 to poly(U), which indicates triple helix formation. L-dA_6 did not show hypochromicity when mixed with poly(G), poly(A), or poly(C). It is particularly noteworthy that L-dA_6 also did not show hypochromicity when mixed with complementary poly(dT). Similarly, L-dA_{12} was prepared and its interactions with poly(dT) and poly(U) were studied. L-dA_{12} interacted strongly with poly(U) (Tm = 53.5°) and weakly with poly(dT) (Tm = 15.5°). On this basis, we hypothesize that *enantio*-DNA generally recognizes complementary RNA over complementary DNA. As anticipated, these *enantio*-oligomers were very stable to nucleases.[2]

These findings led us to prepare DNA containing unnatural 2-deoxy-L-erythropentoses and natural D-2-deoxyriboses in alternating sequence, that is, *meso*-DNA.[2] The *meso*-dodecamer consisting of D-dA and L-dA in alternating sequence (denoted as LD-$[dA]_{12}$) showed a flat CD spectrum, as expected, and was resistant to both bovine spleen phosphodiesterase (BSP) and snake venom phosphodiesterase (SVP). Ultraviolet mixing curves for interaction of LD-$(dA)_{12}$ with homopolynucleotides showed that LD-$(dA)_{12}$ interacts with both poly(U) and poly(dT) as natural D-$(dA)_{12}$: the maximum hypochromicity was approximately 30% at a molar ratio of approximately 1:2 (35% LD-$[dA]_{12}$) of LD-$(dA)_{12}$ to poly(U)/poly(dT), which indicates triplex formation. The melting temperature (Tm) curves for interaction of LD-$(dA)_{12}$ with homopolynucleotides were investigated. The Tm profiles demonstrated that LD-$(dA)_{12}$ interacts with both poly(U) and poly(dT), with Tm values being, respectively, 58.5°C and 48.0°C. As regards the Tm values, the affinity toward complementary polynucleotides of LD-$(dA)_{12}$ seems to be comparable to that of natural D-$(dA)_{12}$.[3]

We conclude that natural DNA recognizes both its complementary DNA and its complementary RNA with nearly the same affinity, whereas unnatural *enantio*-DNA recognizes its complementary RNA with greater selectivity than it does its complementary DNA, and unnatural *meso*-DNA is close to natural DNA in its selectivity.

REFERENCES

1. Fujimori, S., N. Iwanami, Y. Hashimoto & K. Shudo. 1992. Nucleosides & Nucleotides **11:** 341–349.
2. Fujimori, S., K. Shudo & Y. Hashimoto. 1990. J. Am. Chem. Soc. **112:** 7436–7438.
3. Fujimori, S., N. Iwanami, K. Shudo & Y. Hashimoto. 1991. Nucl. Acid Res. (S): 143–144.

Intracellular Oligo Targeting

A Theoretical Dissertation

T. T. CHUANG

Department of Physiopathology of Receptors
Istituto di Ricerche Farmacologiche Mario Negri
Consorzio Mario Negri Sud
66030 S. Maria Imbaro (Chieti), Italy

Antisense technology is potentially a powerful therapeutic method. A major obstacle to its application is the high ratio of antisense to sense sequences required to achieve significant inhibition of gene expression, possibly a reflection of the low efficiency of oligos in reaching their targets. This may be due to two factors: (1) intracellular oligos have a fixed life time due to degradation by nucleases; therefore, oligos may be degraded before having the "luck" of colliding with their target mRNA. On the other hand, (2) it has been demonstrated that intracytoplasmically injected oligos rapidly accumulate in the nucleus (within 60 seconds) to specific, discrete intranuclear sites, where they are bound to unidentified nuclear proteins.[1] This apparently specific route of nuclear uptake may represent that of cytoplasmically synthesized deoxyribonucleotides into the nucleus for DNA synthesis.[2] If so, the intranuclear sites of oligo accumulation may be part of the DNA synthesis machinery and would likely act as quenching sites of oligos. The present paper discusses the feasibility of intracellular oligo targeting as a way to reduce the ratio of antisense to sense sequences needed to achieve significant inhibition.

Small nuclear ribonucleoprotein particles (snRNPs), which are involved in splicing of pre-mRNA, are localized to discrete and previously unidentified nucleoplasmic sites where accumulate also labeled pre-mRNA synthesized *in situ*[3] or injected intranuclearly,[4] probably forming spliceosomes. Thus, targeting of oligos to these snRNP-concentrated sites is sensible because oligos and target mRNAs would become confined to small, defined areas, hence providing maximum possibility for antisense-sense hybridization. Spliceosome targeting would likely involve two major steps, namely, nuclear uptake of oligo and subnuclear localization of the oligo to spliceosomes. The former is likely to be taken care of by natural oligo uptake into the nucleus.[1] The latter has to await the elucidation of a subnuclear localization signal for snRNPs; the answer may lie within intron sequences, because they have been shown to be crucial in the colocalization of intranuclearly injected RNA with snRNPs.[4] This signal, once elucidated, can be included in the oligo sequence to enable targeting to snRNPs. Additionally, accessible sequences on U1, U2, and U5 snRNAs, which are involved in binding to pre-mRNA during splicing, near the sequences complementary to the intron splice site can act as docking receptors for oligos, thereby bringing the antisense and sense sequences face to face. Therefore, a spliceosome-targeted oligo can be envisaged to contain, in addition to the antisense sequence, the snRNA-localization signal and the docking sequence for snRNA. The spliceosomes would act as an "oligoport" where outgoing mRNA and incoming oligo interact, thus economizing on strayed oligos. Obviously, whether spliceosome targeting would interfere with normal snRNP splicing activity is a key factor.

In summary, with increasing knowledge of the highly specific intracellular traffic of various molecules, it may become possible to target antisense oligos to sites of mRNA accumulation. This strategy should reduce dramatically the quantity of oligos needed for antisense inhibition.

REFERENCES

1. LEONETTI, J.-P., N. MECHTI, D. DEGOLS, C. GAGNOR & B. LEBLEU. 1991. Proc. Natl. Acad. Sci. USA **88:** 2702–2706.
2. REICHARD, P. 1988. Ann. Rev. Biochem. **57:** 349–374.
3. SPECTOR, D. 1990. Proc. Natl. Acad. Sci. USA **87:** 147–151.
4. WANG, J., L.-G. CAO, Y.-L. WANG & T. PEDERSON. 1991. Proc. Natl. Acad. Sci. USA **88:** 7391–7395.

Cellular Uptake, Distribution, and Degradation of Phosphorothioate Oligonucleotides in *Euplotes crassus*

L. L. THOMAS AND C. M. PRICE

Chemistry Department
University of Nebraska
Lincoln, Nebraska 68588

Genetic manipulation in eukaryotic systems generally involves transformation. Currently, no working transformation method is available for hypotrichous ciliates because of the arduous task of developing a vector. The difficulty in developing a vector is due to the lack of knowledge of both the promoter region and the replication origin. Also, selectable drug-resistant markers are unavailable.

The feasibility of using antisense oligonucleotides to study gene regulation in the ciliate *Euplotes crassus* was evaluated. To minimize nucleolytic degradation problems, phosphorothioate oligonucleotides were used.

RESULTS

Phosphorothioate oligonucleotides were 3′ end labeled with ^{32}P or ^{35}S to investigate cellular uptake and degradation. This labeling procedure introduces a phosphorous that is not readily removed by a phosphatase. Fluorescein-labeled phosphorothioates were employed to determine cellular distribution.

Cellular Uptake. Cells were incubated with ^{32}P-radiolabeled phosphorothioate + 1 μM unlabeled phosphorothioate. At 0, 4, 8, 24, 36, and 48 hours the cells were microfuged briefly (3 seconds), supernatant was removed, and cells were rinsed twice with 1 ml of sea water. The cell pellet and supernatant were Cerenkov counted (FIG. 1).

Degradation. Degradation experiments were performed to ensure that the phosphorothioates were not being degraded within the cell. ^{35}S was used as the radiolabel, instead of ^{32}P, to provide an all sulfur backbone. Cells were treated as in the uptake experiments plus the DNA was isolated and analyzed to determine the extent of degradation (FIG. 2).

Cellular Location. A fluorescein-labeled phosphorothioate was employed to perform the uptake experiment as just described. The phophorothioate was fluorescein labeled according to the method by Leonetti *et al.*[1] Cells were incubated in 0.02 μM fluorescein-labeled phosphorothioate + 1 μM unlabeled phosphorothioate for various times. Cells were spun, rinsed, fixed, and viewed with a fluorescent microscope. To highlight the macronucleus, cells were treated with Hoechst stain after incubation with the fluorescein-labeled phosphorothioate (data not shown).

CONCLUSIONS

Preliminary results show a steady uptake of the phosphorothioate. Degradation is slow with a trace of undegraded oligonucleotide remaining even after 48 hours.

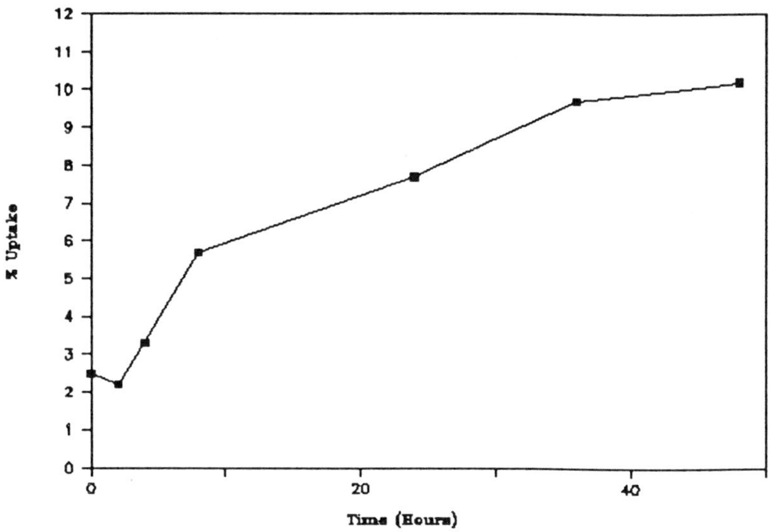

FIGURE 1. Percentage uptake of ^{32}P-labeled phosphorothioate over 48 hours.

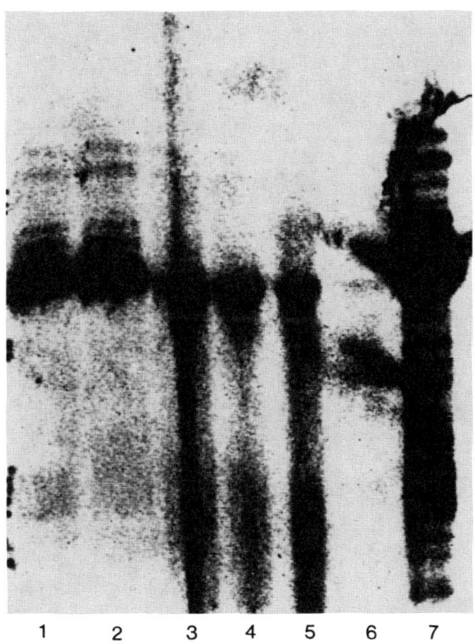

FIGURE 2. Autoradiogram of 8% polyacrylamide denaturing gel showing the extent of degradation. *Lanes 1 and 2* are controls in which the (^{35}S)-labeled phosphorothioate was added to 1 ml of sea water and analyzed for degradation after 12 and 24 hours of incubation, respectively. *Lanes 3, 4, 5, 6,* cellular DNA at 0, 12, 24, and 48 hours of incubation, respectively. *Lane 7* is the (^{35}S) phosphorothioate only.

Cells incubated in fluorescein-labeled phosphorothioate also show a gradual uptake of oligonucleotide. The initial distribution of the oligonucleotide appears to be cytoplasmic through 8 hours. The phosphorothioate then appears to diffuse to the macronucleus sometime during the 8–24-hour interval.

REFERENCE

1. Leonetti, J. P., N. Mechti, G. Degols, C. Gagnor, & B. Lebleu. 1991. Intracellular distribution of microinjected antisense oligonucleotides. Proc. Natl. Acad. Sci. USA **88:** 2702–2706.

Sequence-Specific Activity of Antisense Oligonucleotides Conjugated to Poly (L-Lysine) Carriers[a]

GENEVIÈVE DEGOLS, JEAN-PAUL LEONETTI, AND BERNARD LEBLEU

University of Montpellier II
Sciences et Techniques du Languedoc
UA CNRS 1191 Génétique Moléculaire
Place E. Bataillon
34095 Montpellier Cédex 5, France

Antisense oligonucleotides (ODN) offer interesting prospects for the specific control of viral gene expression, as demonstrated for HIV-1 and other virus-infected cells *in vitro*. Their efficacy is limited by several factors including metabolic stability, cellular uptake, and intracellular bioavailability (reviewed in ref. 1). The chemical conjugation of poly(L-lysine) (PLL) to the 3′ end of antisense ODNs considerably increases their antiviral activity against the vesicular stomatitis virus without loss of sequence specificity[2]; increased cellular uptake and protection against phosphodiesterase degradation presumably contribute to the increased efficacy of these ODN-PLL conjugates.[3]

Synthetic oligonucleotides have been used as well for the control of HIV-1 infection in several T-cell lines *in vitro* (as reviewed in ref. 4), but little or no sequence specificity has been documented. The most active compound is a (dC)28 phosphorothioate derivative (EC50 = 0.5 μM). Nonconjugated and PLL-conjugated 15-mer ODNs complementary to the translation initiation region of tat mRNA have been compared in acutely infected (1,000 TCID 50/ml) MT4 lymphocytes (TABLE 1). A strictly sequence-specific and dose-dependent inhibition (EC50 = 0.15 μM) has been obtained with PLL conjugates; little sequence recognition and a lower efficiency (EC50 = 20 μM) are seen with free ODN, in agreement with published data.[4] The lack of sequence recognition probably results from nonspecific inhibition of HIV reverse transcriptase and/or virus adsorption. Whether and how PLL conjugation prevents it cannot be ascertained; however, PCR analysis of viral DNA synthesis has confirmed differences in the intracellular targets of free and PLL-conjugate ODNs (not shown).

The antiviral activity of ODN-PLL conjugates is only transient and is limited by the toxicity of PLL for T lymphocytes. The association of these ODN-PLL conjugates with sulfated polyanions, such as heparin, is beneficial in several aspects, as sulfated polyanions (1) interfere with early steps in the HIV multiplication cycle, thereby diminishing initial virus load; (2) strongly reduce PLL cytotoxicity as documented by trypan blue exclusion; and (3) modify the release of the conjugated material.

Such ternary complexes between ODN-PLL conjugates at doses as small as 75 nM and heparin (50 μg/ml) completely suppress HIV production when given in a single addition immediately prior to virus inoculation. As a matter of comparison, AZT (0.5 μM), heparin (50 μg/ml), or their combination only transiently suppresses

[a] This work has been supported by CNRS and the Agence Nationale pour la Recherche sur le SIDA.

TABLE 1. Antiviral Activity of Free and PLL-Conjugated Antisense Oligonucleotides[a]

	Percentage RT Inhibition			
	0.5 μM ODN		40 μM ODN	
ODN	HIV-1 BRU	HIV-1 ELI	HIV-1 BRU	HIV-1 ELI
tat BRU-PLL	93	21	ND	ND
tat ELI-PLL	31	89	ND	ND
tat BRU	5	ND	95	78
tat ELI	ND	0	92	90

[a] MT4 cells (3×10^6 cells/ml) were infected with the HIV-1 BRU, or HIV-1 ELI isolate (1,000 TCID50/ml) and incubated 30 minutes at 4°C. Cells were washed, diluted to 3×10^5 cells/ml, and incubated at 37°C in the presence of PLL conjugated or unconjugated oligomers. tat BRU($^{5'}$CTAGGATCTACTGGCTrA$^{3'}$), tat ELI($^{5'}$TTAGGATCTACTGGATrA$^{3'}$) and VSV($^{5'}$CATTTTGATTACTGTrA$^{3'}$) oligomers were synthesized on a riboadenosine-derivatized support using a Biosearch Cyclone automatic DNA synthesizer and purified by reverse phase chromatography. Covalent linkage to PLL through an *N*-morpholine ring was achieved by periodic acid oxidation and borocyanohydride reduction of the 3′ end ribose, as previously described.[3] tat BRU and tat ELI have been designed to hybridize fully to the translation initiation region of tat mRNAs in the corresponding isolates. Culture samples were removed 4 days after virus inoculation for HIV-1 BRU and 5 days after infection for HIV-1 ELI to determine cell count, viability, syncytia formation (not shown), and reverse transcriptase activity, according to established procedures.

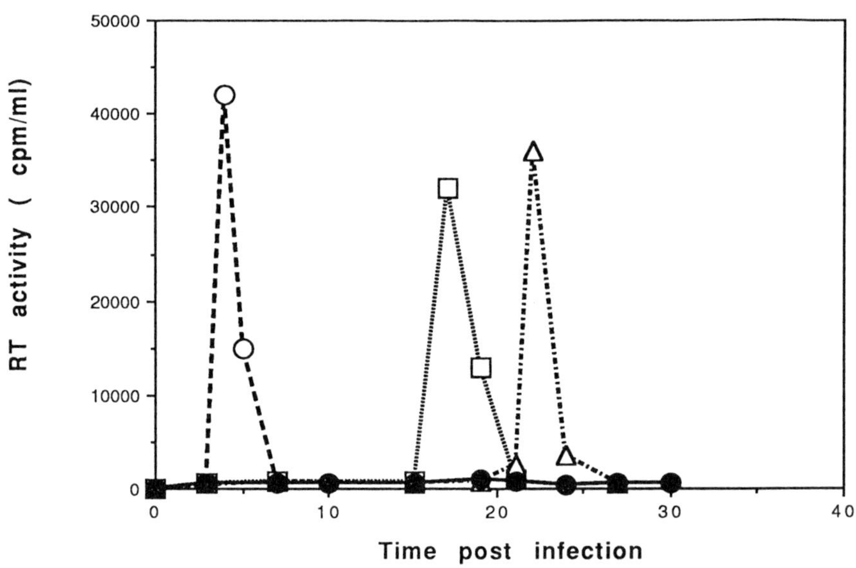

FIGURE 1. Potentiation of antisense-PLL conjugate activity by sulfated polyanions. MT4 cells were left untreated or were treated with various combinations of antiviral agents at the indicated concentrations. The cells were then infected with HIV-1 BRU, and viral production was monitored as described in TABLE 1. Time post infection refers to days. (○----○), untreated cells; □-----□, cells treated with heparin (50 μg/ml) and a control (complementary to VSV N protein mRNA initiation site) oligonucleotide-PLL conjugate (0.5 μM); △----△, cells treated with heparin (50 μg/ml) and AZT (0.5 μM): ●----●, cells treated with heparin (50 μg/ml) and tat BRU-PLL conjugate (0.5 μM).

HIV-1 production in this model system. Interestingly, sequence specificity is maintained in this formulation (FIG. 1 and data not shown).

Additional studies are required to understand the underlying mechanisms, exploring the relative efficacy in this model of various ODN chemistries and targets in the infected cells. Whether comparable inhibition can be attained in chronically infected T lymphocytes or in macrophage reservoirs will also be of interest.

ACKNOWLEDGMENTS

The authors are indebted to Dr. C. Devaux for helpful discussion and access to appropriate containment facilities.

REFERENCES

1. LEBLEU, B., G. DEGOLS & J. P. LEONETTI. 1991. Antisense oligonucleotides as pharmacological modulators of genes expression. *In* Targeted Drug Delivery. R. L. Juliano, ed.: 329–354. Springer-Verlag. Heidelberg.
2. LEMAITRE, M., B. BAYARD & B. LEBLEU. 1987. Specific antiviral activity of a poly (L-lysine) conjugated oligodeoxyribonucleotide complementary to vesicular stomatitis virus Nprotein mRNA initiation site. Proc. Natl. Acad. Sci. USA **84:** 648–652.
3. LEONETTI, J. P., G. DEGOLS & B. LEBLEU. 1990. Biological activity of oligonucleotide-poly (L-lysine) conjugates: Mechanism of cell uptake. Bioconjugate Chem. **1:** 149–153.
4. MATSUKURA, M., H. MITSUYA & S. BRODER. 1991. A new concepts in AIDS treatment: An antisense approach and its current status towards clinical application. *In* Prospects for Antisense Nucleic Acid Therapy of Cancer and AIDS. E. Wickstrom, ed.: 159–178. Wiley-Liss, Inc. New York.

Inhibition of the Friend Retrovirus by Antisense Oligonucleotides

Indirect Evidence for the Necessity of Forced Cellular Uptake

C. ROPERT,[a] M. LAVIGNON,[a] J. L. IMBACH,[b]
P. COUVREUR,[c] AND C. MALVY[a]

[a]*UA 147 CNRS*
U140 INSERM
Institut G. Roussy
Rue Camille Desmoulins
94800 Villejuif, France

[b]*Université des Sciences et Techniques*
34060 Montpellier, France

[c]*Faculté de Pharmacie*
Chatenay Malabry, 92290 France

Resistance to degradation and uptake in cells are two key elements in the efficiency of antisense oligonucleotides (oligos). Alpha anomeric oligos are resistant to degradation but do not trigger the action of RNase H. (For a review see ref. 1.) When a cellular target is used that does not require the action of RNase H and electropulsation to ensure the uptake of oligos, the measured antisense effect has so far been moderate.[2] One possibility is that internalized oligos are not well distributed or do not reach a sufficient concentration. Recent reports indicate the usefulness of liposomes to greatly increase the action of antisense oligos.[3]

We therefore undertook experiments with liposomes and nuclease-sensitive β oligos which are more readily available than α oligos. We chose a β antisense oligo targeted against the AUG initiation codon of Env mRNA of the Friend murine retrovirus. It inhibits the synthesis of Env protein in a wheat germ extract (WGE) (100% inhibition at 10 μM). However, this oligo has no activity on Friend murine retrovirus proliferation (measured by focus immunoassay) in Dunni cells (mouse fibroblasts) even when using electropulsation to facilitate its uptake by cells. Degradation of this oligo in the culture medium is complete in 2 hours. When protected in 3′ by derivatization with ellipticine, this oligo is still intact after 23 hours and displays the same effect in WGE as does the nonfunctionalized one and a moderate effect on retrovirus proliferation (20% inhibition at 1 μM compared to the control oligo). However, it has no effect when electropulsation is used. The nonfunctionalized oligo has then been encapsulated in pH-sensitive liposomes, which should prevent degradation of oligos by lysosomes[4] and which were shown to increase by 1,000 times the biological efficiency of poly(rI)-poly(rC) (P. Milhaud and J. Philippot, Montpellier, personnal communication). Encapsulation of oligonucleotides was successful with the following composition: dioleylphosphatidyl ethanolamine, oleic acid, and cholesterol in a ratio of 10:5:2.[5] The action of nonfunctionalized oligos is then considerably increased (80% inhibition at 0.1 μM; no effect of control oligo[5]). Experiments with non-pH-sensitive liposomes indicate a lower but noticeable activity.

These experiments show that in our system, forced uptake of oligos, in addition to their resistance to degradation in the extracellular medium, is necessary to obtain interesting antisense activity. Delivery through electropulsation is less efficient than that by liposomes. Experiments with α oligos encapsulated in pH-sensitive and pH-insensitive liposomes may indicate the role of the balance between intracellular degradation and RNaseH activity in the global activity of oligos.

REFERENCES

1. RAYNER, B., C. MALVY, J. PAOLETTI, B. LEBLEU, C. PAOLETTI & J. L. IMBACH. 1989. *In* Oligodeoxynucleotides. Cohen, J. ed. Macmillan Press. New York.
2. LAVIGNON, M., J. R. BERTRAND, H. PORUMB, B. RAYNER, J. L. IMBACH, C. PAOLETTI & C. MALVY. 1991. Nucl. Acid Res. Symposium series No. **24**117–119.
3. LEONETTI, J. P., P. MACHY, G. DEGLOS, B. LEBLEU & L. LESERMAN. 1990. Proc. Natl. Acad. Sci. USA **88:** 2702–2706.
4. HUANG, L. & C. Y. WANG. 1989. Biochemistry **28:** 9508–9514.
5. ROPERT, C., M. LAVIGNON, C. DUBERNET, P. COUVREUR & C. MALVY. 1992. Biochem. Biophys. Res. Com. **183:** 879–885.

Structure-Function Relation of the Hepatitis Delta Virus Genomic Ribozyme

MARTA BLUMENFELD,[a] GILBERT THILL,[a]
FRANCK LESCURE,[a,b] AND MARC VASSEUR[a,b]

[a] *GENSET*
1, Passage Etienne Delaunay
75011 Paris, France

[b] *Laboratoire de Virologie Moléculaire*
Université Paris VII
2 Place Jussieu
75251 Paris Cédex 05, France

Hepatitis delta virus (HDV) contains a circular, single-stranded RNA genome of 1.7 kb that can undergo site-specific self-cleavage and ligation.[1] Although some HDV features are similar to those of certain pathogenic RNAs of plants, sequence analysis of HDV self-cleaving domains shows no homologies with virusoid ribozyme motifs such as hammerhead or hairpin-like structures.[2] The smallest RNA subfragment from genomic HDV that can self-cleave was postulated to contain only 1 nt upstream of the break site and 84 nt downstream.[3] We were interested in further analyzing the structure-function relation of the HDV genomic ribozyme by performing internal deletions and mutations on the minimal self-cleaving fragment.[4]

With the RNAFOLD program to evaluate the potential secondary structures of the putative minimal ribozyme Rz85 (−1 to +84), we obtained several alternative conformations, with G_0 values ranging from −37.4−−32.4 kcal. (FIG. 1B). A stable GC-rich stem-loop extending from positions +44 to +73 was systematically observed in all alternative structures. This hairpin was also present in the pseudoknot-like structure[5] proposed by Perrotta and Been (stem IV in FIG. 1A).

To analyze the role of stem IV and to test if this structure was part of the catalytic core of the HDV ribozyme, we studied the cleavage activity of an RNA molecule selectively deleted in this region.

We synthesized Rz-71, a 71-nt long molecule, deleted from positions +51 to +68. *In vitro* transcription at 37°C of HDV71 template gave rise to a faster migrating RNA and a slower transcript (Rz-71-2 and Rz-71-1, respectively; data not shown). Sequence analysis of both Rz-71-1 and Rz-71-2 showed that Rz-71-1 was the bona fide full-length ribozyme, whereas Rz-71-2 appeared to be the 66-nt 3′ cleavage product. These data then suggested that Rz-71 was able to self-cleave during transcription. This finding was further confirmed by performing transcription under self-cleavage inhibition conditions, such as transcription in the presence of ITP or at 4°C. In addition, Rz-71-2 was easily phosphorylated with (γ-^{32}P)ATP even without previous phosphatase treatment, whereas efficient end-labeling of Rz-71-1 required prior dephosphorylation. These observations indicated that Rz-71-2 had a 5′OH and Rz-71-1 a 5′-P, as expected for a 3′ cleavage product and a full-length precursor, respectively.

To conclusively demonstrate that Rz-71 was able to self-cleave, we prepared 5′ end-labeled Rz-71 ribozyme obtained by transcription at 4°C and incubated the

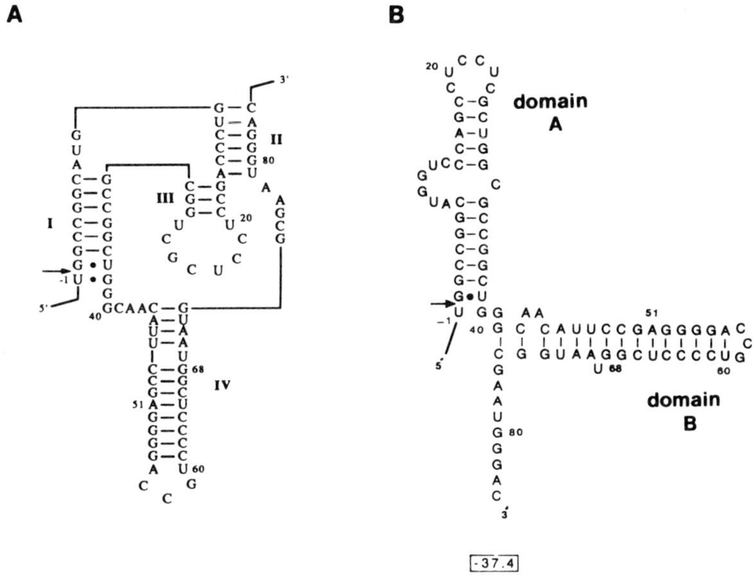

FIGURE 1. Potential secondary structures for HDV genomic self-cleaving RNAs. Sequences are numbered from the cleavage site (*arrow*). Watson-Crick base pairs are indicated by dashes, wobbles by dots. **A** represents the pseudoknot-like structure proposed by Perrotta and Been[3] for an 85-nt long HDV genomic ribozyme (−1 to +84); I to IV indicates the stem-loop regions. **B** depicts the lowest free energy structure of Rz-85 ($G_0 = -37.4$ kcal) obtained by the RNAFOLD program.

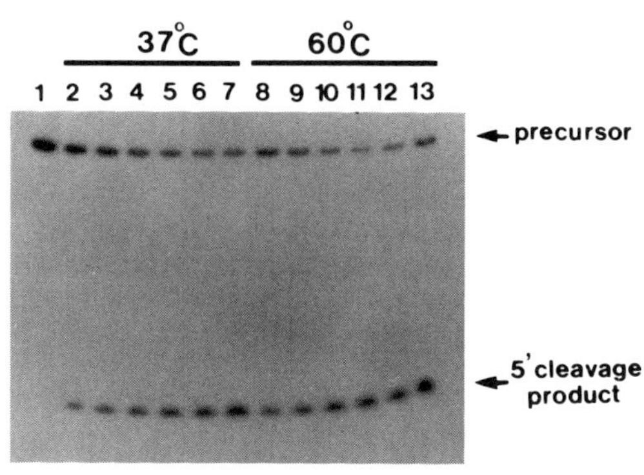

FIGURE 2. Self-cleavage of Rz-71 at 37°C and 60°C. 5′ end-labeled Rz-71 was denaturated at 95°C for 1 minute in 1 mM EDTA, snap cooled on ice, supplemented with 40 mM Tris HCl, pH 8.0, 10 mM $MgCl_2$ and then incubated at 37°C (*lanes 2 to 7*) or 60°C (*lanes 8 to 13*). Aliquots were removed at 1 minute (*lanes 2 and 8*), 2 minutes (*lanes 3 and 9*), 5 minutes (*lanes 4 and 10*), 10 minutes (*lanes 5 and 11*), 20 minutes (*lanes 6 and 12*), or 60 minutes (*lanes 7 and 13*); in the control (*lane 1*), the aliquot was removed before the addition of $MgCl_2$.

purified (^{32}P)-RNA under cleavage conditions. FIGURE 2 shows the time course corresponding to the incubation of Rz-71 with 10 mM $MgCl_2$ at 37°C. Under this condition, Rz-71 ribozyme self-cleaved with a $t_{1/2}$ of 3 minutes (FIG. 2, compare lanes 2 to 7). The reaction could still be enhanced by partial denaturing conditions, as at 60°C and 10 mM $MgCl_2$ the observed $t_{1/2}$ was less than 2 minutes (FIG. 2, lanes 8 to 13).

Therefore, deletion of 18 nt in a very stable GC-rich stem-loop (stem-loop IV) results in a 71-nt long ribozyme that self-cleaves under nondenaturing conditions with a $t_{1/2}$ of 3 minutes. Under the same conditions, the nondeleted ribozyme cleaves to 50% in less than 15 seconds. According to these results, stem-loop IV seems to play a structural role, not being directly involved in the catalytic core of the HDV ribozyme. *Trans*-cleavage reactions further confirmed these results.

REFERENCES

1. TAYLOR, J. M. 1991. Curr. Top. Microbiol. Immunol. **168:** 141–166.
2. SHARMEEN, L., M. Y.-P. KUO, G. DINTER-GOTTLIEB & J. TAYLOR. 1988. J. Virol. **62:** 2674–2679.
3. PERROTTA, A. T. & M. D. BEEN. 1990. Nucl. Acids Res. **18:** 6821–6827.
4. THILL, G., M. BLUMENFELD, F. LESCURE & M. VASSEUR. 1991. Nucl. Acids Res. **19:** 6519–6525.
5. PERROTTA, A. T. & M. D. BEEN. 1991. Nature **350:** 434–436.

Antisense Oligonucleotides as Antiviral Agents[a]

JUDY COULSON AND ALAN D. B. MALCOLM

Department of Biochemistry
Charing Cross and Westminster Medical School
London, W6 8RF, UK

Antisense oligodeoxynucleotides (ODNs) have shown tremendous potential as antiviral agents. Many problems, however, remain to be solved. Of these, delivery to the affected tissue is one of the most important. With epitheliotropic viruses such as the human papilloma viruses (HPV), problems of delivery are greatly reduced. We report here our studies in the development of antisense oligodeoxynucleotides as anti-HPV agents.

HPV only divide in differentiating epithelial tissue. The DNA of the genome is a supercoiled circular molecule of just under 8 kb in length. Many different strains of HPV have been identified. HPV 1 appears to be principally responsible for ordinary skin warts, HPV 6 for genital warts, and HPV 16 and 18 for genital lesions (warts) that show a high frequency for conversion to the malignant state.[1]

The HPV genome contains nine ORFs. Seven of these are expressed during the early part of the life cycle of the virus and two during the late part. Two of the early genes in particular have been shown to play a major role in virus-induced tumorigenesis. E6 affects transformation by binding to and causing degradation of the tumor suppressor gene p53,[2] whereas E7 binds to the retinoblastoma gene product p105.[3] These genes are frequently found to be integrated into the cellular genome in biopsies from cervical carcinomas, and E6 and E7 gene products are detectable. In HPV16 E7/EJ *ras* transformed cell lines, the continued expression of E7 is required to maintain the transformed state.[4] We have therefore chosen the E6 and E7 sequences as targets for our antisense work.

We have cloned the E6 and E7 genes of HPV16 into the transcription vector Bluescript and used these as templates for *in vitro* transcription. This was carried out using T3 RNA polymerase to produce sense RNA and T7 RNA polymerase for antisense RNA synthesis. A 5′ 7meGppp5′G cap analog was included in the reaction mix and is used to initiate transcription, yielding RNA that is efficiently translated in a rabbit reticulocyte lysate.

Unmodified antisense oligonucleotides HPVAS1, HPVAS2, and HPVAS3 (FIG. 1) have been synthesized and tested for their effectiveness in causing translational arrest of E6 and E7 RNAs *in vitro.* The first two are 16 and 32 mers, respectively, directed against the initiation codon region of E7; the third is a 16 mer directed against the initiation codon region of E6.

Micrococcal nuclease-treated rabbit reticulocyte lysate was used to translate the E6 and E7 RNAs. Synthesized proteins were labeled by the incorporation of (^{35}S)-methionine.

In translational reactions containing capped sense E6 RNA in the absence of antisense, a discrete band of labeled protein is seen on an SDS polyacrylamide gel.

[a] This work was supported by Stiefel International. J. C. holds a studentship from SERC.

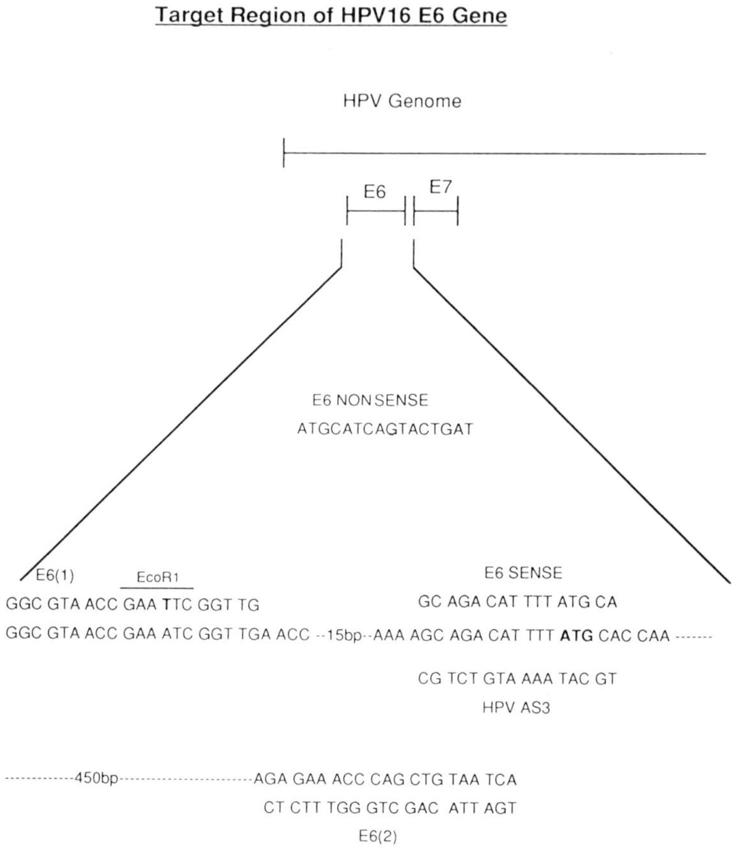

FIGURE 1A. The E6 coding region showing the target site of the antisense ODN HPV AS3, along with sequences for the sense and nonsense control ODNs and the PCR primers: forward sense primer E6(1) and reverse antisense primer E6(2) used in the construction of pBS-E6.

This runs at a size slightly larger than that of globin (16 kD). The E6 band migrates at between 17 and 21 kD, the actual position being concentration dependent. An additional band is seen in the E6 lane which is likely to be a dimer of E6, induced by high concentrations and not denatured in this system; it cannot be due to the translation of longer transcripts, as these are not seen on a glyoxal RNA gel.

In the presence of 25 μM HPVAS3 the translation of the E6 RNA is inhibited. The higher molecular weight band also disappears in the presence of the antisense oligo, supporting the hypothesis that it is a dimer of the E6. In the presence of 25 μM sense oligonucleotide, the E6 bands remain unaffected, showing that the effect of the antisense oligo is specific. The antisense RNA, transcribed from the construct using T7 RNA polymerase, also results in the disappearance of the HPV specific bands, lending further support to the sequence-specific translational arrest induced by HPVAS3.

Similar results have been obtained using HPVAS1 and HPVAS2 to modulate the *in vitro* translation of the E7 gene in a cell-free system.

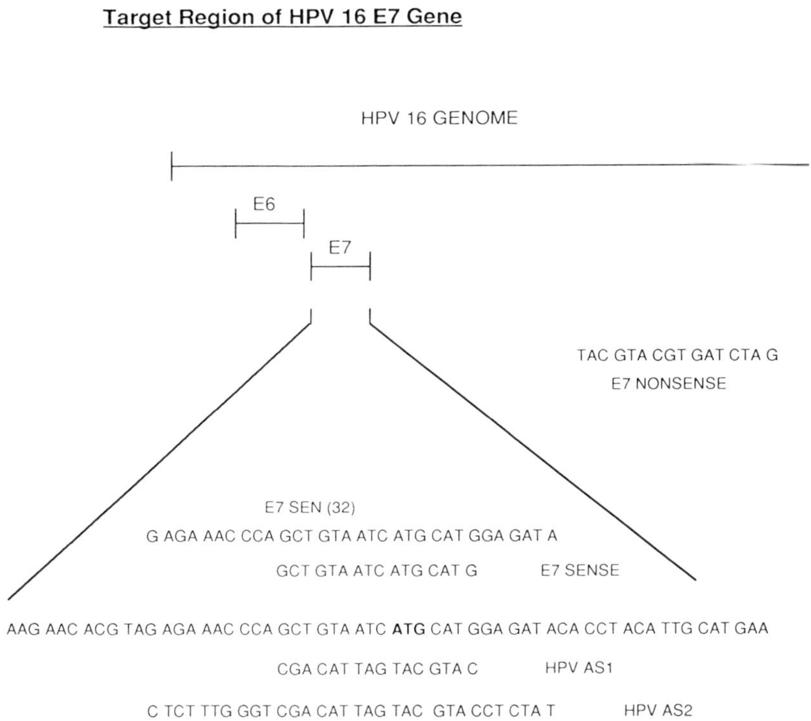

FIGURE 1B. The E7 coding region showing the target site of the antisense ODNs HPVAS1 and HPVAS2 along with sequences for sense and nonsense control ODNs.

ACKNOWLEDGMENT

We thank Professor Mike Clemens for providing the rabbit reticulocyte lysate.

REFERENCES

1. Arends, M. J., A. H. Wyllie & C. C. Bird. 1990. Human Pathol. **21:** 686–698.
2. Scheffner, M., B. A. Werness, J. M. Huibregtse, A. J. Levine & P. M. Howley. 1990. Cell **63:** 1129–1136.
3. Dyson, N., P. M. Howley, K. Munger & E. Harlow. 1989. Science **243:** 934–937.
4. Crook, T., J. P. Morgernstern, L. Crawford & L. Banks. 1989. EMBO J. **8:** 513–519.

Conference Summary

DAVID T. DENHARDT

Department of Biological Sciences
Rutgers University
Piscataway, New Jersey 08855-1059

Exciting advances in several areas, such as reversal of the oncogenic phenotype, dissection of signal transduction pathways, and antiviral therapy, were reported at this conference. The ability to manipulate gene expression by means of exogenously administered oligodeoxynucleotides ("oligos"), or longer lived analogs thereof, complementary to specific sequences in the genome is clearly going to impact on many fields of biology and medicine. Also, we are beginning to realize the extent to which naturally occurring endogenously generated antisense RNA species are important in regulating gene expression. Although most presentations dealt with the interaction of antisense (AS) oligos or RNA with a putative single-stranded RNA target, there was also some discussion that appropriate oligos can form triplex structures with DNA and inhibit transcription (Claude Helene, Edith Postel).

Our understanding of what is necessary for a successful AS experiment remains fragmentary. Each system has its own characteristics. For example, David Knecht sought to identify that part of the myosin heavy chain II gene that was most effective in suppressing expression when used to generate AS RNA in *Dictyostelium discoideum.* A variety of factors were studied (size, location of the AS RNA in the RNA transcript, portion of the gene that was targeted, type of promoter/vector used), and the predominant factor determining success in this case was that the AS sequence be targeted against the 3′ 3.7-kb end of the gene. The AS RNA reduced the target mRNA level, but how this was accomplished and why the 3′ end was the most effective target were not evident.

Several investigators noted that AS oligos targeting either regions immediately upstream of and containing the AUG translation initiation codon or regions incorporating intron sequences, especially splice donor or acceptor sites, were most likely to be effective. Hybridization of an oligo to a message can block ribosome function and translation of the target mRNA. There was general agreement that the precise sequence of nucleotides, typically 15–20 bases in length, is critical to the success or failure of the experiment. Reasons for this presumably have to do with the accessibility of the target to the AS oligo. Factors affecting accessibility include secondary structure in the RNA or protection by associated proteins. Given this unpredictability it is wise to test oligos directed at a variety of different sequences in the target RNA.

AS RNA molecules generated endogenously may be subjected to processing, transport, and degradation just as normal transcripts are. Although for high level production of an AS RNA an efficient promoter would appear desirable, it is equally important that the resulting transcript have characteristics that allow it to persist in an active form in the cell; a weak promoter can generate high levels of a stable RNA. AS RNA experiments can be successful, that is, the expression of the targeted gene can be reduced, even when the apparent steady-state abundance of the AS RNA is less than that of the sense RNA. It is not necessary that the concentration of AS RNA be higher than that of the target RNA in the cell to modulate its abundance. Also important is the relation between the abundance of target mRNA and the phenotype it controls. If a small change in the amount of the gene product can produce an effect, then AS-mediated down-regulation will be successful. If, on the other hand, there is considerably more mRNA and protein in the cell than is required to achieve a given end, then even a substantial reduction in the expression of the targeted gene may not yield an observable consequence.

Introduction of a functional gene into a cell (gene therapy) is an intriguing challenge. For some purposes, retroviral vectors provide a useful way to introduce new DNA into a cell. Some cell types are efficiently transfected with DNA of moderate molecular weight coprecipitated with calcium phosphate or complexed with cationic molecules such as DEAE, polybrene, or polylysine. Max Birnstiel described a procedure for exploiting cell surface receptors to take up DNA complexed with polylysine to which is bound the cognate ligand, transferrin for example, for the receptor. During receptor-mediated endocytosis the DNA is initially taken into the endosomal compartment, from which its release in functional form may be problematic. Release from endosomes can be promoted by certain viruses, and Birnstiel reported that when the transferrin-polylysine-DNA complex was covalently linked to the adenovirus capsid by transglutaminase, very efficient transient expression of a marker gene could often be obtained. Although stable transfection was not efficient, this kind of approach may be an effective way to introduce high molecular weight DNA into cells.

Cells in culture, or in the body, will, with greater or lesser efficiency, take up oligos from the medium, possibly by pinocytosis in a protein kinase C-stimulated process (Cy Stein). Uptake is a function of the cell and how the oligo is presented; encapsulation of oligos into liposomes (Alain Thierry) or conjugated with poly(L-lysine) (Geneviève Degols) can increase the efficiency of the process. Oligonucleotide analogs such as methylphosphonates, phosphoramidates, and phosphorothioates, though more expensive to make than the standard phosphodiesters, are more resistant to degradation; the nucleotide analog can confer much of this resistance even if present only in terminal positions. The neutral methylphosphonate may exhibit an enhanced ability to penetrate the cell and anneal with the intracellular target. Muthiah Manoharan reported preliminary observations on the use of oligonucleotides linked to cholic acid to promote uptake into the cell.

Ribozymes are RNA molecules that possess a catalytic center able to cleave an appropriately based paired RNA substrate. Via an AS sequence the "hammerhead" ribozyme can be targeted to a specific mRNA, and under optimal conditions can act catalytically to inactivate multiple copies of the mRNA. John Zaia reviewed some of the properties of ribozymes and factors that influence their activity, and he described the advantages of using adeno-associated virus (AAV) as a vector for introducing AS sequences during gene therapy.

It is well established that the malignant phenotype of cancer cells is the consequence of increased expression of several positively acting oncogenes, usually coupled to the reduced expression of one or more tumor suppressor genes. K. Lawlor, R. Narayanan, and colleagues reported that diminished expression of the DCC (deleted in colorectal carcinoma) gene in Rat-1 cells provoked by antisense RNA or phosphorothioate oligonucleotides led to the loss of cell adhesion and a more transformed phenotype. For those cancers in which one can identify a specific function that must be expressed for the cell to exhibit its malignant behavior, it is not unreasonable to seek ways to block that expression. Antisense oligos may turn out to be an important weapon to accomplish this end. In an initial *in vivo* application of an AS strategy, Len Neckers used AS oligos designed to inhibit N-MYC expression in cells from neural crest tumors to effectively retard tumor growth. N-MYC expression in the tumor was inhibited and tumor growth reduced by some 50% in this mouse model of neuroendocrine tumors. Edward Shillitoe reported that antisense sequences targeting the E6 and E7 genes of human papilloma virus-18 in HPV-transformed cells partially suppressed the transformed phenotype of the cells.

The use of AS oligos to define the importance of the c-*kit* and c-*myb* protooncogenes in hematopoietic malignancies was reviewed by Alan Gewirtz. Normal bone marrow cells were insensitive to these AS oligos in a colony cluster assay, whereas the proliferation of many lymphoid and myeloid leukemia cells was blocked. C-*myb* AS oligos suppressed colony formation in over half the primary acute and chronic myelogenous leukemias studied; c-*kit* oligos, in contrast, were more effective against CML than AML. Interestingly, a PCR analysis of the CML colonies surviving the AS oligo treatment revealed that most, if not all, of the cells expressing *bcr-abl* had been eliminated. This suggests that the bone marrow of cancer patients might be purged of tumor cells by AS oligos.

Investigations with AS strategies have enhanced our understanding of signal transduction pathways and the control of cell growth. Bruno Calabretta researched the roles of c-*abl,* N-*ras,* and c-*myb* in normal hematopoiesis using 18-mer oligos. The power of this approach is that it yields clues to the functions of these and other genes in the complex differentiation pathways that give rise to normal circulating blood cells. C-*abl* encodes a tyrosine kinase on the inner cell membrane and is expressed as part of a *bcr-abl* chimeric mRNA/protein in many cases of CML. Inhibition of c-*abl* with AS oligos suppressed myeloid colony formation from CFU-G and CFU-GM progenitor cells but not erythroid colony formation from BFU-E and CFU-E progenitors. AS-mediated inhibition of expression of MYB protein, a transcription factor, implicated it in the non-lineage-specific proliferation of the mid-late stage progenitor of many cell types. N-RAS function was similarly shown to be required by normal marrow mononuclear cells to respond to certain growth factors (e.g., IL-3, GM-CSF, or M-CSF), but not others (e.g., G-CSF or erythropoietin). Evidence that casein kinase II is a mediator of the mitogenic effect of EGF was provided by Walter Pyerin's observations that antisense oligos to CKII inhibited the ability of EGF to stimulate proliferation of human fibroblasts.

A fascinating consequence of exposure to AS oligos *in vivo* was reported by Sonoko Ogawa. She described experiments in which lordosis and proceptive behavior of female rats were reduced by the appropriately timed administration into the hypothalamus of AS oligos targeting the progesterone receptor mRNA. Previous work had implicated progestin binding to neurons in the hypothalamus as being important in reproductive behavior. Stephen Epstein reported that the proliferation and migration of smooth muscle cells could be inhibited by AS phosphoramidate oligos targeting the c-*myc* mRNA, and he discussed the potential of this approach to prevent restenosis after coronary angioplasty (paper not included in this volume). An interesting use of antisense oligos to desensitize the receptor to luteinizing hormone was described by Brian Cooke; the take-home message was that oligos targeted to regions in the mRNA encoding the cytoplasmic COOH-terminal portion of the receptor promoted production of a truncated protein insensitive to down-regulation after exposure to phorbol esters or cAMP. Thomas Caskey reviewed research on (1) the use of the *Cypla-1* promoter (responsible for the induction in liver of aryl hydrocarbon hydroxylase by polycyclic hydrocarbons) to drive gene expression in transgenic mice, and (2) the effectiveness with which various portions of the HPRT gene were able to inhibit expression of the gene. Important conclusions were that the *Cypla-1* promoter/enhancer was strongly inducible in the liver of transgenic animals and that a 172-bp portion of the first intron was as effective as larger portions of the HPRT gene, including the entire cDNA, at reducing cellular HPRT levels.

Antisense oligos are certain to be effective in combating persistent viral infections. Overt viral infections can often be treated successfully with selected antiviral agents, which often act on a viral enzyme necessary for replication of the viral genome. Covert, latent infections are much more difficult to treat because fewer virus-specific functions are amenable to attack. The beauty of the antisense strategy is that it can "go after" the viral nucleic acid directly. Paul Zamecnik, one of the pioneers in the use of AS strategies to combat viral infections, reviewed the work of his group on the use of oligonucleotide analogs to inhibit the replication of several viruses, including Rous sarcoma virus, influenza virus, and human immunodeficiency virus (HIV).

Paul Ts'o reviewed earlier research by him and his colleagues. They demonstrated that oligomethylphosphonates (MATAGEN™ [MASKING TAPE FOR GENE EXPRESSION]) complementary to particular *Herpes* virus intron-exon junctions when incorporated into a polyethylene glycol cream, could cure a viral infection in the mouse ear. Ts'o presented a detailed analysis of factors that determine the selectivity and specificity with which oligonucleotides, or derivatives thereof, could inhibit gene expression. Appropriately chosen sequences could effectively distinguish between *Herpes simplex 1* and *Herpes simplex 2* viral infections. There seems no impediment (aside from cost perhaps) to the use of this approach to ameliorate a variety of virus-induced lesions in human beings.

Particularly significant is the evidence that the AS strategy is going to be of importance in the treatment of AIDS. HIV, like any other virus, is vulnerable to inhibition by AS nucleic acids. The remarkable specificity of AS sequences for the virus and the ability to target multiple sites in the viral genome at one time are what make this approach special. Properly targeted complementary nucleic acids are likely to have minimal side effects, and it is improbable that the virus will mutate to a resistant form at several sites simultaneously. Posters displayed at the meeting presented sketchy information that hinted at intense activity by a number of investigators to exploit ribozyme technology to enhance the effectiveness with which an antisense-based strategy could inhibit virus proliferation.

Index of Contributors

Agrawal, S., 2–10, 318–320, 321–323
Alder, H., 64–69, 209–218
Ansorge, W., 295–297
Aurelian, L., 159–177

Banerjea, A. C., 271–273
Barger, S. W., 219–230
Barja, P., 288–290
Baserga, R., 1, 64–69
Belinsky, M. G., 277–279
Bennett, C. F., 306–309
Bergot, B. J., 310–312
Birnstiel, M. L., 136–153
Blumenfeld, M., 336–338
Bradley, M. O., 124–135
Britton, D. H. O., 265–267
Brocard, M.-P., 77–87

Calabretta, B., 117–123
Caracciolo, D., 117–123
Carr, B. I., 315–317
Caskey, C. T., 154–158
Chang, E., 159–177
Chatterjee, S., 95–106
Chen, C.-J., 271–273
Cho, K. R., 283–285
Chuang, T. T., 326–327
Cohen, J. S., 315–317
Cook, P. D., 306–309
Cooke, B. A., 291–292
Cosenza, S. C., 231–239
Cotten, M., 136–153
Coulson, J., 339–341
Couvreur, P., 334–335
Cowsert, L., 306–309
Crisell, P., 274–275
Curiel, D. T., 136–153

Degols, G., 331–333
Delohery, T., 313–314
Denhardt, D. T., 70–76, 280–282, 342–346
Dinter-Gottlieb, G., 277–279
Dritschilo, A., 300–302

Ecker, D. J., 306–309
Elkins, D., 95–106

Fahmy, B., 37–44
Feng, B., 280–282
Ferrari, S., 11–26
Fields, S., 313–314
Freire-Moar, J., 288–290

Ge, L., 263–264
Gervasoni, J., 313–314
Gewirtz, A. M., 178–187
Ghosh, K., 315–317
Graeven, U., 293–294
Grande, A., 11–26
Green, D. R., 193–203
Grunwald, D., 77–87
Guinosso, C. J., 306–309

Haglund, K., 271–273
Harel-Bellan, A., 27–36
Harmison, G. G., 271–273
Helene, C., 27–36
Herlyn, M., 293–294
Ho, S. P., 265–267
Holt, J. T., 88–94
Homann, M., 268–270

Imbach, J. L., 334–335
Inagaki, M., 315–317

Jacobs-Lorena, M., 204–208
James, W., 274–275
Jimenez, G., 107–116
Johnson, L. K., 306–309

Khochbin, S., 77–87
Knecht, D. A., 45–56
Koizumi, M., 276
Kraynak, A. R., 124–135
Krishna, S., 313–314

Lavignon, M., 334–335
Lawlor, K. G., 283–285
Lawrence, J.-J., 77–87
Lebleu, B., 331–333
Ledwith, B. J., 124–135
Leonetti, J.-P., 331–333
Lescure, F., 336–338
Lin, D., 209–218
Lin, J.-C., 107–116
Lorenz, P., 295–297

Malcolm, A. D. B., 339–341
Malvy, C., 334–335
Manam, S., 124–135
Manfredini, R., 11–26
Manoharan, M., 306–309
Matthieu, J.-M., 188–192
McGee, D. P. C., 306–309
Mercer, W. E., 209–218
Metelev, V., 321–323

Miller, P. S., 159–177
Misiura, K., 321–323

Narayanan, R., 283–285
Neckers, L., 37–44
Nichols, W. W., 124–135
Nishikura, K., 240–250

Ogawa, S., 298–299
Ohtsuka, E., 276
Olazábal, U. E., 298–299
Osborne, M. P., 283–285

Pagano, J. S., 107–116
Pawlita, M., 268–270
Pepperkok, R., 295–297
Pestka, S., 251–262
Pfaff, D. W., 298–299
Phillips, S., 136–153
Pietrzkowski, Z., 64–69
Plank, C., 136–153
Postel, E. H., 57–63
Prasad, Y., 277–279
Price, C. M., 328–330
Pyerin, W., 295–297

Raab-Traub, N., 107–116
Ramasamy, K., 306–309
Reiss, K., 64–69
Rhodes, A., 274–275
Rittner, K., 268–270
Roach, A., 188–192
Rodeck, U., 293–294
Ropert, C., 334–335
Rosolen, A., 37–44
Rossi, J. J., 95–106
Rotenberg, S., 313–314

Sarin, P. S., 2–10
Schaapveld, R. Q., 283–285
Scherczinger, C. A., 45–56
Schubert, M., 271–273
Sczakiel, G., 268–270
Seitz, A. W., 263–264
Shi, Y., 193–203
Shields, M. T., 209–218
Shillitoe, E. J., 286–287
Shudo, K., 324–325
Skorski, T., 117–123
Smith, J. B., 277–279
Soprano, D. R., 231–239
Soprano, K. J., 231–239
Springer, R. H., 306–309
Steele, C., 286–287
Stein, C. A., 313–314
Stein, S., 303–305
Steinlein, P., 136–153
Sun, D., 2–10
Sung, N. S., 107–116
Surmacz, E., 64–69
Szczylik, C., 117–123

Tang, J. Y., 2–10, 318–320
Taub, R., 313–314
Taylor, N. R., 95–106
Telang, N. T., 283–285
Temsamani, J., 318–320
Thierry, A. R., 300–302
Thill, G., 336–338
Thomas, L. L., 328–330
Thuong, N. T., 27–36
Togawa, K., 315–317
Tonkinson, J., 313–314
Torelli, U., 11–26
Tosic, M., 188–192
Ts'O, P. O. P., 159–177
Tsutsumi, A., 288–290
Tung, C.-H., 303–305

Ullrich, S. J., 209–218

Valtieri, M., 117–123
Van Eldik, L. J., 219–230
Vasseur, M., 336–338
Venturelli, D., 117–123
Vickers, T., 306–309
Vogelstein, B., 283–285

Wagner, E., 136–153
Weinmann, R., 293–294
Welsh, M. J., 219–230
West, A. P., 291–292
Whitesell, L., 37–44
Wong, K. K., 95–106

Yates, A. A., 45–56
Yumet, G., 231–239

Zaia, J. A., 95–106
Zamecnik, P. C., 2–10
Zatloukal, K., 136–153
Zhang, L. M., 313–314
Zheng, H., 193–203
Zon, G., 310–312

Subject Index

Adeno-associated virus (AAV), in anti-HIV therapy, 101–104, 343
Adenovirus, in transferrinfection, 137–141, 143–152
AIDS, 107, 193, 268, *See also* HIV
  antisense strategy and, 345, *See also* HIV
  antisense treatment and, 2, 8–10
  oligonucleotide analogs in, 318–320
cAMP production, antisense DNA and, 291–292
Anticode strategy, *See* Antisense strategy
Antisense antiviral activity, 2, 5–10, 268–270, *See also* Antisense DNA; Antisense oligonucleotides; Antisense RNA
Antisense cloning strategy, 90–92
Antisense DNA, 1, 209, 212–217, *See also* Antisense oligonucleotides; Antisense RNA
  as anti-HIV agent, 2, 8–10
  as antiviral agent, 339–341
  to bFGF, 293–294
  in bone marrow purge, 178–186
  and cAMP production, 291–292
  and cell cycle inhibition, 64–68
  and cell cycle regulation, 231, 233–239
  and cell proliferation, 233–239
  chemotherapeutic activity of, 1
  and CKII, 295–297
  and DNA synthesis, 231, 234–235, 237–239
  in gene characterization, 11–12
  and growth stimulation inhibition, 295–297
  and HIV replication, 2, 5–10
  HL60 cells, interaction with, 313–314
  in human leukemia therapy, 178–186
  and IGF-1 receptor role, 64–68
  and LH receptors, 291–292
  liposomal delivery of, 300–302, 343
  nuclease-resistant, 2
  and protooncogenes, 37–43, 231–239
  reliability of, 1, 67–68, 101
  and S100β, 220–228
  and T-cell gene function, 193–202
Antisense inhibition, *See also* Antisense DNA; Antisense oligonucleotides; Antisense RNA
  of DNA synthesis, 65
  of β-galactosidase, 252–261
  of gene expression, 42–56, 160–161, 172–175, 188–190, 204–207
  gene portion effect on, 47–48
  of HIV, 2, 5–10, 268–270
  of oncogenes, 11–14, 37–43
  of phenotype transformation, 286–287
  reversion of, 49–50
  of RSV, 9
  sense orientation transformants and, 50–54
  sequence size effect on, 48–49
Antisense mRNA, 255–260
Antisense oligodeoxynucleotides, *See* Antisense DNA
Antisense oligomers, *See* Antisense oligonucleotides
Antisense oligonucleotides, *See also* Antisense DNA; Antisense RNA; Matagens; Oligonucleotide analogs
  antiviral activity of, 331
  cell cycle effects of, 212–217
  in EBV episome disruption, 108–111
  Friend retrovirus inhibition by, 334–335
  as gene expression inhibitors, 42–43
  in hepatoma cell physiology, 315–317
  and HPV, 339–341
  improving uptake of, 306–309
  intracellular targeting of, 326–327
  and latent EBV infection, 107–108
  modifications of, 306–309
  and neuroectodermal tumors, 37–42
  oncogene inhibition by, 37–43
  and p34 protein function, 209–217
  to PR, 298–299
  sequence-specific activity of, 331–333
Antisense oligoribonucleotides, *See* Antisense RNA
Antisense oncogene, in transformation reversal, 124–134
Antisense promoter mapping, 88–90, 92
Antisense RNA, 157–160, 209, 342, *See also* Antisense DNA; Antisense oligonucleotides
  anti-HIV activity of, 2, 5–10
  and DCC gene, 283–285
  in fibroblast transformation, 283–285
  and β-galactosidase synthesis, 252–261
  history/perspective, 251–261
  and p53 regulation, 77–87
  and phenotype transformation, 286–287
  phosphorothioate analogs of, 2–10
  and PKC isoenzymes, 288–289
  dsRNA unwinding/modifying activity and, 248–249
  mRNA translation blockage by, 11–12, 251–252
  splicing inhibition by, 70–72
  synthesis of, 2–3, 7–10
  and target RNA levels, 73–75
  and T-cell activation, 288–289

and TIMP expression, 70–72
variables affecting activity of, 47–49
Antisense RNA inhibition
of β-galactosidase, 255–260
of gene expression, 204–207
of genes, 45–56, 188–190
of HIV, 268–270
mechanism of, 70–76
of protooncogene expression, 128–132
and target RNA concentration, 73–75
of TIMP gene, 280–282
at transcription level, 70–72
at translation level, 72–73
at transport level, 72
of tumorigenicity, 127–128
Antisense sequences, 1
and cell proliferation, 12–14
and gene inhibition, 11–14
Antisense strategy
and AIDS, 2, 8–10, 318–320, 345
in antiviral treatment, 101–102, 164–167, 339, 341, 345
and cell cycle control, 64–68
cloning method in, 90–92
cornerstones of, 160–161
and differentiation, 154–158
in gene characterization, 1, 11–14, 16–22, 154–158
gene expression control by, 27–35
and glial cell function, 219–228
and molecular problems, 1
oligonucleotide backbone construction, 161–163
in oncogene suppression, 167–168
in TCR studies, 193–201
and viral gene function, 164–167
and virus replication, 345
Antisense treatment
and AIDS, 2, 8–10, *See also* HIV
for EBV, 107, 111–115
of transformed cells, 124–134
Antiviral research
anticode mechanism in, 166
anticode specificity in, 164–166
Matagens in, 164–166
Apoptosis
gene control of, 14
protooncogenes and, 19–20, 194–197, 201
of T cells, 193–197

Basic fibroblast growth factor (bFGF), 293–294
Bone marrow purge
antisense DNA in, 178–186
efficiency of, 185–186
protooncogenes in, 185
Burkitt's lymphoma, 107, 109

Cancer, multidrug resistance in, 300–302
Casein kinase II
in cell growth stimulation, 295–297
and EGF mitogenic effect, 344
Cell cycle, *See also* Cell proliferation; Differentiation; Hematopoiesis
antisense DNA and, 64–68, 231, 233–239
antisense strategy and, 64–68, 212–217
genetic regulation of, 231–239
Cell proliferation, *See also* Cell cycle; Differentiation; Hematopoiesis
antisense DNA and, 233–239
antisense sequences and, 12–14
and differentiation potential, 14–16
IGF-1 receptor role in, 64–68
and myeloid differentiation, 14–16
oncogenes and, 12–16
protooncogenes in, 12–13, 124

Deleted in colorectal cancer (DCC) gene, 283–285, 343
Differentiation, *See also* Cell cycle; Cell proliferation; Hematopoiesis; Myeloid differentiation
antisense strategy and, 154–158
genes in, 11–22
in leukemic cell lines, 14–16
in neuroectodermal cells, 37–43
p53 regulation in, 77–87
and proliferative capacity, 14–16
protooncogenes in, 16–22, 124
DNA, 42, *See also* Antisense DNA
duplexes, 4, 159–160
enantio-/meso- forms, 324–325
receptor-mediated endocytosis of, 136–138, 343–344
transfection with, 343
in transferrinfection, 136–138, 141–152
in triplex formation, 1, 27–35, 342
DNA synthesis
antisense DNA and, 231, 234–235, 237–239
inhibition of, 65
Duplex formation, 255–256
DNA in, 4, 159–160
oligonucleotides in, 4–5, 159, 172–175
RNA in, 12, 240–249

Endocytosis
DNA importation via, 136–138
Epstein-Barr virus (EBV)
antisense treatment for, 107, 111–115
disease associations, 107–108
latent infection, 107–108
replication of, 108
Epstein-Barr virus (EBV) episome
disruption of, 108–111

in EBV-immortalized cells, 107, 111–115
and latent infection, 108

Fibroblast growth factor, *See* Basic fibroblast growth factor
Fibroblast transformation, 283–285

β-Galactosidase synthesis
anti-mRNA inhibition of, 255–260
antisense blockage of, 252–261
Gene characterization
antisense DNA in, 11–12
antisense strategy in, 1, 11–14, 16–22, 154–158
Gene expression
antisense inhibition of, 42–56, 160–161, 172–175, 280–282
antisense RNA inhibition of, 188–190, 204–207
antisense strategy in, 27–35
Genes, *See also* Oncogenes; Protooncogenes
antisense inhibition of, 11–14, 72–73
in apoptosis, 14
in cell cycle regulation, 231–239
in differentiation, 11–22
RNA virus, 247
role in myeloid differentiation, 11–14, 16–22
Gene therapy, 343
Glial cell protein (S100β), 219–221
function, 222–228
Glioblastoma cells, p34 protein expression in, 210–212

Hematopoiesis, 180–181, 195, *See also* Cell cycle; Cell proliferation; Differentiation
leukemic, 182–185
protooncogenes in, 117–122, 180–185, 344
Hepatitis delta RNA, 277–279
Hepatitis delta virus (HDV) ribozyme, 336–338
Hepatoma cell physiology, antisense TGF-β in, 315–317
Herpes simplex virus (HSV), 164–167
Matagens and, 172–175, 345
HIV, *See also* AIDS
AAV therapy for, 101–104
antisense inhibition of, 2, 5–10, 268–270
antisense RNA and, 2, 5–10
ribozyme cleavage of, 265–267
ribozyme inhibition of, 101–104
HIV replication
antisense DNA and, 2, 5–10
antisense RNA inhibition of, 268–270, 274
gene inhibition of, 274–275
ribozyme inhibition of, 101–104, 271–275
Hoogsteen, *See* Triplex orientation
HPLC
in analysis of oligonucleotides, 321–323
in separation of oligonucleotide analogs, 310–312
HPV-18, *See* Human papillomavirus
Human immunodeficiency virus, *See* HIV
Human papillomavirus (HPV)
antisense oligonucleotides and, 339–341
and neuroendocrine tumors, 37–43, 344
in phenotype transformation, 286–287
strains of, 339

IGF-1, *See* Insulin-like growth factor I
Inhibin
regulated expression of, 263
ribozyme and, 263
Insulin-like growth factor I (IGF-1), 64–68

Leukemia, chronic myelogenous, 182
Leukemia therapy
antisense DNA in, 178–186
protooncogenes in, 178–186
Leukemic cell lines, *See also* Murine erythroleukemia; Myeloid leukemic cells
differentiation in, 14–16
in hematopoiesis, 182–185, 344
Liposomal delivery, of antisense DNA, 300–302, 343
Luteinizing hormone (LH) receptors, antisense DNA and, 291–292, 344

Matagens, 159, 163, 345, *See also* Antisense strategy; Oligonucleotide analogs
in animal experimentation, 166–167
anti-HSV specificity of, 164–166, 172–175
mechanism of, 166
potency enhancement, 168–172
in triplex formation, 159, 172–175
Melanoma cells
phenotype reversal in, 78–87
transfection of, 293–294
Murine erythroleukemia (MEL) cell lines, 77, 79, 81–87
Myeloid differentiation, *See also* Differentiation
cell proliferation and, 14–16
c-fes in, 16–21
oncogenes in, 11, 16–22
PKC in, 20
protogene products in, 20–22
Myeloid leukemic cells, 344

Neuroendocrine tumors, antisense oligonucleotides and, 37–43, 343

Oligodeoxynucleotides, *See* Antisense DNA
Oligonucleotide analogs, *See* also Antisense DNA; Antisense oligonucleotides; Antisense RNA; Matagens
  in AIDS chemotherapy, 318–320
  anti-HIV activity of, 2, 5–10, 318–319
  backbone construction of, 161–163
  in chemotherapy, 1
  in duplex/triplex formation, 28–32, 159, 172–175
  HPLC separation of, 310–312
  nuclease-resistance of, 163, 310, 343
  RNA duplexes with, 4–5
  synthesis of, 4
Oligonucleotide phosphorothioates, *See also* Oligonucleotide analogs
  in AIDS chemotherapy, 318–319
  cellular uptake/degradation of, 328–330
  end-capped, 318–319
Oligonucleotides, *See also* Antisense DNA; Antisense RNA; Matagens
  HIV inhibition by, 9–10
  RSV inhibition by, 9
  triple helix formation by, 27–32
Oligoribonucleotides, *See* Antisense RNA
Oncogenes, *See also* Protooncogenes
  antisense inhibition of, 11–14, 37–43
  antisense strategy and, 167–168
  and cell proliferation, 12–16
  in myeloid differentiation, 11, 16–22
  in phenotype reversal, 124–134
Oncogene transcription
  antisense inhibition of, 88–92
  triplex formation and, 57–63

p34 protein
  antisense oligonucleotides and, 209–217
  in glioblastoma cells, 210–212
p53 protein
  antisense RNA and, 77–80
  in B16 cells, 78–87
  and differentiation, 77, 81–87
  in MEL cells, 77, 79, 81–87
Papilloma virus-transformed cells, 286–287, *See also* Human papillomavirus
Phenotype reversal, 124–134
Phenotype transformation
  antisense RNA and, 286–287, 343
  human papillomavirus in, 286–287
Phosphorothioate, *See* Oligonucleotide phosphorothioates
PKC, *See* Protein kinase C
Polycythemia vera, 182
Progesterone receptor (PR) mRNA
  antisense to, 298–299
  and rat mating behavior, 298–299
Protein kinase C (PKC)
  in myeloid differentiation, 20
  in oligonucleotide uptake, 343
Protein kinase C (PKC) isoenzymes
  antisense RNA and, 288–289
  in T cells, 288
Protein synthesis inhibition, 255–261
Protooncogenes, *See also* Oncogenes
  in antileukemia therapy, 178–186
  antisense DNA and, 37–43, 231–239
  and apoptosis, 19–20, 194–197, 201
  in bone marrow purge, 185
  in cell cycle regulation, 231–239
  in cell proliferation, 12–13, 124
  in differentiation, 16–22, 124
  in hematopoiesis, 117–122, 180–185, 344
  in vitro suppression of, 37–41
  in vivo suppression of, 41–42
  in leukemic hematopoiesis, 182–185
  in macrophage suppression, 14
  in neuroectodermal cell differentiation, 37–43
  ribozyme inhibition of, 276
  transcription modulation, 57–63
  in transformation reversal, 124–134
  triplex formation and, 57–63

Receptor-mediated endocytosis, 136–138, 343–344
Retinoblastoma, 37
Ribonuclease mimics, 303–305
Ribozymes, 1, 27, 251
  efficiency of, 98–101
  hammerhead, 97–101, 343
  of HDV, 336–338
  and HIV, 265–267
  HIV inhibition by, 101–104, 271–273
  and inhibin cleavage, 263
  and mRNA cleavage, 263, 276
  protooncogene inhibition by, 276
  types of, 95–98
RNA
  messenger, *See* mRNA
  synthetic ribonuclease cleavage of, 303–305
RNA
  antigenomic, 277–279
  double-stranded, *See* dsRNA
  hepatitis delta, 277–279
  oligonucleotide duplexes of, 4–5, 7
  oligonucleotides complementary to, 1
  self-cleaving, *See* Ribozymes

RNA duplexes
dsRNA unwinding/modifying activity and, 240–249
in translation blockage, 12
dsRNA
in protein synthesis blockage, 255–260
in vivo modification of, 240–243
dsRNA unwinding/modifying activity
and antisense RNA, 248–249
biological function of, 245
cloning of, 248
distribution of, 244–245
in mammalian cells, 240
nuclear localization of, 243–244
and mRNA, 247–248
dsRNA alteration by, 242–243
and dsRNA duplexes, 240–249
and RNA virus genes, 247
specificity of, 242
mRNA
antisense RNA blockage of, 11–12, 251–252
and antisense RNA gene inhibition, 54–55
dsRNA unwinding/modifying activity and, 247–248
and p53 regulation, 77–87
in protein synthesis inhibition, 255–261
ribozyme cleavage of, 263, 276
dsRNA unwinding/modifying activity and, 247–248
TIMP, 70–72
translation blockage, 251–252
Rous sarcoma virus (RSV), 9, 102, 345

S100β, *See* Glial cell protein
Small cell lung carcinoma, 37

T-cell receptors (TCR), 193–197
antisense study of, 197–201
T cells
antisense DNA and, 198–202
antisense RNA activation of, 288–289
apoptosis of, 193–197
TGF-β, *See* Transforming growth factor β
TIMP, *See* Tissue inhibitor of metalloproteinases
Tissue inhibitor of metalloproteinases (TIMP) gene
expression of, 70–72
inhibition of, 280–282
Transfection, *See* also Transferrinfection
antisense RNA in, 124–125
and antisense RNA production, 159
with DNA, 343
of HeLa cells, 142
in macrophage suppression, 14
of melanoma cells, 293–294
of mouse hepatocytes, 142
of tissue culture cells, 151
Transferrin conjugates, 136–139, 141–147, 150–152
Transferrinfection
adenovirus in, 137–141, 143–152
DNA in, 136–138, 141–152
in gene construct expression, 142–152
Transformation reversal, 124–134
Transforming growth factor β, 315–317
Triple helix, *See* Triplexes; Triplex formation
Triplexes, 1
gene expression control by, 27–35
transcription control by, 32–35
Triplex formation, 28–32, 342
base triples, 28–29
and c-*myc* transcription, 57–63
DNA in, 1, 27–35, 342
in vitro, 57–58
in vivo, 58
kinetics of, 30–31
Matagens and, 159, 172–175
oligonucleotides in, 28–32, 159, 172–175
and oncogene transcription, 57–63
protooncogenes in, 57–63
specificity of, 30
stabilization of, 31–32
third strand orientation of, 29–30
Triplex orientation, 28–32
Tumor suppressor gene, *See* Deleted in colorectal cancer gene

Viral infections, *See also* specific viruses
antisense DNA and, 2, 8–10, 339–341
antisense oligonucleotides and, 101–102, 108–111, 334–335, 339–341, 345
antisense RNA and, 2, 5–10, 268–270
antisense strategy and, 164–167, 318–320, 331, 345
antisense treatment for, 2, 8–10
Matagens and, 164–166, 172–175

Wilms' tumor, 37